Data Network
Design

McGraw-Hill Series on Computer Communications

ISBN	AUTHOR	TITLE
0-07-019022-4	Edmunds	*SAA/LU6.2 Distributed Networks and Applications*
0-07-054418-2	Sackett	*IBM's Token-Ring Networking Handbook*
0-07-004128-8	Bates	*Disaster Recovery Planning: Networks, Telecommunications, and Data Communications*
0-07-020346-6	Feit	*TCP/IP: Architecture, Protocols, and Implementation*
0-07-005075-9	Berson	*APPC: Introduction to LU6.2*
0-07-005076-7	Berson	*Client/Server Architecture*
0-07-012926-6	Cooper	*Computer and Communications Security*
0-07-016189-5	Dayton	*Telecommunications*
0-07-016196-8	Dayton	*Multi-Vendor Networks: Planning, Selecting, and Maintenance*
0-07-034242-3	Kessler	*ISDN*
0-07-034243-1	Kessler/Train	*Metropolitan Area Networks: Concepts, Standards, and Service*

Other Related Titles

ISBN	AUTHOR	TITLE
0-07-051144-6	Ranade/Sackett	*Introduction to SNA Networking: A Guide for Using VTAM/NCP*
0-07-051143-8	Ranade/Sackett	*Advanced SNA Networking: A Professional's Guide to VTAM/NCP*
0-07-033727-6	Kapoor	*SNA: Architecture, Protocols, and Implementation*
0-07-005553-X	Black	*TCP/IP and Related Protocols*
0-07-005554-8	Black	*Network Management Standards: SNMP, CMOT, and OSI*
0-07-021625-8	Fortier	*Handbook of LAN Technology, 2/e*
0-07-063636-2	Terplan	*Effective Management of Local Area Networks: Functions, Instruments, and People*
0-07-004563-1	Baker	*Downsizing: How to Get Big Gains from Smaller Computer Systems*

To order or receive additional information on these or any other McGraw-Hill titles, in the United States please call 1-800-822-8158. In other countries, contact your local McGraw-Hill representative.   MH92

Data Network Design

Darren L. Spohn

McGraw-Hill, Inc.

New York   San Francisco   Washington, D.C.   Auckland   Bogotá
Caracas   Lisbon   London   Madrid   Mexico City   Milan
Montreal   New Delhi   San Juan   Singapore
Sydney   Tokyo   Toronto

Library of Congress Cataloging-in-Publication Data

Spohn, Darren L.
    Data network design / Darren L. Spohn.
        p.   cm.—(McGraw-Hill series on computer communications)
    Includes index.
    ISBN 0-07-060360-X (alk. paper)
    1. Computer networks.   2. Computer network protocols.   3. Computer
networks—Standards.   4. Computer network architectures.   I. Title.
II. Series.
TK5105.5.S66   1993
004.6—dc20   92-45053
                                                                       CIP

    4 5 6 7 8 9 0   DOC/DOC   9 9 8 7 6 5 4

ISBN   0-07-060360-X

*The sponsoring editor for this book was Jerry Papke, the editing super-
visor was Stephen M. Smith, and the production supervisor was
Suzanne W. Babeuf.*

Printed and bound by R. R. Donnelley & Sons Company.

This book is printed on acid-free paper.

AppleTalk, Macintosh, and LocalTalk are trademarks of Apple
Computer; AIX, AS/400, IBM, NetView, and RISC are trademarks of
International Business Machines Corporation; ARCnet is a trademark
of Datapoint; ARPANET was developed by the Department of Defense;
AT&T and UNIX are trademarks of American Telephone and Telegraph;
DEC, DECnet, VAX, and MicroVax are trademarks of Digital Equipment
Corporation; Ethernet and XNS are trademarks of Xerox Corporation;
MAP is a trademark of General Motors; X-Windows is a trademark of
Massachusetts Institute of Technology.

This book is dedicated to my parents, Karl and Regina, and to my best friend, Becky

ABOUT THE AUTHOR

Darren L. Spohn started his communications career as a noncommissioned officer in the Pennsylvania Air National Guard. In 8 years of service, Mr. Spohn spent the majority of his time on national and international telecommunications projects in both the United States and Germany.

During college, Mr. Spohn worked in various LAN and WAN consulting positions in the Washington, D.C. area. He co-founded Design Consultants, a small consulting firm that manufactured and sold personal computers, as well as providing LAN/WAN consulting.

After graduating from Capitol College in Laurel, Md., with a B.S. in Telecommunications Engineering, he joined MCI Communications. Since then he has held a number of engineering and management positions and has led the design efforts of four intelligent national data networks, including an NET intelligent multiplexer network and the frame relay/cell switch network that forms the technology platform for MCI's next-generation Virtual Private Data Services (VPDS).

Mr. Spohn is now Data Marketing Product Manager of Packet Services Development, and previously was Manager of Data Network Engineering and Design, with responsibilty for MCI's internal data networks, as well as customer-switched and multiplexed network design. He serves on various planning committees for MCI's future data-switched services.

Active in many standards and professional organizations and forums, Mr. Spohn has added to his bachelor's degree an M.S. in Engineering Management from Southern Methodist University, where he is currently pursuing a joint Doctorate in Engineering and MBA.

Contents

Part 4   Requirements, Planning, and Choosing the Technology   477

Chapter 15. Requirements Definition   479

Preface

Purpose of This Book

The primary objective of this book is to present a broad scope of data and computer communications standards, architectures, hardware, software, protocols, technologies, and services as they relate to designing data networks. Emphasis is placed on technologies playing a predominant role in the '90s. This includes X.25 packet switching, frame relay, IEEE 802.6/DQDB (cell relay), SMDS, ATM, B-ISDN, and SONET. This book is designed to walk the reader through the process of a data network design using these technologies. It is the author's intent to provide a broad overview of these topics with insight into practical design aspects of each, allowing the reader to perform an end-to-end design. Standards and references are given if the reader wishes to pursue more detailed study.

The logical and physical design of hardware and software is not the only process in a network design. Network designs encompass many aspects, including making the business case, compiling the requirements, choosing the technology, planning for capacity, completing the vendor selection, and weighing all the issues, before the actual design begins. After these efforts have produced a workable design plan, there are additional issues which must be addressed, including operations and maintenance and support structures. While resolution of many of these additional tasks often falls to the project manager, many other people must be involved in the many processes affecting the integrity of the design, and must assess the impact of each decision as it relates to the overall network design.

This book will concentrate on the science and procedure involved with a data network design, and will include detailed operational descriptions of:

- X.25 packet switching
- Frame relay
- IEEE 802.X (including 802.6 DQDB)
- SMDS
- FDDI
- ATM
- B-ISDN
- SONET

A detailed design book, resplendent with addressing schemes, detailed bus structures and discrete circuit operation, and protocol specifics, would take volumes. In fact, one book could be published on each protocol and technology. The primary thrust of this text, however, is to provide a global view of these technologies and data network design as a whole.

Intended Audience

This text is designed for data communications novices through advanced senior engineers. The basics are provided in the beginning chapters. The text presents design material at a high level, assuming the reader has access to resources or colleagues with a background in basic data communications and a working knowledge of transmission basics. Although this skill level is assumed, some discussion on hardware and protocol basics is provided. This book serves as professional reading as well as a desktop reference guide.

Overview of This Book

Data Network Design is divided into seven parts.

Part 1 provides the business case that drives the need for a data communications network. Chapter 1 defines data network design through the evolution of voice to data and the recent global data revolution. Chapter 2 presents the needs that drive data network design and the move toward switched and virtual data networks, and provides an in-depth analysis of outsourcing.

Part 2 provides a broad overview of the standards that define data networking: organizations, architectures, circuits, services, hardware, and protocols. Chapter 3 presents national and global industry standards organizations, current forums, standard protocol architectures, the seven-layer OSI Reference Model, and the move toward OSI. Chapter 4 presents standard circuit types, topologies, services, and network hardware. Chapter 5 provides a thorough coverage of the first two layers of the OSI Reference Model. Chapter 6 then provides coverage of layers 3 through 7 of the OSI Reference Model, as well as bridge and router protocols.

At this point the reader will understand the basics.

Part 3 presents a detailed study of the major present and future switching technologies. Chapter 7 compares multiplexing techniques and hardware with circuit switching and packet switching. Chapter 8 is devoted to the principles and application of packet switching. Chapters 9 and 10 present the definition, standards, protocols, and transmission theory; and the operation and service provided by frame relay, respectively. Chapter 11 presents the IEEE 802.6 MAN standard, along with DQDB bus theory. Chapter 12 then presents the theory and application behind SMDS. Chapter 13 provides an overview of B-ISDN service in relation to the ATM standard. Chapter 14 then presents SONET. Each of these technology chapters in Part 3 explains protocol structure, format, interfaces, and theory. They also contain hardware and design evaluations.

Part 4 steps the reader through defining data network design requirements and performing the traffic analysis and capacity planning process, and finishes with a comparison of circuit-, packet-, frame-, and cell-switched technologies. Chapter 15 assists in analyzing the complete set of user requirements. Chapter 16 then provides the traffic analysis calculations that turn these requirements into a capacity plan. These two chapters lead directly into the access design. Chapter 17 provides numerous comparisons of every aspect of each switched technology and service.

Part 5 deals with choosing vendors. Chapter 18 provides guidelines for the RFI and RFP process as well as criteria for choosing a vendor and the future relationship with that vendor.

Part 6 contains the actual network design portion of this book. Chapter 19 starts with the access design, the point when the user accesses the network or service. Chapter 20 continues with the backbone design. The backbone is often a switched network service. This chapter also contains some valuable practical insights on address and router design techniques. Chapter 21 presents operations and maintenance issues, as well as providing a thorough review of network management protocols and techniques. Chapter 22 closes with a study of design tool capabilities and a short study on the data communications job market.

Part 7 contains an international network design focus and ends with two case studies. Chapter 23 presents a review of the international data communications market, its leading issues, and a study of international private and public data networking. Chapter 24 presents two industry-leader design tools: MAKE Systems' Netool and IBM's Intrepid.

This book also contains several appendixes. Specifically, Appendix A lists the major acronyms and abbreviations used in this book. Appendix B provides a reference of national and international standards sources. Appendix C gives CCITT and ANSI standards for the advanced technologies discussed. Appendix D provides a summary of Class "C" IP addressing. And, finally, the Glossary defines common terms associated with the technologies, architectures, services, and protocols encountered throughout this book.

Introduction

More bandwidth, cheaper bandwidth, flexible bandwidth! The cry goes out for larger and more efficient data pipes over which to transport a plethora of user traffic. As data transfer bandwidths for text, video, voice, and imaging traffic increase exponentially, so do the networks required to support that traffic. We are now living in a distributed data world. Everyone needs access to everyone else's data. Private lines are quickly becoming the exception as switched public and private data networks span the globe. Computers need to talk to one another the same way people pick up the phone and dial anyone in the world. Because of this need, the market for high-speed data transport is exploding. The age of gigabit-per-second data transport is here. LANs, MANs, and WANs have already crossed the 100M bps range and are moving toward FDDI, DQDB, and ATM technologies as platforms for multimegabit services such as SMDS and B-ISDN.

LANs have become an integral element of almost every major corporation. The move is toward visual-oriented end-user interfaces in computer software packages which are becoming icon-based through the use of graphical user interfaces (GUIs). As the number of LANs continues to grow, so do their interconnectivity requirements. The low-speed private line bridge solutions are maxing out, and cost-effective, flexible higher bandwidth is required. Frame relay has emerged to provide users with the first high-speed bandwidth-on-demand sevice and competes directly with private lines. LAN-to-LAN is not the only technology best served by frame relay. Many businesses have an imbedded SNA traffic base. Much of this traffic is also ideal for frame relay transport.

Many businesses' bandwidth requirements are exploding, such as medical institutions which transfer multimegabit imaging files and film-making industries which store and transport video images recorded directly into a computer and digitized. The most important aspect of these styles of networks is their ability to store and retrieve large image files. Switched Multimegabit Data Service (SMDS) has become the first cost-effective, viable option over private lines. SMDS provides a switched public network which handles speeds up to 155M bps, with

plans for even higher bandwidth support. FDDI is hot on its heels, and in many cases ATM will provide the longer-term solution.

In the search for new technology to provide data transport of this scale, we begin to rely on versions of older packet technology. Packet switching has moved into the '90s, in the form of frame relay and cell switching. Even as these two new leading technologies vie for the private and public data market, ATM fast packet vendors are alpha testing products for quick release. As services such as frame relay and SMDS are gaining increasing support from equipment vendors, Telcos, and interexchange carriers, the user, as usual, seems to be the driving factor as to which technology will succeed.

Fiber Distributed Data Interface (FDDI), a 100M bps token-ring-based technology, is serving as a short-term solution for interconnecting LANs within a local campus area. FDDI is a standard of the ANSI X3T9.5 Committee. The need to expand services past the geographical constraints of FDDI are being met by standards such as IEEE 802.6 DQDB networks running SMDS. Synchronous Optical NETwork (SONET) standards are now providing the gigabit optical, virtually error-free transport method for these high-bandwidth services, and ATM hubs and switches will soon appear on the market.

Designing a data network to handle a diverse user base is a complex task. Capacity requirements begin to boggle the mind as LAN users project peak traffic periods of up to 50:1. With X.25 packet switching, the network provided many services which made the user rely on the network for error correction of data and end-to-end link level data integrity. Frame and cell networks push the requirements to the end user CPE and the devices running the applications, and require the end user to apply much more intelligence when transporting data. Now, computer users who before had only to worry about how long it would take to move or access data are faced with data transport protocols which promise to deliver a majority of the data, leaving the discovery of lost data and subsequent retransmission to the application. Two very different ways of viewing data communications are merging—the computer user who looks at data transport in view of delay, and the telecommunications user who views data transport in terms of acceptable levels of lost data, error correction, and retransmission.

The data network designer must now bridge the gap between data transport services and user services. This means building a network which involves both local-area and wide-area data communications. The new developments discussed will bridge the gap between these historically separate domains and provide the end user with a wide-area communications potential. Large-scale wide-area communication networks are becoming economically feasible for smaller businesses, and the LECs and IXCs are building public data networks to capture this market. They are providing connectivity which is flexible and easy to install, provides a low amount of delay and high throughput, and offers pricing that is both fixed-based (like with private line service) and usage-based (like with IP servcice). This allows for cost efficiencies while providing higher transport speeds and larger data volumes.

After learning the technologies and services available, many users ask the classic question, "which service do I use?" This book will show that the answer to this question is based on many factors and may have multiple answers. There is rarely a single solution, and the decision on which technology and service to use generally comes down to what is best for the application and what can be afforded—price versus performance—as well as what means the least change for the user and best positions him or her for future expansion.

How to Use This Book for Courses

Chapters to be taught in a basic architectures, protocols, technologies, and services course (PT1) are Chapters 1 through 14, with stress on Chapters 5 through 14.

Chapters to be taught in an advanced protocols and technologies course (PT2) are Chapters 8 through 23. The student should have a minimum working knowledge of the material contained in Chapters 1 through 7.

Chapters to be included in a pure design course (ND1) are Chapters 15 through 23. The student should have a minimum working knowledge of Chapters 1 through 7, and some knowledge of the material covered in Chapters 8 through 12.

The material in this book can be taught and covered in two or three semesters. Three semesters' worth of study, along with the suggested course outlines and guidelines for selecting the course material, are outlined above. There is some overlap, and the recommended progression is from the basic course PT1, to the advanced course PT2, and finally to the pure design course ND1, with both design tool and test scenario hand calculated network design labs. PT2 and ND1 should begin with the overview chapters shown to reaffirm a working knowledge of basic protocols and their operation, since the advanced protocols are modifications or perturbations of simpler protocols. Labs should contain design problems based on the cumulative knowledge gained from the class reading and outside reading assignments (recent technology updates). The exercises should involve multiple design tool exposure and problems. Special or final exams (all or a portion) should include a network design problem. Students should be encouraged to use the text as a "working document," noting any changes as the standards are revised and updated. The author plans to publish updated editions of this book as appropriate technology changes may warrant. Supplemental documentation and instructional tools can be obtained from the author at extra charge.

Author's Disclaimer

Accurate and timely information is provided up to the date of publication. Many of the standards used in this book were drafts or recommendations at the time of writing and are assumed to have become standards by publication. At times, the author will present material which is practical on a large-scale design, but is simply not possible in the normal small business communications environment. Also, in many cases, examples are presented on a larger scale. The presented material must be "scaled down" on a case-by-case basis. Many data communications networks operate, and will continue to run, quite well on dedicated private lines, but eventually the economics of switched technologies and services, even on the smallest scale, are worth investigating. Please excuse the blatant assumptions that the user is ready for these advanced technologies—in many cases it will take some time before they can be implemented. Also, please excuse any personal biases which may have crept into the text.

Acknowledgments

Many people have helped prepare this book. They have provided comments on various drafts, information on products and services, and other value-added services. In particular, I would like to thank Mr. Carl Geib; Mr. Arthur Henley; Mr. Gary Kessler of MAN Technology Corporation; Dr. David McDysan; Mr. Paul Metzger; Dr. James F. Mollenauer of Technical Strategy Associates, Chair of the IEEE 802.6 (MAN) Subcommittee; Mr. Gene Wahlberg; and other colleagues over the last several years, who have shared their knowledge and expertise. They have helped me develop a greater understanding and appreciation for data network design.

This book does not reflect any policy, position, or posture of MCI Communications. A caveat should be added that this work was not funded or supported financially by MCI Communications or by MCI Communications resources. Ideas and concepts expressed are strictly my own. Information pertaining to specific vendor or service provider products is based upon open literature freely provided and modified as required. My friends and associates at MCI Communications did support the project in spirit, especially my management team of Mr. Lance Boxer and Mr. John Nitzke, and are hereby thanked.

Also, special thanks go to Mr. Herb Frizzell, Sr. for the English style and syntax review as well as select submissions; Mr. Paul Metzger for graphics submissions, special subject reviews, and ideas for textbook and classroom use; Ms. Beverly Dygart of MAKE Systems and Mr. Charan Khurana and Ms. Margot Peterson of IBM for assisting in the design tool documentation; Mr. Edward Brounston of General DataCom, MCI Communications, IBM, MAKE Systems, Netrix, and Wellfleet Communications for providing figures and graphics; and finally Gary Kessler for portions of the appendixes and Glossary.

And thanks especially go to Becky Thomas—her never-ending support throughout helped me accomplish this project.

The combined support and assistance of all these people has made this book possible.

Darren L. Spohn

Data Network Design

1

The Business Case — The Need for Data Communications Networks

How does one determine when a data communications network is needed? Or that an existing data or computer communications network requires a new technology to meet its ever-growing needs? Some may answer, "we need to perform a business case analysis." Others may say, "we know our needs — they are driving us toward one — and we need it now!" Data communications managers and engineers need to solve the questions of why, what, when, and how, and then make business-wise decisions on future data and computer communications.

Part 1 of this book will take you through many of the initial decision processes to determine the actual needs driving a data network. The dollars and cents "business case" can be accomplished only after the network designer or manager knows and understands the requirements. Part 1 will help define the statement of work needed to provide a network design analysis. Past history provides insight into the future. This is also true in data communications. Outsourcing is also important, and a discussion of factors driving the need for outsourcing of data networks will be discussed.

1

What Is Data Network Design?

Data network design is a broad field of study encompassing many communications fields — architectures, topologies, standards, services, protocols, data transport techniques, analysis/planning, and other sciences. It goes beyond simply understanding the technical details of protocols connecting one element of hardware or software to another to facilitate data transfer, and the knowledge of standards and basic telecommunications. Specialization by engineers and managers in these fields is no longer enough.

Past technologies were engineered for the transport of voice traffic. The entire infrastructure of communications was based on voice (analog) technology, and data communications had to adapt to the analog world. A myriad of developments in analog data transport technologies sallied forth the importance of the information era. Digitization of transport networks was the one most significant factor in the evolution and acceleration of the information era. Digital technology has increased network, switch, and computer performance. The same technology has created a transport network capable of provisioning the subscribers with all modes of communications for both voice and data.

Thus, our charter is to understand all aspects of network design in order to make comprehensive decisions in the design of networks. A working understanding of these principles enables the design engineer or manager to successfully design a data communications network from conception through implementation. This chapter focuses on defining data network design by providing a brief high-level overview of the history of voice and data communications. The trend from voice to data affects the network design engineer and manager. The effect of emerging broadband data upon both the industry and network design in general is also critical. We will discuss the data revolution, as well as the move toward computing power decentralization and its effect upon data communications structures and architectures to transport that data. Lastly, we will look at data network engineering and managerial psychology, and discuss their emerging roles and responsibilities.

1.1 WHAT IS NETWORK DESIGN ?

When beginning a network design, the engineer or manager must clearly comprehend several fundamental concepts to understand what network design encompasses. Change is accelerating as we move into the twenty-first century. Data transport formats change, speeds increase at exponential rates, vendors strive to produce products to meet standards under development or create new standards, and technologies become more efficient at a rate which makes equipment seem almost obsolete by the time it is employed in the network. We need to understand how data is quickly replacing voice. Today's engineer must become a data specialist to survive. Studies on recent trends in both the data communications industry and corporate staffing make it apparent that businesses are turning more and more toward data gurus who drive the network design for their company.

Data network design is a difficult term to define. It encompasses and interacts with almost every portion of the business. Data is defined as information from which conclusions are drawn. It is often used in the plural to refer to multiple collections of information. Data can reside in microcomputers, minicomputers, mainframes, storage devices, and even the human brain (although this is sometimes questioned). Data units are grouped into bits and bytes, and these units, in multiples, form data streams. The transportation of data streams over a medium is called data transport. The medium used is called a network. Figure 1.1 depicts a sample data communications network. It is oversimplified to demonstrate that the network extends from the output port on a user device to the input port on a remote destination user device, and may even include user premises equipment.

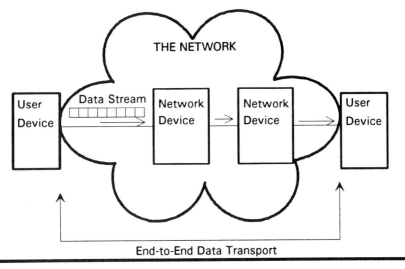

Figure 1.1 Data Communications Network

The true network portion deals with the interconnectivity of multiple devices in a network using passages of data pipes to provide bandwidth. A network is also defined as a snare or catching device, that once built, attracts and "catches" many applications not originally destined to be supported. The "snare" concept will become more evident in subsequent chapters.

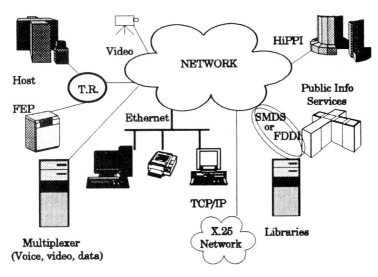

Figure 1.2 Data Communications Network with User-to-Network Interfaces

A data communications network, when designed efficiently, can and does become the lifeblood of a business organization. It can also be extremely detrimental to the business if designed incorrectly. Figure 1.2 depicts another example of a data communications network, with more detail than the former, showing a few of the many types of user-to-network elements interfacing with the network.

This book deals primarily with network design relating to computer and data communications networks. A data communications network design is defined as a complete communications system-level design defining user access (ingress and egress), the transportation medium and data transport elements, and all internal and external factors which affect, manage, or interact with the communications medium. Figure 1.3 below shows a high-level view of the domain of the data communications network. This domain includes the hardware, software, management, and any other subset of the communications network. This list is far from comprehensive but includes the basic elements. We will focus on the network at both a micro and macro level — from detailed architecture and application design to systems level design. Thus, the data network design will encompass many aspects not readily apparent to the design engineer.

Figure 1.3 Data Communications Network Domain

1.2   THE TELEPHONE NETWORK

History repeats itself, and voice and data communications history is no exception to the rule. Voice communications has a long history of competition, starting with the telegraph and box phone, through the extension of long lines business across the country, and into modern day divestiture. This evolution, a by-product of Congress, has been cyclic — competition, regulation, monopoly, deregulation, and again competition. With the advent of a digital network, data communications also has had competition in the T1 Carrier business. Now for a trip through the history of the United States telephone network. It took many competitive efforts to achieve what we have today in the United States. AT&T, once plagued by independent telephone competition at the beginning of the century, succeeded in building a monopoly only to lose it again during divestiture.

1.2.1 The United States

The telephone was patented on March 7, 1876 by Alexander Graham Bell. The first box telephone was patented the very next year, on January 30, 1877. That same year, over six hundred telephones were connected to each other via point-to-point private lines. Bell also carried his invention to Europe in 1877 and demonstrated its operation to Queen Victoria and other world dignitaries. Thus, the information era of telecommunications was born. On January 28, 1878 the first telephone exchange was opened in New Haven, Connecticut, and users were switched manually at "central" by an operator. The number of exchanges spread until the New England Telephone Company was formed on February 12, 1878. Throughout the next three years many companies changed hands. Companies like Western Union, the National Bell Telephone Company, and the American Speaking Telephone Company competed over telephone rights. In November 1879 an agreement was reached that Western Union would remain solely in the telegraph business and the National Bell Telephone Company, which became the American Bell Telephone Company, would remain solely in the "telephone" business.

The next major development was the formation of the American Telephone and Telegraph Company (AT&T) in 1885. This company was formed as a subsidiary of the American Bell Telephone Company to handle the nation's long lines business, but AT&T grew, and in 1900 absorbed the Bell Company. Later, in 1911, AT&T organized territorial divisions called Bell Associated Companies, which in turn paid licensing fees to AT&T for their patents. Throughout the early part of the century AT&T fought a hard battle against independent telephone competition, buying out those they could and refusing to connect to others. The Kingsbury commitment was made by AT&T in 1913, allowing independents to interconnect with AT&T. In 1921 the Graham Act was passed by Congress, legitimizing AT&T as a "natural monopoly". From this point until divestiture, AT&T enjoyed the status of a monopoly.

The AT&T monopoly comprised AT&T Headquarters, 23 Bell Associated Companies, Western Electric Company, and Bell Telephone Laboratories. This monopoly was broken up during what is called "divestiture". The Modified Final Judgment (MFJ) was issued by Judge Harold Greene of the United States District Court in Washington, D.C. for AT&T to divest itself of all of the Bell Operating Companies (BOCs). Figure 1.4 shows each of the Regional Bell Operating Companies (RBOCs) and their area of coverage. These seven RBOCs are now called:

- Pacific Telesis
- U S West
- Southwestern Bell
- Ameritech
- BellSouth
- Bell Atlantic
- NYNEX

Another major decision at divestiture was to not allow RBOCs to manufacture Customer Premise Equipment (CPE). One hundred and sixty Local Area Transport Areas (LATAs) were designated for local access. Independent telephone companies could elect to come under the LATA plan or not; most larger ones did. The RBOCs could provide service within designated LATA areas only, leaving the long distance companies — AT&T, MCI, Sprint — to provide the inter-LATA connectivity. AT&T was split into AT&T Communications and AT&T Technologies and forbidden to use the "Bell" system, and thus AT&T's new "death star" logo.

Data communications networks evolved in much the same way as voice networks. Dedicated private lines connected users to centralized switches, and eventually to national and worldwide switched networks with classes of access and hierarchical backbone designs.

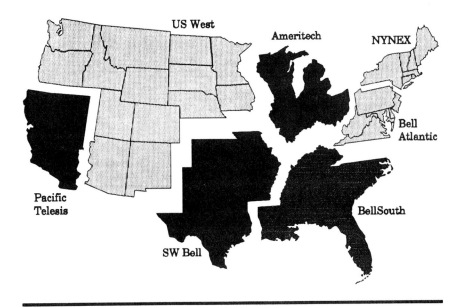

Figure 1.4 Regional Bell Operating Companies

1.2.2 Europe and the Pacific Rim

The European and Pacific Rim telephone industry has evolved in a similar manner as the United States. Most predominant telecommunications carriers are owned by government entities through what are called postal telephone and telegraph (PTT) agencies. In the United States, communications entities are privately owned but regulated by governmental agencies. In Europe and the Pacific Rim the PTTs are both operated *and* regulated by the government. This system is rapidly changing today in the face of increased international competition for communications services. Many of the PTT monopolies are being realigned into two organizations — traditional services such as local dial tone and competitive services such as virtual private network services. The multinational (transnational) companies helped force this change. The competitive environment prevailing in the United States after the AT&T breakup is now happening internationally.

1.2.3 Formation of the T1 Carrier System

As data and voice communications proliferated in the early '60s, the Bell system saw a need to develop a digital transmission network. The early telephone systems were based on frequency division multiplexing (FDM) and relied on multiple signals to be stacked in frequency bands. The new digital networks were based upon time division multiplexing (TDM) techniques. Bell Labs developed the first digital transmission system called the T1 Carrier System. The theory for TDM came in the early 1940s, but not until the invention of the transistor by Bell Labs did the technology catch up with the theory. This new digital transmission system was designed to operate using TDM time slot multiplexing techniques. Since their deployment in 1962, digital transmission systems now form the majority of every common carriers' network, with some carriers such as MCI Communications and Sprint providing a 100 percent digital network. Digital transmission facilities have even extended to the customer premises for service offerings such as Integrated Services Digital Network (ISDN). Digital transmission speeds range from DS0 (64K bps) through DS5 (see Table 1.1).

TABLE 1.1 Digital Transmission Speeds

Digital Multiplexing Level	Number of Voice Channels	Bit Rate (Mbps)		
		N.America	Europe	Japan
0	1	0.064	0.064	0.064
1	24	1.544		1.544
	30		2.048	
	48*	3.152		3.152
2	96	6.312		6.312
	120		8.448	
3	480		34.368	32.064
	672	44.376		
	1344*	91.053		
	1440*			97.728
4	1920		139.264	
	4032	274.176		
	5760			397.200
5	7680		565.148	

*Intermediate multiplexing rates.
Source: Kessler, Gary, *MANs*, McGraw-Hill, 1991.

The European and Pacific Rim digital infrastructure is not as well developed as North America's. Digital data networks are not as predominant and coverage is limited in many European countries. However, this is changing as the PTTs diversify, privatization increases, and competition drives the need for better digital transmission facilities.

1.3 THE EVOLUTION FROM VOICE TO DATA

Data communications not only gives business the competitive edge, it puts them on the "leading edge". This edge will offer both benefits and risks. Once a business experiences effective data communications, it becomes dependent on its own network. The data and computer communications network becomes the lifeblood of the company. A company needs voice network communications, but a voice network can only become so large. It is limited to the population size of the company. Voice communications can be forecasted, and follow common distributions. Data communications traffic characteristics can be a different ball game with new rules. Once a company uses a data network, there is no limit to its potential. Data growth is occurring at a factor of 25 percent per year, far outpacing voice. Fax, groupware, electronic mail, local area networks, wide area networks, PBXs, mainframe, terminal, and computer communications are just a few examples of how data communications far exceeds voice communications in pure traffic volumes. Figure 1.5 shows the growth of data services over roughly the past thirty years.

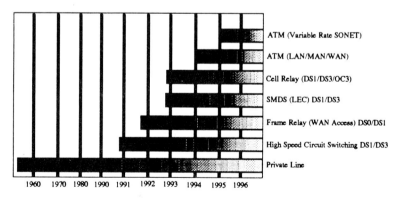

Figure 1.5 Data Services Growth

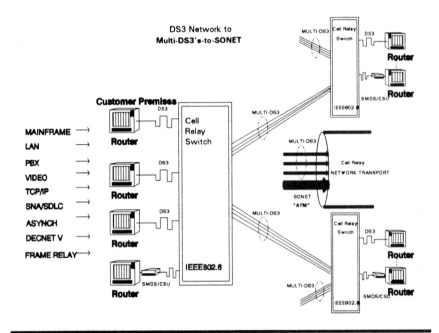

Figure 1.6 The Network as a High-Bandwidth Solution

Figure 1.6 above shows an 802.6/SONET network providing connectivity to many high-bandwidth applications now driving large data communications networks. Many types of traffic are transmitted over a typical data communications network, including:

- Voice
- Data
- Video
- Teleconferencing
- Bit mapped imaging
- Multimedia
- Facsimile

These traffic types transport many distributed processing applications such as:

- Real-time access to files
- Shared high-resolution graphics
- Compound database applications
- Remote database access
- E-mail and messaging systems

- Desktop publishing
- CAD/CAM/CAE/CIM
- Medical imaging
- LAN interconnect
- Bulk file transfer

- Inventory control systems
- Video-conferencing
- Distributed electronic publishing
- Electronic funds transfer (ETP)
- On-line databases and catalogs

- Industrial design
- Product testing
- Check processing
- Financial modeling

The list continues to grow. As fast as the bandwidth becomes available, new applications develop to take advantage of it. This is an evolutionary process. How did we move so quickly from voice networks using analog transmission facilities to data networks riding digital transmission facilities? With this question in mind, we move on to a discussion of the evolution of data communications.

1.3.1 History of Voice to Data

The public telephone network was built on a narrow band to broadband hierarchy. Figure 1.7 depicts the five-step public telephone network hierarchy. Customers receive narrow-band access, typically twisted pair voice. A "pair" is two wires that are "twisted" to minimize analog transmission troubles such as cross talk and interference. In the voice world, these twisted pairs are bundled into cables and then aggregated at the central office, where they become part of the local loop. If they traverse local access transport areas (LATAs), they are aggregated into even larger bandwidths for a long haul over microwave, copper, or more predominantly, fiber. Users typically access the local exchange carrier (LEC) through a Class 5 switch, and occasionally through a Class 4 or 3. These smaller switches then connect to larger switches, such as the Class 3 and 2. The largest is the Class 1 switch, which switches circuits at the largest aggregate level. The larger the class, the greater the throughput and number of users switched. Data communications networks follow the same general hierarchy, with switches in the T1 Carrier System becoming larger as they service greater bandwidths.

Voice channels are time division multiplexed. When compared to data, the main difference is that voice calls are circuit switched, whereas data can be either packet or circuit switched. All of these methods will be discussed later in the text. The primary commonality is that data and voice transport seem to be evolving in the same direction, toward larger-bandwidth circuits. These circuits resemble water pipes, where the larger the user (e.g., a hotel connected to a mainframe), the larger the required pipe size. For example, a single-family home resembles a personal computer, where one or two water pipes would suffice, whereas a hotel would resemble a mainframe, where many water pipes (or one huge pipe like a water main) would be needed to accommodate the entire structure. New technologies such as frame relay and cell switching allow for even better utilization of existing pipes, similar to a water saver on a shower head.

1.3.2 History of Data

Data communications first began with the invention of the telegraph by Samuel F. B. Morse in the year 1846. It was a simple(x) device, allowing one-way transmission of start and stop signals over a large distance. The key of the telegraph is electrical technology applied to telecommunications that for the first time in 5000 years of history *time* and *space* (distance) was spanned at the speed of light. The first telegraph message was from Baltimore to Washington, D.C. and the first message was: "What hath God wrought!" Morse won a $30,000 Congressional prize for this effort. While it was a significant invention, we will dwell on the telegraph concept: transmitting coded pulses over large distances via a transmission medium or carrier. This method is used today even with the most current advanced technology, such as fiber optics transmissions of light pulses over great distances. We have become much more adept at the transmission of data as we will see in future chapters.

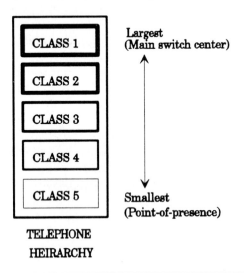

Figure 1.7 Public Telephone Company

The first telecommunications company in the United States was Western Union, founded in 1856. The first data communications protocol was the Morse code (using binary "dots" and "dashes"), the first data communications interface was the tap set or the telegraph operator, and the first data communications service offered was instantaneous overland telegraphy. This discrimination of protocol, interface, and service is very important when discussing data communication principles, and we will see that data communications history repeats itself over and over.

The military was one of the first users of data communications networks. The telegraph played a significant role in the Civil War by northern armies. Many data processing and early computer systems were developed during World War II, when systems integration was necessary, and after the war all of the command and control centers, weapons and sensor systems, voice networks, and the computers which ran these systems needed to be centrally controlled within one interconnected communications systems. This was the beginning of the Department of Defense (DOD) telecommunications architecture. Today, the United States DOD architecture comprises one of many data communications platforms in use. The latest fad in protocols at DOD is GOSIP — government OSI protocols that are now a requirement in Request for Proposals (RFPs).

The next major advance by the DOD was the establishment of Advanced Research Projects Agency Network (ARPANET). AR-PANET was established in 1971 as the first "packet" switch network. This data network connected both military and civilian locations, as well as universities. In 1983 a majority of ARPANET users, including European and Pacific Rim contingents, were split off to form the Defense Data Network (DDN) — also referred to as MILNET. Some locations in the United States and Europe remained ARPANET, and are now merged with the DARPA Internet, which provides connectivity to many universities and national telecommunications networks.

Host-based networks accessed by local and remote terminals evolved through the use of private networks and packet-switched services. The primary example is the IBM Systems Network Architecture (SNA). This architecture provides the platform for many dumb terminals to communicate with an intelligent host or mainframe in a hierarchical fashion. This hierarchy developed because intelligence was at the host, while terminals had little resident intelligence (high cost) and depended upon host resident applications.

Local area networks (LANs) were the next major development in the computer communications networking environment with the advent of Ethernet by Xerox in 1974. The advent of client/server architectures and distributed processing brings us to modern day communications. It behooves us to study data communications history in the recent history of computing, since this is much more relevant at this point. Thus ends History of Data 101 — now for some recent computing history.

Personal Computers (PCs) such as the Apple computer were born in "garage shops". They were known as "toys" for games and other amusements. However, as applications were developed in software and speeds and memory increased as costs dropped, the larger computer manufacturers began to see PCs on the desks of users that previously only had a "dumb" terminal to the host. The big boys woke up and legitimized the PC by IBM's announcement of their PC in 1983. The original personal computer (the one with the big blue IBM logo on the front) hit the market ten years ago. Since 1983 the PC has been the industry standard for corporate and government microcomputing. In that short period of time, the number of professionals using PCs has jumped from 25 to 75 percent. These figures are shown in Figure 1.8. This trend is likely to continue, with a PC on every desktop (and possibly every wrist!) by the year 2000. The estimated size of the personal computing market for 1991 was $35 billion.

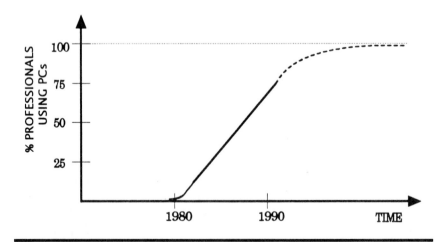

Figure 1.8 Data Communications Trends

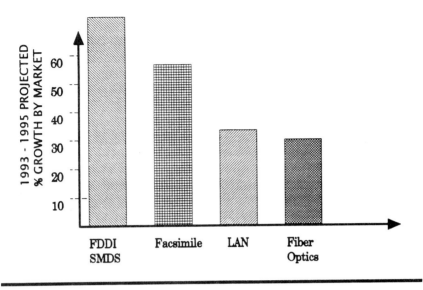

Figure 1.9 Market Growth for LAN/MAN Services

The PC has provided the user with the device for global access to the world of data. Paper has not disappeared from the desktop, and the age of the "paperless office", which some futurists saw as an inevitability, has not yet arrived, but mass storage of information has shrunk to a fraction of its original size and cost. The growth rate of telecommunications hardware such as facsimile, LANs, and fiber optic hardware is estimated at between 20 and 50 percent. Figure 1.9 shows some of these trends and increases. The cost of computer hardware and increased software writing tools has been dropping faster than the cost of transmission facilities.

The challenge that many corporations now face is the flattening of networks and the management of these PC networks which are proliferating at astounding speed. Thus begins the evolution of the corporate data network. What started as a PC for home use fad has now become a corporate must. It was a natural evolutionary choice for the visionary network design engineer to take these islands of information — the distributed LAN and the centralized mainframe arenas users — and create a common environment to facilitate communications. The network design engineer must bridge the gap between different groups of network users hungry for high bandwidth transport from the desktop, and the management information systems (MIS) professionals who have historically networked mainframe technology with hierarchical networks. Emerging cost-effective router solutions of the LAN world are expanding. Voice professionals, who are scrambling to come up to speed in the data market or be swept under the data evolution tide, are evolving as well. What has caused this slide is the decrease in the cost per MIP, now decreasing about 50 percent every two years. Couple this with the distributed processing to handle bandwidth-intensive applications and we see the inevitable repositioning of the mainframe and the desktop computer/workstation.

Be prepared to battle the MIS oligarchic organizations. They still exist. The older more innovative mainframe proponents are now changing their views toward the distributed dynamic bandwidth LAN/MAN/WAN architecture environments of the PC or data communications products now providing multi-MIPS processing power on the desktop. It is the PC which has driven the control from the hands of the MIS executive to the user's fingertips. This is a "power of the people" movement. Just as minicomputers invaded mainframe turf when the cost of minis fell to departmental budget approval levels and the department level users did not have to go to corporate MIS for budget approval, so, also, the router phenomenon has exacerbated LAN-WAN connectivity due to falling cost. Thus cost and control are more in the hands of the end user. Why conform to corporate MIS dictates when you can build your own departmental LAN and handle 90 percent of your data processing needs? The MIS manager is thus forced to integrate his or her VAXs and IBM mainframes into the corporate distributed data structure, which is now being run by entrepreneurial LAN managers. These LAN managers are born from change, and provide the users with the hand-holding, customized design, application development and customer service they have been denied and deprived of for years.

MIS should focus attention on the design and maintenance of the host and front-end processor systems. Bringing these devices into the new mainstream of LAN/WAN/MAN networking is essential, and if done wrong could be fruitless and costly. While these devices still provide valuable resources and applications, they must be integrated by area networks with foresight and precision. In later chapters, the pros and cons of each approach and how to effectively design a data communications network to span both worlds will be presented.

1.3.3 Voice Is Low Speed Data

A standard voice-grade channel can be sampled at 64K bps (or 56K bps). This equates to a bandwidth of 4000 Hz, the typical voice frequency sampled in accordance with Nyquist's theory. Nyquist's theory states that the number of samples taken from any signal needs to be at least *twice* the bandwidth. So the minimum sampling rate for a 4000-Hz voice call would be 8000 samples per second. In fact, many digital encoding techniques now enable a voice channel to be transmitted at speeds as low as 8K bps.

From this analysis, it is readily apparent that voice is indeed just low-speed data when transmitted over digital facilities. It is treated that way for data transmission. The only caveat is that low-speed data for voice is delay sensitive. Users will not tolerate "appreciable delay" during a full duplex or half duplex conversation, nor will they allow their sentence flow to be garbled by the misplaced arrival of words or letters. Satellite delay, which most people have experienced, is an example. Certain new technologies such as Switched Multi-Megabit Data Services can handle delay-sensitive voice transmission when other technologies such as frame relay cannot.

1.4   THE DATA REVOLUTION

The '90s have yielded a broadband data revolution with the rallying cry of "more bandwidth" and "smarter bandwidth" — or more accurately "bandwidth on demand"! Broadband switched data communications is fast becoming the prevalent market in the '90s. This decade may also be called the era of internetworking, simply because of the widespread use of these broadband technologies and services and their interaction with LANs. Key services and switching technologies representing the emerging data market include:

- Frame relay
- SMDS (DQDB)
- MANs
- FDDI / CDDI
- B-ISDN
- ATM
- SONET

The data industry is being revolutionized by the allure of new technologies and services — offering much desired enhanced user control such as bandwidth-on-demand, huge bandwidth pipes, and the integration of multimedia — with reduced network access elements. These services are depicted, along with their bandwidth offerings, in Figure 1.10. In this figure, the timeline goes from left to right, and distinguishes between switched services and private line services. The Y axis shows the characteristic of burstiness, or how bursty the traffic is during transmission. The X axis shows the bandwidth throughput available with the service. The first services to emerge were low-speed access circuits (LSACs) and packet switching. The most recent advances are the Asynchronous Transfer Mode (ATM) protocol and Synchronous Optical NETwork (SONET) transport technology.

Service providers are offering networks with high availability and good reliability. This is true even for corporations and governments which are building enterprise networks. The switch from conventional private lines to broadband switched services is exacerbative, but there will always be a need for dedicated private lines. The incremental reliability of the carriers' network data error rates and the reduced price of switched data services, based on the "economies of scale" of the carriers' frame and cell-based infrastructures, has made switched data services more appealing in comparison to dedicated private line services. And the drive is on for global and transnational data and voice communications networks. Corporate global enterprise networks are proliferating at an astonishing rate as international circuit costs plummet. Does this point to a new world order for data?

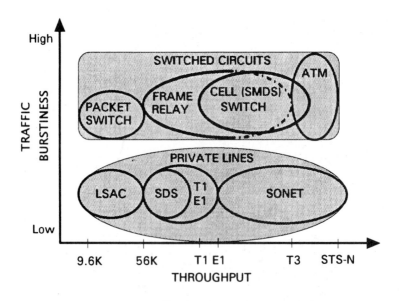

Figure 1.10 Switched and Private Line Services Comparison

1.4.1 A New World Order?

It seems destined that frame relay and cell switching will revolutionize private and public networking in the '90s. Are ATM and SONET the ultimate solutions to multimegabit and gigabit data transport? It is clear that many intelligent applications such as imaging, groupware and multimedia networking, and distributed data processing are driving the need for broadband services. As these networks come of age, the transport networks these services use must also evolve. Intelligent broadband networks, while somewhat fault-tolerant, rely even more on their fiber transport to be both error and outage free. Redundancy alternate restoral must be observed at every step in the design. Since many of these new technologies do not provide error correction, redundancy and alternate pathing to advent broken/disrupted transports is critical. Network survivability must be engineered into every part of the network. There are tradeoffs between the technologies, costs, applications, and time that often drive a decision as efficiently as technology. Either way, a new world order for data is emerging, one which quickly takes on a global view.

Corporate, government, and university internetworking has also exploded. When the corporate network of company A is tied to the corporate network of company B, via either private lines or switched services, internetworking takes place. This is often the case when two companies need to share vital information such as engineering CAD/CAM files, databases, groupware, and other applications which require multiple on-line users to coordinate file sharing. This is, in effect, connectivity between two corporate, government, or university private networks through public network services. Internetworking works well as long as security precautions are taken. Internetwork resources can be local, national, or international (Figure 1.11).

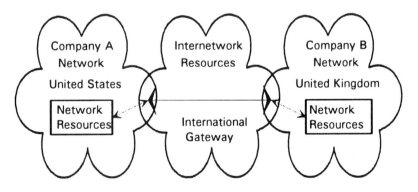

Figure 1.11 Internetworking

1.4.2 The Global View

Data communications has taken on a global view in many facets. Standards such as IEEE 802.6 (SMDS), ATM, and SONET promise to provide interface requirements for worldwide interconnectivity. As these standards emerge, vendors, standards bodies, and users strive to agree upon global interconnectivity issues. While each interest group has its own agenda, the trend is still moving toward global connectivity. International private line connectivity is growing faster than transoceanic cables and fiber can be installed. Figure 1.12 shows the installed and planned installation base of trans-Atlantic and trans-Pacific transoceanic cables. International switched services are blooming, with product offerings such as MCI's Global Communications Service (GCS) providing a broad range of switched services and outsourcing needed by international corporations and government entities. These changes are accelerated by decreasing international transmission costs, which a few years ago were too cost-prohibitive for many customers.

Recent political, technological, economic, and regulatory changes occurring worldwide have spurred international data network interoperability. Primary growth has been in the United States, Europe, Pacific Rim, and Southeast Asia. Markets in South America, New Zealand, Australia, Russia, and countries once part of the Eastern Bloc and the now-defunct Soviet Union are slower to emerge, but are gaining momentum through the development of infrastructures based on state-of-the-art technology. PTT monopolies are realigning and open market competition is making advanced data communications a reality. Many international businesses seek to establish ties and form agreements with the PTTs to remain strong. These are the communications markets of the '90s. Data network design engineers must maintain global views when designing computer and data communications networks.

The '90s are also the age of mergers and strategic partnerships. Every day the sun rises on a new international merger or partnership between carriers, hardware vendors, PTTs, governmental agencies, and small companies who will fill niche markets in the '90s. Many joint ventures have sprung up both nationally and internationally. These range from the computer vendors trying to beat out the smaller clone vendors to the large interexchange and international carriers who vie for entrance into foreign markets. This trend will continue to increase.

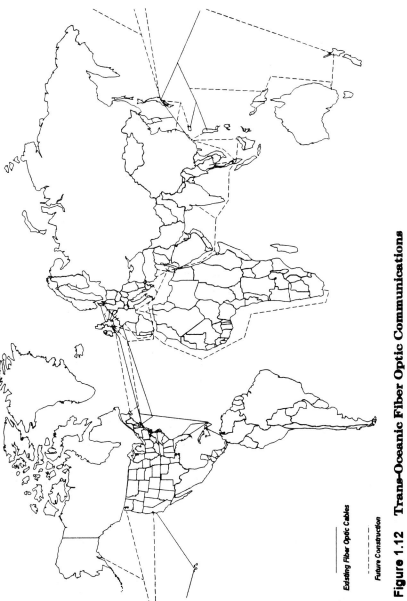

Existing Fiber Optic Cables

Future Construction

Figure 1.12   Trans-Oceanic Fiber Optic Communications

1.5 THE NEW WORLD DATA ENGINEER OR MANAGER — THINK DATA SERVICES!

Data services is the new philosophy for data communications. In an age where customer service and quality reign over price, the real winner is the company or individual who provides quality network services such as good user response, well-planned operations activities, user-friendliness, and an entrepreneurial answer to customer needs. The service providers who accomplish this through deployment of intelligent network technologies will emerge the true winners. Equipment cost is usually one of the last considerations, as multiple vendors are capable of providing products at close to the same price. The communications industry is now competitive for both technology and services. Four main keys to provide data services are:

- Fast and efficient user response
- Efficient and productive operations
- User-friendliness (provider and user defined)
- Entrepreneurial answer to customer needs

Emerging data technologies are never at a loss for vendors. This is evident in the large number of vendors offering products before standards are published. These vendors offer versions of a current recommendation, and announce plans to upgrade to the "official standard" when indeed it becomes a published standard. Often, even users drive standards in test labs before either the vendors or the standards committees can come to agreement and publish the officially accepted standard. Customers have real-world business problems that must be solved today.

Faster and larger networks are definitely within the corporate plan, but these networks are only being built to a size that will efficiently utilize the resources available and be manageable from an operations viewpoint. The engineer of the '90s designs "smart" networks, not just faster and better networks. This new smart network becomes an efficient machine, run by smart data services. The data transport of the '90s is becoming less intelligent and counts on the user to provide the intelligence and network services. The network provider has to offer an efficient transport network to allow intelligence to reside with the user. Network providers must also offer the alternative intelligent network to extend to the user premises. If both services are not offered, the user will go elsewhere.

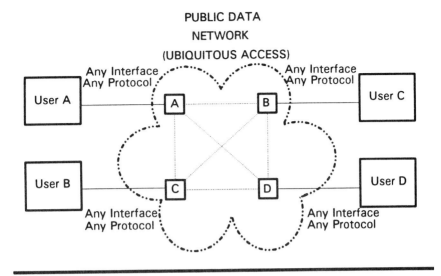

Figure 1.13 The Public Data Network

1.5.1 Decentralization and Ubiquitous Access

Coping with decentralization of computing power and providing the user ubiquitous network access are challenges network service providers now face. Decentralization of computing power makes this transition from network intelligence to user intelligence more palpable. Fully distributed network environments are replacing mainframes and minicomputers and are causing this migration. As the MIPS float out to the desktop, the network intelligence (or capability thereof) goes along with them. Thus, when network equipment is purchased, the need to connect highly intelligent devices at extremely high bandwidths must be considered. Ubiquitous connectivity to an intelligent transport network spanning both the nation and the globe now becomes the rallying cry of the user. User-to-every-user connectivity is the wave of the '90s. Networks must be designed using this major consideration. Users want to access a network cloud and be able to talk to any other user connected to that cloud — true ubiquitous access. Users want public data network services which combine the flexibility of the private data networks with the ubiquity and survivability of the public data networks.

Figure 1.13 simplifies this view and shows one network where any one user can connect with any type of interface and protocol and talk to any other user on the network. While there are many complications and protocol conversions involved in this process, this overall cloud of users who can connect via multiple interfaces and protocols does provide a picture of the future data network service.

There is a trade-off between intelligent networks and intelligent users. Many factors, driven by global industry standardization and the development of technology, will influence these decisions on where the network intelligence resides. The market is both technology-driven and user-driven. International providers want the net intelligence to reside in international gateway nodes. National carriers want intelligence to reside in carrier point-of-presence (POPs). Local exchange carriers (LECs) or PTTs want intelligence in the serving central office (CO). Customer premise equipment (CPE) vendors only want "pipes" from the carriers and the intelligence to reside in the premise's CPE. The customer needs to mix and match all of the above options for the best cost and functionality advantage. They work hand in hand to define new services. Both the network users and network providers must work together to ensure that the technology being used meets business goals. This is critical for continued business operation. It is the responsibility of the network design engineer and manager to take up the flag and ensure that intelligence is used to bridge the technology with the user. Relating these responsibilities to the strategic business objectives of the company will guarantee success.

1.5.2 Planning

Planning is one of the most important functions a data engineer or manager can perform. Proper planning saves money, time, and resources. As data networks continue to grow in size and complexity, so do user demands upon that network. Planning becomes the key factor to network survivability, and thus job security. The data engineer must be able to adapt to the changing user and technology environments. Therefore, an ongoing process of redesign and optimization is a prerequisite for an efficient and cost-effective network.

One example is the recent growth of LANs. Five years ago, the network engineer who built upon proprietary multiplexer solutions to support high-bandwidth traffic could not foresee the growth and change in wide area networking. Soon after, it became quite evident that there would soon be a need for multimegabit LAN data transport. A data manager with foresight laid plans for a network which could support this transport, such as a router or cell switch solution. The manager who did not is now scrambling to throw bandwidth at a proprietary multiplexer solution while quickly coming up to speed with new technologies, as network users are clamoring for these new technologies and services. The recent trend in planning is toward hybrid networks which provide ubiquitous access for a broad range of applications and network support systems. Networks are hybrid at various levels: voice/data, private/public, CO based Centrex-like services/PBX, etc. This book is designed to help you understand these services and technologies.

1.5.3 Dependence of Business on Data Communications

Most business and corporate managers have little idea of the level of dependence they have on data communications networks. This is evident by the small amount of resources usually dedicated to networking in relation to the funding which pays for networks (in fact, data networks produce the greatest revenue and knowledge base to make a company successful, yet executive benefits get more funding than corporate communications budgets). Studies do show that a major portion of a business's office budget is dedicated to data communications. From this dedication, it is intuitively obvious that data communications networks are fundamental to the successful operation of any business. Computers, terminals, modems, facsimile machines, security systems, and even most telephone systems transit some form of data communications network. In fact, we have become more reliant upon data communications than many people realize. (As an experiment, remove the user, or shut down his or her network, LAN, or even PC, and watch to see who screams. It will clearly show user appreciation for the network).

Cost also plays a major role in determining network needs, but relates more toward services provided by the vendor than toward the actual cost of the equipment. Often services supplied by carriers/ vendors are tax-deductible expenses, whereas the purchase of equipment follows the accounting deferment of a capital expenditure. The equipment expenses often turn out to be a very small part of the total expenses of building and running a network. Ongoing support, especially people, are a disproportionate cost compared to equipment costs. While costs are a driving factor, they should be analyzed in detail only when they exceed the project funding limit.

Many companies planning to offer data communications services to customers must first demonstrate that it works internally in their company. A customer is apt to first ask the question: "Do you use the same service to transmit your own critical traffic?" This important consideration must be answered in planning to provide any service to users. As is often the case, the company offering the service becomes the test bed for the service before it is sold to a customer, thus becoming its own best customer.

1.5.4 Psychology

Psychology plays a major role in data network design, though it is hard to quantify. When the two major disciplines of computer communications and telecommunications come together, entire departments can overlap. Support structures, departmental responsibilities, and even functional group responsibilities are duplicated, while each side tries to come up to speed with the other. This can cause emotional and psychological battles which have nothing to do with technological issues. While these are management issues, they should be anticipated. There will be a need to cross these traditional boundaries. The data network manager must clearly define the roles and responsibilities for each individual on the staff. MIS and telephony have traditionally been separate functions. Only recently have they begun merging — often with the MIS/data computer groups absorbing the telecom entity. The MIS telephony group is now more or less strictly involved with voice, as are the data groups that are probably handling voice much more efficiently. It is necessary to get the telecommunications manager and the computer communications manager together to clearly document responsibilities. Cross training should begin immediately. The designer must eliminate or at least minimize the turf battles which result between telecommunications wiring closets and computer server locations, communications room patch panel and cross connect access, and modem pool and bypass access.

There also exists a sort of technology struggle between the old style of brute force mainframe and front-end processors against the new style of desktop devices with RISC and CISC processors (and even the 386, 486, and 586 PCs) capable of outperforming some of the older mainframes and front end processors in price and performance. Each has its place in the new hybrid computer-telecommunications network. The important thing to understand is the interaction between each of these interest groups and to develop the mind set of a data network engineer who understands both worlds. Above all, think data!

1.5.5 The Data Engineer and the Data Manager

A distinction must be made between the voice engineer and the data engineer. The voice engineer deals primarily with analog communications, which include standard telephony elements such as telephones, wiring closets and punch-down blocks, phone lines, cross connects, PBXs, and other communications equipment which interface to the digital portion of the communications network. The data or computer communications engineer looks at the world from ones and zeros to the big picture of voice and data systems integration. It is not quite that simple, and many data designs are much more complex than analog designs, but there is a major distinction. The data engineer deals with both analog and digital communications because all analog data is changed to digital data before it is transferred across a network. This is especially true with the new virtual private and public data services (VPDS). Thus, a knowledge of both analog and digital is essential to the data engineer. As voice circuits are converted to low-speed digital circuits, it becomes apparent that the future of data communications lies in the digital fiber optics arena.

Analog networks can include digital communications. One example is a modem. Modulator-demodulators (modems) convert a *digital* computer or terminal signal to *analog* tones for transport across an analog network. A data network has the capability of transmitting analog signals across digital facilities by using methods of sampling, coding, and quantizing. These digitizing conversions are explained in basic books on communication theory and are not discussed here, but the principles are similar. While there is still a call for skilled voice engineers, data engineers will dominate the market by the end of the decade, if not before then. Many businesses are realizing the need for effective data communications, and are integrating their voice communications with data communications. Integrated services digital network (ISDN) services are indicating this change.

The data network engineer and manager must have at least a working knowledge of available technologies and strive to attain a broad base of knowledge to be able to provide a working solution to any networking problem. It is good if the data manager has served some time as a data engineer. There are conflicting views on this subject. However, the view to which this book subscribes is that a good engineer with proper management training and project management experience will make a better engineering manager than a general-purpose manager who is later given some technical training. This assumes that the engineer has education in general-purpose business strategies, and that the educational level is achieved through higher education. This statement is based upon two major assumptions. It is easier to train a technical mind to perform administrative management duties than it is to teach an administrative management mind the technical aspects of complex technologies. Also, network project engineers pick up management skills early in their careers by virtue of the engineering environment.

1.6   REVIEW

In this chapter we defined data network design beginning with a history of both voice telecommunications and data computer communications, and how both have evolved and combined to provide a new hybrid data network for the '90s. These new networks provide access for many types of traffic and applications, many of which require large amounts of bandwidth and an intelligent network to provide ubiquitous access through bandwidth-on-demand services and technologies. This data revolution has affected the data network design engineer and manager, from both a national and global view of data services and technologies. The importance of planning in network design was covered. The chapter finished with insight into the roles and responsibilities of the new world data design engineer and manager.

2

The Driving Need
for a Data Network

What drives the need for a data network? Data and computer communications are an integral part of any major business. There comes
a time when the business must cope with the changing needs of the
users. An efficient strategic business plan often provides too high a
level of detail to address these needs, and the tactical operating plans
generally only include voice transmission services, leaving the data
communications to the MIS/data departmental managers to deal
with as best they can. This chapter discusses some perspectives of
the corporate need for data communications, as well as addresses the
business needs driving the requirement for data transport networks.
Outsourcing is analyzed in a objective manner, allowing users to
determine if outsourcing is a viable alternative. Next a discussion of
switched networks versus private lines, and a discussion of virtual
data networks in both the public and private network environment.
To conclude, cost factors of data networks will be explained along
with a quick overview of information services. The new buzz word,
"interoperability", and what it means to today's data networks, will
be looked at as well.

2.1   DEFINING THE NEED

The first step in the process of corporate data identity is to define the need for data communications services and to prioritize this with respect to other needs in the organization. The economic cost justification is for later sections and will concentrate only on the technological factors driving this need and the options available to the network manager.

Let us take a look at the most likely scenario. The business has expanded, creating more positions, more departments, more information flow, and thus computing power continues to grow. As computers proliferate, so does the need for data communications and data transport. The business becomes reliant on its data communications, and the "network" becomes a force with which to be reckoned. The corporate communications budget becomes larger each year, and the costs of available data services decrease daily. Maybe LANs grow throughout departments, or mainframe connectivity and applications move to the workstation. There is an infinite number of scenarios, but each leads to one conclusion: an explosion of data traffic needs to be moved around the business and distributed through computing to the desktop.

How do we define these increased data requirements and the phenomenal growth of user applications, processing power, and the sheer need for connectivity and bandwidth? Let us now view some perspectives on recent data communication growth.

2.1.1 Perspectives

Data communications, including text, video, and imaging, will grow from a $50 billion market at the start of the '90s to a $200 billion market by the middle of the decade, a 300 percent increase in five short years! This exponential curve gets steeper as time goes on. The number of million instructions per second (MIPS) sitting on the average desktop would have filled a medium office building 20 years ago. Twenty-five percent of these MIPS are interconnected today, and predictions show this by the middle of the decade to exceed 50 percent. New distributed processing applications are driving this requirement to interconnect. And not only are MIPS more distributed, but more traffic is being generated. The cost of PC-based MIPS is now about $1000, compared to the $100,000 cost of each mainframe MIP. But this is not the only driving factor for decentralized computing. It is the centralized computing power which is now done on the desktop, as well as the need for everyone to share everyone else's information through applications like groupware, shared databases, and electronic mail. Again, applications drive the technology.

The real rivalry is between the user applications designed to transfer large files and databases between desktops, and the transport needed to efficiently support that traffic. It is a battle of the desktop MIPS against the network transport megabits per second. As always, the network bandwidth is never large enough. The hard part is providing the wide area network transport to effectively utilize and distribute the desktop MIPS. But new technologies such as Fiber Distributed Data Interface (FDDI) and Switched Multimegabit Data Service (SMDS) now have the capability to provide LAN bandwidths over wide geographic areas.

Network architectures and vendor proprietary architectures, such as IBM's System Network Architecture (SNA), are becoming more distributed. New applications are constantly emerging from these vendor and standard distributed environments, enough to easily fill the bandwidth provided by the network. Bandwidth is like money. How much does it take to satisfy? More! The other major development driving the data and computer communications explosion is the movement of voice traffic off of private networks and back to public switched networks. Network services is another category. Virtual private network (VPN) services emulate private networks but ride on public network service platforms. This migration of voice to VPNs has caused private networks to increasingly support data, thus increasing the number of hybrid private/public data communications networks. A customer can balance the cost savings of a private data network, accessing a public data network and the services it offers, or a combination of both. Carriers, with their network nodes providing large "economies of scale", are now seeking to lure private network data traffic to their public platforms by using public switched data service offerings.

Most existing private wide area networks use one of the following:

- X.25 packet switching (the predominant international technology)
- High-speed circuit switching (HSCS)
- IBM Systems Network Architecture (SNA)
- TCP/IP

Voice continues to use circuit switching. This is now changing with the new technologies we will soon discuss.

2.1.2 Network Perception

One important factor to address when performing the business case is the executive perception of the network need. Not only must company executives understand what the data engineers and managers are trying to accomplish, but they must also understand the basic technology at hand. This begs the question — "How familiar are executives with the technology being used?" Familiarity begins with the executive using a PC in his or her daily tasks, assuming that the executive does not just use his or her PC for electronic mail and word processing. This could take the form of a desktop, laptop, or even a terminal using an multimedia applications. If these fundamentals are not used by the executive making the decisions for data communications systems, it will be difficult for the executive to relate to more broad-based computer communications decisions. In some cases, one must use highly developed consultative skills to cost justify network expansion via efficiency studies. Corporate executives handle decisions at a strategic level. Engineering management handles decisions on an operational or tactical level. Executives can best relate to the strategic importance of a communications network if they understand the underlying computer technology. They must understand how the technological decisions will affect their organization's strategic mission.

The technical aspect of the business case is just as important as the financial aspects. What may look good in numbers may position the company for technical disaster or obsolescence. One will also find trade-offs in the initial design, either integrating the most critical applications or lowest-priority applications. Make sure that both the technology and the financial, as well as the strategic and operational, perspectives are studied before a business decision is made.

2.1.3 From Centralized to Distributed Networks

Computer communications networks have quickly evolved from centralized power computing using mainframes, through the minicomputer era and now into the personal microcomputer and distributed processing era. The first data computer communications networks resembled a hierarchical star topology. All access from remote sites was homed back to a central location where the mainframe computer resided (usually an IBM host or DEC VAX). Figure 2.1 shows this centralized, hierarchical computer concept.

Today, more and more computing is accomplished through distributed processing. Most sites have the intelligence and capability to communicate with many other sites directly on a peer level, rather than through a centralized computer, although the host can be appraised of what took place and can maintain the database from which management reports can be generated and operations and maintenance issues addressed, such as IBM's NetView. Also, the actual processing of information is distributed among many of the user sites, rather than done solely at a centralized location.

Figure 2.2 shows a computer communications network with distributed processing. Many local, metropolitan, and wide area networks are built with the concept of distributed processing in mind.

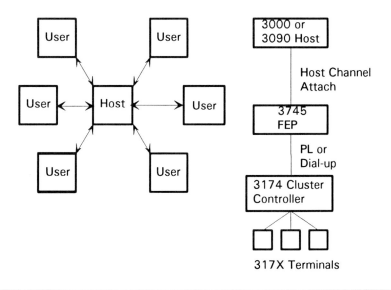

Figure 2.1 Centralized Hierarchical Topology

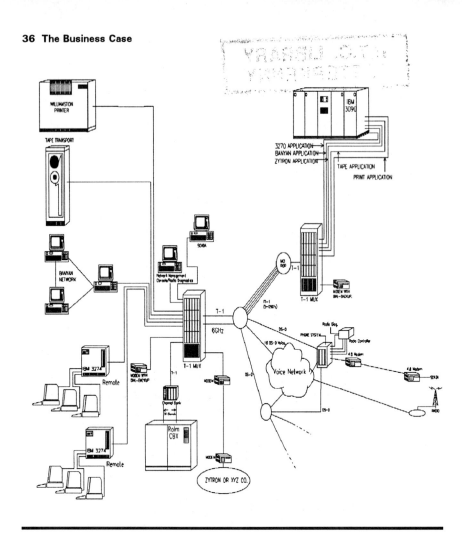

Figure 2.2 Distributed Processing Network

2.1.4 The Need for LAN/MAN/WAN Connectivity

Previously, two scenarios were outlined: the corporation now needs a LAN to allow internetworking between distributed computing devices; and the corporation's LANs have exceeded the bounds of their local existence and now need connectivity across a larger geographic area. The geographically dispersed LANs now have a choice of connectivity ranging from dedicated circuits to transport via a switched wide area or metropolitan area network. The decision of technology and services is based upon many factors other than cost.

In this section we are concerned mainly with validating the need to connect these disparate LANs.

Computer networking has been defined in many terms. Four will be used throughout our discussion:

LAN — Local Area Network — less than 3 km in distance; provides local connectivity typically within a building

MAN — Metropolitan Area Network — less than 50 km; provides regional connectivity typically within a campus or small geographic area

WAN — Wide Area Network — no limit of distance; provides national connectivity

GAN — Global Area Network — no limit of distance; provides global connectivity

The trend is moving away from the majority of traffic originating on the mainframe system to a majority originating on the LAN (or file server) itself. Distributed and work group computing are growing into enterprise and global computing. Typical LAN interconnect speeds range from DS0 to T1, but are now moving quickly toward T3, FDDI, and SMDS. Once the LAN has been established and is operational, there are many factors which drive the LAN to expand its physical and logical size. Data traffic is now growing at rates close to 30 percent per year, and bridges, routers, and gateways make expanding LANs easily and in a cost-effective manner. Even worldwide LAN access is not out of the question. The need to expand local area networking usually falls into one or more of the following categories:

- Increased LAN internetworking speeds
- High transmission rates
- Applications functionality
- Cross-domain access capabilities
- Bypass the boundaries of the local area into the wide area networking arena
- Low installation costs
- Easy expansion

These benefits of internetworking LANs are driving hybrid private/public WAN services, with a strong emphasis toward bandwidth-on-demand. The LAN industry has planned revenue growth from $8 million, at the beginning of the '90s to over $30 million by the mid-'90s, with a majority being made in support services as opposed to the product revenues of hardware and software. Users are spending much more on service and support budgets than hardware and software budgets. This statistic shows the decreasing cost of the equipment as opposed to the support systems required to run the LAN. The use of high-bandwidth circuits and services to support these networks doubles each year. All of these statistics point toward a drastic increase in the need for LAN/MAN/WAN interconnectivity.

2.1.5 The Accelerating Bandwidth Principle (Need for High-Bandwidth Pipes)

The accelerating bandwidth principle is defined by the author as: "users will require an increase of twice the bandwidth that they currently use every three years". This equates to about a 25 percent growth factor. This factor may become a conservative number based on the emergence of bandwidth-intensive applications such as imaging, EDI, and groupware. Again, this value applies to the average business. Some businesses will have a smaller or larger growth factor.

Applications driving these high-growth-factor numbers include:

- LAN/MAN/WAN internetworking
- Data center channel-to-channel connectivity
- Mainframe and Front-end processor communications
- Video
- Multimedia
- Graphic intensive engineering applications (e.g., CAD/CAM)

Users want these LAN/WAN/MAN services bandwidth-on-demand.

It can be charted that the cost of providing data communications decreases as the amount of bandwidth being transported increases. The economic justification point is reached (see Figure 2.3) when the data communications cost of bandwidth and the increasing bandwidth needs of the user meet. This is where it is economically feasible to provide the user X amount of bandwidth.

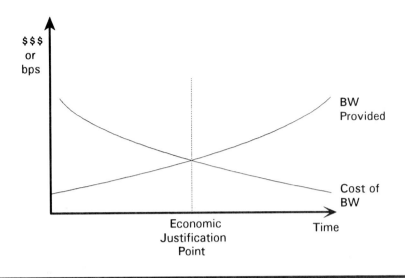

Figure 2.3 Economic Justification Point

2.1.6 Transmission Media and Protocol Enhancements

Two other factors have also served to accelerate the process toward the data network: enhancements to protocols that guarantee the effective transmission of data and to the media transporting these protocols.

Discussed previously was the decentralization of computing power to the desktop. While these desktop machines have the processing power from the host to run the applications, they also have more control of how the information is passed, employing a larger range of controlling protocols such as TCP/IP, DECnet, and other network and transmission protocols. These enhancements have pushed network intelligence out towards the user, allowing the network to employ less intelligent transport protocols but also employing less overhead.

Since network protocols are becoming less intelligent and relying more on higher-level protocols residing at the application, they need to rely on a more reliable transport media. The requirement for intelligent transport protocols is a result of dramatically more efficient (error-free) communications over digital facilities versus analog facilities of the past. This is where reliance on fiber optics enters the picture. Fiber optic transmission allows for low error rates and reliable transport as required by frame relay and SMDS. Fiber optics are now prevalent in many sectors of the communication industry, from the customer premises, through the local loop, and on to the interexchange carriers' backbone networks. We will see later, as we move toward the new technologies such as frame relay and cell switching, the impact fiber optic enhancements have made on new technologies.

2.2   OUTSOURCING vs. INTERNAL DESIGN

Now that we have determined the need for a new data network, should it be designed and maintained internally, or contracted out? Why has outsourcing become a buzzword of the '90s? The question of outsourcing versus internal design is one which spans every aspect of the business. One of the most important decisions for the network manager is whether to build a network for the business owned and managed by company employees, or to outsource network needs to a third party, turning over various elements of control from simple monitoring to ownership and control of the entire network (including people).

This decision is often one of business policy, based on many factors including corporate resource availability, sensitivity of the data, return on investment analysis, skillset and reliability of the vendor, cost factors, retention of control, and the business charter of the company. Other factors, such as how long and how extensive the support is required, the contract stipulations between user and vendor, and the loyalty to existing company employees, also play a major role in the decision. Since a company's computer and communications system is the lifeblood of the organization, outsourcing is one of the most important decisions an organization can make. What is outsourcing? What issues are associated with an outsourcing contract?

2.2.1   What Is Outsourcing?

Outsourcing is the method of contracting an outside vendor to develop and implement a solution for a company's communication needs. This may include, but is not limited to, planning, designing, installing, managing, owning, leasing, operating, and controlling a communications network. Full responsibility for some portion of the company network communications assets will be transferred to or assumed by the outsourcing vendor. Many business aspects could be affected by outsourcing, including resources such as staff and existing investments, questions on skillsets and reliability of the vendors performing the outsourcing function, the cost savings or eventual loss, control of the hardware, monitoring, the types of traffic to be outsourced, and finally the future of the company to either continue outsourcing or to bring the business back in house. Each one of these will be explored in detail. As with any business case analysis, all expenses of an outsourcing deal should be analyzed in detail for proof of validity of cost savings versus expenditures. What needs to be outsourced?

2.2.2 Understand What Needs to Be Outsourced

The organization considering outsourcing must clearly define what portion of the business needs to be outsourced. This could range from simple network management to the entire communications network. Taking a global business view, this could even extend to other areas of the business which are not communications oriented. Besides analyzing the cost ratios, a complete study of user traffic must be performed (similar to that covered in Chapters 15 and 16). Only after the scope of computing and communications resources and systems that need to be outsourced is analyzed can a comparison and business case be completed. This outsourcing study should be accomplished at least every other year to optimize internal efficiency and to determine if the in-house systems are operating at maximum efficiency.

2.2.3 Resources

When a company decides to outsource, there will either be a reduction in staff if they already had one or an avoidance of the need for network staff. If there were individuals assigned to the network, they will either be reassigned or cut. Nine out of 10 times they are cut, or forced to subcontract to the outsourcing company. They then can become hostile and eventually resign. This can translate to cost savings for the company in head count or the ability to use these people for other job functions, or it may have a detrimental effect to efficiency. The outsourcing vendor can become an extension of the networking staff, allowing the network professionals within the company to change the focus of their efforts. The outsourcing deal must be closely monitored for increasing vendor fees, a degradation of performance, and a loss of priority, which may cause potential problems down the road. If the outsourcing goes sour, rehiring good people who were cut in favor of the vendor could be quite expensive and difficult.

Outsourcing can also provide an increased value-added service to companies who only wish to outsource a portion of their network operations, such as monitoring, disaster recovery, or data transport. Thus, the vendor supplies a service which can best be provided internally while control over network resources can be maintained.

2.2.4 Skillsets and Reliability

Will the vendor you have outsourced to have the same or better skillsets and reliability than your own staff? Does your existing staff lack the efficiency and organization needed to manage the network? Measuring the trade-offs of existing skillsets to those gained from outsourcing is a necessity. The critical cost factors in outsourcing are people, not computers. Skilled professionals are difficult to find and expensive to retrain. It takes a large investment to keep them on the edge with constant technological updates and training.

Researching the vendor's past experience with customers is a necessary task before choosing a vendor. Contact current customers for opinions and determine if the vendor has managed a network such as yours before. Experienced data communications managers and staffs are difficult to find, especially those who know and understand the business and the user community. It will take time before you can rely on the vendor for these skills.

A benefit of outsourcing is that it enables your experienced staff to work on other strategic projects. The routine everyday repetitive tasks are sometimes best outsourced to relieve existing staff and free up resources to focus on mission critical applications. These freed up resources can enhance the long-term strategy of the business by avoiding the more mundane network management tasks. One such task discussed is the migration of applications from mainframe platforms to distributed processing platforms of mini- and micro-computers networked by LANs. Many outsourcing firms hire technology experts which are expensive resources to keep on staff, yet are packaged with the outsourcing deal. Outsourcing can also be the catalyst to help combine departmental resources from both telecommunications and computer communications departments into one department.

2.2.5 Monetary Impacts

Monetary impacts vary between corporations, but the bottom line is how outsourcing affects your departmental and corporation expenses. Cost still seems to be the driving factor in the decision to outsource. It is a consensus that outsourcing can save a corporation money, and that large capital expenditures can be minimized (mostly incurred by the outsourcer — but shared with other contracts) but all cost savings and additional costs incurred (including opportunity costs) must be measured in the business case. Costs associated with corporate networking must be clearly defined and compared to the portion of these costs outsourcing will eliminate or augment. An outside vendor can achieve cost savings by providing network services such as access to their switched networks, which can save the corporation currently using private lines a considerable sum of money. The true savings can be achieved when the vendor supplies all network services down to the user, and the cost of outsourcing can be directly compared to the cost of existing or planned operational costs.

Some companies with financial difficulties or shortages of capital prefer outsourcing to take care of communications while they concentrate on building the core business. This rationale should only be pursued for the short term and has an addictive effect upon the user, who eventually may want to bring the network in-house, or in the case of large outsourcing companies like EDS, being purchased by the outsourcing company.

2.2.6 Control

There is always the fear of losing control of communications to the vendor. This is a valid fear. If you control the intelligence of a company, you control the company. Loss of the company computing resources or communications can cause the company to stop dead in its tracks, and possibly even financially ruin the business. Imagine a bank who could not communicate between its branches across the country, or the New York Stock Exchange not being able to communicate the hundreds of billions of dollars in trade each day! A corporation always wants control of its mission critical applications which are key to its success, and full outsourcing will limit this control. Privacy, security, and availability become critical questions for the vendor. How do you manage the vendor's effectiveness in handling your traffic? This is part of the trust relationship which must be built between the user and the vendor. If the corporation retains its network manager, which it should, they are now in charge of making sure the vendors fulfill their outsourcing objectives. If that manager loses his people and high-level control, the company loses his or her allegiance and this will cost the company even more. The same proactive attitude which would prevail internal to the corporation should also be shown by the vendor.

2.2.7 To What Extent?

How high on the OSI reference model do you outsource? Outsourcing can range from private line physical layer transport of data to complete outsourcing services which include all seven layers of the model. The more layers that are outsourced, the higher the level of expertise needed, and the farther the outsource extends to the user. Typical outsourcing for switched services ranges from the physical layer (Layer 1) to the transport or internetworking layer (Layer 4). In this scenario, almost all of the application control remains with the user.

2.2.8 How Long?

There is no predetermined length of time for outsourcing, but, in general, contract periods should not extend more than five years before renewal. This cycle can be reduced down to as much as two years with reviews and cancellation clauses, depending on the shelf life of the technology. This prohibits locking the organization into one vendor or one technology cycle. The half-life for technology is shrinking year by year, and extended outsourcing contracts can lock the user into an obsolete or non-cost-effective technology platform — ultimately impacting the overall business. Price structures should be renegotiated even more often. Beware locking into and becoming dependent upon one vendor who uses this reliance to raise the price of outsourcing accordingly. One way to avoid this time trap is to always use technologies that strictly adhere to international standards and provide a clear migration path to future technologies. Also, some corporations have divided up outsourcing into at least two parts to let competing vendors vie for their business.

2.2.9 Vendor/User Relationships — The Contract

A very close relationship exists between user and vendor. This relationship is built during the vendor proposal stage. A good idea is to observe how the vendor handles methods of measurement for performance, diversity, availability, reliability, technology revisions and upgrades, and backup plans for their other large accounts. This is probably how you will be treated as well. Always remember to treat the vendors with respect, as they will be your partner for the length of the contract. Be aware, however, that some of corporate resources will be required to "manage the vendor", and a lack of vendor control or understanding can easily damage this relationship.

2.2.10 Loyalty

Loyalty takes many forms, and the most important is employee loyalty. An employee is as loyal to the company as the company is to him or her. The staff affected by outsourcing should be informed as soon as the deal is being considered. Outsourcing should be portrayed as an opportunity, not a career stopper. Try to provide growth and promotion opportunities for employees working under outsourced networks. After an outsourcing deal has been signed, reassign, or restructure as many people as possible. Transfer to the outsourcing company is also a possibility (who better knows your own network than you!). Retrain the network managers to become vendor managers and operations managers, managing the operations and outsourcing agreements, maximizing the knowledge and resources of the vendor.

2.2.11 Summary of Benefits and Drawbacks

The primary *benefits* of outsourcing are:

+ Use vendor experience / specialists
+ Focus on core business rather than running networks
+ Outsource management and retain design and improvement
+ Tap a good source of quick network resources
+ Reduce cost
+ Augment existing work force with skilled workers
+ Reduce computing / communications staff
+ Combine computing and communications departments
+ Develop applications and business practices

The *drawbacks* of outsourcing are:

  - Loss of control
  - Possible loss of resources if retain networking operations
  - Loss of expertise in-house
  - Alienation of users
  - Possible sacrifice of technology flexibility
  - Risk of impact to critical systems if vendor fails

These drawbacks can be countered with good planning, smart management, and proper choice of vendor. Unfortunately, these factors are not always controllable.

2.2.12 The Future

The market for outsourcing continues to grow at a rapid rate. Outsourcing is clearly here to stay, and it is clearly having a major impact on the computer and communications industries. There are obviously many short-term benefits from outsourcing, but what about the long-term effects? Many people who disagree with outsourcing say that the long-term expenses outweigh the short-term gains. If you lose your ability to manage the network internally (control), and the external vendor source fails to perform, how hard is it to rebuild the internal networking department? What has been lost, mainly skilled people, which cannot be regained? When outsourcing is chosen as the alternative, a strategic plan must be implemented with contingencies for each of these possible scenarios. Turning over your network to the vendor in increments until you can judge his or her capabilities, called selective outsourcing, is one answer. Regardless of which solution is chosen, a clear-cut plan must be in place vis-à-vis the vendor for a minimum of two years. Define the vendor's plan for updating technology and workforce. Make sure the vendor is able to adapt to your business needs, as well as his or her own, to maintain your future growth and competitiveness. Alternate outsourcing vendors must also remain an option.

If the design is performed internally, the network engineers must include both good strategic planning as well as innovative techniques. They must know when to use the resources available and know the time to contract services out when the skillsets do not exist to complete certain tasks. After completing the analysis of outsourcing and deciding not to outsource, it is time to analyze the systems and technologies available to optimize the in-house operation. Most companies will shop around before outsourcing, and as they do multiple vendor outsourcing may become a viable alternative. Portions of the organization are outsourced, while others are retained in-house. It seems now as if most companies are outsourcing systems integration, while voice and data integration and data network management come in last.

During the planning stages, a balance must be maintained between designing a network to accommodate internal applications and designing the internal applications to accommodate the network. It is most common after a network is built for both the network and the applications which ride it to develop and grow together. Part of the capacity planning process, discussed later, is critical at this point. Applications are analyzed on a case-by-case basis, as well as together, to determine the best network solution. The network design engineers will have various levels of knowledge at their disposal, ranging from city pairs and projected traffic bandwidths to a complete user profile for each application.

Once the decisions concerning outsourcing have been made, the requirements are compiled and two scenarios are compared — the private line and switched network configurations. This step goes hand in hand with the requirements analysis covered later. Cost factors are usually the driving factor. The successful network designer must also be a consultative cost analyst.

2.3   PRIVATE LEASED LINES vs. SWITCHED NETWORKS

Developing an understanding of the need for high-performance, high-bandwidth data communications helps us understand the general types of networks available today. These include private line or leased line networks, switched networks, and hybrid designs accommodating both. Two primary alternatives to data transport networks are available. Dedicated lines, also called private or leased lines, are dedicated circuits between two or more user devices. This type of circuit represents a dedicated private portion of bandwidth between two or more ports on the network, hence the term private line. This private circuit is available for a set fee to a customer 24 hours a day. High volumes of traffic justify this type of circuit. No other customers use it since it is dedicated to one user only. These circuits are also leased by carriers to the public, hence the term leased lines.

The second alternative is switched network access. This can range from simple circuit switching, in which users dynamically select from a pool of multiple public service lines with fixed bandwidths, to intelligent ubiquitous switched access networks where bandwidth is only allocated and used when needed.

Corporate communications usually become a hybrid network employing both solutions. The circuits requiring dedicated bandwidth to accommodate predictable volumes of constant-bandwidth traffic use lines and circuits requiring one to many connectivity, bandwidth on demand, and flexible access use switched networks. These decisions are also influenced by other factors such as burstiness of data, traffic patterns and bandwidth maximums, minimum delay, etc., which affect performance. The three types of network services which match the access types above include; private line services dedicated to one customer, virtual private services which look like private services but rides on a public network platform, and public network services.

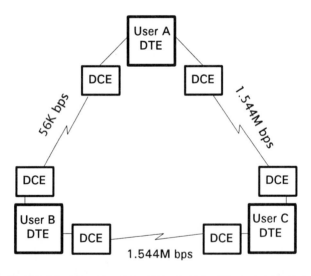

Figure 2.4 Leased Line Network

2.3.1 Private (Leased) Line Networks

Private lines are the simplest form of point-to-point communications. Leased lines leased from a carrier are a form of private line. Private lines provide a dedicated circuit between two points. Figure 2.4 shows three user devices connected via private lines. User A has a dedicated 56K bps circuit to user B, as well as a dedicated 1.544M bps circuit to user C. Users B and C have a dedicated 1.544M bps circuit between them. Private line bandwidths will vary, but typically follow standard speed conventions of 9600 bps, 19.2K bps, 56K bps, and 1.544M bps. A user will generally lease a private line when they want the assurance that it will always be available between two points of choice. They do not want to share this bandwidth with anyone else.

Leased lines come in many grades and speeds. The most basic traditional service available is analog leased lines. These lines require a modem for data-to-voice conversion and transmission. This is the poorest quality of service, yet the least expensive. Digital data service (DDS) is the next higher grade of digital transmission offered by the major IXCs and regional holding companies both inter-LATA and intra-LATA, respectively, as well as by alternate access providers such as Teleport and Metropolitan Fiber Systems. DDS is a private line digital service for data transmission, providing access speeds of 2400, 4800, 9600, and 19,200 bps. DDS is generally more expensive and more reliable than analog leased lines.

Options for higher bandwidth access include subrate digital multiplexing (SRDM), fractional T1, and dedicated T1, T2, and T3. SRDM offers the same access speeds as DDS, but allows aggregation of many low-speed channels into a single DS0 for cost savings. Fractional T1 offers the same type of service, but at a DS1 level. Dedicated T1, T2, and T3 offer just what they state, a single, dedicated high-bandwidth circuit to access the carrier. If a user wants additional functionality, reliability, and availability, switched services are the alternative. In later chapters, the emergence of Synchronous Optical Networks (SONET) will show the technological quantum leap that will change high-speed user access in the late '90s.

While dedicated private line circuits provide the benefit of guaranteed available bandwidth, they are typically nonredundant. If the private line should fail or be taken out of commission (e.g., a fiber cut) the users on each end cannot communicate (unless they have dial backup). Thus, the user must decide what level of availability is needed for communications between facilities. A decision for switched services is predicated on these trade-offs.

Many options are available for assuring high availability in private lines. Automatic Line Protection Service (ALPS) can provide a hot-swap capability between two circuits of the same speed. When employing ALPS, two 1.544M bps trunks are used. If one goes down, the ALPS system automatically switches traffic to the remaining 1.544M bps trunk with minimal service interruption of typically 50 milliseconds. Figure 2.5 depicts the use of an ALPS system between User A and User C. If circuit #1 were to fail, the ALPS service would switch user traffic to transport across circuit #2. Circuit #1 could be brought up and switched back into service after the problem which took the circuit out of commission is corrected.

2.3.2 Switched Networks

Switched networks can range from simple circuit switching to advanced packet and cell switching, and include new technologies such as asynchronous transfer mode (ATM). The main characteristics of switched networks include:

+ Addressing capability
+ Multiple protocol and interface support
+ One-to-many connectivity
+ Network intelligence above the physical transport layer

Circuit and packet (including frame and cell) switching are two major types of switched network techniques. Each of these techniques will be defined in detail later. The ultimate method of achieving high availability will prove to be a switched network. Two emerging examples of new switched service offerings are Frame Relay and switched multimegabit data service (SMDS).

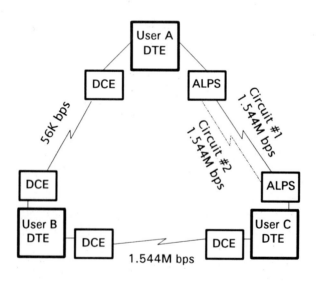

Figure 2.5 Leased Line with ALPS

2.3.3 Hybrid Networks

Hybrid networks consist of some measure or mixture of private line and switched network access services, or even traffic which transits a private line before entering or after exiting a switched network. Either way, the important considerations are hardware, software, and protocol compatibility (or transparency). Network management becomes more difficult when multiple network elements are crossed, as does network support. Since switched networks offer significant cost savings and concentration benefits over leased lines, the driving factor should be from leased lines to switched network access. Also, many of the services offered on switched networks will not extend to leased lines.

Today, the big choice in switched data is between frame relay and SMDS, and in dedicated lines or (leased) private lines. Ubiquitous network access, quickly changing technologies with enhanced reliability, and the apparent economics of access costs are driving many network designs to the switched network solution. Network support and management are also driving factors. Each of these decisions should be accompanied by a cost comparison of separate networks to a single network, or a private line to switched solution.

Multiplexer networks are somewhat of a hybrid between a private line and a switched network. Intelligent multiplexers have the capability to take multiple low-speed user inputs and provide multiple high-speed output paths. If one path should fail, there are alternate paths to reroute the traffic. While this achieves the desired redundancy, there is little or no capability for intelligent switching and true routing functions (except with alternate path routing capabilities in integrated switching multiplexers such as Netrix and Stratacom). Many vendors have recognized the need for multiprotocol, multifunctional customer front-end equipment capable of interfacing to both private lines and the large variety of switching architectures and protocols. Router vendors have been leaders in this evolution, even providing switching capability in hardware which is a fraction of the cost of broadband hardware.

These trends toward switched data and intelligent distributed networks and services leads to a discussion of the evolution of virtual networks.

2.4 EVOLVING VIRTUAL DATA NETWORKS

This section delves into the evolution of virtual data networks. They are increasing in popularity as users and businesses strive to out-source their networking but still maintain virtual control. To the user, a virtual data network is a self-contained network. To the carrier of a public network with many virtual network offerings, the virtual data network is a "network within a network". Carriers tout and advertise their VPN services to the user as having the benefits of a private network but also providing the economies and cost savings of switched services. The virtual network is a subset of the carriers' larger network, but provides the image of a complete data network to the user. Figure 2.6 demonstrates this concept, where Users A, B, and C are partitioned as Virtual Data Network #1 and Users D, E, F, and G are partitioned as Virtual Data Network #2. Users H, I, J, and K access the common public data network in the same manner, but they are not dedicated to a virtual data network. All users actually access the same public data network, but some do not see the big picture. Thus, the user sees only the extent of the network they are allowed to see, and control of user-owned elements can remain in the hands of that user. The virtual data network typically has all of the inherent qualities and capabilities of the larger net-work, but control of network resources is often limited. Customers can also be denied access into other customers' VPNs or virtual data networks.

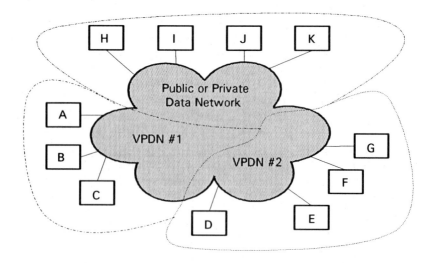

Figure 2.6 Virtual Networking

The availability of switched broadband data services has lured users away from private point-to-point networks and onto the public networks offered by the interexchange carriers. The main line has been the increased reliability of the carriers' public data network platforms and cost savings vis-à-vis private line operations due to the economies of scale of the larger carriers' backbone, as well as the built-in alternate routing in a large carrier network. This switched alternative to dedicated facilities can be provided in DS1 and DS3 speeds, as well as 155.52M bps voice, data, and video switched B-ISDN services.

2.4.1 The Virtual Network

Virtual networking is another buzzword of the '90s. Many interexchange carriers are building or have built large public voice and data networks, and offer portions of these to customers as virtual voice or data networks. MCI Communications offers a virtual voice service called VNET, or Virtual NETwork, and AT&T offers SDN, or Software Defined Network, which allows customers to customize voice networks with the effect of each customer having their own virtual voice network. This same theory of virtual services is now being applied to data. With the bandwidth sharing capabilities of new technologies such as frame relay and cell switching, virtual networks are even more attractive alternatives to private lines.

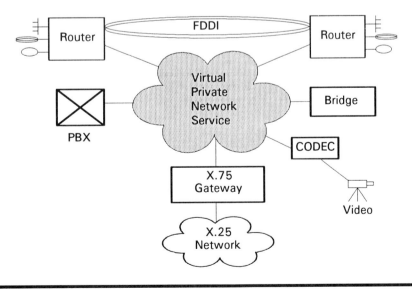

Figure 2.7 Virtual Private Data Network

Figure 2.7 shows a virtual private data network connecting multiple access devices and protocols.

Virtual networks are a conglomerate of private line access and switched services, providing an attractive method of connectivity for a geographically diverse user base. The primary advantage is that the user can access the virtual network without having to purchase multitudes of private lines. (This does not alleviate the fact that typically two-thirds of the total cost is in the local loop and one-third is in the IXC.) Virtual networks often combine dedicated private lines such as DDS, public switched services such as packet and switched T1, and a variety of other hybrid solutions that maximize user connectivity and minimize cost. The virtual network is often run and managed by software, making reconfiguration simple and still allowing for technology transparency when needed. Users are provided the capability of altering their virtual private network to accommodate individual changing traffic patterns. This provides them with the level of control not available through common switched services.

Some benefits of virtual networks are:

- ❖ Increased network control for users
- ❖ Network redundancy of the switched network being accessed
- ❖ Cost reductions based on least cost routing algorithms
- ❖ Intelligent routing and protocol support
- ❖ Inherent fault tolerance
- ❖ Real-time user control
- ❖ Bandwidth management
- ❖ Integrated access
- ❖ Network management
- ❖ Security
- ❖ Value added use of public switched network service

Virtual networks can either be customer premises (CPE) based or central office (CO) based. In CPE-based virtual networks, processing is done at the customer premises, and transmission facilities are just dumb pipes, adding no intelligence for data transmission. Management and control of the network can either be centralized or distributed. Network access is wherever the customer decides to base a network element. For carrier-based virtual networks, the processing and network intelligence is based in the central offices and switch junction sites. The data transmission facilities have additional intelligence, and the network management and control is centralized by the provider. Network access is via designated network access gateways. Also, the provider can bundle many value-added services with virtual networks, which the CPE devices would have to replicate in a CPE-based virtual network.

Recently, virtual networks have a new resource: pay per use, bandwidth-on-demand switched data services offered by the IXCs. These services include access rates ranging through T1 and into multimegabits. Many IXCs are also pushing control and network management of these services to the customer premises. Connectionless services such as frame relay and SMDS are opening up new possibilities for virtual networks. Soon, SONET pipes will be part of the virtual network services offered. And this is only the beginning, as these services will be ready for international deployment shortly.

The virtual network also offers the platform for voice and data integration. While virtual networks are evolving for data transport, it must be noted that many businesses use virtual networks for voice traffic as opposed to data (see above discussion on VNET). This is likely to change and move in the opposite direction, eventually toward an all digital packet or cell format for combined voice and data transmission. The spread of high-bandwidth services along with their respective transmission facilities will aid in this migration. The entire public network will provide addressing for access on a global scale.

MCI Communications is now offering two virtual network services; VPDS 1.0 which incorporates switched T1 and T3 services and VPDS 2.0 for frame relay switching and true bandwidth on demand, and will soon be offering SONET based cell-to-cell services (HyperStream). These services provide more than just the switched network advantages listed above. Support services consistent with the virtual network voice offerings are also available, along with a wide variety of interconnect possibilities including LAN/MAN/WAN, FDDI, CDDI, frame relay, SMDS, and ATM. Many other carriers are now following suit, offering switched services of varied types.

2.4.2 Public Network vs. Private Network

As discussed, the trend of the '90s is toward private virtual networks within the public data networks. Private data network growth remains greater than public network growth, but this may change. Historically, why have private networks become so valuable? The mass confusion of divestiture, who bundles what, caused large customers to try to get control of their destinies and networks. The intelligence was in CPE switching equivalent to do the job. Many users chose private networks ten years ago because of the falling transport prices of the IXCs. Divestiture had allowed these long-distance bandwidth prices to drop, making private networks a viable solution for voice and data. Many users built intelligent multiplexer networks to aggregate traffic and more efficiently use the carrier bandwidth (higher utilization). The monopoly carriers held back T1 offerings from customers until the new carriers offered T1 bandwidths and even AT&T was forced to provide T1 capabilities to the DOD as the first customer. Private networks also provided efficient, reliable interoffice communications. Efficiencies in voice compression also helped to keep down cost.

Long-distance carriers provided private virtual networks with voice services for years, allowing users to manage virtual private networks within the carriers' public voice network. This is creatively marketed as services over and above private line usage. Now, virtual private data networks are being offered in the same manner. The cheap voice rates offered by carrier voice VPNs has made migration to carrier-based VPNs irresistible. The T1 multiplexer networks that were cost-justified a few years before, saw voice bleeding off to VPNs while data remained to provide approximately 80 percent of the fill in T1 private multiplexer networks. Data currently accounts for over half the bandwidth transported by private networks, and most private networks are now dominated by data, which has become their primary investment.

Today many users have the capability of building a private data network, but with that capability comes many support issues. Administration, network support services, and vendor dealings are but a few of these issues. High-bandwidth-use applications such as LANs and imaging are driving the need for private or virtual private network capabilities. The issues can be solved in part or completely by using carrier-offered virtual private data network offerings. Private networks can be built to provide a dynamic environment for application and traffic change, and a larger mix of data services, and are supplemented by carrier service offerings.

Public network services are now competing with the advantages of private networks. Some carriers are building large public data networks, or modifying their voice networks to carry data, and selling virtual private data networks. Common carriers are legal carriers which can provide public communications services to specific geographic areas. Value-added carriers (such as CompuServe) provide additional services. Private network vendors are also making the move into public network technologies, driven by the market. One of the biggest challenges to router and multiplexer vendors, who are now marketing broadband interconnectivity services and hardware, is to move from a private network environment to a public network posture.

The engineer or data manager needs to choose the correct virtual private data networking solution which provides services in a timely manner, while dealing with complex network technologies that emerge in private and public networks even before standards are finalized. ISDN, SS7, and the other networking standards often have several versions. Even ESF has different versions among LECs. Then there is the risk of buying into a technology which may be rejected by the users as inadequate or become obsolete by the emergence of even newer technology. Thus, the virtual data network must improve its operating efficiency, save costs, and provide users with a platform for future expansion.

Public networks offer many advantages over private data networks. The competitive transmission environments of the IXCs are yielding new service offerings which are pulling private networking back into the IXCs domain. Much of this is accomplished through the use of software defined networks and services. Public networks now employ super switches that provide DS1, DS3, and 155M bps access into SMDS, B-ISDN, and ATM switches with routing over SONET lines. Users can choose the most cost effective service offerings from among these new high speed switched services offered by the IXCs — mixing voice, data, and video traffic, each with changing traffic patterns. Custom virtual network designs can be created within the public data network. Hybrid private and virtual data networks with hybrid topologies are also commonplace. The carriers advertise the economies and dependability of public networks while providing the user with the advantages of private networks — control, CPE-based intelligence options, etc.

An added advantage to public networks now being used quite often is for the user to tap the knowledge of the public network engineers and managers. Customers of public network services gain access to public network assets, such as engineering personnel, experience, resources. If done correctly, data consultation can be obtained at the lure of additional voice and data business. This is like having an in-house consulting service. But it is a two-way street. At the same time the provider is allowed to learn the customer's business and applications and can sell into this customer base.

2.4.3 Virtual Networks as Outsourcing

Virtual networking is in many ways a type of outsourcing. By providing advanced voice, data, and video services with customized network management, billing, and support systems to back these services, many interexchange carriers are providing a form of outsourcing. This outsourcing takes the form of resource reduction for systems and services which otherwise would have to be accomplished in house. Other benefits of virtual private networks which parallel outsourcing include carrier-provided network management, bandwidth management, fault tolerance, intelligent bridging and routing, security, order entry and order processing, and integration and standardization. Virtual data networks can also be used as an overflow technique or as a complete replacement for existing networks. The network manager needs to decide what portion of the network resources to retain or replace, keeping in mind the costs of access to the public network, not just end-to-end circuits as in a private network. These advantages are additional bonuses of switched services. If anything, the network provider becomes a free consultant, offering experience, knowledge, and sometimes even manpower to potential or existing customers. The rules of outsourcing previously discussed also pertain to virtual private data networking and switched public data network services.

2.5   COST FACTORS

While the technical business case may call for a new network, the economics may paint a different picture. The network must be an economically viable investment moved in at some point early in its life-cycle. How much will the network cost compared with how much the network will save the company? Savings could include straight cost avoidance savings, productivity increases, headcount decreases, and future cost avoidance. Cost should primarily be weighed against reliability and performance, with support running a close third. The network should be implemented with minimal adaptation costs, multiple uses and capabilities, with a strong return on investment for the business.

Network operations usually constitutes a major portion of network costs, thus outsourcing of network operations could be a viable alternative. While networks are the focus of this book, many other support systems are required to make the network design run. These systems include network management, billing, administration, order entry and tracking systems, and many other enablers to the business operations of the company. These services are used to manage and maintain the network and can constitute a considerable portion of the entire network expense. The price of procuring and installing CPE and end-user equipment, which use high-bandwidth data services, is also a consideration. It can be seen that the network designer must consider many costs other than those associated with the data communications transmission and hardware procurement.

2.6 INFORMATION SERVICES

The RBOCs have been lobbying to provide information services for years. Judge Greene has recently ruled in favor of the RBOCs providing information services, but placed a stay on the order. At the time of writing, this stay was being contested. Information services first became an issue in 1987, when it was ruled to keep information services away from the RBOCs. That restriction was loosened in 1988 by allowing the RBOCs to provide voice processing and information gateway services. The RBOCs failed to produce marketable services. There are obvious advantages to this ruling, such as the growth of the information services market and its impact on the users, as well as improvement of the public access network. Some of the services which could be offered are highlighted below. Along with advantages, there are always disadvantages, such as increased rates passed on to the user and even a lack of rate decreases for local access (which should run parallel with long-distance rate decreases of the past decade). Do not underestimate the power of the user, however, who will have the final say on the success of information services.

Possible information services (IS) offered by RBOCs are:

- Electronic messaging
- Credit check and validation
- Public electronic yellow and white pages
- Video on demand
- Shared data base services
- Audiotex audio information
- Real estate listings
- On-line retrieval services (dial-in databases)
- Cable television

The RBOCs can fund ventures such as these information services by raising rates on the long-distance providers, by increasing the access rate charge to other information service providers, and by passing along these expenses to end users. The RBOCs would like to provide information services such as multimedia and advanced entertainment television services. However, the RBOCs may soon be challenged by alternate access competition. These bypass providers will compete with the RBOCs and LECs to provide cost-effective information services. These entities are going after the data transport business with a fervor, but paying little attention to the applications themselves.

Some users are requesting services from the inter-exchange carriers including billing, network operations control, order entry and order processing, and network design and planning. The IXCs are performing some alternate bypass of the LECs to users through alternate access carriers, which are taking business from LECs. There are other alternate access bypass carriers bypassing the local exchange carriers and offering some switched services, such as frame relay, directly to the user. These services are often packaged with long distance service, which in many instances is provided by one of the big three carriers.

The carriers tend to introduce new information services to reap profits in new revenues, while the RBOCs usually introduce them as cost reductions. Another plus for the IXCs is that users seem to want to rely upon the skilled resources of the carriers. The information services provider who offers the following is most likely to win the game:

★ Innovative pricing
★ Services differentiation
★ Services provided through broadband switching systems and digital cross connects

These services will be offered through intelligent CPE devices which provide for multiple access types to be aggregated into a broadband device for transport over broadband switched services. Lately, IXC-based networking services have become more predominant, especially when international connectivity is a major concern.

2.7 INTEROPERABILITY

Interoperability in data communications is defined as the communication and intelligent interaction between dissimilar network architectures, protocols, and systems linked by some common medium. This medium can take the form of an operating system, logical connection, or physical connection. The medium may even be as simple as network interface cards which support the same protocol.

Some difficulties encountered in providing interoperable networks include the large investment, multiple standards and protocols to support and convert, changing business requirements, lack of network management and design tools, and simple network inconsistencies. Open systems becomes increasingly important as the number of protocols and applications interacting in the wide area network increase. The next chapter will cover the multiple industry standards through which the network designer achieves true interoperability.

2.8 REVIEW

We have determined already what constitutes a network design, the history behind voice and data networking, and where data and computer communications is heading. We took a detailed look at the actual need for data and computer communications networks. We defined this need through industry, business, and user perspectives or perceptions. The move from centralized to distributed networks was explored, and its impact reviewed on high-bandwidth on-demand connectivity. We showed the two factors directly influencing this data explosion: transmission media and protocol enhancements. An objective look at outsourcing showed the alternatives to designing and building part, or all, of your data network. Next, a comparison of private line and switched networks showed the evolution to virtual voice and data networks. Discussions of cost factors influencing network design were laid out as well as an overview of the state of information services and interoperability. Now that we have discussed the building blocks of networking we will move to Part 2 of the book: standards in the industry. Standards will be the connecting link that defines the measurements to making dissimilar devices conform as much as possible to some common standard.

2

The Basics

We have now determined the genuine need for a data communications network. A study of data and computer communications is now warranted. Seven major architectures dominate and shape the protocols of computer and data communications. Each major architecture will be covered and explained. This includes an overview of the primary standards organizations, committees, and forums which are influencing many of the technologies discussed in this book. Also the many protocol architectures prevalent in the industry today will be compared and contrasted to the International Standards Organization Open Systems Interconnect (OSI) Reference Model. A discussion of the basic circuits, services, and hardware types which comprise networks will be covered as well. Finally, an overview of common protocols and interfaces is required. Only when the circuit, services, and hardware resources for data networks are understood, along with the standards and definitions of protocols and interfaces, can an advanced study be done of switching technologies and services.

3

Understanding
the Standards

Seven major architectures dominate and shape the protocols of computer and data communications, and each warrants a summarized discussion. A detailed description of the more popular protocols of each architecture will be given later. We will overview the primary standards organizations, committees, and forums and how they shape today's computer and communications industry standards. How standards organizations interface is even more fascinating than how standards themselves develop. An explosion of standards forums has occurred, primarily driven by vendors, users, and service providers. A few of the most recent forums are influencing many of the high-bandwidth services such as frame relay and SMDS. Since the International Standards Organization Open Systems Interconnect (OSI) Reference Model dominates the future of data communications, and many protocols are referenced to this well-known seven-layer OSI model, a description of its layers and functions will be helpful. Interoperability among standards will wrap up this discussion of standards.

3.1 WHY STANDARDS?

Perhaps the single most important driving factor for standards is the user. The user is affected most by the lack of standards. Some of the most important questions a user can present to a vendor are "does it conform to industry standards, which ones, and how?" Standards play a critical role in an age where national and international interoperability between standards is the key to success in data communications. Standards are defined as the written word which defines the accepted method of technology, whether it be hardware or software, systems or processes, for vendor, user interconnectivity, and service provider. These standards provide many benefits to vendor, service provider, and user alike.

In the past, standalone systems (e.g., CPU, terminal, printer, etc.) worked well together for one application. In fact, at one time IBM had over 50 operating systems which worked just fine as standalone units. The interconnection of these systems by users created a de facto standard called Systems Network Architecture (SNA). In fact, ASCII was developed by non-IBM companies so as not to get "locked" into IBM's EBCDEC coding, for example. Then ANSI blessed ASCII. The exponential rate of technology development and technological advances seems at times to outpace even the ratification of standards.

3.1.1 Vendors

Vendors live and die by standards. If they do not design products around standards, and provide proprietary implementations, users may take their business elsewhere. Vendors are becoming more concerned with meeting industry standards. Vendors who remain completely proprietary, or try to dictate the standards with their offerings, are finding that users are unwilling to risk their business on proprietary systems. Beware vendors who provide lip-service standards and only support them when forced to by the user community.

Vendors can also drive standards, primarily by de facto industry standardization. This is when a vendor is the entrepreneur in the industry and wants to associate the new technology to their name. One example of this is Hayes, whose leadership in modem interoperability made "Hayes compatible" the de facto standard. Sometimes the vendor is not the only one in the market with the product, but their product quality or market share makes them the standard which other vendors must design around. IBM is given credit for being the early de facto standard for microcomputing, as the PC was an IBM product brand name.

3.1.2 Users

Users do best when they purchase equipment conforming to industry standards. A certain comfort level exists in knowing that the equipment you stake your business on has the ability to interface with equipment from other vendors. In the context of international interconnectivity, standards are paramount. Users play a key role in developing standards since their use of standard equipment (as well as vendor acceptance) determines the success or failure of the standard. Take for example the ISDN standard. Universality is part of a standards success, however. They say: "We will provide it when customers sign up." Customers say: "We will sign up when it is universally available at the right price, unless we see something else better and cheaper". Standards are used by users on a voluntary basis. Users play a very important part in the standards bodies. The main reasons why many corporations send representatives to the standards meetings are to benefit in the long term through standardization and eventually interoperability and to assure the standard is "friendly". This will be discussed in detail later.

3.1.3 Providers

Providers of network services also need to adhere to standards while simultaneously participating in the standard-making process. Service providers are more driven by hardware vendors, but service providers prefer vendors that adhere to industry standards. This does not lock them into one vendor's proprietary implementation. Providers must not only make vendor hardware interoperable within their network cloud, but must also assure industry standard interfaces over which they provide their value-added services.

3.2 STANDARDS ORGANIZATIONS

The vehicles for driving industry standards, and the rallying points for standardization and interoperability, are the standards organizations. These standards bodies, along with the users and vendors, play the most important part in deciding what actually becomes a standard. Users can also drive standards with a need for a specific technology and a desire to incorporate it as soon as possible. The standards organizations provide common ground between users and vendors while providing guidelines for the industry that define the interoperability not only between computer communications, but between computer and user communications. While standards organizations are comprised of both users and vendors, they seek to remain objective about the standardization of technologies which could have a drastic impact on the businesses of both.

Standards organizations have increased their visibility and importance in recent years. With this increased visibility comes increased bureaucracy. As more organizations participate in this process, the following grows. Users, service providers and vendors alike view the chance to participate in the standard-setting process as an opportunity to both express and impress their views upon the user community. This is a double-edged sword, as while participation is necessary, biases are brought to the committees which tie up decision making and bog down the process of making standards. Thus, in some ways standards organizations have become political quagmires. A specific vendor or carrier will argue the point of a word meaning until the point is won through attrition. The larger vendors have the deep-pocket financing resources to "attrit" the others. While standards organizations take their time to publish standards, many vendors take the lead and build equipment which they either design around a proposed standard or partially issued standard. They promise compliance with the standard when it is published. If these limitations are not enough, technology seems to accelerate at an exponential rate, causing standards to take longer to come to fruition because of their complexity, and thus forcing users to implement proprietary short-term solutions. Customers have business problems today that can be solved by proprietary or semiproprietary implementations of standards — if they wait until the perfect standard is designed, they may be out of business.

Taking this into account, there is no doubt that the standards process has a dramatic effect on the data network industry. As users have become active participants in setting standards, many niche markets are eliminated and competition has been enhanced among vendors. Users are unquestionably in control, or are they? Vendors play the game by supporting their proprietary solutions. They wish to make a standard before their competitors proprietary solution becomes a standard. This alone can draw out a standard process for many months or even years. Thus, many points of contention between self-interest factions comprise a standards committee.

The process of standardization is a long and painful process. Proposals are formulated by committees, grouped into documents, debated for lengthy periods of time until all members of the committee agree by vote, and then passed to a higher organization for blessing and publication. This process often takes a minimum of one year from when the proposal is made, and can take many years if the subject matter is a much debated issue. The danger in delay is that many vendors will develop their own proprietary standards in the interim, and at times (as with standards such as the IBM one cited above) become more widely used and endorsed than the official standard as set forth by the committee. Standards developed must be broad enough to cover all contingencies, yet detailed enough to leave no dispute nor further deliberation or interpretation. At times, standards may be superseded by user or vendor implementations, or vice versa. And even standards have lacuna or holes that are undefined and left to the vendors for implementation (which are different). RS232 is a "standard" but manufacturers do not use the 25 pin leads in the same manner.

The major standards organizations in the United States are the American National Standards Institute (ANSI) and its main communications committee the T1 Standards Committee, the Institute of Electrical and Electronics Engineers, Bellcore, and the United States branch of the International Telegraph and Telephone Consultative Committee (CCITT). The international standards bodies covered in detail here include the International Standards Organization (ISO) and International Electrotechnical Commission (IEC), as well as the International Telegraph and Telephone Consultative Committee (CCITT). Although the Federal Communications Commission (FCC) is not a standards-setting organization, it is a major regulatory power in the United States and deserves mention. Each standards bodies will be covered in detail, followed by a discussion of how standards bodies are influenced and how they interface with one another.

3.2.1 American National Standards Institute (ANSI)

The American National Standards Institute (ANSI) acts as the North American primary standards body, as well as the official interface to all international standards bodies. To ensure that standards sanctioned by ANSI are impartial to vendor, user, and service provider alike, contributions pending standardization are contributed from many voluntary nonprofit, nongovernmental organizations, including the Institute of Electrical and Electronic Engineers (IEEE), the Electronic Industries Association (EIA), and the Computer and Business Equipment Manufacturers Association (CBEMA). ANSI is also a member and active participant in the International Standards Committee, and the United Nations sits at the top of the standards bodies hierarchy.

ANSI standards define both electronic and industrial standards for both the national and international communities. ANSI defines national standards such as American National Standards. These standards are published with the following number scheme:

ANSI/NNNN XXXX-19XX - Standard Name

where **NNNN** is the name of the contributing organization (e.g., IEEE)

XXXX is a letter and/or number combination signifying the field of study and the reference number of the standard within that field.

19XX is the date the standard was officially published.

Standard Name is the name of the standard

ANSI defines international standards sanctioned by the ISO in a similar manner:

ANSI/ISO XXXX-19XX - Standard Name

where the fields are the same as for national standards.

3.2.2 T1 Standards Committee

The T1 Standards Committee on Telecommunications is one of the most important standards bodies dealing with data communications and telecommunications in the United States. The committee, established in 1984, is sponsored by the Exchange Carriers Standards Association (ECSA) and accredited by ANSI. Its primary charter, through the use of subcommittees, is to:

> develop technical standards and reports supporting the interconnection and interoperability of telecommunications networks at interfaces with end-user systems, carriers, information and enhanced-service providers, and customer premises equipment (CPE).

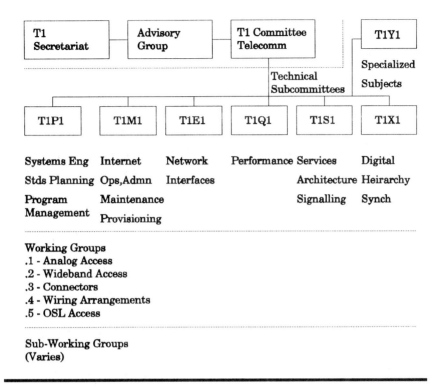

Figure 3.1 T1 Telecommunications Committee Levels

The T1 advisory group has seven subcommittees which report on recommended standards to the main committee.

Figure 3.1 shows the three levels of the T1 Telecommunications Committee: the Technical Subcommittees, Working Groups, and Subworking Groups. Each level addresses more detail than the previous level.

The T1E1 subcommittee deals with digital internetworking and interconnectivity standards. The T1M1 subcommittee deals with maintenance and operational issues. The T1P1 subcommittee deals with personal communications. The T1S1 subcommittee deals with ISDN standards and works with the U.S. National Committee for the CCITT.

When reading technical recommendations published by the T1 Committee, it is useful to understand the document numbering plan. These numbering plans shown below are consistent across all ANSI technical documentation. Figure 3.2 shows the general document numbering plan format.

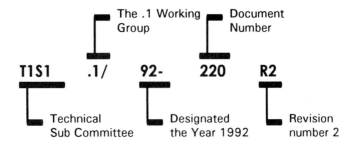

Figure 3.2 T1S1 General Numbering Plan

Two other organizations which adopt ANSI standards are the Federal Information Processing Standards (FIPS) group which are used as government procurement standards and the Department of Defense (DOD), which dictates mandatory standards for governmental use.

3.2.3 Institute for Electrical and Electronics Engineers (IEEE)

The Institute for Electrical and Electronics Engineers (IEEE), Inc. was formed in 1963 by a merger of the American Institute of Electrical Engineers (AIEE) and the Institute of Radio Engineers (IRE). It is now the world's largest professional engineering society. Standardization has always been the core activity of these two organizations. Through an international membership, the IEEE carries on that function by developing and disseminating electrotechnical standards which are recognized worldwide. Standard guides, practices, and reference manuals are developed by the IEEE to define these standards. *The IEEE is truly a member-driven standards body which represents the engineering community.* Anyone can become a member of the IEEE. Standards are approved through the IEEE Standards Board. Many publications, journals, and newsletters are published by members under the IEEE title, containing various proposals and technology trends.

The primary IEEE standards discussed here are the IEEE 802.X standards, which include 802.1 through 802.6 and define such technologies as Ethernet (802.3), Token Bus (802.4), Token Ring (802.5), and metropolitan area networking (802.6). The IEEE 802 standards are developed by the IEEE LAN Standards Committees.

3.2.4 International Standards Organization/International Electrotechnical Commission (ISO/IEC)

The International Standards Organization (ISO) is a voluntary international standards body made up of over 90 countries and chartered to cover many subjects. The organizational structure is based upon technical committees made up of subcommittees, which in turn are made up of Working Groups. Standards are published first as Draft Proposals (DP) from Working Groups. Once approved, they become Draft International Standards (DIS) for ballot by the Technical Subcommittee. Upon approval they become an international standard with the same DP number originally assigned by the Working Group.

Data communications standards, more commonly called information technology standards, are handled by the ISO Technical Committee 97. The International Electrotechnical Commission (IEC) and the ISO formed the Joint Technical Committee 1 (JTC 1) to produce joint ISO/IEC standards for information processing. The ISO/IEC JTC 1 has participants of standardization bodies from twenty-two countries who submit standards to the ISO. It is interesting to note that ANSI administers the secretariat of the JTC 1, as well as four of its subcommittees. Three of these four primary standards subcommittees defining our subject matter here are telecommunications, open systems interconnection, and text and office systems.

There is also a task force developed for routing protocol standardization. The ISO/OSI Routing Task Force is now working on the development of link state algorithms to quickly communicate topology and network state changes to all elements in a routed network. The importance of developments such as this one becomes clear in the next few chapters.

3.2.5 International Telegraph and Telephone Consultative Committee (CCITT)

The International Telegraph and Telephone Consultative Committee (CCITT) publishes standards called recommendations. The CCITT is chartered by the International Telecommunications Union (ITU) to produce telegraphy and telephone technical, operating, and tariff issue recommendations. The CCITT is also a United Nations sponsored treaty organization. The U.S. voting member in the CCITT is a representative of the U.S. Department of State, and includes technical advisors through the U.S. National Committee for the CCITT.

The CCITT publishes approved recommendations every four years in the form of various colored books. The book sets have been colored red, blue, white, green, orange, yellow, red, and blue for the years 1960, 1964, 1968, 1972, 1976, 1980, 1984, and 1988, respectively.

3.2.6 United States National Committee for CCITT

United States participation in the CCITT is channeled through the U.S. National Committee for the CCITT. This department, administered by the U.S. Department of State, contains subcommittees which directly participate in the CCITT international study groups. They provide the transfer of standards information between the United States and the international community.

There are four U.S. CCITT study groups which directly interface with CCITT study groups. The A study group concentrates on services, operations, and tariffs and contributes to the CCITT study groups I through III and IX. The B study group concentrates on ISDN specifics and contributes to the CCITT study groups XI, XVIII, and all matters relating to ISDN. The C study group concentrates on the Telephone Network and contributes to the CCITT study groups IV through VI, X, XII, and XV. Finally, the D study group concentrates on data communications and contributes to the CCITT study groups VII, VIII, and XVII.

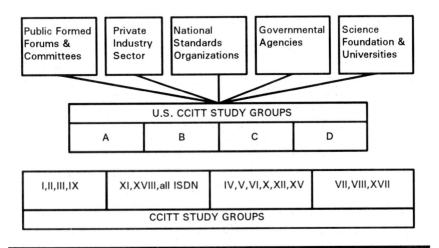

Figure 3.3 Interest and Study Group Interactions

The U.S. CCITT study groups receive input from many groups of U.S. businesses. These include public forum and committees such as the Frame Relay Forum (FRF), private industry sector groups such as businesses who have membership in the CCITT, national standards organizations such as the T1 Committee, government agencies such as the Federal Communications Commission (FCC), and science and educational sectors such as the National Science Foundation (NSF).

Figure 3.3 shows the interactions between the U.S. interest groups, the U.S. CCITT study groups, and the CCITT study groups.

3.2.7 Bell Communications Research (Bellcore)

Bellcore is more of a development center than a standards organization. The original intent of Bellcore was to act as a center for coordination and research for the regional holding companies (RHCs) in the aftermath of divestiture. It since has developed into a very powerful driving force for the Regional Bell Operating Companies (RBOCs). Historically, Bellcore has centered its research on the operations and support systems (OSS) for the BOCs. This included hardware and software research, development, and even design of these systems. Always setting it sights on the public network, Bellcore has recently started to expand its consulting services and participation in standards recommendations. The Switched Multi-Megabit Data Service (SMDS) Bellcore Recommendations, discussed in later chapters, are evidence of this move into new technologies becoming standards.

Bellcore technical publications are divided into Bell System pre-divestiture (before January 1, 1984) and post-divestiture documentation. The predivestiture documentation included compatibility bulletins (CBs), information publications (IPs), technical descriptions (TDs), and technical references (PUBs). Since many of these documents have been revised and reissued by Bellcore under the post-divestiture format, we will concentrate on those new formats.

Bellcore's post-divestiture documentation is divided into:

Framework Technical Advisories (FAs). These contain preliminary early Bellcore views with very generic requirements, and act as catalysts for the telecommunication industry. FAs are often superseded by TAs.

Technical Advisories (TAs). These contain preliminary views of proposed generic requirements for interfaces, products, and new services and technologies. TAs are often superseded by TRs.

Technical References (TRs). These contain the developed and mature views of TAs.

Family of Requirements (FRs). These contain complete sets of related TRs.

Special Reports (SRs). These contain documentation of other technical interest not mentioned in previous documents.

Science and Technology (STs). These contain highly technical information which needs conveyance to the telecommunications industry.

Figure 3.4 shows these document types and their evolution cycle.

3.2.8 Federal Communications Commission

The Federal Communications Commission (FCC) is not a standards organization, but it does serve as the regulatory authority for radio, television, wire, and cable communications within the United States. The FCC has the charter to retain regulatory control over interstate and international commerce concerning communications. The FCC strives for competitiveness in these markets, which relates to the public benefits of marketplace fairness, up-to-date and quality communications systems, and a broad range of communications offerings for the consumer. The FCC also has responsibilities for reviewing the rate and service change applications for telegraph and telephone companies, reviewing the technical specifications of communications hardware, and setting reasonable common carrier rates of return. These goals extend to the regional holding companies (RHCs) and the interexchange carriers (IXCs), and to any technical and regulatory policy issues by which they are affected. One of the recent changes that the FCC has implemented is the local exchange carrier (LEC) price cap regulation, designed to drive the LECs to improve profitability by improving efficiency. This financial incentive forces the LECs to offer more attractive products and services, rather than raising the price of existing services. This may serve to push the LECs faster toward implementing user services such as ISDN. The FCC has also recently shown its commitment to open network architecture (ONA) and the OSI Reference Model.

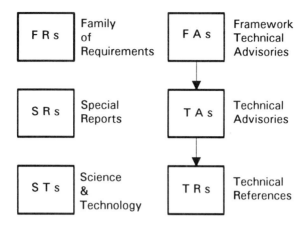

Figure 3.4 Bellcore Postdivestiture Documentation

3.2.9 Additional National Standards Bodies

Some other national standards organizations include the Computer and Business Equipment Manufacturers Association (CBEMA) which sets standards for computer and business hardware. The Electronic Industries Association (EIA) and Telecommunications Industries Association set standards on the communications, computers, and electronics fields, with an emphasis on interconnectivity between these fields. Both represent the manufacturing community, and the EIA is active as a trade organization and as an ANSI-accredited standards organization. The Exchange Carriers Standards Association (ECSA) was incorporated in 1983 and sponsors the T1 Standards Committee.

3.2.10 Additional International Standards Bodies

Some additional international standards organizations include the European Computer Manufacturers Association (ECMA) which establishes manufacturing computer standards and the European Telecommunications Standards Institute (ETSI) which focuses upon solving data communications and telecommunications issues for data processing hardware manufacturing and marketing within Europe. The member body of the ISO and IEC for the Netherlands is the Nederlands Normalisatie-instituut (NNI), for Saudi Arabia it is the

Saudi Arabian Standards Organization (SASO), for Sweden it is the Standardiseringskommissionen i Sverige (SIS), and for Finland it is the Suomen Standardisoimisliito (SFS). The group dedicated to establishing European standards as national standards is the Comite Europeen de Normalisation (CEN). The CEN is also active in the ISO. Members of both the CEN and ISO include the Danish standards organization Dansk Standardiseringsrad, the National Standards Authority of Ireland (NSAI), and the British Standards Institution (BSI). The French national Association Francaise de Normalisation (AFNOR) supports the ISO activities, as does the German Deutsches Institut fur Nromung (DIN). Canada has the Canadian Standards Association (CSA) as their primary standards association, and the primary Japanese standards body is the Japanese Industrial Standards Committee (JISC).

3.2.11 How Are Standards Influenced?

Many external factors cannot be controlled by the standards organizations, yet they have a pronounced influence upon them. We have already discussed the influence of the people who attend the standards committees, as well as the vendor proprietary development which goes on outside the standards organizations. The primary cause for this vendor development is the exponential growth in technology. Vendors need to standardize their designs to get products out the door before their competitors do. This often causes the vendor to create standards of his or her own. The best standard is one that builds upon these proprietary creations of vendors to produce the solution which has the least impact to both the vendor and installed user community and still defines a workable standard for future use.

Another major factor influencing the standards industry is the political rulings which effect both the service provider industries and the hardware provider markets. This can span from FCC or government rulings to policy issues pronounced on a national or international level. These factors can have a direct effect on standards organizations and their current focus. One example is the resulting affect of allowing the RBOCs to offer widespread information services, causing committees to be established or focused to set standards for a flood of information service offerings.

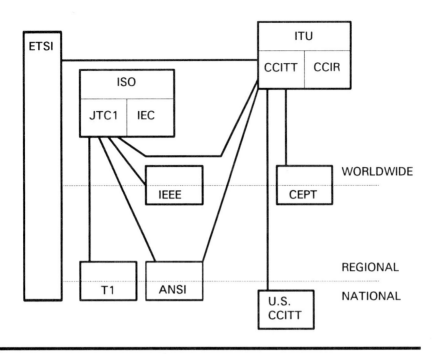

Figure 3.5 Standards Organizations

The final factor is the interaction between organizations and committees on both a national and international level. Standards set by organizations who fail to anticipate all of the impacts can either be accepted or shunned by other rivals of nonnational organizations, causing dual standards to be implemented, thus defeating the purpose of standardization. Now for a discussion of the interface between standards organizations.

3.2.12 How Do They Interface?

Figure 3.5 shows the standards organizations discussed and how they interact. The figure is layered, showing organizations which span the United States, a given geographic region (such as Europe), and those which span the globe.

Table 3.1 shows a summary of the standards organizations discussed, along with their areas of participation and influence.

In an age where global connectivity and communications is becoming the keystone for international business, the existence of global communication standards is imperative. The industries of both national and international business must drive these standards bodies to develop standards which are as consistent as possible. Interfaces, protocols, regulation, network management, and many other support technologies must become interoperable.

TABLE 3.1 Standards Bodies

Standard Body	Standards Issues Covered	Sphere of Influence
ANSI	Electronics and industrial standards	National
T1 Standards Committee	Data communications	National
IEEE	General engineering, computer and communications, LANs and MANs	International
ISO/IEC	Data communications, NM, information technology	International
CCITT	Telecommunications, data communications, telegraph and telephone	International
U.S. CCITT	Telecommunications, data communications, telegraph and telephone	National
BELLCORE	RBOC research & development, SMDS	National
FCC	Regulatory for communications	National
CBEMA	Computer & business hardware	National
EIA/TIA	Communications, computers, electronic fields	National
ECMA	Manufacturers' computer standards	European
ECSA	Telecommunications, data communications	National
ETSI	Data communications, telecommunications issues	European
NNI, SASO, SIS, SFS, CEN, DS, NSAI, BSI, AFNOR, DIN, CSA, JISC	Varies	International

Heading down the road toward OSI, it is the responsibility of standards organizations to push for a global standard for telecommunications, and that standard seems to be OSI. From the music of the spheres in the "seventh" heaven, to Dante's rings of descending layers to the bottom-most depths of hell, man can relate to the degrees or "layers" that connect things together. Thus is the seven-layer OSI model. OSI defines the development of hardware and software, systems and media, as well as communications services and processes. Vendors and service providers who define hardware and systems based upon these standards will gain a larger piece of the world data communications market share. Governments, IXCs, and PTTs can also play a major role in shaping these standards. But this global standardization process does not stop at organizations such as the CCITT and JTC1, however. On the contrary, it typically starts at user group and forum levels. Following is a discussion of the user's groups and forums helping to shape these standards both nationally and internationally.

3.3 CURRENT FORUMS

A new style of "jump starting" the standards process is emerging in the '90s: forums and committees. These are not formal standards committees and forums, but independent groups formed by vendors, users, and industry experts who want standards for interoperability, but who do not want to wait for the exorbitant lag times for formal standards groups to publish the "official" standard. Sometimes these forums and consortiums offer valuable contributions to standards organizations, speeding along the acceptance of a protocol while providing more from the user's perspectives and needs. It involves more than just writing and publishing the standards to make standards stick, and the vendor-user agreements built in these forums are often essential to a standard's success.

One critical aspect influencing standards acceptance often overlooked is the development of services and applications to accompany the standards. An example of this is ISDN basic (BRI) and primary (PRI) rate interfaces, where the technology is fully developed but is still at a loss for applications. ISDN was basically a PTT/telephone company phenomena to upgrade, digitize, and put the latest technology into the utilities' networks for maximum efficiencies — the end user is not a significant factor in the equation. Little wonder end users do not perceive its immediate value. Acceptance by the vendor community also plays a key role in the success or failure of standards. The forum which keeps both the standard and the applications for that standard in mind will likely cause the standard to live long and prosper, or at least live longer than those who concentrate primarily on the development of standards. User trial communities and university test beds are other methods employed by these forums to help speed up the testing and acceptance of these new technologies. The Internet protocols are an example of university (academia) lead standards work.

What recent forums have made sizable impacts on technology development and standardization? Most of these forums provide training centers, seminars, interoperability tests in test labs, and strictly adhere to parallel standards development by national and international standards bodies. Standards development seems to take place best in this "free market" of standards just as the "free market" of ideas has stimulated Western culture.

3.3.1 Frame Relay Forum

The Frame Relay Forum was formed in January 1991, as a non-profit organization dedicated to promoting the acceptance and implementation of frame relay based upon national and international standards. The group originated when the "Gang of Four" (cisco Systems, Inc., Digital Equipment Corporation, Northern Telecom, Inc., and StrataCom, Inc.) developed the frame relay Local Management Interface (LMI) specification, made it voluntarily available to those who wanted to join (about 40 to date) and then saw a need to carry on this work using the vehicle of the Frame Relay Forum. The coalition is now over 40 members strong, and includes participation from Pacific Rim and European firms. Issues currently under current study include flow control, encapsulation and translation, and multicast capabilities, as well as marketing of products and user acceptance of frame relay. The organization is divided into three groups: technical, organizational, and marketing. The Frame Relay Forum (FRF) submits all of its standards work to the international standards organizations, ensuring worldwide interoperability.

While the FRF is not a standards body, it is a forum for frame relay users to discuss implementation issues and drive the standards bodies such as ANSI and CCITT to implement what the users (and vendors whose equipment relies on the user version of the standard) really require. One example of their efforts is their assistance to ANSI in implementing the LMI specifications. The committee also drives for interoperability between implementations of frame relay, and thus between users, as well as developing testing and certification standards.

3.3.2 SMDS Users Group/SMDS Interest Group

The SMDS Interest Group and SMDS Users Group are chartered with developing guidelines for interoperability between SMDS hardware, services, and applications. Participation involves primarily SMDS hardware vendors, service providers including the seven RBOCs, and the major interexchange carriers. As with many user groups, the SMDS Interest Group works with the major standards bodies and other user groups to ensure compatibility among standards.

3.3.3 Internet Society

The Internet Society is chartered with speeding the evolution and growth of the Internet communications network as an international research network. Within this group is the Internet Activities Board, comprised of users of Internet, which was the old governing board for the Internet. The Internet Activities Board has established such standards as TCP/IP. The Internet Society remains the administrative body for the Internet, providing functions such as database administration, user training, and Internet interoperability among its user community. The Internet Engineering Task Force (IETF) establishes the standards for Internet engineering, while the Internet Research Task Force (IRTF) pursues ongoing research.

3.3.4 Switched Digital Services Applications Forum

The Switched Digital Services Application Forum (SDSAF) is dedicated to promoting the use of switched services from 56K bps to T3, including frame relay. Membership is comprised of various vendors and carriers, each with a vested interest in the success of switched services. The thrust of their charter is to address the issue of vendor interoperability in the switching world. The SDSAF will attempt to drive the development of applications to suit these types of switched services; video, FAX, and LAN traffic. This provides applications for switched services to expand upon their primary use for private line redundancy. The SDSAF goal is that if applications are developed using switched services, and interoperability and private networking issues can be solved and more publicity may be given to switched services.

3.3.5 ATM Consortium

The charter of the Asynchronous Transfer Mode (ATM) Consortium is to focus on acting as an industry catalyst to speed final acceptance of the ATM protocol. They address the need for multimegabit, multimedia transmission across the WAN. Many in the industry have a vested interest in the development of ATM standards, and most of the members of this consortium, as with other consortium participants, are vendors who will produce the hardware and software to run and support the soon-to-be ATM standards.

The ATM Consortium is similar to the ATM Forum, which has also been formed by major customer premises equipment (CPE) vendors and central office providers. The charter of the ATM Forum is similar to that of the ATM Consortium, namely developing interoperability specifications for ATM hardware and services based upon emerging ATM standards. The ATM Forum is now addressing physical ATM interfaces, and will move to implementations for standards to support continuous and various bit rate services. The focus of both groups is toward both private and public networking.

3.3.6 Internet Engineering Task Force (IETF)

The Internet Engineering Task Force is the standards-setting body for the Internet. The IETF developed the Simple Network Management Protocol (SNMP) and provides provisions for it to be integrated into the OSI architecture. The IETF is involved with measuring performance characteristics for routers and bridges, as well as providing recommendations for TCP/IP networking standards and developing router protocols. They also shared in the development of OSPF as an IGP interior gateway protocol, which they historically supported as the standard router protocol, but now plan for an OSI standard, IS-IS protocol.

The Internet community is composed of many user networks tied together through an internetworking protocol (TCP/IP). This includes ARPANET, NSFNET, NYSERnet, and thousands of others.

3.3.7 Additional National Forums

Another group formed recently is the Network Management Consortium. This consortium will speed the integration and interoperability of various vendor SNMP implementations of systems and agents. The members are primarily network equipment vendors who employ products with SNMP capabilities.

A North American ISDN User's Forum has been formed, chartered with establishing a commercial infrastructure for ISDN in the United States. Also, the Open Systems Interconnect Network Management (NM) Forum is composed of many vendors developing network management interfaces to the OSI architecture. The OSI NM Forum has developed the System to Management System standard which will speed interoperability between proprietary vendor hardware.

3.3.8 International Forums

The Pacific Telecommunications Council (PTC) is a forum composed of vendors, users, government agencies, and research institutes from over 23 Pacific Rim countries. Formed in 1980, the PTC serves as a common forum for pooling network experience and interconnectivity and standards adherence. The PTC acts more as a convention for common information sharing, and does not produce any standards or recommendations.

3.4 STANDARD COMPUTER ARCHITECTURES

Seven major computer architectures have shaped and standardized the computer networking industry. These include the International Standards Organization's Open Systems Interconnect Model (OSIRM), the Institute for Electrical and Electronics Engineers Local / Metropolitan / Wide Area Network (IEEE 802.X), Integrated Services Digital Network (ISDN), International Business Machines' Systems Network Architecture (SNA), Digital Equipment Corporation's Digital Network Architecture (DNA), General Motors' Manufacturing Automation Protocol (MAP), and Boeing Computer Services' Technical and Office Products Systems (TOP). We will lastly cover the new IBM move toward OSI protocol stacks, the System Application Architecture (SAA), which defines the IBM equivalent of layers three through seven of the OSI Reference Model. While we will explain the origins of each of these architectures, only the

OSIRM, and IEEE 802.X, and portions of the ISDN, as it relates to frame relay, will be covered in great detail in this text.

3.4.1 Open Systems Interconnect Reference Model (OSIRM)

The Open Systems Interconnection Reference Model (OSIRM) defines the functions and protocols necessary for international data communications. This model was developed by the ISO discussed in the previous section. Work began on a standard worldwide architecture in 1977 with the formation of ISO Technical Subcommittee 97 (TC97), Subcommittee 16 (SC16), and was officially documented in 1983 under ISO 7498. The basic reference model is shown in Figure 3.6 showing a source, intermediate, and destination node protocol stack. The layers are represented from layer one, which interfaces to the physical media, to layer seven, which resides on the user end device and interacts with the user application. Each of these seven layers represents one or more protocols which defines the functional operation of communications between user and network elements. All communications between layers are peer-to-peer. These layers are defined in greater detail in the next section. Standards are finally emerging which span all seven layers of the model. The term "open system" refers to a standard which conforms to the OSI standards.

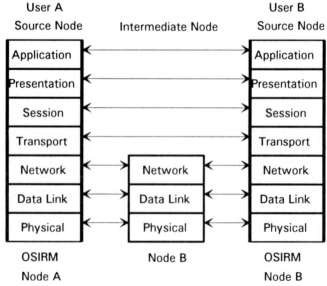

Figure 3.6 OSI Reference Model

3.4.2 IEEE Local/Metropolitan/Wide Area Network (LAN/MAN/WAN) 802.X Series

A Local Area Network (LAN) is defined as a network providing high-speed connectivity to a small geographic area, such as a building or floor, and providing data connectivity for users to access intelligent devices (file servers, print servers, hosts, etc.), peripherals, and other users. The Institute of Electrical and Electronics Engineers (IEEE) has played the major role in the development of LAN standards.

There are other terms for the extension of LANs. A metropolitan area network (MAN) provides high-speed connectivity (up to 100M bps) in a geographically close metropolitan area, and typically allows users the same resources as a LAN. The MAN opens users to an array of other transport, such as video, imaging, and even voice. A Wide Area Network (WAN) provides a similar service as the MAN, but over greater distances, such as between cities, regions, or even countries. The WAN provides transport similar to the MAN, but at lower speeds (typically T1 and some T3).

The IEEE 802 reference model defines three layers and two sublayers of operation. Figure 3.7 shows these three layers of the IEEE 802 model. Layer one is the physical layer, layer two is composed of the medium access control (MAC) sublayer and the logical link control (LLC) sublayer, and layer three is the network layer. The standards for the network layer are still under development.

The Logical Link layer operates the same for all LAN architectures, but not for the MAN architecture (a completely different beast!). The logical link control layer and physical layer operate differently for each of the local and metropolitan area network architectures. The LLC layer manages the call establishment, data transfer, and call termination through three types of services: connection oriented, unacknowledged connectionless oriented, and acknowledged connection oriented. All three will be covered in the next chapter.

The three major LAN architectures are:

Ethernet - IEEE 802.3 and Ethernet common specifications form what is called the Ethernet standard. The first Ethernet products appeared in 1981, and now sales for Ethernet outpace the other 802 architectures. Ethernet works primarily in the link layer, where users contend for bus resources and send data through CMSA/CD and Token Passing. The interface is 10M bps.

Token Bus - IEEE 802.4 forms what is called the Token Bus standard. The Token Bus has never been very popular as a standard, probably because its standard specified an operating range that was designed to operate at what later turned out to be the peak performance which could be achieved. Token Bus defines both physical and data rate options which use analog signaling.

Token Ring - IEEE 802.5 Token Ring architecture was developed by IBM development labs in Zurich, Switzerland. The first Token Ring products appeared in 1986. Since then, Token Ring has been slowly gaining on Ethernet as the popular LAN standard. Token Ring operates on the 802.2 Logical Link Control (LLC) layer with IEEE 802.2 Type 1 protocol and 802.5 Media Access Control (MAC) Token Passing Protocol.

There is one major Metropolitan Area Network Architecture: the Distributed Queue Dual Bus (DQDB) (IEEE 802.6 defined) architecture. Chapters 11 and 12 are devoted entirely to this architecture and the Switched Multimegabit Data Service (SMDS) which operates over the DQDB architecture.

Figure 3.8 shows a comparison of the IEEE 802 standards to the OSI Reference Model. This comparison will be carried through this section to show the relative layering of all other architectures defined in this section.

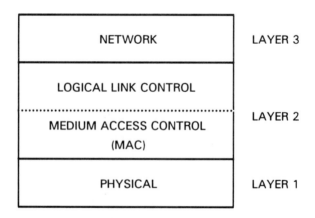

Figure 3.7 IEEE 802.X Architecture

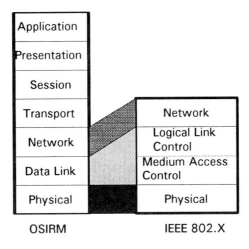

Figure 3.8 IEEE 802.X Architecture Compared to the OSIRM

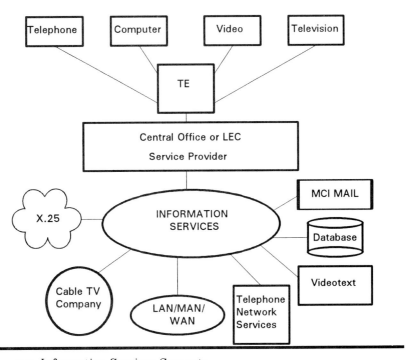

Figure 3.9 Information Services Concepts

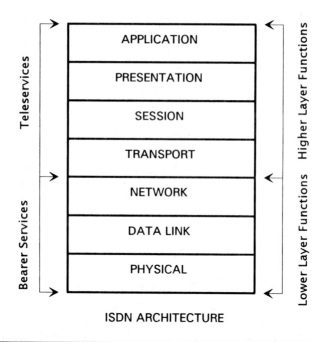

ISDN ARCHITECTURE

Figure 3.10 ISDN Architecture

3.4.3 Integrated Services Digital Network (ISDN)

Integrated Services Digital Network (ISDN) standards were begun by the CCITT in 1972, with the first standards documents published in 1984. The original intent of ISDN was to provide a conversion of telecommunications transmission and switching techniques to a digital architecture, providing end user to end user digital service for voice, data, and video. But ISDN standards have been used for much more. ISDN standards are also at the root of frame relay standards. These roots have enabled frame relay to become an overnight international standard, helped by the fact that ISDN is designed to provide a standard international user interface. Figure 3.9 shows the ISDN architecture concept of multiple device connection through an ISDN network termination device (called a TE) and into the central office environment, where information services are provided.

The ISDN architectural structure is composed similar to the OSI Reference Model. The ISDN architecture layout is shown in Figure 3.10. While all seven protocol layers are the same as the OSIRM, the physical, data link, and network layers define the lower layer functions, which include the bearer services, such as frame relay. These layers define physical connectivity and transmission; data link management, flow, error, and synchronization control; and network addressing, congestion control, end-to-end call establishment, routing or relaying, and switching, respectively. The transport, session, presentation, and application layers define the higher-layer functions, including the teleservices which define services such as messaging, telephone, and telex. Figure 3.11 shows a comparison of the ISDN architecture to the other architectures defined in this section.

3.4.4 IBM's Systems Network Architecture (SNA)

The introduction by IBM of Systems Network Architecture (SNA) in 1974 signaled the beginning of a vendor proprietary architecture which continues to rule the standards for computer architectures even today. Even though many users and vendors (even IBM!) recognize OSI as the future standard architecture, the dominant computer architecture remains SNA. SNA was IBM's method of creating a computing empire through standardization, and revolved around the mainframe and front end processors. The point of IBM attempting to standardize among its own products, and not other vendors' equipment and protocols, is sometimes forgotten. By providing a hierarchy of network access methods, IBM created a network which could accommodate a wide variety of users, protocols, and applications, while retaining ultimate control in the mainframe host and front end processors. The move from centralized to distributed processing has had pronounced effects on SNA, but even IBM sees SNA as too strict a protocol structure for true distributed processing.

OSIRM		ISDN
Application		Application
Presentation		Presentation
Session		Session
Transport		Transport
Network		Network
Data Link		Data Link
Physical		Physical

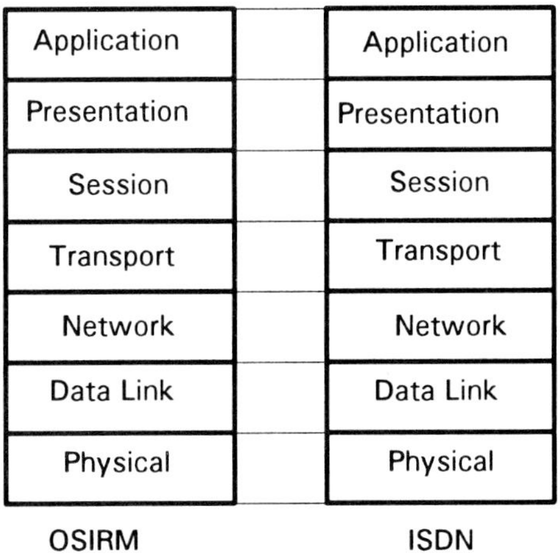

Figure 3.11 ISDN Architecture Compared to the OSIRM

The SNA architecture layers are similar to the OSI Reference Model. Figure 3.12 shows the SNA architecture model. The network control functions reside in the Physical, Data Link, Path, Transmission, and Data Flow control layers. The Physical and Data Link layers define functions similar to the OSIRM, with serial data links employing the SDLC protocol and channel attachments employing the S/370 protocol. The Path Control layer provides the flow control and routing between point-to-point logical channels on virtual circuits through the network, establishing the logical connection between both the source and the destination nodes. The path layer provides paths without addressing, which forces OSI protocols to bridge SNA traffic rather than to route or switch. The Transmission layer provides session management and flow control over SNA sessions which it establishes, maintains, and terminates. The Transmission layer also contains some routing functions. The Data Flow layer provides services related to the actual user sessions, with both layers operating at times in parallel. The network services functions reside in the Presentation and Transaction layers. These two layers combined are also called the Function Management layer. The Presentation layer formats and presents the data to the users, as well as performing data translation, compression and encryption. The Transaction Services layer provides network management and configuration services, as well as a fully functional user interface to the network operations. Figure 3.13 shows a comparison of the SNA architecture to the other architectures defined previously.

3.4.5 Digital Equipment Corporation's Digital Network Architecture (DNA)

DECnet, or Digital Network Architecture, was created in 1975 by Digital Equipment Corporation. DECnet Phase V is the most recent publication of the architecture. The architecture is again similar to the OSI Reference Model, but with the addition of a network management layer.

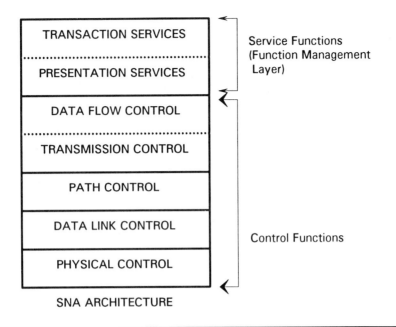

Figure 3.12 SNA Architecture

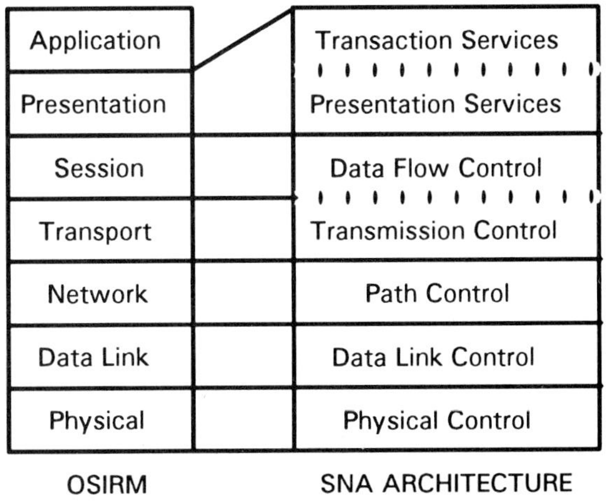

Figure 3.13 SNA Architecture Compared to the OSIRM

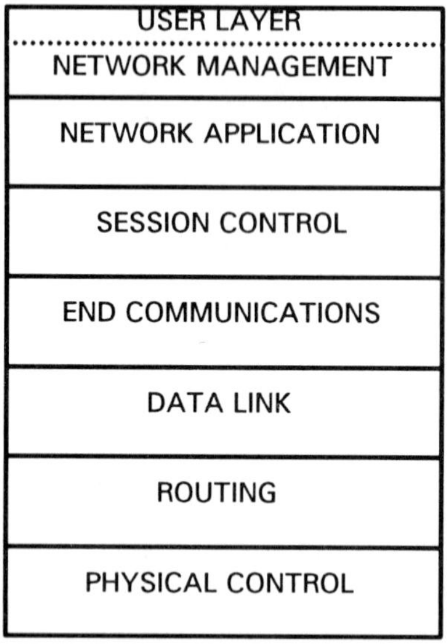

```
┌─────────────────────────────┐
│         USER LAYER          │
│·····························│
│     NETWORK MANAGEMENT      │
├─────────────────────────────┤
│     NETWORK APPLICATION     │
├─────────────────────────────┤
│       SESSION CONTROL       │
├─────────────────────────────┤
│     END COMMUNICATIONS      │
├─────────────────────────────┤
│          DATA LINK          │
├─────────────────────────────┤
│           ROUTING           │
├─────────────────────────────┤
│       PHYSICAL CONTROL      │
└─────────────────────────────┘
```

Figure 3.14 DNA Architecture

Figure 3.14 shows the DNA architecture model. The Physical layer is similar to the OSI Reference Model, as is the Data Link layer, which uses X.25, Ethernet, and DDCMP protocols. The Routing layer provides sort of a network-wide datagram service, and the End Communications layer provides the end-to-end communications service to the higher functional layers, as well as performing logical channel multiplexing. The Session Control layer provides the higher layers processes with system-dependent communications functions through sessions. The Network Application layer provides user services to the Network Management and User layers. The User layer contains user-defined application programs. The Network Management layer provides management for all other seven layers and distributed network management throughout nodes in the network.

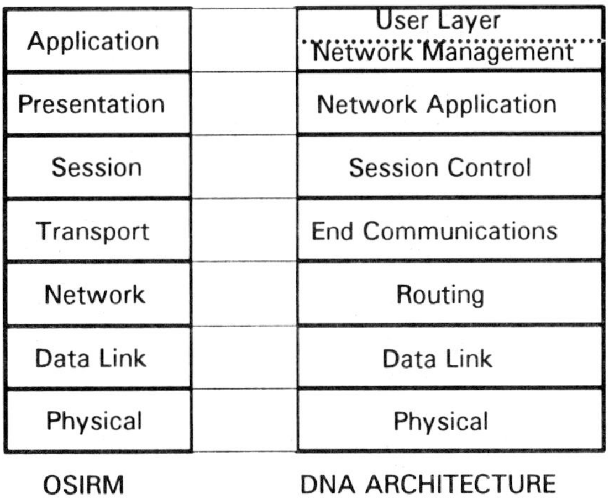

Application
Presentation
Session
Transport
Network
Data Link
Physical
User Layer
Network Management
Network Application
Session Control
End Communications
Routing
Data Link
Physical
OSIRM
DNA ARCHITECTURE

Figure 3.15 DNA Architecture Compared to the OSI Reference Model

Figure 3.15 shows a comparison of the DNA architecture to the OSI Reference Model.

APPLICATION
PRESENTATION
SESSION
TRANSPORT
NETWORK
DATA LINK
LLC
MAC
PHYSICAL
Same as
IEEE
802.4

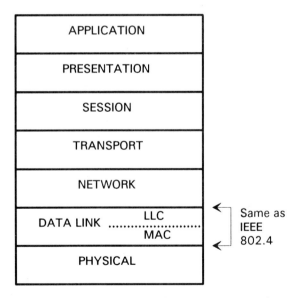

Figure 3.16 MAP Architecture

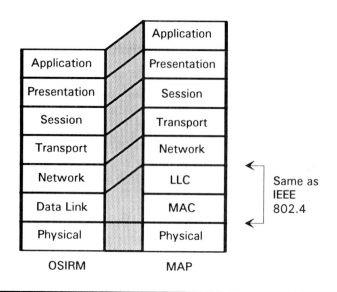

Figure 3.17 MAP Architecture Compared to the OSIRM

3.4.6 General Motors' Manufacturing Automation Protocol (MAP)

The Manufacturing Automation Protocol architecture was developed by General Motors as a conglomerate of existing protocols and the OSI Reference Model. MAP uses an architecture structure similar to the OSI Reference Model, with the lower layers utilizing the IEEE 802.4 Token Bus Protocol and the higher layers using the standard OSI protocols previously defined. General Motors is also building implementation profiles to interface the MAP architecture to the OSI architecture. Figure 3.16 shows the MAP architecture model. Figure 3.17 shows a comparison of the MAP architecture to the OSIRM.

3.4.7 Boeing Computer Services' Technical & Office Protocol (TOP)

The Technical and Office Protocol (TOP) was developed by Boeing Computer Services through a user group which needed computer communications standards for vendors serving Boeing to meet stringent architectural guidelines. The architecture structure is again similar to the OSI Reference Model, with the lower layers using IEEE 802.3 Ethernet or 802.5 Token Ring protocols. Boeing is also building implementation profiles to interface the TOP architecture to the OSI architecture. Figure 3.18 shows the TOP architecture structure.

Figure 3.19 shows a comparison of the TOP architecture to the other architectures previously defined.

3.4.8 IBM's Systems Application Architecture (SAA)

Systems Application Architecture (SAA) was announced by IBM in 1988, providing a new architecture for computer communications which incorporates most of the OSI Reference Model functional layers into existing IBM-defined protocols and new computing platforms. This was IBM's response to the migration away from centralized mainframes and front end processors and toward distributed processing. It provides more peer-to-peer connectivity and interoperability among IBM products but still allows the mainframe to control the activity - or at least receive network management information on activity for tracking and report generation.

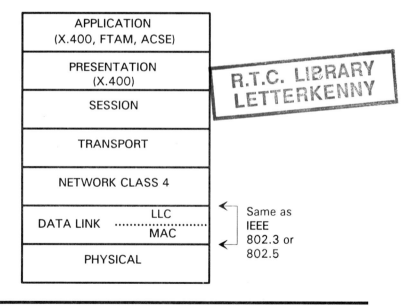

Figure 3.18 TOP Architecture

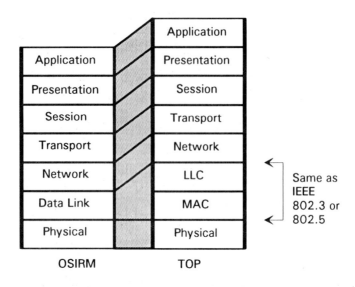

Figure 3.19 TOP Architecture Compared to the OSIRM

The Systems Application Architecture focuses on the top five layers of the OSI Reference Model: the Applications through Network layers. The Physical and Data Link layers are the same as for SNA, and also contain the LAN protocols of the IEEE 802.5 Token Ring. Other LAN protocols can also be supported through IBM 3172 Interconnect Controllers and the IBM Open Systems Interconnect / Communications Subsystems (OSI / CS). The Network layer is served by the X.25 protocol for connectionless-oriented services (called CLNS) and portions of the LAN protocols, as well as Internet protocols for connection-oriented services (called CONS). The Transport layer provides three classes of service: Class 0 defines simple, Class 2 defines multiplexing, and Class 4 defines error correction and recovery. The Session layer services support both Version 1 and 2 of the ISO session protocols (still in the standards process). The Presentation layer services supported include both the Kernel and Abstract Syntax Notation 1 (ANS.1). Finally, the Application layer services support Common Management Information Protocol (CMIP) and Common Management Information Services (CMIS), File Transfer, Access and Management (FTAM), X.400 message handling and X.500 directory services, and Association Control Service Element (ACSE) protocols. Figure 3.20 shows the SAA architecture model, paralleling the OSI Reference Model and showing the protocols supported at each layer.

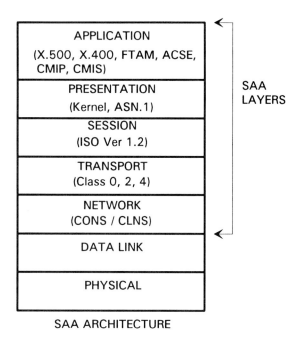

SAA ARCHITECTURE

Figure 3.20 SAA Architecture

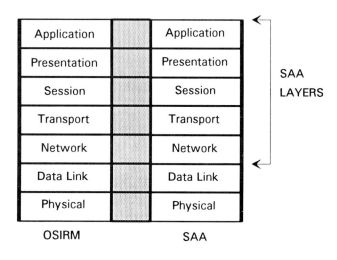

Figure 3.21 SAA Architecture Compared to the OSIRM

IBM is only implementing these protocol suites on the System/370, System/390, Application System/400, and Personal System/2 platforms, as well as their new router products. The OSI/CS is the method of running the SAA architecture protocols on these IBM hardware. This form of OSI implementation keeps IBM in the proprietary arena, while providing the capability to interface to non-IBM architectures and protocols. Sometimes SAA implementations to other vendors' equipment and protocols is cumbersome and restricted to gateway-like functionality while most IBM products are more tightly coupled and integrated. So IBM can "support" standards while pursuing customer lock-in via proprietary implementations. Figure 3.21 shows a final comparison of the IBM SAA to the OSIRM.

3.4.9 Architecture Joint Ventures

Some major architectures have blended to produce cross-architecture architectures. One example is the Office Document Architecture (ODA), developed as a joint effort between ISO, CCITT, and ECMA. This architecture provides a common format for document interchange that can be used for all word processors. Many of the major vendors with proprietary architectures mentioned above now support the ODA architecture. Similar efforts are taking place with ISDN standards, in particular for frame relay service which spans many standards and architectures.

3.5   LAYERS OF THE OSI REFERENCE MODEL

Many computer networking architectures utilize the basic structure of the seven layer Open Systems Interconnect Reference Model. Each layer of the OSIRM will now be covered in detail. The OSIRM presents a layered approach to data transmission; seven layers, with each layer adding a value-added service to data transport. Data flows down from layer seven to layer one, where it is transmitted across layer 0 through the network, and back up to the level seven to the remote user. Not all seven levels are always used, this is dependent upon the application and user needs. Also, all seven layers are not well defined, particularly the Application, Presentation, and Session layers. All seven layers will be covered starting with the layer closest to the physical transmission medium, and assume a physical communications media, layer 0 (discussed in Chapter 5). In reality, the user data originates and terminates at layer seven. Examples provided will be discussed in later chapters.

3.5.1 Physical Layer

The first layer encountered after the physical medium is the Physical layer (layer one). This layer manages the physical connection of network elements, including electrical voltages and currents, mechanical, basic signaling through pin connections, and signaling formats. The physical media can either be wire or wireless, and the interface can be monitored by a simple breakout box. Examples of the physical layer include EIA-RS-232C, EIA-RS-449, CCITT X.21/X.21bis, CCITT V.35, IEEE 802 LAN, ISO 9314 FDDI, and the new HSSI interface. One example of a wireless physical interface is the move toward wireless LANs.

3.5.2 Link Layer

The second layer encountered is the Link layer, sometimes called the Data Link layer. The Data Link level manages the flow of data between user and network, or between Data Termination Equipment (DTE) and Data Communications Equipment (DCE). It is the second layer's responsibility to assure reliable and efficient data transfer, data formatting, detection, correction and recovery of data errors, data transparency, and some forms of addressing. The Link layer also contains frames or packets, and includes the Medium Access Control (MAC) sublayer used by the IEEE 802.X LAN protocols such as Ethernet, Token Ring, and FDDI. Some of the new services such as Frame relay and SMDS use only the first two layers of the OSI Reference Model, and rely heavily on enhancing the link layer services. Examples of the Link layer include ISO 7776/CCITT X.25 LAP/LAPB & LAP-D, CCITT HDLC, and MAC Layer protocols such as the ISO 9314-2 FDDI Token Ring MAC.

3.5.3 Network Layer

The third layer encountered is the Network layer. The Network layer manages the details of transmitting data across the physical network between network elements, as well as between networks. It is the third layer's responsibility to manage flow control, define data call establishment procedures for packet and cell switched networks, and manage the segmentation and assembly of data across the network. The Network layer is the most protocol intensive portion of packet networks. Some examples of protocols used in the network layer are CCITT X.25 and X.75 packet level and gateway protocols.

3.5.4 Transport Layer

The fourth layer encountered is the Transport layer. The Transport level controls the quality and methods of data transport across the entire network. This can be independent of the number or types of networks through which the data must transit. It is the fourth layer's responsibility to manage end-to-end control for complete message transmission, retransmission, and delivery. The Transport layer also assures that the packet/message segmentation and reassembly process is complete. This service has become increasingly more important since it provides a higher-level error correction and retransmission protocol for new services such as Frame relay and SMDS. Examples of the Transport layer include CCITT X.224 OSI Transport Protocol and TCP (although TCP is part of the ARPANET protocol stack, it is commonly used in combination with IP as the protocol of choice for WAN connectivity). An OSI-IP standard (ISO 8473) has been issued as an improved version of the DOD-IP.

3.5.5 Session Layer

The fifth layer encountered is the Session layer. The Session layer manages the user to network interactive sessions and all session-oriented transmissions. These sessions are communications between users, typically a user community such as terminals or LAN workstations, and a central processor, typically a host or front end processor. Many of these functions of user software interaction are handled via the top three layers together, but each layer has a subtle but distinct functionality still being ironed out in the standards process. Some examples of the Session layer are terminal to mainframe log-on procedures, transfer of user information, and setting up information and resource allocations. The ISO standard for the session layer is the ISO 8327/CCITT X.225 connection-oriented session protocol.

3.5.6 Presentation Layer

The sixth layer encountered is the Presentation layer. The Presentation level determines how data is presented to the user. Official standards are now complete for this layer. Also, many vendors have implemented proprietary solutions. One reason for these proprietary solutions is that the use of the Presentation layer is very "equipment dependent". Some examples of the Presentation layer protocols are video and text display formats, data code conversion between software programs, and peripheral management and control, using protocols such as CCITT X.400/X.410, and CCITT X.226 OSI connection-oriented protocol.

3.5.7 Application Layer

The seventh layer encountered is the Application layer. The seventh layer manages the program or device generating the data to the network. The Application layer is an "equipment-dependent" protocol, and lends itself to proprietary vendor interpretation. Examples of Application layer protocols include CCITT X.500 - X.520 Directory Management, ISO 8613/CCITT T.411-419 Office Document Architecture (ODA), and the ISO 10026 Distributed Transaction Processing (TP).

3.6 INTEROPERABILITY — THE MOVE TOWARD OSI

While many users have a large investment in one of the proprietary vendor architectures, such as IBM SNA, DEC DNA, and TCP/IP, they are still moving toward Open Systems. The market once dominated by IBM's Systems Network Architecture is now moving with increasing speed toward global standards through Open Systems Interconnect. Even IBM is moving toward OSI. Why has IBM moved toward Open Systems with their new Systems Application Architecture? As discussed before, IBM detects their proprietary market for computer communications eroding, along with the move toward distributed processing and the need to produce products that interoperate with international networks employing OSI standards.

Open Systems Interconnect is the international standard by which open systems communicate. OSI is an any-to-any open system of data communications via a common set of protocols, allowing internetworking and interoperability between enterprise networks and multivendor products. The move toward OSI standardization depends on both vendor and user implementation. Users are moving from vendor proprietary architectures such as SNA and TCP/IP toward OSI, the U.S. government is moving from TCP/IP toward the U.S. Government Open Systems Interconnection Profile (GOSIP), and the European and Pacific Rim countries are moving toward OSI at an even faster rate than the United States. GOSIP version 1 has now been followed by GOSIP version 2 that adds that X.400 and FTAM products should support the transport of office document architecture (ODA) — the international standard for compound documents. GOSIP version 3, now being drafted, seeks to combine industry and government requirements into one specification.

3.6.1 Vendor Support of OSI

Most vendors have moved toward support of the upper layers of the OSI Reference Model. This support has made significant advancements toward interconnecting of disparate communications systems. This shows where true intelligence lies in a data communications network. Lower layer implementations usually support X.25 for wide area networking and the 802.X standards for local and metropolitan area networking. One of the primary endless loops is the vendor who is waiting for the user to start buying OSI products before producing large quantities of OSI compatible products, while the users are waiting for the vendors to sell OSI products en mass before they will buy.

The compromise position seems to be making investments in hardware and software to support workable protocols such as TCP/IP and SNA now, and migrating to OSI based networks in the future. Vendors need to support OSI for it to be a success. One sign of vendors moving to support OSI is through the new OSI IS-IS routing protocol. IS-IS allows routers for the first time to communicate network and internetwork routing information between devices. This will be a major breakthrough for the router market and for interoperability among vendors. Support for the new X.400 Message Handling and X.500 Directory Services standards are also pushing vendors toward OSI. The vendor support for OSI is becoming more predominant through internetworking products such as routers and gateways, which provide routing and switching, encapsulation and translation, and mix OSI and proprietary protocols.

3.6.2 User Support of OSI

Many trade shows, conventions, and the Open Systems Forum are pushing OSI. X Windows, a new OSI application, has covered many trade show floors, and is a shining example of the commitment toward OSI applications. The best user indication of support of OSI standards is through participation in the standards committees. Since the OSI standards development cycle held little or no research community participation, users do not fully understand these standards. Once the user begins to participate, OSI standards will better reflect the needs of the user.

Users can begin to implement OSI products as slowly or as quickly as they wish. Start slowly by replacing proprietary network elements with OSI based elements. Start quickly by building test networks replace core backbone networking structures when they meet your needs. Users can begin by implementing one layer of the OSI protocol stack, and eventually communicating over all seven layers. Profiles containing portions of proprietary architecture interacting with the OSI architecture provide another implementation solution. Either way, the shift toward OSI networking is on. Catch the wave.

3.6.3 TCP/IP and OSI

TCP/IP and OSI have both been developed to internetwork multiple diverse data networks through vendor proprietary solutions. LAN operating systems are also developed with proprietary solutions in mind, but all tend to sway toward TCP/IP implementations rather than OSI solutions. In the past, focus has been placed upon improving interim protocols such as TCP/IP, which has proven sufficient in today's networks. Users want the TCP/IP that works today as a reasonable solution to immediate business problems rather than waiting for the perfect pie-in-the-sky OSI solution that has yet to appear and is thus hard to plan and budget for. Some development has also been oriented toward running OSI at higher layers over TCP/IP. Most likely, there will be a temporary coexistence of TCP/IP and other proprietary standardized protocols and OSI for the remainder of the decade, with TCP/IP remaining the dominant internetworking protocol until the end of the decade. This is true by the example of the Internet, where TCP/IP and UNIX reign.

3.6.4 The United States Government

The U.S. government also needs to make the move away from TCP/IP toward OSI. The U.S. Government Open Systems Interconnection Profile (GOSIP) is the U.S. government's solution to OSI. GOSIP solves the problem of OSI by mandating any purchase of communications systems to be OSI compliant, if possible. Sometimes the word "compliant" however has a rubber-band definition. GOSIP also uses an implementation profile which defines the specific set of OSI protocols and features supported. A few other driving forces for OSI are the Corporation for Open Systems (COS) Mark and OSINET. COS Mark is an attempt at an OSI multivendor warranty for vendor products and OSINET is an association made up of OSI product vendors and suppliers. Together with GOSIP, these interest groups and the others mentioned above are driving the OSI standards needed for interoperability.

The Federal Information Processing Standards (FIPS), specifically FIPS 146, defines what portion of the OSI standard is used as government agency procurement standards. The National Institute of Standards and Technologies (NIST) releases the GOSIP guidelines, and is currently on the GOSIP 4 release.

3.6.5 OSI — The True International Standard

The European and Pacific Rim countries are standardizing on OSI. Since their networks are developing at a slower rate, with less proprietary vendor competition, they are slowly building data communications networks which comply to the OSI standards and support both voice and data. This move is both government and end user driven. Here, all hardware and software is required to comply with OSI standards. This dedication of the world data communications market toward OSI shows that one true international standard architecture is desired.

3.6.6 Advantages and Drawbacks

Some *advantages* of implementing OSI include:

- ⊕ Interconnectivity
- ⊕ Interoperability
- ⊕ Vendor independence
- ⊕ Fewer compatibility problems
- ⊕ Investment protection
- ⊕ Tap vendors with systems integration experience
- ⊕ Ideal for client-server architecture and distributed processing
- ⊕ International standard architecture
- ⊕ Reduce cost of hardware and software
- ⊕ Eliminate multivendor platforms
- ⊕ Improve interoperability among network elements
- ⊕ Reduce training costs
- ⊕ Increase vendor competition which will lower prices
- ⊕ Seamless, concurrent computing

Some *drawbacks* of implementing OSI include:

⊗ OSI may not be needed
⊗ OSI products often inferior to proprietary solutions
⊗ OSI products are more expensive than proprietary solutions
⊗ Many proprietary protocols are sold bundled with other products and services
⊗ May change to operations of the company
⊗ Stranding different generations of technology
⊗ Many vendors still experimenting with OSI
⊗ Standards gridlock on key business affecting standards
⊗ OSI products are vendor-oriented, not user-oriented
⊗ Steep learning curve
⊗ Too little, too late

3.7 REVIEW

We have covered most aspects of what constitutes a standard — de facto and de jure. We gained a basic understanding of what standards are and why they are needed by vendors, users, and service providers. Each of the major North American standards committees and forums, along with the major international organizations, were introduced. The seven major computer architectures were explored, as well as IBM's new Systems Application Architecture (SAA). Each layer of the OSI Reference Model was covered in detail, citing examples. The move toward OSI and the vendor, user, and service provider support required to make this a reality is clear to see. Thus a brief review of each aspect of standards, with references for further study, shows that the standards listed here must be understood and addressed. Next we will cover circuits, services, and hardware evolving from these standards.

4

Circuits, Services, and Access Hardware

Now we will discuss transmission types and circuit definitions, which have their roots in two- and four-wire voice transmission and grow into the branches of data transmissions. In data communications, many of the same basic rules apply for both voice and data, and a discussion of duplex transmission options will explain these principles. A high-level description of common network topologies is provided as a primer for later design chapters. Coverage of the two major types of data services is given: connection oriented, and connectionless data transfer. These definitions lay the groundwork for discussing packet, frame and cell switching. An overview of access hardware classifications will tie the various concepts together. Both local area networks (LANs) and wide area networks (WANs) access devices tie access to transport services. Multiplexing devices will be covered in a subsequent chapter. As technology marches on, hardware which was at one time specialized in a given function is now performing many functions through an increasing demand upon software.

4.1   TRANSMISSION TYPES

Transmission of information can be defined in two levels: the actual communication circuit operation defining data flow direction and the method in which information is transmitted over these circuits. Three methods of circuit operation exist: simplex, half duplex, and full duplex. Two modes of data transmission prevail: asynchronous and synchronous.

4.1.1 Simplex, Half Duplex, and Full Duplex

When talking about circuits, we must reference their means of signal transfer.   When dealing with voice, simplex defines a two-wire connection between two points over which voice can travel in both directions but only in one direction at a time. Half duplex defines a two-wire connection between two points where the voice can be transmitted in both directions simultaneously. Full duplex defines a four-wire connection between two points over which two simultaneous transmissions can occur on a pair of wires.

When dealing with data communications, these concepts are similar. For example, point-to-point ISSI SMDS service resembles half duplex, because two DS3s are required for simultaneous transmission of data in both directions. While two wires are used in both analog and data communications, for data these wires (media) are often fiber optic cables, capable of carrying gigabits of data, as opposed to low-bandwidth voice signals.

4.1.2 Asynchronous/Synchronous Transmission

Asynchronous transmission is characterized by the absence of a clock in the transmission media. Data is transmitted in pieces or as individual characters, and each character is synchronized separately by information in its header (start) or trailer (stop) bits — self-contained. The access device is not synchronized with the network device, but the speeds of transmission must be the same. Figure 4.1 shows an asynchronous transmission stream with each character framed in start and stop bits and variable delay of time "t" between each character transmitted.

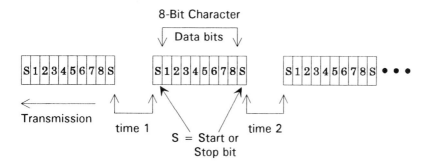

Figure 4.1 Asynchronous Transmission Stream Framing

With synchronous transmission the access device and network device share a common clock and transmission rate, and all transmissions are synchronized. Data flows in character streams, called message-framed data, and each transmission has a specific start and stop sequence. Thus, the access and network devices need to be synchronized to ensure that the entire message is received in the order transmitted. Figure 4.2 shows a typical synchronous data stream. The message begins with two synchronization (SYNC) characters and a start-of-message (SOM) character. The control (C) character(s) denote the type of user data or message following. The data follows next. The cyclic redundancy check (CRC) character checks the data for errors, and the end of message (EOM) character signals the end of the transmission stream. Synchronous transmissions vary in format but retain this same general structure.

Synch	Synch	SOM	Control	Data	CRC	EOM

Synch - Synchronization

SOM - Start of Message

Control - Specifies User Message

Data - User Data Stream

CRC - Cyclic Redundancy Check

EOM - End of Message

Figure 4.2 Asynchronous Transmission Stream with Framing

Asynchronous transmissions may operate at low speeds (typically 1200 bps) or, in the case of technologies like asynchronous transfer mode, at very high speeds (typically OC-3 and above). Synchronous transmissions operate at speeds 1200 bps and higher. Asynchronous interfaces include RS232-C and D, as well as X.21. Synchronous interfaces include V.35, RS449/RS-442 Balanced, RS232-C and D, and X.21.

4.2   CIRCUIT TYPES AND SERVICES

When taking a microlevel look at circuit connectivity, it is necessary to analyze the methods of connectivity available to users. There are four major methods of circuit connection available: point-to-point, dial-up, private (leased), and multidrop. These fundamental configurations are found in most multiplexing and switching architectures.

Many of these types of circuits are actual service offerings of the Regional Bell Operating Companies (RBOCs), independents, and interexchange carriers (IXCs). They are offered by the IXCs in conjunction with hybrid network solutions such as switched services and virtual private and public networks. In some cases they are more attractive than dedicated or switched services. Either way, these circuit types form the fundamental set of connectivity for all types of data communications.

4.2.1 Point-to-Point

Point-to-point circuits are defined as a single connection or link between two users. This connection can be physical, virtual, or logical. The physical media can be two wire, four-wire, coaxial, fiber optics, or a variety of other proprietary and standard interfaces. The point-to-point circuit can be permanent or temporary, dedicated or dial-up. Figure 4.3 depicts a simple point-to-point circuit between User A and User B running simplex. Figure 4.4 shows the same point to point circuit using half duplex, and Figure 4.5 depicts full duplex communication over a point-to-point circuit. Arrows are added to show the direction of transmission for two- and four-wire use.

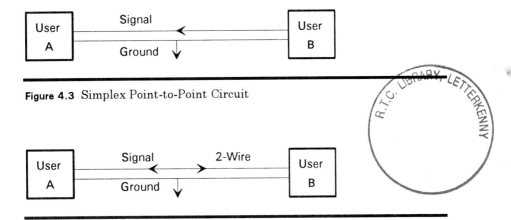

Figure 4.3 Simplex Point-to-Point Circuit

Figure 4.4 Half-Duplex Point-to-Point Circuit

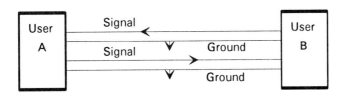

Figure 4.5 Full-Duplex Point-to-Point Circuit

4.2.2 Dial-Up Lines

Dial-up lines are similar to point-to-point circuits, only they are not guaranteed to be dedicated and available when accessed. The user relies on the service provider for availability of the dial-up lines, and many users share dial-up lines as alternate access during emergencies. While dial-up lines are used primarily by modems, the network designer will most likely use some form of dial-up circuit for network redundancy. Switched 56 and T1 services provide a means of dial-up lines for users who have dedicated data circuits and require additional redundancy. Figure 4.6 depicts User A with a dial-up circuit to User B, being used as a redundant circuit for the failed point-to-point leased line circuit. Occasional users sometimes rely exclusively on dial-up data services. Frequent and heavy users usually employ dedicated (leased) private lines.

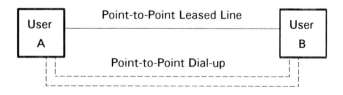

Figure 4.6 Point-to-Point Redundancy

4.2.3 Private Lines

A private line, or leased line, is a dedicated circuit which has been leased from a carrier for a predetermined period of time. This private line is also referred to as a dedicated line, where the user pays extra for a defined quality of service, such as conditioning to ensure a given error rate minimum. As carriers install all-fiber networks, digital private lines are replacing the old voice-grade analog circuits, at the same or even cheaper cost.

Private lines are purchased based upon continual 24 hour a day utilization. Private lines are now mostly used for "mission critical" applications where switched service or other dial-up alternatives are not sufficient, or as backbone high-bandwidth trunks. Thus, a user who purchases a private line must justify the set cost against a service which would offer usage-based billing. Services such as frame relay and SMDS allow users to be billed based upon their usage, as well as their connect fee. This creates a virtual leased line of appropriate bandwidth only when needed. Most users only use the full bandwidth of their leased lines a small percentage of the time. Such bandwidth-on-demand services will replace many private lines and allow the user greater cost savings when the virtual leased line is not in use. MCI's frame relay and Switched Multimegabit Data Service (SMDS) services offer these benefits.

Leased lines are also used for access to the local telco, and are then called access lines. These access lines feed or funnel the user's data to carrier-based services that are often switched. Access lines can be purchased through either local Telco or as alternate access. Access lines from alternate access carriers are generally cheaper than the Telco price. But, of course, the alternate access carrier usually "cream-skims" the voluminous and lucrative traffic and leaves the "skimmed milk" for the LEC to serve smaller, more remote, or occasional users. The average price for T1 access through an alternate access carrier is around $900 for the one-time installation fee and about $500 per month after that. This price is only for approximately a "last mile" tail circuit, and prices for access increase as the distance increases. As a comparison, the average Telco price would be $1300 for installation and $600 per month after that. Prices continue to drop but the ratio will remain the same.

Another form of T1 circuit is the High-rate Digital Subscriber Line (HDSL). HDSLs eliminate the need for repeaters every 5000 feet as in a standard T1 line, are not effected by bridge taps, are less than half the cost of standard T1s, and can support switched or dedicated services. The drawback is that they need to be within 12,000 feet of the serving central office.

Private lines in Europe and Pacific Rim countries are still very expensive, and transoceanic fiber access is limited. A carrier also must make an agreement with the party at the other side of a fiber to offer the transoceanic service. European pricing for private lines to the United States ranges from as low as $4000 per DS0 for the United Kingdom to as much as $9000 per DS0 for Italy. These prices are dropping, but they show the significant investment in small amounts of bandwidth which is often taken for granted in the United States. This is why multiplexers have proven so popular internationally.

4.2.4 Multidrop

When one user, typically the originator of information, needs to communicate simultaneously with multiple users, a multidrop, or multipoint circuit is used. Figure 4.7 shows a two-wire multidrop and Figure 4.8 shows a four-wire multidrop. When using multidrop circuits, there is a master-slave relationship between User A and Users B, C, and D. Each transmission between these users must first go through User A.

In the data world, multidrop circuits work the same way. Take SMDS for example. The SMDS Subscriber Network Interface (SNI) allows multiple users to interface with a Metropolitan Area Network (MAN) switch via one SNI interface. Thus, the MAN switch acts as user A, and users B, C, and D are connected via one SNI. See Figure 4.8 (except that the transmit and receive are one circuit instead of two). This connectivity will be explained in greater detail later, but provides a good example of multidrop data functionality.

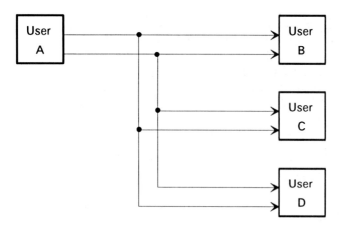

Figure 4.7 Two-Wire Multidrop

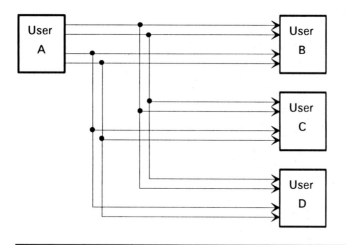

Figure 4.8 Four-Wire Multidrop

4.3 NETWORK TOPOLOGIES

There are five major network topologies for computer and data communications networks. These include: point-to-point, multipoint or common bus, star, loop or ring, and meshed. We will briefly outline each topology, and in later design chapters will demonstrate their implementation via new technologies. The term "node" will be used to designate a network transport element such as a router, switch, or multiplexer.

4.3.1 Point-to-Point

Point-to-point connectivity is the simplest topology, providing a single link between two nodes. This link can be composed of multiple physical and logical circuits. Figure 4.9 shows three examples of point-to-point links. The first example shows a single link between Node A and Node B with a single physical and logical circuit, the second depicts a single link between Node A and Node B with multiple logical circuits riding over a single physical link, and the third depicts a single link between Node A and Node B with multiple physical circuits, showing that each physical circuit could have multiple logical circuits.

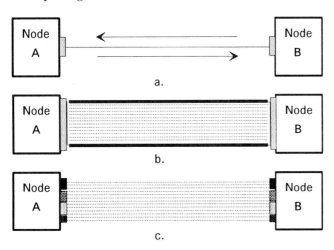

Figure 4.9 Point-to-Point Link Examples

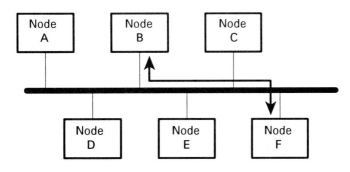

Figure 4.10 Multipoint Topology

Point-to-point configurations are the most common method of circuit connectivity. Many of the services described in this section: point-to-point, dial-up lines, and private lines, use point-to-point topologies. Almost every user access to the many types of network architectures uses some form of point-to-point topology.

4.3.2 Multipoint (Common Bus)

In the multipoint, or common bus, topology all nodes are physically and logically connected to a common bus structure. These nodes are often local to the bus. Figure 4.10 shows the multipoint topology, where Nodes A through F communicate via a common physical and logical bus. The IEEE 802.4 Token Bus and IEEE 802.3 Ethernet use a common bus topology, as do many other vendor architectures.

4.3.3 Star

The star topology was developed during the mainframe era, when many computer communications were centrally controlled by a mainframe. It also has its analog in the voice world with a PBX that provides one central switching CPU but serves hundreds of thousands of users that radiate networks in a star-like fashion from the CPU or PBX switch to telephones on people's desks. All devices in the network are connected to the central node, which usually performs the processing. Nodes communicate with each other through point-to-point or multidrop links off of the central node.

Figure 4.11 shows a star topology, where Node A serves as the central node and Nodes B through E communicate with each other through Node A. One example of a star topology is when accessing a host processor from remote terminal locations. Each terminal must access the host directly, and all processing capability lies within the host. These terminals are often called "dumb" terminals since the processing power and intelligence is host-resident. Another example is a LAN wiring hub which we will see later in this chapter.

4.3.4 Loop (Ring)

The loop, or ring, topology is used for networks where communications data flow is unidirectional. A ring is established, and each device passes information along the direction of the ring. Figure 4.12 shows a ring network where Node A passes information to Node C via the ring and past (or through) Node D. Node C returns information to Node A via Node B. Examples of the ring topology are the IEEE 802.5 Token Ring and the IEEE 802.6 Distributed Queue Dual Bus (DQDB) architecture, both of which will be discussed later.

4.3.5 Mesh

Most switched networks employ some form of mesh architecture. Meshed networks connect many nodes via multiple point-to-point circuits. Figures 4.13 and 4.14 show two types of meshed networks. Figure 4.13 shows a partially meshed network where Nodes A, C, F, E, and G are well meshed with at least three paths to any other node, while Nodes A and D have only two paths to the network.

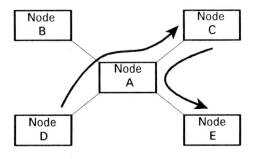

Figure 4.11 Star Topology

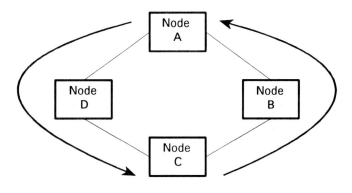

Figure 4.12 Ring or Loop Topology

Figure 4.14 shows a fully meshed network where each node has a point-to-point circuit to every other node. Almost every major computer and data communications network uses a meshed topology to give alternate routes for backup and traffic loads, but few use a fully meshed topology primarily because of cost factors associated with having a large number of point-to-point circuits. We will see later how to design based upon these topologies.

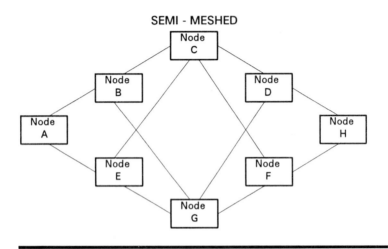

Figure 4.13 Semi-Meshed or Partially-Meshed Network

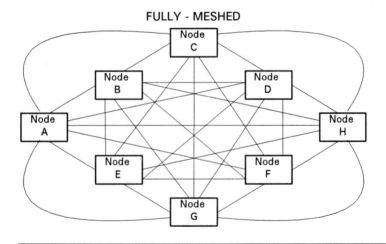

Figure 4.14 Fully-Meshed Network

4.4   DEFINITION OF NETWORK SERVICES

Data network service offerings are categorized as either connection oriented or connectionless oriented. Connection oriented defines a service which depends upon an established connection between end points, physical and virtual, and sometimes logical, for the transfer of data. Connectionless services, on the other hand, provide end-to-end logical and physical connectivity, but do not establish virtual circuits. Connection-oriented services are generally used in wide area networking, where connectionless services are used primarily in local area networking.

4.4.1 Connection-Oriented Network Services (CONS)

Connection-oriented services establish a connection between the origin and destination before transferring data. The data is transferred over a connection established through a service provider, and this same connection is then taken down after the transmission is complete. All connections are first established, data is transferred, and then the circuit is taken down. The service provider monitors the user connection and can control the information flow between the users. While connection-oriented services add large amounts of overhead through the use of complex protocols, the process ensures that information is exchanged in a reliable manner. When the call is first placed, the destination address is determined from the sending node information. All data delivery is via a predefined order regulated by protocols which allows for detection of errors in transmission.

The voice telephone call was the first connection-oriented network service. Most wide area networking protocols operate in connection-oriented mode. This makes more efficient use of the circuit and takes it down when not needed. Virtual circuits, such as X.25 Permanent Virtual Circuits (PVCs) are one example of a connection-oriented service. Virtual circuits are logical and physical paths established through the network from origin to destination. Most of the OSI architecture is based on connection-oriented services. Connection-oriented services operate best when establishing somewhat constant data flows which require constant bandwidth allocation. Private lines and high-speed circuit switching are a few examples of connection-oriented services provided by the interexchange carriers, whose primary service are based on connection-oriented services. Frame relay is sort of a blend of connection and connectionless service, with the first service offerings tending more toward being connection-oriented. The connection-oriented service is provided by frames transmitted via DLCIs across the network in permanent virtual circuits.

4.4.2 Connectionless-Oriented Network Services (CLNS)

In connectionless-oriented services, no prior end-to-end connection is required for data transmission. Thus, there is no predefined path through the network which the data travels. The originating node transmits a message encapsulated into a packet or cell, and allows for transmission to the destination via the best path. Each node in the transmission path interprets the address information in each packet as it processes it, and based on this node's position in the network and its alternate paths to the destination, passes on packets not terminating at that node, and processing those which do terminate at the node. The delivery of the packet is not guaranteed, so the service relies on higher-level protocols to perform the end-to-end circuit data management and to ensure data integrity. The flow control is minimal, and the service often does not provide for error detection or correction. Bandwidth efficiency and message sequencing is sacrificed for high performance through fast routing of packets, and there is less overhead incurred through call establishment and management. Connectionless service is sometimes called datagram service.

There are two modes of connectionless services: unacknowledged and acknowledged. Unacknowledged connectionless service operates per the standard definition of connectionless service as defined above, where the transmission contains little or no intelligence for message transmission. Acknowledged connectionless service provides for the originating node to perform connection management while the destination nodes act as unacknowledged nodes.

Connectionless service is the predominant mode of communications in local area networks because it is well suited to the bursty traffic found on LANs. Beside the common LAN technologies such as Token Ring, Ethernet, and FDDI, we are now seeing a new emergence of connectionless data services with technologies such as Asynchronous Transfer Mode (ATM). ATM can operate with the restrictions of connectionless communications because of its reliance on fiber optic systems to provide virtually error-free transmission. This is an important principle. High throughput and processing of packets, or datagrams, takes place, but the error rate is kept low by the power of the fiber optic medium of transmission. Other technologies such as 802.6/DQDB can operate in either connection or connectionless mode.

4.5 HARDWARE DISTINCTIONS

Data networking topologies are the roads by which data can travel. Now to explore the wide variety of LAN and WAN access devices which utilize these roads. Multiplexer access will be considered later. The access technologies discussed complement and, in some places, replace the conventional multiplexed and switched networks. As the fine line between channel connectivity and LAN/WAN connectivity starts to disintegrate, devices such as the bridge, router and gateway become an integral part of data communications networks. These three devices — bridge, router, and gateway — are the primary access devices to interconnect local area subnetworks to wide area networks. These devices provide connectivity among LAN subnetworks and also to wide area networks.

Critical item: before bridges and routers were available, the functions these devices now serve were performed in mainframes and front-end processors.

As personal computing arrived on the scene, and as MIPS moved to the desktop during decentralization, the bridging and routing functions, traditionally accomplished at the host/FEP complex, migrated over toward the desktop as well. The advent of personal computing, along with local and wide area networking, made routing outside the mainframe environment a necessity.

Perhaps the greatest driving factor was the LAN. Due to its diverse markets, technologies, and protocol suites, there evolved a need to make diverse LANs speak one language (or at least provide a translation between similar languages on similar types of LANs — e.g., Ethernet to Ethernet, Token Ring to Token Ring). When bridges and routers first came along, they were designed to deal with lower-speed local area networking. Now the functions of both begin to merge, and with the use of increased processor speeds and technologies [i.e., Reduced Instruction Set (RISC) processors], with reduced costs, they begin to support the local and wide area networking from low access speeds of DS0 up to T1 and even DS3 speeds. And these routers are protocol rich to deal with dissimilar protocols.

The three main categories of LAN/WAN interconnect hardware are bridges, routers and gateways. Each provides a different set of functionality which can either be provided separately or together in one piece of equipment. Each provides protocol support for certain levels of the OSI reference model, as well as other architectures. For simplicity of discussion, and since the OSI reference model seems to be the common architectural point of reference, the protocol support for each hardware device will be given in reference to the seven-layer OSI reference model. Repeaters and brouters also come into play, as subsets and conglomerations of these categories, as well as other access devices such as channel service units (CSUs) and data service units (DSUs). LAN hubs that provide LAN concentration and even some bridging and routing will also be covered in detail. Each of these new hardware technologies offer specific advantages and disadvantages depending upon user applications, protocols, addressing, and data transport needs. The network designer must understand each of these hardware to ensure successful LAN-MAN-WAN connectivity, interoperability, and integration.

4.5.1 Repeaters

Repeaters are inexpensive distance extension devices, providing physical distance extension for point-to-point circuits. This allows a network to extend the distance between network devices, similar to an extension cord for electricity, while providing electrical isolation during problem conditions. Repeaters possess very little intelligence. They are commonly used as signal regenerators, protecting against signal attenuation without affecting signal quality. Due to this lack of intelligence, repeaters add value by maintaining the integrity of all data being passed. One drawback to using repeaters is possible network congestion caused by the overhead they add due to repeating. Repeaters use only the physical layer of the OSI reference model. Figure 4.15 portrays User A and User B communicating via a repeater and the relation to the OSI reference model. Figure 4.16 shows the OSI reference model layer used by repeaters.

4.5.2 Line Drivers/Limited-Distance Modems

Line drivers, also called limited-distance modems (LDMs), are used to extend the distance of the physical circuit. Basic telecommunications courses teach that modems provide modulation/demodulation between analog and digital data. LDMs provide the same functionality, but in the form of a repeater.

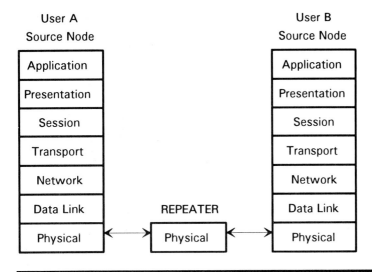

Figure 4.15 Repeaters and the OSIRM

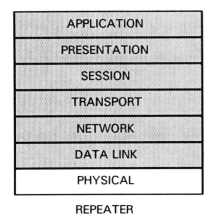

Figure 4.16 Repeater Layer of the OSIRM

Figure 4.17 shows the OSI reference model layer used by line drivers and limited-distance modems.

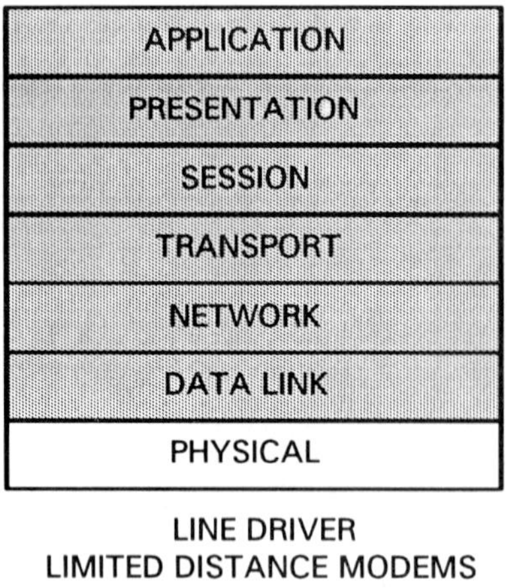

APPLICATION
PRESENTATION
SESSION
TRANSPORT
NETWORK
DATA LINK
PHYSICAL

LINE DRIVER
LIMITED DISTANCE MODEMS

Figure 4.17 Line Driver/Limited-Distance Modem Layer of the OSIRM

4.5.3 Channel Service Unit (CSU)/Data Service Unit (DSU)

The terms Channel Service Unit (CSU) and Data Service Unit (DSU) are often used interchangeably. These two devices perform separate functions, but the functionality of both can often be found in one box, called a Digital Data Set (DDS). The DSU is the lower-speed device, providing signal format and protocol translation, as well as acting as the termination point for digital circuits. The CSU is a higher-speed device which can also be used at lower speeds. The CSU terminates digital circuits the same as a DSU, but it also provides many feature functions not provided by the DSU, such as filtering, line equalization, line conditioning, signal regeneration, circuit testing capabilities, and error control protocol conversion (i.e., B8ZS). Some CSU/DSUs also have the capability for Extended Super Frame ESF monitoring and testing, and some have the capability to multiplex traffic from multiple input ports into a single point-to-point or multidrop circuit.

DSUs come in many speeds and with many different functions. There are five major categories of DSUs. Fixed-rate DSUs operate at speeds of 19.2K bps and below (subrate) or at the fixed speed of 56K bps. Multirate DSUs can be purchased which operate at variable speeds at or below 56K bps. These two types of DSUs can also be obtained with a secondary channel for network management. The fourth type of DSU is the switched 56K bps DSU, which operates with switched 56K digital services. T1 DSUs are also available, with switched T1 DSUs on the technology horizon. Standard CSUs provide a T1 circuit interface, and can have properties similar to the DSUs mentioned above.

Many CSU/DSU vendors are now marketing DS-3/E-3 products which provide the High-Speed Serial Interface (HSSI) for direct DS-3 connectivity. A few of the vendors, such as Wellfleet, Digital Link, and ADC Kentrox, are also offering Switched Multimegabit Data Service (SMDS) support. But "let the buyer beware", for many of these SMDS support functions are proprietary between the CSU/DSU vendor and a particular hardware vendor.

With the emergence of broadband services such as SMDS, the CSU takes on an entirely new function apart from its normal functionality. Some SMDS CSUs actually perform some of the required protocol conversion and cell segmentation required in SMDS, going far beyond their original role in life. These CSU/DSUs actually take the high level L3_PDU frame and segment it into L2_PDU cells, performing part of the SMDS protocol function within the CSU/DSU! The CSU/DSU then interfaces to the SMDS network through a Data Exchange Interface (DXI).

DSU/CSU interfaces include 56K bps, T1, V.35, and HSSI (on DS-3 models). T1 DSUs can be purchased for less that $1500, and DS-3 models can be acquired for less than $10,000. CSU/DSUs operation is similar to that shown in Figure 4.15. Figure 4.18 shows the OSI Reference Model layer used by CSUs and DSUs.

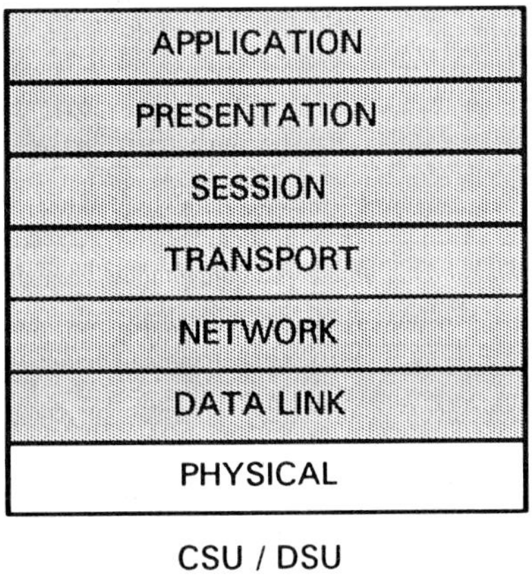

| APPLICATION |
| PRESENTATION |
| SESSION |
| TRANSPORT |
| NETWORK |
| DATA LINK |
| PHYSICAL |

CSU / DSU

Figure 4.18 CSU/DSU Layer of the OSIRM

4.5.4 Intelligent LAN Hubs

LAN hubs are a product resulting from the move away from bus LAN topologies to star topologies. LAN hubs have been classified into three generations. The first-generation LAN hub started to appear in 1984 and acted as a repeater for a single type of LAN connectivity. These hubs provided the function of a LAN concentration point, supporting a single bus which provided physical connectivity for multiple ports on multiple LANs operating on the same architecture. This function was similar to that of a combination patch panel and repeater. The second-generation LAN hubs provided the same bus architectures, but accommodated different LAN architectures over multiple ports, such as Ethernet and Token Ring. Additional features such as local and remote network management and configuration capability were also added.

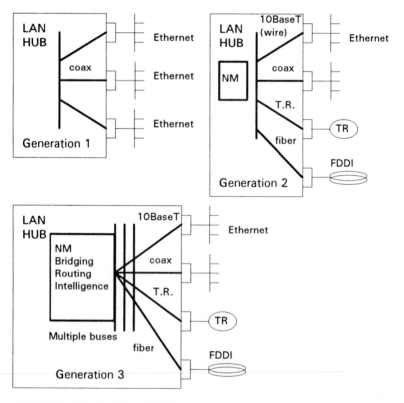

Figure 4.19 Three Generations of Wiring Hubs

Third-generation LAN hubs provide multiple buses for connectivity similar to the second generation, but also add bridging and some rudimentary routing functions. The range of physical media support is much wider, and often multiple buses are in the hub architecture. These multiple buses range from 4M bps Token Ring buses to 800M bps synchronous fast packet buses. Third-generation LAN hubs also have additional network management features, and are sometimes called "smart hubs". Many support Simple Network Management Protocol (SNMP) along with new developing standards (such as IEEE hub management).

Figure 4.19 shows the three generations of LAN hubs along with their support at each stage. Hub configurations tend to model the star topology, with the hub as the center and each LAN device which is attached directly to the hub.

Figure 4..20 shows a building using a LAN hub to connect multiple LANs of the same type, let us say Ethernet to Ethernet, between multiple floors. The following sections on bridges and routers will explain these techniques from the protocol and OSI reference model point of view. Figure 4.21 shows the OSI reference model layers used by LAN hubs. Notice that hubs use the physical, data link, and part of the network layers.

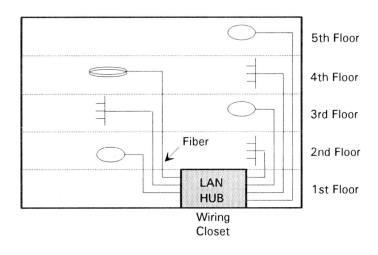

Figure 4.20 LAN Hub Building Plan

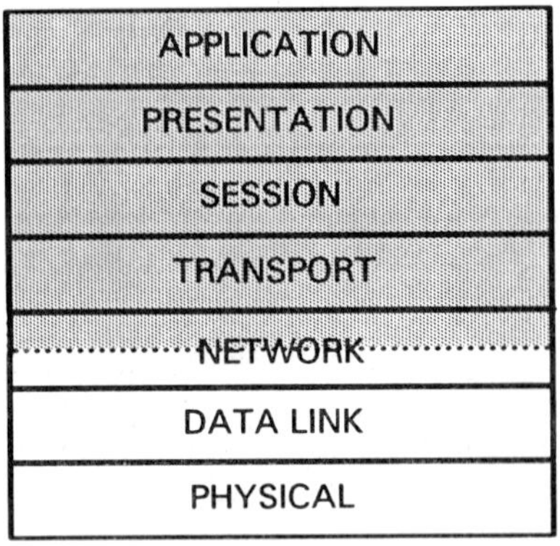

```
┌─────────────────────────────────┐
│            APPLICATION           │
├─────────────────────────────────┤
│            PRESENTATION          │
├─────────────────────────────────┤
│             SESSION              │
├─────────────────────────────────┤
│            TRANSPORT             │
├·················NETWORK···········┤
│            DATA LINK             │
├─────────────────────────────────┤
│            PHYSICAL              │
└─────────────────────────────────┘
```

OSIRM of LAN HUBS

Figure 4.21 LAN Hub Layers of the OSIRM

4.5.5 Bridges

Bridges provide connectivity between local area networks of like architecture, forming one of the simplest of local area and wide area network connections. A bridge uses a minimal amount of processing and thus is less expensive to link LANs using the same physical and link layer protocols. These LANs can be across the hall or across the country. Bridges can also connect devices using physical and link layer protocols to devices using the higher level IEEE 802.X protocol suite (including FDDI). Since bridges are protocol transparent, they do not provide flow control or recognize higher-level protocols. They use only the physical and link layers of the OSI reference model (see Figure 4.22), and support both the logical link control and the media access control layers of LAN transmission. Figure 4.23 portrays the same User A and User B now communicating via a bridge over the physical and link layers of the OSI reference model. Bridges operate at the media access control (MAC) layer of the OSI data link layer. Both users are implementing the same protocol stack for layer one and two, and the bridge does not modify the information flow in any way (except a possible MAC layer conversion in translation bridging). The bridge supports linking at the physical and link level, but provides no addressing or switching functionality. Thus, the user provides all addressing and protocol translation. Bridges simply pass traffic from one network segment to another based on the destination address of the packet being passed.

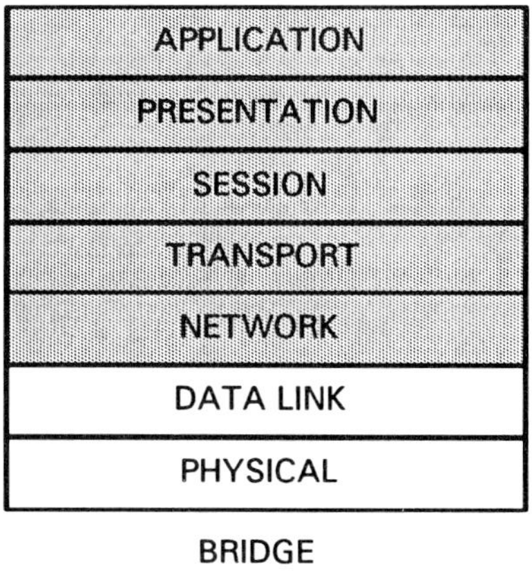

APPLICATION
PRESENTATION
SESSION
TRANSPORT
NETWORK
DATA LINK
PHYSICAL
BRIDGE

Figure 4.22 Bridge Layers of the OSIRM

User A
Source Node
User B
Source Node
Application
Presentation
Session
Transport
Network
Data Link
Physical
BRIDGE
Data Link
Physical
Application
Presentation
Session
Transport
Network
Data Link
Physical

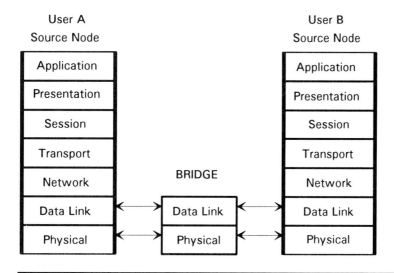

Figure 4.23 Bridge Communications via the OSIRM

Bridges will store and forward packets between bridges similar to a packet switch, but the bridge cannot act as a switch. Each packet is sent to a remote user based upon a destination address. Bridges can recognize either a fixed routing table scheme or, for more expensive bridges, a dynamic learning routing scheme. Bridges can "learn" the network through use of intelligent routing schemes, and some bridges are able to dynamically update their routing tables. Bridging protocols will be discussed later. Another key capability of bridges is their ability to filter data, but a major drawback to bridges is that they cannot forward data if operating at their maximum filtering rate. When using source route bridging, there is also a limit of seven imposed on the number of bridges which can be linked together (due to hop-count restrictions). Bridges deployed in a network do not have knowledge of the network to which they are attached. They are blind to devices other than those that attach to their logical path structure. The flexibility of bridge connectivity will be discussed further when we cover bridge protocols. There are four major types of bridges: transparent, translating, encapsulating, and source routing. Each provides different functionality for the various LAN architectures.

When operating in transparent mode, bridges at both ends of a transmission support the same physical media and link layer (MAC-level) protocols from the IEEE 802.X suite (or possibly FDDI), but transmission rates may vary. From the point of view of the network node, transparent bridges take no part in the route discovery or selection process. The higher-level protocols (OSI layer three and higher) need to be the same or compatible for all connected applications, because bridges are transparent to protocols at, or above, the network layer. Figure 4.24 shows examples of transparent bridging between two local Ethernet LANs and between two local Token Ring LANs. Remote bridging between two Token Ring LANs is also shown — across the street or across the world.

When operating in translation mode, bridges at both ends of the transmission can use different physical media and link (MAC-level) protocols. Translating bridges translate from one media format to another — manipulating the frame structure associated with each media type. Protocols in the network layer and higher still need to be compatible.

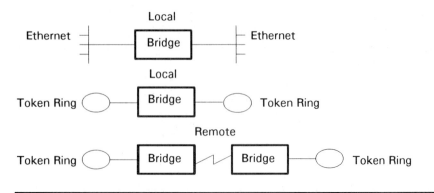

Figure 4.24 Transparent Bridging

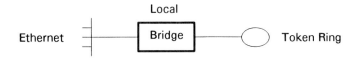

Figure 4.25 Translation Bridging

Figure 4.25 shows an example of translation bridging between local Ethernet and Token Ring LANs. Translation bridges do not provide segmentation services, so the frame sizes of each LAN host must be configured for the same supportable length.

One type of bridge, the Media Access Control (MAC) bridge, can be used as an inexpensive method of connecting Ethernet to Token Ring LANs without having to use routers, gateways, or intelligent hubs. These bridges convert packets between the two frame formats, as well as provide the standard forwarding and filtering of packets. These bridges also require the user of each LAN architecture to employ the same upper layer protocol suites (transport layer and higher).

When operating in encapsulation mode, bridges at both ends of the transmission must use the same physical and link layer (MAC-level) LAN protocols, but the transmission network between the bridges can provide a similar or different physical media and MAC-level protocol. Encapsulating bridges provide a network interconnection/extension by placing received frames within a media-specific "envelope" and forwarding the encapsulated frame to another bridge for delivery to the destination. This is common when multiple Ethernets are served by a Token Ring or FDDI backbone. The backbone then serves as the wide area network protocol.

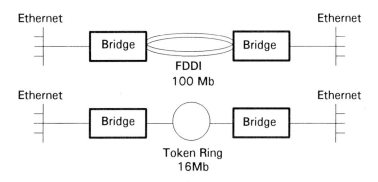

Figure 4.26 Encapsulation Bridging

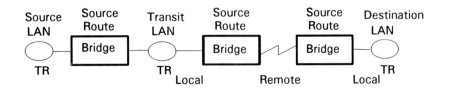

Figure 4.27 Source Route Bridging

Figure 4.26 shows two examples of encapsulation bridging where two remote 10M bps Ethernet LANs are bridged via a metropolitan 100M bps FDDI network and a 16M bps Token Ring network. The concept of using a Token Ring as the backbone WAN for multiple Ethernet LANs is attractive and efficient from the viewpoint of packet size and segmentation. The maximum frame size for Ethernet is 1500 Bytes, whereas the Token Ring frame size can be up to 4500 Bytes.

The fourth type of bridging is through source route bridging. Figure 4.27 shows a source route bridging scheme between two remote Token Ring LANs and three source route bridges. The third Token Ring LAN is only used for transit. Source route bridging will be discussed in detail in the next chapter.

Bridges are best used in small, geographically concentrated networks which do not require a large customer addressing base and are needed to connect a fairly static network design. Bridging speeds vary, supporting subrate through T1 to T3, and even supporting FDDI's 100M bps, and higher, bridging speeds. These high speeds are needed to support the high-speed LANs connected to the bridge, such as 10M bps Ethernet and 16M bps Token Ring. Bridges provide either local, remote, or both local and remote support.

Careful future planning is required when deploying a bridged network solution. The engineer who employs a bridge solution may find that very soon his or her bridge solution will resemble the wood and stone bridge — built in 1850 and designed to accommodate a horse and carriage. Soon there will be a need to drive not only a car but trucks over the bridge, but one year later rather than one-hundred years later. Thus, bridges can be good solutions for networks only utilizing one protocol and architecture with no plans to change, or for very static network designs with multiple protocols and architectures which have close local control.

Some major disadvantages are associated with bridging. Bridges are susceptible to multicast or broadcast "storms". These "storms" occur when a network bridge floods the network with repeated useless information, regenerating messages planned for one user so that many retransmissions occur to all bridges in the network. The broadcast propagates throughout the network until response time and available bandwidth become unmanageable. This problem increases with the size of the network and number of users attached.

To minimize the problem, smart bridging techniques can provide some level of traffic isolation. Some bridges cope with broadcast storms by segmenting the bridged network into domains which restrict broadcast storms to a limited area. This containment method coupled with a multicast traffic ceiling effectively controls broadcast storms. Bridges are also limited in both address retention and memory. They are designed to retain a limited amount of information and can handle only limited network changes. The more changes occurring in the network, the greater the traffic passing between routers to update routing tables — thus an unstable network would occur.

Due to these disadvantages and limited capabilities, bridges should not be used in network designs calling for multiple protocol support, dynamic networks requiring frequent changes, or large networks of greater than 50 nodes. For networks with these requirements, more intelligent and robust devices will provide much of the bridging functionality and additional routing intelligence, as well as eliminating the disadvantages of bridging. Enter the router.

4.5.6 Routers

Routers are the next generation of computer internetworking devices. Routers have emerged into the marketplace as the hottest thing since multiplexers, with much more intelligence than bridges and multiplexers. Routers provide interconnectivity between like and unlike devices on the local and wide area network, as well as extending the LAN into the metropolitan area networking arena. Multiprotocol routers provide support for multiple protocols simultaneously. Figure 4.28 shows two routers providing connectivity to multiple Ethernet and Token Ring LANs, as well as providing a WAN FDDI link between them and an SMDS link to a switched network.

Routers are protocol sensitive, and can either bridge or route a large suite of network-layer and higher-layer protocols. Thus, they support varied LAN devices which also employ a variety of networking protocols and addressing schemes. Routers understand the entire network, and will route based on many factors to determine the best path. Routers use the physical, link, and network layers of the OSI reference model to provide addressing and switching functionality. Routers route packets from node-to-node based on the packet-defined protocol information and factors such as least-cost routing, minimum delay, minimum distance, and least congestion conditions.

Figure 4.29 shows the relation of the router to the OSIRM. Both users may exercise the same protocol stack up to layer three, but the protocols above layer three may also be the same. A router's main functionality resides in the data link and network layer protocols but also uses the physical layer. Applications at both ends of the transmission do not need to support the same LAN protocol from the IEEE 802.X suite, or protocols up through OSI level three, but they do need to have the same protocol from the fourth through seventh layers of the OSIRM (or at least the intelligence at the user end to provide the gateway functionality if needed). Figure 4.30 shows the OSIRM layers used by routers.

Routers use their own internetworking protocol suite. Through the use of routing tables and routing protocols such as OSPF, routers retain artificial intelligence called "dynamic knowledge" of the entire network. They can discover network topology changes and provide rerouting based upon these dynamic routing tables. Routers can limit the number of hop counts by their intelligent routing protocols. Routers employ large addressing schemes sometimes up to four bytes worth of addresses in a logical network. Routers also support large packet sizes. For example, frame relay uses a maximum packet size of around 8000 bytes. Internal bus speeds are also much higher, such as Wellfleet's speeds of up to one gigabit per second. The other major advantage of routers is their ability to perform these functions primarily through the use of software, which makes future revision and supportability upgrades much easier.

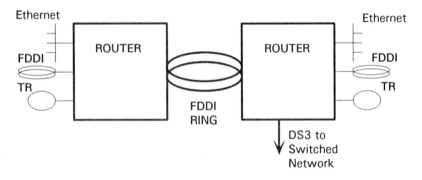

Figure 4.28 Router Connectivity Example

User A
Source Node
Application
Presentation
Session
Transport
Network
Data Link
Physical
ROUTER
Network
Data Link
Physical
User B
Source Node
Application
Presentation
Session
Transport
Network
Data Link
Physical

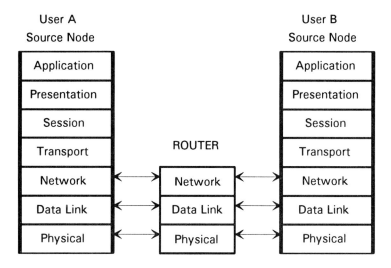

Figure 4.29 Router Communications via the OSIRM

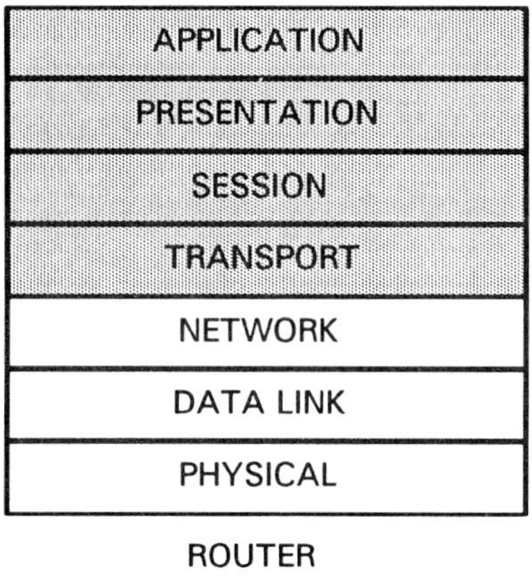

APPLICATION
PRESENTATION
SESSION
TRANSPORT
NETWORK
DATA LINK
PHYSICAL
ROUTER

Figure 4.30 Router Layers of the OSIRM

Additional advantages accrue to using routers rather than bridges. Routers provide a level of congestion control not present in bridges, thus allowing the router to dynamically reroute traffic over the least congested paths. Routers eliminate the broadcast storm danger by providing segmentation capability within the network. Thus, the network designer can build a hierarchical addressing scheme and design smart routing tables which operate somewhat similar to the filtering capabilities of bridges, but with the additional flexibility to define virtual network subsets within a larger network definition. Routers differ from bridges in that they provide protocol translation between users at the link level, while bridges just pass information in a store and forward mode between devices of similar protocol structure. Greater detail will be forthcoming when we discuss router access network design. Routers which utilize IP routing schemes can solve packet fragmentation problems caused by technologies such as X.25 and FDDI. Packet fragmentation occurs whenever two protocols with different size packets are used. Routers also have the ability to translate between medium access control (MAC) layers. Routers also can be isolated and routed around when network problems exist, unlike bridges. Routers contain a level of investment protection over less intelligent devices.

Now for the few disadvantages to routers. Routing algorithms, discussed in great detail later, incur more overhead than bridges. This is because of additional intelligence needed in the routing protocols and the various congestion control techniques implemented. Many router vendors are implementing multiple processors within the network interface card and faster platforms and processors (such as RISC machines) to eliminate throughput problems caused by these increased traffic loads. The market growth for routers has exploded in the past few years, far exceeding the growth of remote and local bridges or gateways. Table 4.1 shows a comparison of bridge and router uses and capabilities.

TABLE 4.1 Bridge-Router Comparison

FUNCTIONALITY	BRIDGE	ROUTER
Data Sources	one source and destination	multiple sources and destinations
Addressing	no	yes
Packet Handling	pass packet transparent	interpret packet
Forward Packets	out bridge	specific destination
Global Network Intelligence	none	knows status of all devices
Priority Schemes	no	yes
Security	based on isolating	based on routing protocol

4.5.7 Brouters

The term "brouter" is a conflate word formed by "bridge" and "router" — thus "brouter". Brouters perform the functions of both bridges and routers. Brouters have the ability to route some protocols and bridge others. Some protocols need to be bridged (such as DEC LAT) rather than routed. The term "brouter" was derived from the need to expand single-port bridges to multiple ports to support IBM Source Routing Protocol and Source Routing Transparent Protocol. The routing done by brouters is transparent to both the network layer protocols and end stations, and is accomplished in the Media Access Control (MAC) address. Thus, brouters do not look at the network-level address. Rather they route based on the MAC header. Router logical functionality is similar to that of Figure 4.29, and the OSI reference model layers of support are the same as Figure 4.30.

4.5.8 Gateways

Gateways provide even greater functionality than routers or bridges. Gateways provide all of the interconnectivity provided by routers and bridges, but, in addition, furnish connectivity and conversion between the seven layers of the OSI reference model as well as other proprietary protocols. Gateways are often application specific and, because of their complex protocol conversions, are often slower than bridges and routers. Some applications use priority schemes not consistent between the OSI layers and proprietary protocol structures.

One example of a gateway function is interfacing a device using SNA with a device using the OSI protocol stack. The gateway will convert from SNA to an OSI protocol structure (as well as the reverse conversion from OSI to SNA). Thus, the gateway's main functionality resides in the role of protocol translator for architectures such as SNA, DECnet, Internet TCP/IP, and OSI. It can also translate between LAN architectures such as Ethernet to Token Ring LANs and vice versa. If protocol functionality is needed in excess of that found in routers, then the gateway is the hardware of choice. Gateways are often based in mini-computers and mainframes, and are considerably more expensive than routers. Some routers have limited built-in gateway functionality. Figure 4.31 shows the OSI reference model layers used by gateways.

Figure 4.32 portrays the same two users yet again, but this time connected via a gateway. This figure also shows the relation of gateways to the OSI reference model. Both users may have different protocol stacks in any of the seven levels with both OSI and non-OSI protocols.

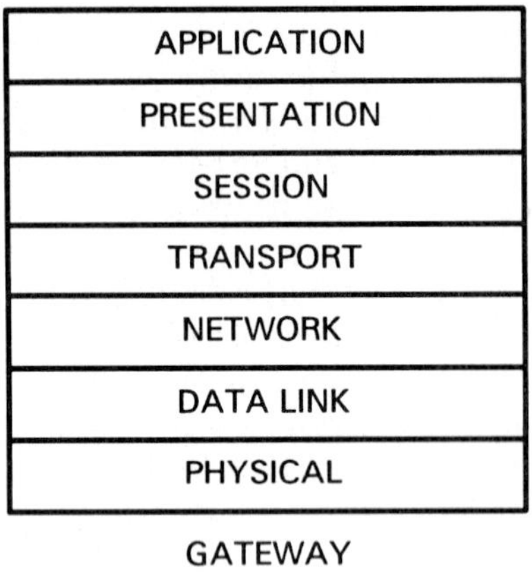

| APPLICATION |
| PRESENTATION |
| SESSION |
| TRANSPORT |
| NETWORK |
| DATA LINK |
| PHYSICAL |

GATEWAY

Figure 4.31 Gateway Layers of the OSIRM

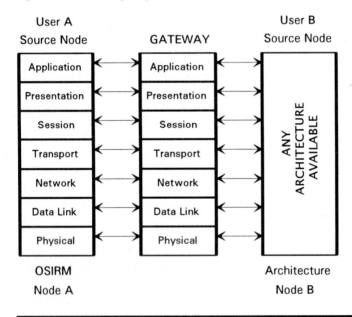

Figure 4.32 Gateway Communications via the OSIRM

There are three major disadvantages of gateways: low throughput during peak traffic conditions, user-to-gateway priority handling, and store and forward characteristics. During periods of peak traffic, a gateway may become the network congestion point, having to spend the majority of its time translating between many protocol suites. Gateways are often store and forward devices forwarding only the information requested by the destination node. In spite of these drawbacks, and the high expense of gateways, there is a growing need for this functionality. As users transition toward the OSI industry standards, gateways will fill an important niche in uniting disparate protocols for many years to come.

4.5.9 The Future of Bridges, Routers, and Gateways

While most bridges and routers primarily support Ethernet, the shift today is toward support of multiple protocols, with Ethernet and Token Ring in the forefront. Ethernet support still predominates over Token Ring. There are very few bridge vendors who support both architectures. FDDI has become a cornerstone for local area networking, and routers and bridges now include FDDI interfaces and can even serve as access points for FDDI LAN backbones. FDDI backbones also serve as a common ground between Ethernet and Token Ring LANs.

Many personal computer manufacturers are building hardware and software to perform bridging and routing functions. Expansion boards plugging into the PC bus may be less expensive than purchasing a standalone unit, but usually at the sacrifice of throughput or delay. Personal computers cannot compete with router vendors introducing internal bus speeds into gigabit ranges, and their processors cannot handle the hundreds of thousands of packets per second processing required by standalone models such as Wellfleet's Backbone Concentrator Node (BCN), using packet processing speeds of 320,000 packets per second. Software can also be implemented on UNIX-based computers to perform bridging and routing, but the same problems are encountered. Advances in processor and backplane bus speeds, hardware architectures, and software coding have enabled some bridges to become more intelligent and perform a level of routing, while some routers are now performing bridging. New dynamic routing protocols such as IS-IS will make routers even more desirable.

Network management for these devices is being accomplished through proprietary platforms or Simple Network Management Protocol (SNMP) until the OSI standard stabilizes. Most vendors provide SNMP agents in their software base, realizing that SNMP has become the network management de facto standard of choice. SNMP is the only standard which enables multiple vendors and hardware platforms to communicate through one network management platform. The network design engineer must review all user network management requirements before making a decision, because additional platform support, such as DECnet DECmcc or IBM LAN Manager, may be required. Each vendor implements a proprietary network management system.

TABLE 4.2 Device Selection Criteria

REQUIREMENT	DEVICE
Extension of similar LAN distance only	r
Higher-speed transmission	B
Less overhead	B
Ease of installation	B
Low operational maintenance	B
Small centralized network w/many point-to-point links	B
All devices using same protocol structure	B
Connecting similar LANs	B
Simple topology	B
Two-port connections	B
Specialized routing restrictions	b
Routing choices based upon least cost	R
Ease of upgrade for additional protocols	R
Large network, large number of users	R
Congestion, flow control required	R
Complex intricate topologies	R
Hierarchical LAN designs	R
High probability of data storms	R
High-speed PPS processing	R
Low data loss during flow control	R
Adapt to constantly changing technology	R
Compatibility with switched services	R
Universal addressing scheme	R
Alternate routing around link failures	R
Various packet sizes requiring segmentation	R
Multiple MAC layer implementations	R
Different addressing schemes employed	R
Many alternate paths required	R
Decentralized users requiring translation of multiple high-level protocols	R

At present, cisco Systems, Inc. holds the largest share (over half) of the router market, with Wellfleet coming in a close second. Many of the major bridge and hub vendors are supplementing their products with bridging and routing capabilities. Even the high-end multiplexer vendors are providing bridge and router capability in their products now. A massive migration from centralized to distributed processing is being accomplished through bridge, router, and gateway devices.

Router devices have become the most cost-effective devices for network designers. While routers have remained price competitive (about $6,000 for the low end) vis-a-vis with bridges, they can provide the same functionality and migration path for future connectivity. The router market has been projected to be worth over half a billion dollars by 1994.

Table 4.2 shows the decision alternatives for choosing the proper technology: repeater, bridge, router, or gateway. Brouters could be substituted for routers in this example. The DEV column stands for device: r = repeater, B = bridge, b = brouter, R = router, G = gateway. Many of the decisions could relate to multiple technologies, and are often cumulative in the order listed.

4.5.10 Recent Trends from Mainframes and Front-end Processors to Routers

Will the router replace the front-end processor (FEP)? Maybe not the local FEP, but most surely the remote FEP. But FEPs are just one piece of the IBM puzzle. Many users are turning toward bridges and routers to solve their IBM SNA local area and wide area connectivity. Users with SNA nets are finding an increasing need to connect non-SNA devices to SNA devices, necessitating a router or gateway functionality. IBM too has to connect via non-SNA connectivity arrangements some of its own IBM equipment which is not SNA compatible. This combines a broad range of hardware, from hosts to FEPs, LANs, and terminals using multi-protocol routers. It also consolidates many dedicated point-to-point circuits, as well as providing a concentration point with dynamic alternate routing using less total bandwidth and thus increasing availability.

IBM is now joining the bandwagon of bridging and routing products, and even providing routing capabilities in front-end processors. But this solution may prove to be far too costly a solution when router and gateway vendors can provide standalone products at a fraction of the cost, as well as guarantee internetworking standardization. It is much less expensive to deploy routers as a substitute for IBM FEPs being used as concentration devices. Is IBM's play for this market a little too late? Are they just revamping their SNA product line through proprietary router offerings? The IBM 6611 router does contain special (proprietary) "hooks" for SNA but significantly uses SNMP NM capabilities. Routers may not be able to replace the FEP functionality needed for mainframe attachment, but they can replace remote front-end processors by exploiting PU 4 emulation, and add the additional functionality of a multi-protocol LAN/WAN concentration translator. For point-to-point SNA connections, routers could replace FEPs used only to support 3270 applications using a single host via SDLC "tunneling" or synchronous pass through. IBM has also made a foray into the router market with the IBM 6611, which uses data link switching. This router is based on IBM's proven RISC technology, a RISC processor on each high-speed adapter card, and Micro Channel® bus technology. The IBM 6611 Model 140 and 170 are shown in Figure 4.33.

Routers can handle the SNA question by two methods: encapsulation (or packetization) of SNA traffic or emulation and routing of PU 2s and 4s in native mode. SNA PU 2 devices are typically 3174 cluster controllers providing a concentration and bridging function locally, and SNA PU 4 devices are front end processors providing routing of native mode SNA across local or wide areas. IBM SNA traffic is routed in the wide area and uses alternative router designs. More later. Many non-IBM routers, moreover, are supporting SNA's LU 4 compatibility.

Many benefits derive from using routers to perform PU 2 and PU 4 functions. Native mode PU 4 routers provide routing control and efficiencies gained by both priority routing through the router network and the avoidance of double packetization of data using native mode. Also, a router network provides the ability to consolidate many support systems under one network. Network management can be provided under SNMP utilizing IBM NetView feeds to the host.

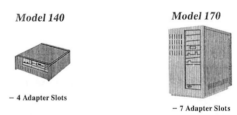

Model 140   *Model 170*

− 4 Adapter Slots

− 7 Adapter Slots

Highlights
- Based on IBM's Proven RISC Technology
- RISC Processors on High Speed Adapter Cards
- Micro Channel® Bus Technology
- Pre−loaded Software

Figure 4.33 IBM 6611 Routers — Model 140 and 170

As router and gateway products evolve, SNA encapsulation will improve and additional capabilities such as IBM NetView feeds, configuration and management capabilities, and solutions for polling overhead will make routers and gateways an even more viable alternative.

4.6 REVIEW

In data communications circuits, services, and hardware types have evolved. The three types of signal transfer: simplex, half duplex, and duplex, as well as asynchronous and synchronous data transfer are now clear. We have also studied the types of circuits and the general distinctions of services offered over them. Five network topologies were shown: point-to-point, multipoint or common bus, star, loop or ring, and meshed. Two classifications of network services were explained: connection-oriented and connectionless. After a detailed discussion of each major type of computer networking hardware (except multiplexers and switches) it is clear that a detailed study of application requirements is needed to determine the type of technology needed for each application. The many types of protocols that provide the intelligence and rules for these devices to communicate will be explored next.

5

Common Protocols
and Interfaces — Part 1

A protocol is similar to a language, conveying meaning and understanding through some form of communications. Protocols are sets of
rules governing communications between data machines and between protocol layers. The definition of protocol also infers documentation, negotiation, and the establishment of rules. For one computer to talk to another, each must be able to understand and translate the other's protocol. Protocols are defined in seven major sections: the physical layer, data link layer, logical link control and
medium access control layers, bridge protocols, router protocols,
network and transport layer protocols, and upper or user layer
protocols. Discussions of both interfaces and protocols will be covered in this chapter. Interfaces play an important part in network
design by providing the physical and logical interface between user
and network equipment. Protocols play a very important part in
data transmissions, for without them we would have islands of users
unable to communicate.

Lower-layer protocols provide the logical link and physical interface to the transport medium. Details concerning physical interface
signaling, timing, and pin level configuration are outside the scope of
this text. Data link layer protocols allow communications with the

physical medium. Included in the data link layer are both the logical link control and the medium access control (MAC) sub-layers, which define local area network communications. Bridge, router, transport, and Internet mid-layer protocols will be covered in the next chapter.

5.1 BASIC CONCEPTS

Interfaces provide boundaries between different types of hardware, while protocols provide rules, conventions, and the intelligence to pass data over these interfaces. Interfaces and physical media provide the path for data flow, while the protocols manage that data flow across the path and interfaces. Both interfaces and protocols must be compatible for accurate data transport. Often, interfaces are referred to only in the physical layer of the OSI reference model. Protocols can also act as interfaces, defined as interface protocols. Such protocols define the interface between the DTE to DCE signaling and transfers that information to higher-layer protocols for user presentation. Many network designs now incorporate multiple levels of protocols and interfaces, from the physical layer to the application layer according to the seven-layer OSIRM.

The concepts behind the use of multiple protocol layers are important. The concepts of physical, data link, and network layer protocols can now be defined on a high level. Less time will be spent on the details of session, presentation, and application layer protocols, since our thrust is toward a design implementing protocols such as frame relay, Distributed Queue Dual Bus (DQDB), Fiber Distributed Data Interface (FDDI), and Asynchronous Transfer Mode (ATM), which have their roots in the lower layer protocols. Some implementations of session, presentation, and application layer protocols will be discussed later in the next chapter.

5.1.1 Physical Layer Concepts

The terms Data Termination Equipment (DTE) and Data Communication Equipment (DCE) refer to the hardware on each side of the communications channel interface. DTE equipment is typically a computer or terminal which acts as an end point for transmitted and received data. DCE equipment is typically a modem or communication device for data transport. The DTE communicates with the DCE, which then interfaces with and provides the DTE with access to the network. Delving into network design, the farther we go into the network, the more these terms blur. DTE and DCE are also used to identify a standard cable configurations, and it is not uncommon to find a DTE-to-DTE cable connecting two devices. This is called a "null modem" connection and is often used to connect two DTE devices such as a terminal and a computer. Figure 5.1 shows a common end-to-end network configuration where DTE 1 talks to DCE1, which in turn formats the transmission for transfer over the network to the end DCE which then interfaces to the end DTE.

The physical layer protocols provide both electrical and mechanical interfaces to the transport medium. The physical layer standards define electrical and mechanical interfaces, as well as procedures and functions of their operation. The physical layer can also activate the transmission media, maintain transmission of data over that link, and deactivate the link upon end of transmission, doing this through interfaces with the data link layer. The physical layer is concerned only with assuring the transmission and reception of a data stream between two devices.

The intelligence managing the data stream, and protocols residing above the physical, are transparent to the physical layer. Physical cable connector characteristics, voltages and currents, timing and signaling are functions of the physical layer. And the physical layer interfaces to many physical network media devices spanning many standards and architectures. Functional specifications define data, control, timing and ground assignments. Procedural specifications define the procedures which govern the activation, use, and interpretation of these interface characteristics.

5.1.2 Data Link Layer Concepts

The data link layer is layer two in most computer architecture models, including the seven-layer OSI reference model. The data link layer interfaces between the first physical layer and the third network layer protocols, interpreting the data flow across the physical media and feeding the network layer protocols information on the outcome of these services. The primary function of the data link layer is to establish a data link across the physical media, manage reliable data flow across this media, and terminate the link after completion of accurate data flow. The data link control functions establish a peer-to-peer relationship across the network. This layer may also provide functions and services such as error control, detection and correction, flow control, framing and character formatting, synchronization, sequencing for proper delivery, and connection control and management functions.

Data link functions and protocols use many circuit topologies including: point-to-point, multipoint, switching, and broadcast. Since computer communications via local area networks utilizes special functions of the data link layer, a section on the medium access control and logical link control layers will be emphasized. The MAC layer protocols deserve special attention since they form the basis of local area network and metropolitan area network standards.

Network devices such as bridges and routers manipulate the protocol transfer of information across many of these layers. While adhering to protocols defined within these two layers, they also contain protocols of their own which perform bridging, routing, management, and other functions. Each bridge and routing protocol will be discussed in detail.

5.1.3 Network and Transport Layer Concepts

Once the physical and data link layers have built the frames of information and placed them onto the physical medium, the network and transport layer protocols take over and pass information through the network in packets and ensure end-to-end reliable delivery. A broad range of network and transport protocols from many different architectures will be explored and explained.

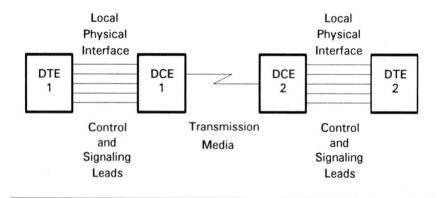

Figure 5.1 DTE to DTE Communications

5.2   PHYSICAL LAYER PROTOCOLS and INTERFACES

There are seven major types of physical interfaces that primarily involve the physical layer of the OSI reference model. The more commonly known interfaces include: RS-232C; EIA-RS-232D; EIA-RS-449; CCITT V.24/V.28; CCITT X.21; X.21bis; CCITT ISDN I.430; CCITT ISDN I.431; T1/E1 and D4/ESF; V.35; High Speed Serial Interface (HSSI); High Performance Parallel Interface (HiPPI); and ESCON.   Some physical interfaces are integrated with link and network layer protocols such as the IEEE 802.X LAN standard interfaces. Selected standards will be discussed. Physical interface standards such as T1/E1 and D4/ESF will also be discussed. The physical interfaces for B-ISDN and ATM will be covered later.

5.2.1   EIA-RS-232C/D, EIA-RS-449, and CCITT V.24/V.28

RS-232-C and EIA-232-D are defined by the Electronic Industries Association as synchronous interface standards for use with the physical layer of the OSI reference model. Adopted at a time when analog transmission was the prevalent industry standard, RS-232-C is probably the most common interface standard, and provides a 25-pin connector DTE interface to voice-grade modems. The EIA-232-D is a recently adopted standard which, though offering a significant upgrade to the RS-232-C standard, has had limited acceptance. The CCITT V.24/V.28 standard is very similar to RS-232-C and provides the international version of the RS-232-C standard.   CCITT V.24 defines the physical interface and V.28 defines the electrical interface. Many of the EIA standards parallel multiple CCITT standards. For example, the EIA-RS-232-C standard contains the electrical, mechanical, and signal definitions for physical connectivity.   The CCITT V.24 (signal definition), V.28 (electrical), and ISO mechanical standards are required to define a similar interface.

Since the "232" standards are over 30 years old and based on the then existing technology, limitations show up on implementation, such as a distance limitation of 15 meters and a speed limitation of 19.2K bps maximum on standard RS-232-C. RS-449 was adopted later to alleviate these distance and speed restrictions and improve performance using a 37-pin connector.   This provides balanced signaling and tighter cabling and electrical specifications.   The maximum achievable speed on RS-449 is 2M bps.

Two other standards were developed to further extend the distance and speed of the RS-232C interface.   The RS-423A offers an improvement of up to 300K bps speed and operates in "unbalanced" transmission mode, while the RS-422A offers speeds up to 10M bps and operates in "balanced" transmission mode.

5.2.2 CCITT X.21, X.21bis

The CCITT developed the X.21 standard in 1972 as a physical inter-face specification for digital networks using digital transmission facilities (unlike RS-232-C which was designed during the analog network era). X.21 also eliminates the restrictions imposed by RS-232-C by using balanced signaling and two wires for each circuit (as opposed to two wires total for transmit and receive in RS-232). This allows for full duplex transmission of both user data, control, and circuit status information. X.21 also adds more logic at the DTE and DCE interfaces. X.21 spans the gap between the physical and data link layers of the OSI reference model, even touching, at times, upon the network layer for circuit switched networks. The call manage-ment capabilities of X.21 will be discussed later. X.21 uses a 15-pin connector and operates only in synchronous transmission. It pro-vides an unbalanced or balanced mode of operation. The major capabilities present in RD-232-C but lacking in X.21 is the ability to pass control information during data transfer, and the separation of transmit and receive signal element timing circuits and signal rate selectors. The X.21bis interim standard was developed as a migra-tion from the RS-232-C, EIA-232-D, and V.24 standards to X.21.

5.2.3 CCITT ISDN I.430 and I.431

The CCITT has defined many standards for Integrated Services Digi-tal Networks (ISDN). Two standards are defined for the physical interface to ISDN: Basic Rate Interface (BRI), or basic access, defined in the CCITT ISDN I.430 standard, and Primary Rate Interface (PRI) defined in the CCITT ISDN I.431. Both standards define the electrical characteristics, signaling, coding, and frame formats of ISDN communications across the S/T (user access interface) refer-ence point. The physical layer provides transmission capability, activation and deactivation of terminal equipment (TEs) and network terminations (NTs), D-channel access for TEs, maintenance func-tions, and channel status indications. The basic infrastructure for these physical implementations, as well as the definition for S/T, TE, and NT, are defined in CCITT recommendation I.412.

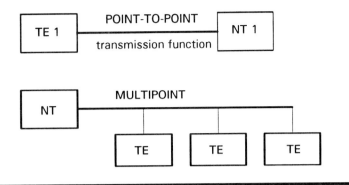

Figure 5.2 ISDN BRI Point-to-Point and Multipoint Configurations

Note that these interfaces are part of the specifications for Narrowband ISDN (N-ISDN). The physical interface in ISDN is one part of the D-channel protocol, and defines a full-duplex synchronous connection between the TE layer 1 terminal side of the basic access interface (TE1, TA, and NT2 functional group aspects) and the NT layer 1 terminal side of the basic access interface (NT1 and NT2 functional group aspects). Figure 5.2 shows both a point-to-point configuration with one transmitter and one receiver per interchange circuit, as well as a multipoint configuration with multiple TE, both for BRI. Both bus distances cannot exceed 1000 meters, except when using a short passive bus as opposed to a extended passive bus in multipoint mode, when the limitation is 180 meters. These bus configurations are explained in Appendix A of the I.430 standard. The bit rate in both directions is 192K bps.

Basic Rate Interface (BRI) — Offers two 64K bps user data (bearer) channels and one 16K bps control, messaging, and network management channel (D channel). This interface is also referred to as 2B+D (the B stands for "bearer" channel, bearing the user's data information). The BRI is primarily used for customer access devices such as the ISDN voice, data, and video phone.

Primary Rate Interface (PRI) — Offers twenty-three 64K bps user data channels and one 64K bps signaling channel. This interface is also referred to as 23B+D. PRI is primarily used by high bandwidth customer devices such as the Private Branch Exchange (PBX), personal computer, and LAN.

The BRI interface consists of two B-channels and one D-channel (2B+D) using an ISO standard 8877 eight-pin, sub-miniature modular plug and an RJ-45 jack. This connector represents one of the first open standard end-user interfaces. The 193K bps transmission rate is accomplished by 4000 frames per second with 48 bits per frame. The ANSI T1.601 standard provides the U reference point connectivity for the BRI local-loop, full-duplex connectivity via twisted pair to the local exchange (Figure 5.3). The ANSI standards also define many other aspects of the interface different from the I.430 standard, due to the two binary, one quaternary (2B1Q) signaling and framing scheme, and may require some protocol conversion at the NT1 device.

PRI provides a single 1.544M bps TI or 2.048M bps E-1 data rate channel over a full duplex synchronous point-to-point channel. Figure 5.4 shows a PRI point-to-point configuration, and depicts the two channels needed for PRI data transfer. The link is established between an NT2 type CPE switch and the Local Exchange Carrier (LEC). CCITT Recommendations G.703 and G.704 define the electrical and frame formats of the PRI interface, respectively. The 1.544M bps rate is accomplished by 8000 frames per second of 193 bits per frame over a standard T1 line. Twenty-four channels of 64K bps each comprise the T1, containing 23 B-Channels at 64K bps each and one D-channel (equivalent to a B-channel) at 64K bps. Twenty-four frames are transmitted in each multiframe utilizing Bipolar 8 Zero Substitution (B8ZS) code.

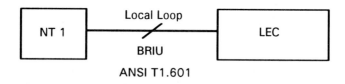

Figure 5.3 Standard ISDN BRIU and Local-Loop Interfaces

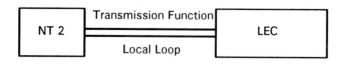

Figure 5.4 ISDN PRI Point-to-Point Connection

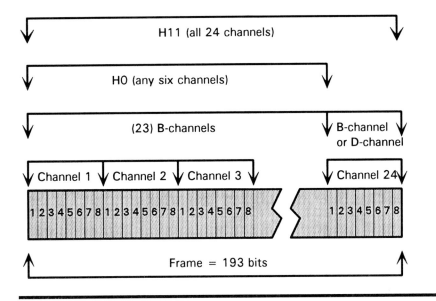

Figure 5.5 CCITT T1 PRI Frame Structure

Figure 5.5 shows the transmitted framing of the T1 PRI interface. The CEPT E1 PRI interface is somewhat different from this, offering 30 B-channels, one D-channel, and a channel reserved for physical layer signaling, framing, and synchronization.

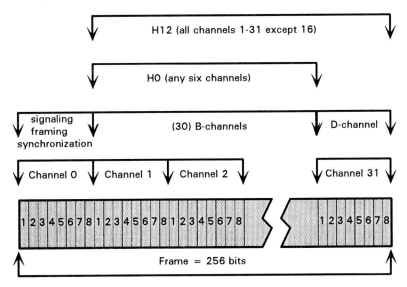

Figure 5.6 CEPT E1 PRI Frame Structure

The frame structure for the CEPT E1 PRI interface is shown in Figure 5.6. These 32 frames are transmitted via high-density bipolar 3 zeroes (HDB3) signaling. The primary attribute distinguishing ISDN service is the concept of common channel signaling, or out-of-band signaling using the D-channel.

H-channels are used in PRIs and defined in I.431. Two types are defined. H_0-channel signals have a bit rate of 385K bps and H_1-channels have a bit rate of 1,536K bps for H_{11} (U.S. channel rate) channels and 1,920K bps for H_{12} (European channel rate) channels. These channels (except for the H_{12} channel implementation) use existing D-channel slots. Also note that the D-channel and B-channel share the same physical connection and medium through TDM.

5.2.4   T1/E1, D4/ESF, and V.35

T1 circuits operate at a speed of 1.544M bps. With overhead they generally offer less than 1.536M bps to the user for data transfer. T1 technology derives from the use of digital processing. With the advent of the transistor effect and the development of integrated circuits, the large-scale telephone utilities began to move this new technology into the telephone transmission network. T1 took the twisted wire pair and allowed 24 voice conversations on it instead of only one. It was little thought at the time to offer this capability to the end users. When the T1 capability was extended to the end user by the advent of CPE multiplexers, private lines from the Telcos were used. Even today, T1 circuits remain the best seller of digital private line services. Each T1 consists of 24 channels with 8 bits per channel with a time frame of 125 microseconds (μS). This adds up to 192 bits per T1 frame, with a framing bit added for a total of 193 bits per frame. The transmission rate of the T1 is 8000 frames per second, and includes an 8K bps overhead channel. E-1 is the European standard of the T1, and differs by offering 2.048M bps bandwidth. The 8K bps framing overhead on each T1 channel can be used to great advantage, depending on what framing protocol is used. The D4 12-frame superframe concept allows the 8K bps overhead on each T1 channel to be used for frame synchronization and signaling. The physical T1 port (often called a T1 framer port) is a DS1 channel as defined under the DSX-1 standard. Framed T1 runs at 1.544M bps over a DSX-1 interface. D4 framing formats can provide non-channelized or channelized (as in DS0) circuits. D4 framing can also be used to access Fractional T1 and DXC services.

There exists an alternate method of using the 8K bps overhead channel. Extended Superframe Format (ESF) is an enhancement to

D4 framing. The ANSI standard for ESF is T1.403. With ESF, both the carrier and the user can nonintrusively monitor the performance of private lines. The ESF divides the 8K bps overhead channel into three network management and reporting functions: 2K bps performs the frame synchronization and signaling, 2K bps provides a Cyclic Redundancy Check (CRC-6) code for providing error detection of end-to-end format and logic errors, while the last 4K bps is the Facility data link (FDL) used as an open control channel.

The obvious advantages to ESF is the capability for remote monitoring and problem detection, without having to take the circuit out of order to test (as in D4 framing). This is called "nonintrusive" monitoring and testing. If ESF is implemented in conjunction with intelligent network equipment, errors affecting performance can be detected and corrected transparent to the user. ESF will play a valuable role with services such as frame relay and SMDS, where network management information provided by ESF can supplement and, in some instances, make up for existing deficiencies or lack of capabilities.

A word on Fractional T1 (FT1). While Fractional T1 is popular, its sales have not come close to reaching those of T1. Users seem to favor using FT1 for tail circuits on complex, large networks. Fractional T1 is built on an interim step between X number of dedicated circuits versus a full T1 (24 channels). It is a cost tradeoff. T3 is still in infancy, but more and more users who have large-bandwidth applications are looking into full and fractional T3 services. Currently, the price breakpoint for T3 seems to be the need for eight T1s or more worth of bandwidth to justify the DS3 facility. This varies based upon carrier tariffs.

DS3 circuits can either use asynchronous protocol multiplexing or synchronous protocol multiplexing. The asynchronous transmission protocols are defined in ANSI T1.107 and the synchronous transmission protocols (SYNTRAN) are defined in ANSI T1.103.

5.2.5 High-Speed Serial Interface (HSSI)

The High-Speed Serial Interface (HSSI) is a physical interface operating at speeds up to 52M bps. The HSSI interface was primarily developed by cisco Systems and T3Plus Networking, Inc. This high-speed interface was designed to become the standard interface between the DS3 rate of 45M bps and the OC-1 SONET interface of 51.84M bps for everything from wide area network connectivity to a DTE to DCE direct-channel interface. It can be used to extend 45M bps DS3 mainframe channels to remote devices, providing a valuable high-speed interface between computer and communications equipment and, in effect, extending the WAN with larger bandwidth pipes. Currently, the HSSI interface is a CPE hardware interface. HSSI is a national standard and is non-proprietary, and is slated for ANSI standardization. The ANSI standard would use a 50-pin tab connector, shielded twisted-pair cabling, similar to the Small Computer Systems Interface II (SCSI-II), and could operate at SONET speeds. Experiments are under way to enable the HSSI interface to be software configured to handle a high-speed frame relay interface.

5.2.6 High-Performance Parallel Interface (HiPPI)

High-Performance Parallel Interface (HiPPI) is a new high-speed broadband parallel point-to-point channel (interface) for super-computer networking. HiPPI was designed by a few scientists in the Los Alamos National Laboratory who required high-speed parallel transfer speeds for graphic-intensive applications. The standard under development is ANSI HiPPI standard X3T9.3. HiPPI operates at speeds of 800M bps on one shielded copper cable or 1.6G bps on two shielded copper cables. The maximum distance is 25 meters. HiPPI circuits can then be connected through multiple circuit switches in a crossbar topology. There are also plans for fiber optic implementations and interfaces into the National Research and Education Network (NREN).

5.2.7 Enterprise Systems Connection Architecture™ (ESCON)

ESCON provides high-speed, direct-channel connectivity for VM-, and VSE-based system processors, direct-access storage devices (DASD), and peripherals. Primary support is for IBM 3990 Storage Control and application software. ESCON transfers data through synchronous, variable length frames as opposed to the older byte-by-byte "Bus and Tag" parallel interface. ESCON operates over fiber optic cable, rather than copper wire in "Bus and Tag". ESCON uses both device and link-level framing protocols, and can operate in either a cached or non-cached mode. Cached transfer operates at 10M or 17M Bytes per second channel speeds. Noncached data transfers will operate at the full device speed.

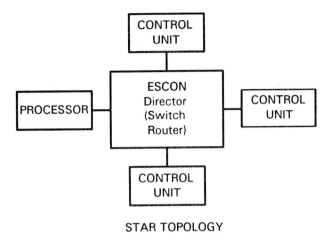

STAR TOPOLOGY

Figure 5.7 ESCON Star Topology with Director

ESCON also uses a device called a "director". A "director" acts as a high-performance switch and router for all attached devices. ESCON topology, when using the director switch/router, resembles a logical star. This configuration is shown in Figure 5.7.

The advantages of ESCON include:

- provides data center local or remote connectivity
- provides DASD extension (9 km versus 400 feet maximum distance between peripherals)
- improves application and resource availability
- increases transfer speed of data
- increases control of data transfer
- increases network and system performance
- provides network management across all DASD and peripherals
- easy configuration
- enables "electronic vaulting" of critical data — disaster recovery
- allows user transparency to system reconfiguration
- lets fiber optic cables replace copper cabling

IBM products will slowly move users toward ESCON, and it will soon become the IBM standard high-end interface. In fact, just now third-party ESCON controllers, switching devices, and converters are appearing on the market. Converter devices are required to convert between parallel "Bus and Tag" to ESCON channels, and make ESCON implementations more cost effective. Two converter models are available: 9034 ESCON Converter Model 1 and the 9035 ESCON Converter Model 2. Although ESCON solves many problems, it is yet a young interface. Many of the DASD and tape control unit peripherals still need to increase their internal speeds to take advantage of ESCON. The industry standards challenge for ESCON will be to map ESCON speeds to ATM and SONET interfaces.

5.2.8 IEEE 802.X Physical Interfaces

Ethernet — The physical interface for IEEE 802.3 Ethernet provides more than just the synchronous interface standard provided by RS-232 or V.35. Recognition of the presence or absence of the control of the Carrier Sense Multiple Access/Collision Detection (CSMA/CD), data transport, collisions, and the translation of signaling from physical to Media Access Control layer (layer two) are provided by the physical layer of Ethernet. These will be discussed in the section on data link control protocols.

There are five types of Ethernet access: 1BASE5, 10BASE2, 10BASE5, 10BASET, and 10BROAD36. Each type defines both the wiring and the device which terminates the end of the wiring. 10Base5 was the thick-cable Ethernet with the drop cable needed by each user to access via a tap. 10Base2 was introduced in the 1980's as thin-wire Ethernet, offering a thinner coax cable with the same method of access. The physical connectors are called BNC and AUI, respectively.

The most current connectivity is via twisted-pair wiring and the 10BaseT standard, where users access the medium through their telephone RJ45 wall jack, no more than 100 meters of 0.5mm gauge telephone twisted-pair into a star topology central hub. Both baseband coax and broadband coax operate at 10M bps, while unshielded twisted-wire pair operates at 1M bps. A transceiver module may be needed to adapt to the 10Base2 thin-wire or 10BaseT twisted-pair Ethernet connection. Other Ethernet proprietary implementations may require an external radio frequency modem or fiber optic inter-repeater link. Another standard in process is the 10BaseF and 10BaseFB. These standards specify fiber optics with Ethernet networks and backbone Ethernet. The distance for Ethernet could be extended to 2 km, the same as for FDDI.

For physical connectivity, cables run from the computer to the LAN and contain multiple wire pairs with 15 pin interface connectors (there is also a RJ-45 connector available). For coax cable wiring schemes, a tapping screw and tap block are needed to pierce the outer shielding and contact the center coax cable conductor and ground on the outer shield. An Attachment Unit Interface (AUI) provides the physical connection to transfer the MAC PDU from the user attached device to the LAN access device called a Medium Attachment Unit (MAU). This MAU attaches to the physical network medium via a Medium Dependent Interface (MDI). All Ethernet interfaces operate at a 10M bps bus speed for IEEE 802.3 and Version 1.0 and 2.0 Ethernet frame format.

Token Bus and Token Ring — There are nine types of Token Bus physical interfaces. These interfaces fit into two categories: the broadband coax and the carrierband coax, both supporting 1, 5, and 10M bps. Physical interface for IEEE 802.4 Token Bus and IEEE 802.5 Token Ring is the dB connector, interfacing the IBM Type 1 shielded twisted-pair cable. Coax cable connections are also available. Token Ring interfaces at either 4M bps or 16M bps through an IEEE 802.5 interface, and uses 802.2 Type 1 LLC support. It also can interface via twisted pair (shielded) at 1, 4, and 16M bps. Figure 5.8 shows an example of Token Ring physical connectivity.

Table 5.1 shows a comparison of physical media based on speed, maximum distance, and signaling. Baseband and broadband LANs are compared in Table 5.2. LANs of the future are predicted to use a new form of transmission; Wave Division Multiplexing (WDM). WDM LANs will operate over single-mode fiber optics theoretically transmitting gigabits of data over terabits of media, possibly even using photonic switching.

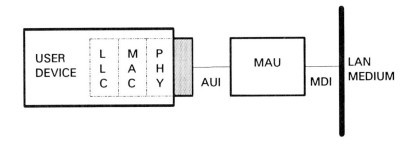

Figure 5.8 Token Ring Physical Connectivity

TABLE 5.1 Physical Media Comparison

PHYSICAL MEDIA	STANDARD NAME	SPEED (Mbps)	MAX DIST. (meters)	SIGNAL MODE
50-ohm Coax	10Base5	10	500	Baseband
50-ohm Coax	1Base5	1	500	Baseband
Thin Coax	10Base2	10	185	Baseband
UTP	1Base5	1	500	Baseband
UTP	10BaseT	10	100	Baseband
STP	Token Ring	4/16	100 (lobe)	Baseband
Broadband Coax	10Broad36	10	3600	Broadband
Fiber Optic Cable	---	16	4500	Baseband
STP (data grade)	Token Ring	16	700(lobe) + 800 inter-MAU	Baseband

TABLE 5.2 Comparison of Broadband and Baseband LANs

ATTRIBUTE	BASEBAND LANs	BROADBAND LANs
Media Type	Coax, UTP, STP, Fiber	Coaxial Cable
# Signals Carried	One	Multiple (FDM)
Type of Signal	Digital (voltage on/off)	Analog (variation of carrier frequency)
Speeds	1M to 16M bps	1M to 5M bps per channel (20 - 30 channels)
Distance Limitation	2000 ft. (then digital repeaters)	6000 ft. (MANs)
Installation & Maintenance Required	Easy Install - Minimum maintenance	Careful design/tuned components
Type of Data Carried	Data Only	Data, Voice, Video
Typical Topical Distribution	Intrabuilding or Campus	Metropolitan Area

5.3   DATA LINK LAYER PROTOCOLS

Asynchronous and synchronous transmission are forms of data link control protocols. Data link control (DLC) protocols have two forms: asynchronous and synchronous. The synchronous data link layer protocols are the more important. "Character-oriented" protocols such as BSC use control characters from character sets such as ASCII or EBCDEC as the control fields set within frames of variable formats. "Byte-count oriented" protocols such as DDCMP from DEC are similar to the character-oriented protocols but use count fields to indicate the number of bytes being transmitted. Finally, "bit-oriented" protocols such as SDLC, HDLC, ISDN BRI and PRI, and X.25 use specific bit *patterns* for frames. The individual bits define the protocol specifics and control the transmission. Bit-oriented protocols are the most important, because they represent the primary data link control protocols. Frame relay DLC layer functions are discussed with frame relay and Asynchronous Transfer Mode (ATM) and Synchronous Optical NETwork (SONET) concepts and will come later.

5.3.1   Binary Synchronous Control (BSC or Bisync)

The Binary Synchronous Control protocol, BSC or Bisync, was introduced by IBM in the mid-1960s as a half-duplex, point-to-point or multipoint character-oriented protocol for bidirectional transmissions of character-oriented data. Each variable-sized frame consisted of control codes such as Start of Text (STX) and End of Text (ETX) to manage transmission of character-coded user information. All codes are derived from a single character set — EBCDEC, modeled after the American Standard Code for Information Exchange (ASCII). Figure 5.9 shows a typical BSC text frame consisting of the following characters: a Frame Pad (PAD), two Synchronous Idle (SYN), a Start of Header (SOH), Header Information, Start of Text (STX), Nontransparent data (denoting a fixed-bit pattern as opposed to a variable one), an End of Text (ETX), a Block Check Count (BCC), and another PAD. An example of Bisync use is the ARPANET data link control protocol, which operates primarily Bisync utilizing transparent text mode, where the bit pattern of the text is variable.

P A D	S Y N	S Y N	S O H	Heading	S T X	Nontransparent data	E T X	B C C	P A D

BSC TEXT FRAME

Figure 5.9 BSC Text Frame Format

5.3.2 DEC Data Communications Message Protocol (DDCMP)

The Data Communications Message Protocol (DDCMP) developed by Digital Equipment Corporation (DEC) provides a byte-count-oriented protocol transmitted either asynchronous or synchronously over half- or full-duplex circuits on point-to-point or multipoint topologies. This protocol is similar to BSC, and uses a Count Field to indicate the number of bytes in the information field, synchronization characters, a flag, Cyclic Redundancy Checks (CRCs) for error management, address and information fields, and the send and receive count fields. Figure 5.10 shows a standard DDCMP frame showing placement of these fields. DDCMP also provides for supervisory and information frames, where this information would replace the Information field. No one coding scheme was standardized because each vendor had a self interest to "proprietize" the coding for their unique products and lock in users and protect the customer base.

5.3.3 Synchronous Data Link Control (SDLC)

In 1973 IBM was the first vendor to produce a bit-oriented protocol, called Synchronous Data Link Control (SDLC). This standard has been modified and adopted by the International Standards Organization (ISO) as the High-Level Data Link Control (HDLC) or ISO4335 protocol and by the American National Standards Institute as the Advanced Data Communications Control Procedure (ADCCP) or ANSI X3.66. The CCITT also has developed two standards based upon SDLC.

The CCITT Link Access Procedure-Balanced (LAP-B) is the X.25 implementation of SDLC, and the CCITT Link Access Procedure-D (LAP-D) is the ISDN and frame relay implementation of SDLC. Since LAP-B and LAP-D are subsets of HDLC, the discussion will center on HDLC. Figure 5.11 shows the relations and progression of various DLCs derived from SDLC.

8	8	8	14	2	8	8	8	16	var	16	bits
S Y N	S Y N	C L S	Count	Flag	RX Count	Send Count	Address	CRC-1	Info	CRC-2	

Figure 5.10 DDCMP Standard Frame Format

The present version of IBM's SDLC primarily uses the unbalanced normal response mode of HDLC together with a few proprietary commands and responses for support of loop or ring topology polling. SDLC operates independently on each communications link, and can operate in either multipoint or point-to-point, switched or dedicated circuit, and full- or half-duplex operation. The primary difference between SDLC and HDLC is that SDLC does not support the extended address field nor the extended control field.

SDLC is steadily replacing the less efficient BSC protocol. Some improvements of SDLC over BSC include: the ability to send acknowledgments, addressing, block checking, and polling within every frame rather than in a separate sequence, the capability of handling long propagation delays, no restrictions to half-duplex, not susceptible to missed or duplicated blocks, not topology dependent, and not character code sensitive.

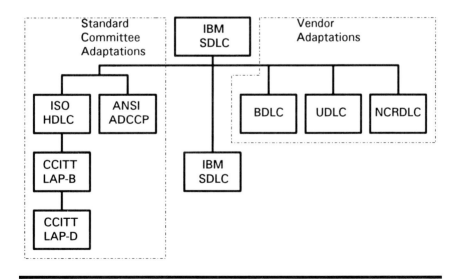

Figure 5.11 SDLC Legacy

5.3.4 High-Level Data Link Control (HDLC)

The High-Level Data Link Control (HDLC) protocol is not only the most popular protocol for data link control implementations, but it also forms the basis for Integrated Services Digital Network (ISDN) and frame relay protocols and services. HDLC is an international standard, adopted under ISO TC97. HDLC is a bit-oriented, simplex, half-duplex, or full-duplex, synchronous protocol passing variable-bit stream lengths over either a point-to-point or multipoint configuration. HDLC can also operate over either dedicated or switched facilities. There are two types of point-to-point link structures: a primary station transmitting commands to and receiving responses from the secondary station, as in Figure 5.12, and as the transmitting and receiving station both acting as a primary and secondary, with the capability of sending either a command or a response as in Figure 5.13. The multidrop link structure with one primary and multiple secondary (unbalanced) is also shown in Figure 5.13. Both configurations may be configured over switched or non-switched facilities.

HDLC has three types of data transfer modes. The two most common types are unbalanced Normal Response Mode (NRM) and Asynchronous Balanced Mode (ABM). NRM is used in multidrop and point-to-point links with the secondary awaiting a poll from the primary. ABM is used in the balanced configuration between combined stations, and allows only one secondary to be active at any time. A third type or method is via unbalanced Asynchronous Response Mode (ARM) which requires the secondary stations in an unbalanced mode to have explicit permission from the primary to initiate transmissions.

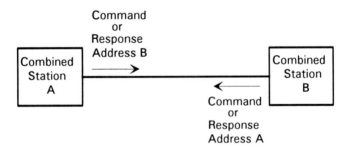

Figure 5.12 HDLC Balanced Mode

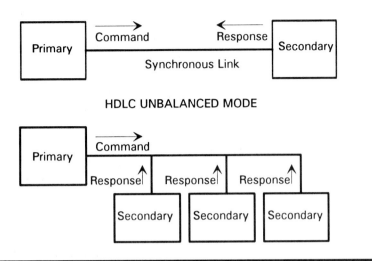

HDLC UNBALANCED MODE

Figure 5.13 HDLC Unbalanced Mode

A standard HDLC frame format is used for both information exchange and control of transmission. This frame is represented in Figure 5.14. This frame format supports both basic and extended control field formats. Two flag fields of proprietary bit patterns encapsulate the frame, an address field provides the address of the secondary station (but is not needed for point-to-point configurations), the information field, of course, contains the data being transmitted, and the frame check sequence (FCS) verifies the accuracy of the fields within the frame. Also included in this frame is a control field to identify one of three types of frames available. The first bit (or two bits with supervisory and unnumbered frames) of each control field is used to identify the type of frame: information, supervisory, or unnumbered. These basic control field formats are found in Figure 5.15, with the 16-bit version of the information, supervisory, and unnumbered frames found in Figure 5.16.

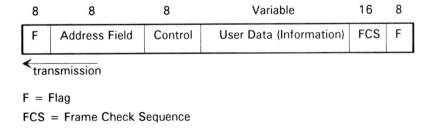

F = Flag
FCS = Frame Check Sequence

Figure 5.14 HDLC Frame Format

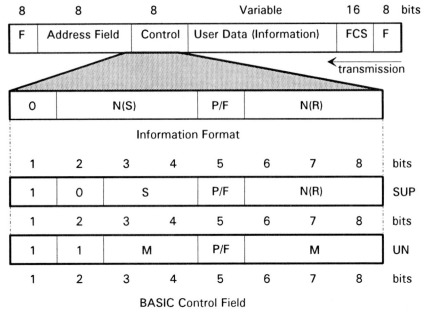

Figure 5.15 HDLC Frame with Control Field Breakouts (8-bit version)

The information frame is used to transport user data between stations. Within this frame, the N(S) and N(R) fields designate a modulo eight send and receive count for the number of frames to be transmitted and received, respectively. The P/F field designates a poll requesting transmission from the secondary station or a final bit indicating the end frame in the transmission sequence. The supervisory frame manages flow control and error control through positive and negative acknowledgments using four modes of operation: Receive Ready (RR), Receive Not Ready (RNR), Reject (REJ), and Selective Reject (SREJ). "S" bits establish one of these modes, and all other fields operate the same as before.

The unnumbered frame specifies a variety of control functions through the mode-setting commands just discussed: NRM, ARM, and ABM. The M modifier bits (M) specify which type of unnumbered frame to use.

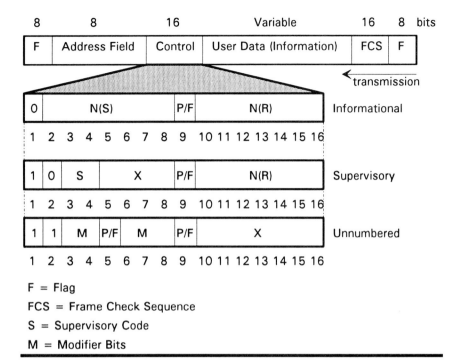

8	8	16	Variable	16	8	bits
F	Address Field	Control	User Data (Information)	FCS	F	

transmission

0	N(S)		P/F	N(R)	Informational

1 2 3 4 5 6 7 8 9 10 11 12 13 14 15 16

1	0	S	X	P/F	N(R)	Supervisory

1 2 3 4 5 6 7 8 9 10 11 12 13 14 15 16

1	1	M	P/F	M	P/F	X	Unnumbered

1 2 3 4 5 6 7 8 9 10 11 12 13 14 15 16

F = Flag
FCS = Frame Check Sequence
S = Supervisory Code
M = Modifier Bits

Figure 5.16 HDLC Frame with Control Field Breakouts (16-bit version)

Other implementations of HDLC are also used. It is interesting to note that HDLC is often used in satellite transmissions, because the window can open between 2 and 127 frames in size before an acknowledgment is needed. This is important due to the 500 millisecond turnaround times for up- and down-link transmissions.

5.3.5 ISDN and LAP Protocols

ISDN provides communications between the end user and the carrier, as well as between carriers. There are three types of Link Access Procedure (LAP) protocols. LAP was the first LAP protocol and was designed based on the HDLC Set Asynchronous Response Mode (SARM) command used in "unbalanced" connections. This mode formed the basis for Link Access Procedure "balanced" (LAP-B), an HDLC implementation that uses balanced asynchronous mode with error recovery to form the basis of the X.25 packet switching protocol.

The next extension of HDLC and LAP was Link Access Protocol over D-channel (LAP-D) standardized by the CCITT in Recommendations Q.920/Q.921 and I.440/I.441 as the Digital Subscriber Signaling System No.1 (DSS 1) data link layer. This implementation of HDLC uses either the basic or extended asynchronous balanced mode configuration and provides the basis for both Integrated Services Digital Network (ISDN) and frame relay services. Figure 5.17 shows a comparison of both LAPB (X.25) and LAPD (ISDN) frame structures. The trade-offs of both will become apparent later.

Integrated services digital network and frame relay protocols and standards span and cover the range of the first three layers of the OSIRM: physical, data link, and network. ISDN physical interfaces in I.430 and I.431 is covered here. Aspects of LAP-D will be explored with frame relay, which has its roots in ISDN protocol suites.

5.4 LLC and MAC LAYER PROTOCOLS

The logical link control (LLC) and medium access control (MAC) layers together make up the data link control layer of the OSI reference model. Figure 5.18 shows this relationship. Together with the physical layer, data link standards comprise the core IEEE 802.X protocol standards. The MAC layer manages the communications across the physical medium, defines frame assembling and disassembling, and performs error detection and addressing functions. The LLC layer interfaces with the network layer through Service Access Points (SAPs), as shown in Figure 5.19. Now to define the MAC and LLC sublayer functions, and how the IEEE 802.3 through 802.5 protocols operate over both.

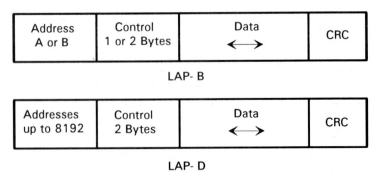

Figure 5.17 LAPB (X.25) vs. LAPD (ISDN) Comparison

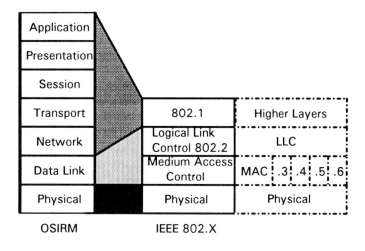

Figure 5.18 LLC and MAC Layer Protocol Stack of IEEE 802.X Compared to the OSIRM

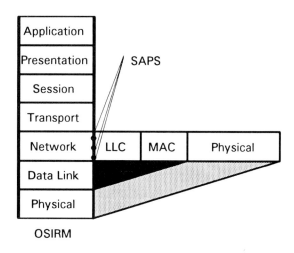

Figure 5.19 LLC and Service Access Points (SAPs)

A detailed discussion of cell-switching protocols as standardized by the IEEE (IEEE 802.6) in 1990 will be covered in Chapters 11 and 12.

While the IEEE 802.2 standard defines the logical link control layer, further developments with the IEEE 802.3 through 802.6 standards define the Medium Access Control (MAC) layer protocols. Multiple MAC protocols can exist under the same LLC. Figure 5.20

shows the physical relationship between LLC and MAC interface points, where a multiple-host application LLC interfaces to a Ethernet LAN is via a single MAC address. A Network Interface Unit (NUI) also provides a single MAC address to the same Ethernet LAN, and supports a terminal and workstation on separate LLC addresses. The host also connects to a Token Ring LAN, while another NUI, with workstation, attaches to the Token Ring LAN.

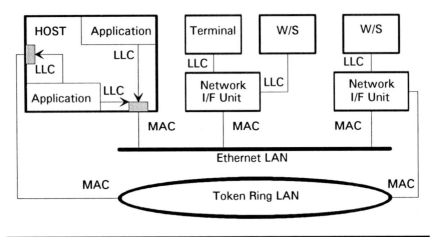

Figure 5.20 Illustration of LLC and MAC Physical Interface Points

5.4.1 Logical Link Control (LLC) Sublayer

The Logical Link Control (LLC) protocols are designed to communicate peer-to-peer over multipoint bus and ring topologies. The IEEE 802.2 standard defines the LLC sublayer within the data link layer. In LLC, primary and secondary stations do not exist — all stations are *common* to the transmission medium, with no intermediate switching. The LLC sublayer protocol allows any 802.3 through 802.6 protocol to carry multiple logical subnetwork traffic of each protocol over the same physical media, such as the local area network (LAN).

The two major modes of service interfacing with the network layer are connection-oriented and connectionless. Connection-oriented service uses the previously mentioned Service Access Point peer-to-peer communications and provides acknowledgments, flow control, or error recovery. There are two classes of connectionless services provided in the LLC: Class 1, or unacknowledged connectionless, which requires both the sending and receiving station address to be contained in each packet, and Class 2, or acknowledged connectionless, which requires the acknowledgment of each individual frame. Both types of connectionless services provide no acknowledgments, flow control, or error control, and rely on higher level protocols to perform these functions.

The LLC and MAC sublayers of the data link layer, along with the other layers of the OSI reference model, are shown in Figure 5.21. From this reference it is clear that the LLC sublayer serves a peer-to-peer protocol function between end users, and that the MAC and physical layers interface to the local area network transport media.

When the logical link layer receives user data in the form of an information field, it adds a header to this field and forms what is called a Protocol Data Unit (PDU). Figure 5.22 shows the formation of the LLC PDU. The information field can be variable in size. The PDU header contains both a destination address and a source address of the origination port for a network hardware device or application for network software. Both are referred to as Service Access Points (SAPs). It is important to note that the size of these address fields determines the number of possible addresses on the network. There are two options, a two byte and a six-byte address field, with each defined as individual or group, with universal or local authority.

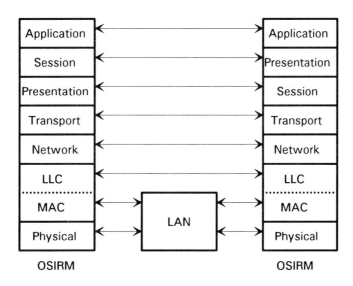

Figure 5.21 LAN Layers of the OSIRM

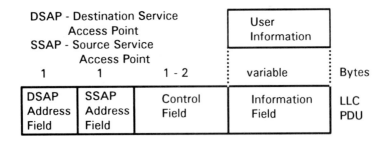

Figure 5.22 LLC Protocol Data Unit (PDU)

The physical and logical link layers of many vendor-proprietary LAN standards follow those of the 802.X standards. As a final note, IEEE 802.1 is defined by the ISO and CCITT as the Higher Layers and Management (HILI) standards. These standards are focused upon bridging protocols, and one example, the spanning tree bridge, will be defined in the next chapter on bridging protocols. The IEEE 802.X standards above 802.6 are left for further study by the reader.

5.4.2 Medium Access Control (MAC) Sublayer

The Medium Access Control (MAC) sublayer in the OSIRM data link layer manages and controls communications across the physical medium, manages the frame assembling and disassembling, and performs error detection and addressing functions. It is the point where distributed processing begins. The four most common MAC layers include:

- 802.3 CSMA/CD Ethernet
- 802.4 Token Bus
- 802.5 Token Ring
- 802.6 Metropolitan Area Networks (MANs)

The following MAC standards still under development:

- 802.7   Broadband Technical Advisory Group
- 802.8   Fiber Optic Technical Advisory Group
- 802.9   Integrated Voice and Data LAN Working Group
- 802.10 LAN Security Working Group

There is also one MAC layer bridge protocol,

- 802.1d bridge (Spanning Tree)

designed to interface any 802 LAN with any other 802 LAN. This protocol will also be covered in the next chapter on bridging protocols.

When the MAC layer receives the Logical Link Control PDU, it adds a header and trailer for transmission across the MAC layer (and physical medium). Figure 5.23 shows the LLC PDU as formatted into a MAC PDU. Each of the specific IEEE 802.X MAC frames conforms to this format and is described in the following sections. This new frame is now called a MAC PDU. MAC addresses identify physical station points on the network. Each station reads this MAC address to determine if the call should be passed to one of the LLC entities. Each Network User Interface (NUI), discussed in the section on Physical interfaces, has its own SAP and address.

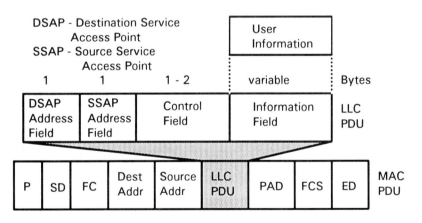

P - Preamble

SD - Starting Delimiter

FC - Frame Control

FCS - Frame Check Sequence

ED - Ending Delimiter

Figure 5.23 LLC PDU Formatted into the MAC PDU

5.4.3 802.3 CSMA/CD (Ethernet)

The Ethernet specifications were developed by the consortium of Xerox, Intel, and DEC in 1980, and the first products for Ethernet appeared in 1981. This standard used a medium access control protocol called Carrier Sense Multiple Access with Collision Detection (CSMA/CD) across a common bus with channel-attached MAC addressed stations. The IEEE later adopted Ethernet as the IEEE 802.3 standard. This is a classical case of vendor-driven standards development. The use of CSMA/CD allows the LLC layer to send data without collision (i.e., CSMA/CD and token passing) at rates theoretically reaching 10M bps. In actuality, it can reach about 3.5M bps under maximum load conditions.

CSMA/CD allows for stations to both transmit and receive data. During a collision the end stations initiate a "back-off" algorithm and follow a mathematical formula to randomize each station's next attempt to retransmit. The medium can be either baseband or broadband. The specific data link functions of Ethernet include encapsulation and de-encapsulation of user data, media access management, such as physical layer and buffer management, as well as collision avoidance and handling, date encoding and decoding, and, finally, channel access to the LAN medium. Each station on the network can attempt transmission and, if the medium is idle, gain the right to transmit. If they receive a busy, they transmit when the medium becomes idle. If they encounter a collision, they stop transmitting and send out a jamming signal to notify all other stations of a collision. The station then waits a random period of time and attempts retransmission.

Figure 5.24 shows the IEEE 802.3 CSMA/CD Ethernet MAC PDU frame.

MAC 802.3 CSMA/CD

7	1	2 or 6	2 or 6	2	variable	var	4	Bytes
P	SFD	Dest Addr	Source Addr	Length	Info	PAD	FCS	

P - Preamble

SFD - Starting Frame Delimiter

FCS - Frame Check Sequence

Figure 5.24 IEEE 802.3 CSMA/CD Ethernet MAC PDU Frame

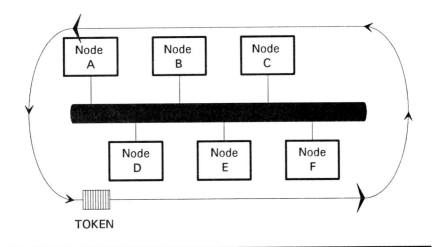

Figure 5.25 Token Bus Token Passing Routine

The preamble field provides synchronization. The Start Delimiter (SD) is a "start of frame" character. The destination and source address provide the MAC layer destination and source address. The length field identifies the length of the data field in bytes. The information field is the LLC PDU. The pad provides extra characters to achieve a minimum frame length value. Finally, the frame check sequence is the standard CRC-32. The maximum size of the 802.3 frame is 1,500 bytes.

5.4.4 802.4 Token Bus

The IEEE 802.4 Token Bus standard defines the MAC protocol for a token passing bus topology. A logical ring is formed on the physical bus, and each station knows only the preceding station on the bus. A token is passed down the bus, from station to station in sequence of the logical ring and by descending station address. This token contains the destination address of the next station. When the destination station receives the token, it can use the medium to transmit information for a limited time before having to turn the token over to the next station on the bus (in sequence). Figure 5.25 shows the Token Bus token passing routine.

Figure 5.26 shows the IEEE 802.4 Token Bus MAC PDU. This is similar to the Ethernet PDU, with the omission of the PAD functionality and the addition of both a Frame Control (FC) character to identify the frame type and an End Delimiter (ED) which indicates the end of a frame.

MAC 802.4 TOKEN BUS

1 +	1	1	2 or 6	2 or 6	variable	4	1	Bytes
P	SFD	FC	Dest Addr	Source Addr	Info	FCS	ED	

P - Preamble

SFD - Starting Frame Delimiter

FCS - Frame Check Sequence

ED - Ending Delimiter

Figure 5.26 IEEE 802.4 Token Bus MAC PDU

5.4.5 802.5 Token Ring

The Token Ring protocol was developed by IBM development labs in Zurich, Switzerland in the late '60s, with the first products appearing in 1986. The IEEE has adopted Token Ring as IEEE Standard 802.5, which works on the IEEE 802.2 logical control layer with IEEE 802.2 Type 1 protocol and 802.5 Media Access Control (MAC) Token Passing Protocol. Basic operation consists of a token which circulates around the physical 'hub' and logical 'ring' topology and provides "priority access" to the network medium. The token is either free or busy. As a free token circles the ring, each station has the capability to seize the token, modify it and load data into it, and send it on to the destination station. If the token is busy (the token contains data destined for a different station), the station regenerates it and passes it on to the next station without modification. Thus, only one station on the ring can transmit data over the common medium at a given time. Priorities can be assigned to the token for specific stations on the ring. At heavy load conditions, the Token Ring protocol is much more bandwidth efficient that other LAN protocols (less idle tokens — as opposed to Ethernet, where there are many collisions during heavy load conditions). Figure 5.27 shows the Token Ring token passing routine.

Figure 5.28 shows the IEEE 802.5 Token Ring MAC PDU. This is similar to the Token Bus PDU, with the omission of the preamble and the addition of the Access Control (AC) field for priority and reservation access control and the Frame Status (FS) character. Also, note that when an empty token is sent, only three characters are needed: the Start Delimiter (SD), Access Control (AC), and an End Delimiter (ED) field in place of the rest of the Token Ring PDU.

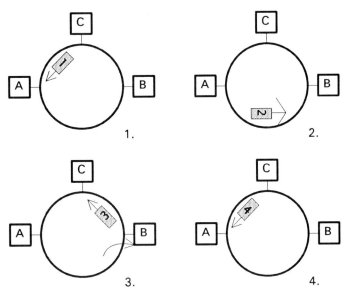

1 - free packet
2 - packet with C destination
3 - packet passes through B node and on toward C
4 - packet received by C and new free token passed to A

Figure 5.27 Token Ring Passing Routine

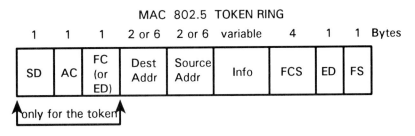

MAC 802.5 TOKEN RING

SD - Starting Delimiter

AC - Access Control

FCS - Frame Check Sequence

ED - Ending Delimiter

FS - Frame Status

Figure 5.28 IEEE 802.5 Token Ring MAC PDU

The maximum frame size for a Token Ring frame using the 4M bps medium is 4,000 Bytes, for the 16M bps medium is 17,800 Bytes.

5.4.6 Fiber Distributed Data Interface (FDDI)

Fiber Distributed Data Interface (FDDI), or FDDI-I, is a local and metropolitan area network standard defined by the American National Standards Institute as ANSI Standard (and CBEMA Committee) X3T9.5 and also recognized as an ISO standard. FDDI operates over both physical and MAC layer protocols, providing a 100M bps transmission over a dual, counter-rotating optical fiber ring between nodes. Although the bandwidth provided on the ring is 100M bps, the actual throughput is usually much lower. Still, this makes FDDI a high-speed LAN technology. Up to 500 dual attachment connection devices can interface to the FDDI ring in series. FDDI rings support a maximum of up to 1000 stations, with a maximum distance between stations of 2 km and a maximum ring total circumference (path) of 100 to 200 km. Many more network stations can be supported by FDDI than lower-speed LAN technologies, leading to better performance curves and less LAN degradation per user. There is also a standard for FDDI protocol over copper — Copper Distributed Data Interface (CDDI).

In an OSI reference model comparison, the FDDI protocols cover the same territory as the Token Ring protocol. In fact, FDDI operation is very similar to Token Ring protocol operation, and Token Ring can be accredited with providing the basics of FDDI. This ANSI standard X3T9.5 specifically defines the Physical Medium Dependent (PMD) (X3.166) layer to define single or multimode operation through full-duplex connectors, optical transceivers, and optional bypass switches. The physical layer of FDDI consists of a Class A Dual Attachment physical interface via the Physical Medium Dependent (PMD) sublayer. The Physical Protocol (PHY) (X3.148) layer implements a Non-return to Zero Inverted 4 bit-to-5 bit (NRZI-4B/5B) encoding/decoding algorithm as well as performing handshaking between each station's PHY protocol. Physical interfaces to the transmission medium are via multimode or single-mode fiber.

The Media Access Control (MAC) (X3.139) layer serves as peer-to-peer communications for the LLC layer and the SMT layer over the ring, as well as routing and traffic allocation. The Station Management (SMT) (X3T9.5/84-48) layer is still under study, and will provide addressing, bandwidth allocation, fault isolation and ring reconfiguration, and initialization of station control functions. This important layer provides the means of inserting and removing stations from the ring. SONET interface and transport is still under study. Figure 5.29 shows these differences, where the Layer Management (LMT) layer provides the station interaction management between physical and MAC layer protocols.

Protocol Data Units (PDUs) similar to those used in Token Ring are formed in the same manner as other LAN protocols, and use the FDDI MAC layer frame format found in Figure 5.30. FDDI fields are labeled in symbols which represent four bits each, and all fields shown in this figure have been discussed. The information field (data packet) ranges from 128 to 4,500 bytes. The address fields conform to the Token Ring standard. The maximum frame length is 9,000 bytes.

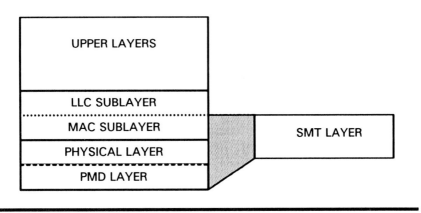

Figure 5.29 FDDI Protocol Stack

16 +	2	2	4 or 12	4 or 12	var	8	1	3 +	Symbols
Pre-amble	SD	FC	Dest Addr	Source Addr	Info	FCS	ED	FS	

MAC FDDI

SD - Starting Delimiter
FC - Frame Control
Info - Information Field
FCS - Frame Check Sequence
ED - End Delimiter
FS - Frame Status

Figure 5.30 FDDI MAC Layer Frame Format

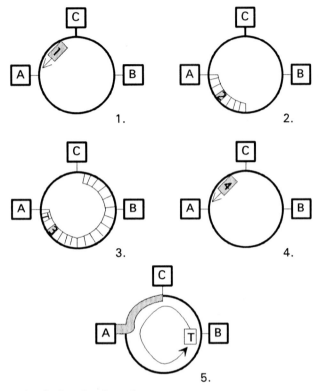

1 - A absorbs the token
2 - A transmits frame to C
3 - A adds new token to end of frame (B ignores token)
4 - C reads in frame, and frame continues to A
    (B has chance to transmit)
5 - A takes frame off network (C has chance to transmit)

Figure 5.31 FDDI Token-Passing Routine

Basic FDDI operation is similar to that described in the section on Token Ring, with the exception of free-token seizing and the release of the token after transmission. Each idle station on the ring has a chance to seize a passing free token. Figure 5.31 shows an example of FDDI token passing. If there is a free token passing by station A, the station can (a) seize the token and (b) transmit a frame to station C. The original token will be taken off the ring, and a new token not released until station A has transmitted its full frame. This is the major difference between FDDI and Token Ring. A new token will be released onto the ring (c) after the completion of frame transmission from station A, even though the destination station (station C) has not received the full frame (d). In this manner, high-speed transmissions can be accomplished without the need for the transmitting station (station A) to clear the original token. The frame which was transmitted by station A will eventually come back around (station C reads the frame as it passes) and station A must purge the frame from the ring (e).

Two classes of stations use the FDDI ring. Figure 5.32 shows both types of stations and their use of the FDDI ring. Class A stations are classified as primary stations and utilize both the inner and outer fiber ring. They are called Dual Attachment Stations (DAS) since they attach to both fiber rings. Class A stations can route around network failures by utilizing a combination of both primary and secondary rings. These stations can also utilize a bypass switch, which enables the ring to remain intact even though the station has lost power. Class B stations cannot provide reroute, and use only the primary (outer) ring. They are called Single Attachment Stations (SAS). Each SAS attaches to the ring via only two fibers, and can be isolated by the hub during failure conditions. These stations are shown connecting to the FDDI LAN through a wiring hub, also called a Dual Access Station (DAS) concentrator. Under normal operation, stations use the primary ring for data transfer.

During a link failure (shown in Figure 5.33) all Class A stations can automatically reconfigure to use the secondary ring. This capability is called self-healing. Class B stations will be off-line because the primary ring they use is inactive during a failure condition. Any station on the link can be taken down without affect to the FDDI ring.

The secondary (inner) ring acts as an on-line backup to the primary (outer) ring. This is due to physical connections on the dual FDDI ring to share a single MAC layer and a single MAC address. FDDI networks can also be configured in a star-wiring arrangement, where a patch panel is used for concentration of workstations. FDDI concentrators are available which can aggregate multiple single attachment devices into a single FDDI ring attachment. Each of these devices must be within 2 km of the FDDI ring.

FDDI can be implemented with either single mode or multimode fiber optic cable. During multimode operation, a 1300-nanometer bandwidth LED light source is transmitted over ANSI-specified 62.5/125 micron (core/cladding diameter) multimode fiber, with a 2-km maximum between stations. During single-mode operation, a laser is used in place of the LED to increase distance.

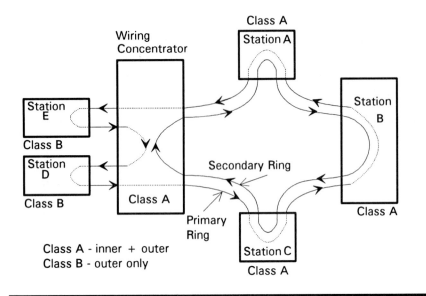

Figure 5.32 FDDI Station Classes and Rings

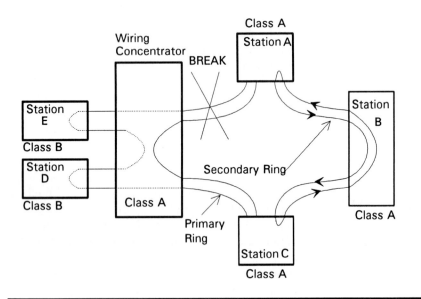

Figure 5.33 FDDI Link Failure Scenario

FDDI can be transmitted over shielded and unshielded twisted-pair fiber distances up to about 100 meters. This is with data grade, not voice grade UTP wire, where it is limited to 50 meters. The Unshielded Twisted Pair Forum (UTF) has developed the shielded twisted-pair standard called Copper Distributed Data Interface (CDDI). FDDI using CDDI can still use the fiber optic backbone, but can provide twisted pair to the desktop. This is then used in conjunction with 32-bit FDDI PC interface cards.

WAN attachment is accomplished through both encapsulating and translating dual attached bridges. Encapsulating bridges are designed to perform encapsulation of the MAC LAN packet into an FDDI packet. This is forwarded through the router network. This is a proprietary approach. Translating bridges bond the MAC LAN packet address to a Subnet Access Protocol-Service Access Point (SNAP-SAP) packet. This packet is then routed between FDDI devices, and is universally understood among many FDDI vendors. Figure 5.34 shows an example of an FDDI Campus wide area network.

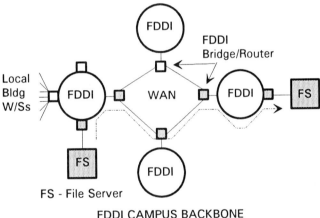

Figure 5.34 FDDI Campus Backbone or WAN

The many advantages to implementing an FDDI LAN include:

+ Accommodate large numbers of users
+ 100M bps bandwidth
+ Large geographic area coverage (long-distance LAN)
+ High reliability through self-healing
+ Fiber immune to copper cable interference — high transmission quality

+ Direct efficient attachment for hosts
+ Supports more stations than standard LAN protocols
+ Interoperability through FDDI standard
+ Built-in network management (SMT)
+ Support multimedia services on LAN

There are also a few disadvantages to FDDI LANs:

− Expensive to install (cost of hardware and fiber runs)
− Need to use existing copper wiring at lower data rates and distances (except CDDI)
− Speed conversion problems

5.4.7 FDDI-II

Development is under way on a mechanism called Hybrid Ring Control (HRC) which will allow FDDI LANs to transport multiplexed asynchronous packet data and isosynchronous circuit-switched data. Figure 5.35 shows the new FDDI-II protocol structure. The physical layer will probably change to SONET, but, as of now, it still uses the old FDDI-I 100M bps interfaces. There is now a hybrid multiplexer layer between the physical and MAC layer. The MAC layer has been split into a Packet Processing layer (from the old FDDI-I standard) and a new isosynchronous MAC layer. Frames are now called cycles. The PMD layer passes a "cycle" onto the FDDI-II ring every 125 microseconds. The Hybrid Multiplexer layer strips off the header and passes the cycle to the appropriate MAC layer. The station management has to be worked out for compatibility between different vendor hardware.

Multiple 6M bps portions of data can be dynamically allocated to T1 channels to support voice, data, and video. This 6M bps (actually 6.144M bps) chunk of data consists of multiple 64K bps channels. Each cycle can carry up to 16 channels, and each channel up to 6.144M bps. Isosynchronous traffic can also use the 6.144M bps channels. This service is good for interconnection of PBX equipment to the LAN and WAN, and combines telecommunications and data communications mixed media on the same fiber. Other planned enhancements include the WBS subchannels, dedicated packet groups, and cycle formatting. Future protocol support may include XTP, HCR, and HiPPI. To date, the deployment of FDDI-II has been stalled by the lack of a vendor to produce a chip-set (due to the high development costs associated with advanced technology chip-set development).

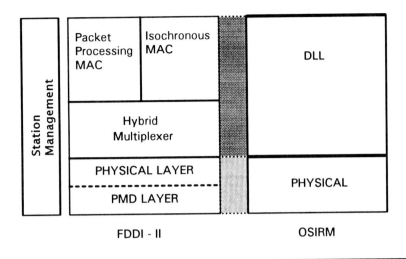

Figure 5.35 FDDI-II Protocol Structure

5.5 REVIEW

We have covered many of the common protocols spanning both telecommunications and data communications. Protocols from the lowest two layers of the OSI reference model have been discussed. Many others were touched upon. It is clear that there are many different physical interfaces, but there is a central logic and progression exhibited among them. When dealing with LANs, the physical interface protocols are intricately tied to the data link layer protocols. For LAN data transport, the data link protocol layer is divided into Medium Access Control (MAC) and Logical Link Control (LLC) sublayers, performing local medium access and managing end-to-end data transport, respectively. Bridge, router, and higher-layer protocols will be explained in the next chapter.

6

Common Protocols
and Interfaces — Part 2

Because many network designs involve bridges and routers, the protocols by which these devices communicate and exchange information should be understood. This chapter also presents a discussion of the transport and internet midlayer protocols and shows how they negotiate the transport, networking, and internetworking of data. SNA, APPC, and XTP protocols are also discussed.

6.1 BRIDGE PROTOCOLS

The IEEE 802.1 Spanning Tree "Learning Bridge" Protocol (STP) and IBM Source Routing Protocol are the two major protocols for network bridging. These protocols operate at the physical and Medium Access Control (MAC) layers to provide a limited form of relaying packets over the local and wide area network medium. True bridging protocols operate as relay points only. They provide a LAN extension similar to the repeater function but with limited additional intelligence. Some bridges used fixed routing tables to control bridging, while others, such as the IEEE 802.1 Spanning Tree Protocol, employ dynamic routing tables. Bridging schemes such as these allow the bridges to dynamically change packet relaying based upon network topology changes. Ethernet relies on the Spanning Tree approach when scanning for address destinations on the WAN. Token Ring uses Source Routing to accomplish the same.

6.1.1 IEEE 802.1 Spanning Tree "Learning Bridge" Protocol (STP)

The IEEE Standard 802.1 Spanning Tree "Learning Bridge" protocol (STP) is a bridging protocol based upon the IEEE 802.1 Transparent Spanning Tree algorithm. STP defines routing table operation for bridges which span multiple networks, and provides the function of frame (packet) forwarding and the capability for the bridge device to "learn" station addresses and network topology changes. Each bridge has a routing table which interacts dynamically and changes to correct routing problems such as routing loops or unavailable circuit paths. A flat common addressing scheme is used to assign a unique data link address to each device attached to a networked bridge, and to eliminate transmission broadcast "loops". Each data frame passing through a bridge is examined and forwarded on through a process called "filtering". If it is destined for the local LAN, it is sent on to the user device. The resulting "tree" spans each subnet and assures that there is only a single route between any two LANs. If the topology is changed, a new spanning tree is created by the protocol. STP performs traffic management by employing packet filters to forward or drop packets based upon their source and destination address, protocol type, and multicast or broadcast address. Some vendors implementing this protocol, such as Wellfleet communications, also accommodate load sharing across multiple physical links.

STP is a true bridging protocol and is inefficient and disadvantageous as a networking protocol. STP is better utilized when the network is made up of many point-to-point circuits. STP does have a few disadvantages. STP elimination of loop paths ties up expensive leased-line resources. Also, spanning tree table building after network failures takes considerable time and introduces long user delays.

There is also a standard called Source Routing Transparent (SRT) by the IEEE which marries the IEEE STP and the IBM SRP into one bit-selective bridging protocol. The selective bit is transmitted to the router within the MAC frame information field. SRT can then interact with the source route bridging of token ring while performing transparent bridging to other LAN implementations. Many bridges achieve forwarding rates of over 14,500 frames per second sustained over a long period of time using this technique. STP is used primarily for LAN-to-LAN and LAN-to-WAN connectivity, but should be used with caution as performance and response time are much poorer than with routing protocols. SRT can also allow SNA Source Routing into not only Ethernet TCP/IP networks but also DECnet networks.

6.1.2 IBM Source Routing Protocol (SRP)

The IBM Source Routing Protocol allows LAN users to specify their routing for each packet transmitted. Thus, each packet transmitted by the LAN device to the bridge contains a complete set of routing information for the bridge to route upon. Thus it becomes a combination of bridging and routing, but the distinction is that the routing is performed at the data link layer, not the network layer. The addressing scheme is hierarchical in nature. Each device makes its ring number part of the data link address, thus creating a hierarchy of addresses for each level of devices. The bridge then routes according the packet instruction, and routing is performed at the data link layer rather than the network layer. SRP has more overhead than SRT, but the processing is reduced for each bridge it transverses. Some LAN operating systems only work with source routing, such as IBM's PC LAN Program.

The information for source routing to perform its function is contained in the routing information field (an extension of the source address field). Figure 6.1 shows the structure of this field, which starts with a 2 Byte control header, and then contains consecutive ring and bridge numbers, each 12 bits and 4 bits long, respectively, to identify the path towards the destination ring. The maximum number of ring numbers the packet can transit is eight, therefore the maximum number of bridges to be "hopped" is seven. Source routing has a seven-hop count maximum. If the routing information field is zero, the bridge does not pass the frame (performs filtering). Any frame with the Routing Information Indicator (RII) bit in the address header set to one is routed to the next bridge in sequence.

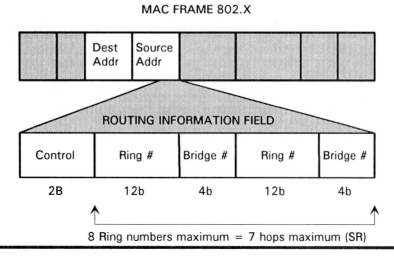

Figure 6.1 MAC 802.X Frame

6.1.3 Extended Source Routing

Wellfleet Communications, Inc. has implemented a new protocol within their router products. Extended Source Routing enables network end stations to support both local source route bridging and internetwork routing, while eliminating the seven-hop count restriction on source route bridging. Routing tables are built dynamically through use of the source route explorer packets. Each router is able to reset the hop count to zero as it passes the source route packet on to the next router. While this method is just another form of encapsulation of the token ring packet, it improves reliability of transmission, eliminates the hop count restriction, and can decreases response time across the network. Other router vendors are implementing similar protocols.

6.2 ROUTER PROTOCOLS

Routers are playing a role of ever increasing importance in data communications. Three forms of router protocols communicate and route information. As discussed, routers perform both *routing* and *bridging* functions. Both methods require the router to perform address translation. There are multiple routing protocols which perform routing with different priorities. Some perform routing based on the shortest path to the destination node, some least cost routing, and others based upon complex algorithms. These routers use a series of algorithms to perform the task of routing, along with dynamic routing tables to manage this routing. Almost all routers support bridging protocols. It is preferable to perform translation bridging with a router as opposed to encapsulation bridging. "Tunneling", a form of encapsulation, should also be avoided if possible. The three forms of protocols routers use include:

Gateway Protocols — router to router communications between like routers between routing tables. Communications can take place between autonomous systems (Exterior Gateway Protocol - EGP) and within autonomous systems (Interior Gateway Protocol - IGP). We will cover RIP, IGRP, EGP, BGP, OSPF, and IS-IS.

Serial Line Protocols — communications over serial or dial-up links between unlike routers. We will cover HDLC, SLIP, and PPP.

Protocol Stack Routing and Bridging Protocols — protocol stacks tell the router whether the information needs to be bridged or routed, and how to do it. We will cover TCP/IP, OSI, SRB, STP, and SRT.

Gateway and serial line protocols work in parallel — the gateway protocol passes the routing table information and keep alive packets, the serial line protocol passes the true user data. Protocol stack routing and bridging protocols operate across the network to manage the flow of frames.

6.2.1 Distance Vector Routing Protocols

The first routing protocols to emerge on the market were based on distance vectors. Figure 6.2 shows an example of the use of distance vector routing determination for routing table exchange.

Distance vector protocol allows router A to talk to its nearest neighbors, router B and router C, for an intermittent time period (i.e., 30 seconds). Router A will constantly exchange its routing tables with router B and router C, as well as vice versa. Thus, router A in effect broadcasts its status to router C who, in turn, broadcasts to router D and router E, and so forth throughout the network. Problems can arise due to convergence, where the transfer of these routing tables between nodes takes thirty seconds or more per hop. In a large network, this can amount to a multiple-minute delay causing a different network status at multiple locations throughout the network.

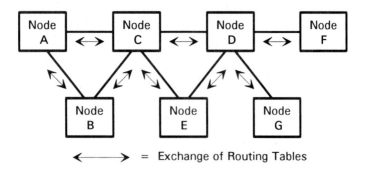

Figure 6.2 Distance Vector Routing Table Exchange

There are many implementations of the distance vector routing protocols. One of the first developed was the Routing Interconnectivity Protocol (RIP), developed for the XNS protocol suite. RIP is the most basic and operates in a connectionless mode at the application layer, interfacing with transport layer protocols through UDP. Its decision for routing is based upon hop count only. RIP computes this hop count distance while ignoring the length of the hop and the capacity available in that and other hops. This can cause problems when a higher-bandwidth path is available and desirable for transport, as in Figure 6.3. Router A would like to send a 500K file to router E. Since the path through router D requires only one hop, it will be chosen (assuming distances for both routes are equal) instead of the higher-bandwidth path through router B and C. RIP also requires an abundant amount of overhead to update neighboring routing tables every thirty seconds. Rather, this is not just an update; for each router in the network transmits its entire routing table to each of its neighbors every thirty seconds. This problem obviously compounds as the network becomes larger. RIP protocol also has a hop-count restriction of 16 hops, and is prone to routing loops (nodes stuck transmitting back and forth to each other because each believes the other has the shortest path to the destination).

Internetwork Gateway Routing Protocol (IGRP) is another type of routing protocol proprietary to cisco Systems routers. This protocol operates similar to RIP, but is superior because it understands bandwidth limitations between hops, as well as time delays. While this protocol improves upon RIP, it also doubles the transmit time of information between nodes, amplifying the convergence problem. Internet, such as in campus research WANs, uses two additional versions of distance vector protocols, Edge Gateway Protocol (EGP) and Border Gateway Protocol (BGP), which work similar to IGRP.

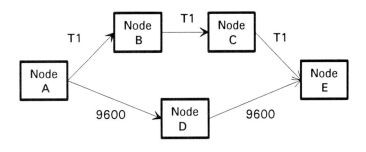

Figure 6.3 RIP Routing Example

6.2.2 Link State Routing Protocols

An even more efficient routing protocol is now employed called link state. Routers using link state routing protocols request changes in link states from all nodes on the network. Thus, each router requests everyone else's routing tables. This makes good use of the routing process and eliminates having various network states running simultaneously throughout the network.

Each node has a view of the entire network. These link state updates are sent using 64-Byte packets in a multicast mode, and require acknowledgments. This protocol will also notify users if their IP address is unreachable. This method is more memory intensive for the router, and requires large amounts of buffers and memory space. More router CPU cycles are also needed, as more complex algorithms must be used.

There are two major implementations of link state routing protocols on the market: Open Shortest Path First (OSPF) and Intermediate System to Intermediate System (IS-IS). OSPF, a standard IGP developed by the OSFP Working Group of the IETF and identified under TCP/IP (DARPA), provides a much more comprehensive view of the network. OSPF uses the Dijkstra, or Shortest Path First (SFP), algorithm for packet routing. OSPF routing is based upon many parameters, including shortest path, bandwidth available, cost in dollars, congestion, number of hops, time delay, and even the bit error rate (BER). All costs for links are designated on the outbound router port. Routers are sectioned into "areas", with each area maintaining its own topological database. These areas are then connected to a backbone area.

OSPF routing supports three types of networks: point-to-point, broadcast, and nonbroadcast. Point-to-point networks join a single pair of routers. Broadcast networks attach more than two routers, with each router having the ability to broadcast a single message to multiple routers through a single address. Nonbroadcast networks, such as X.25, attach more than two routers but do not have broadcast capability. OSFP also supports subnetting and filtering, which we will discuss in the chapter on backbone design. OSPF is only useful with TCP/IP networks. OSPF also supports bifurcated routing; the ability to split packets between two equal paths.

IS-IS is an emerging standard, and is yet to be fully defined. IS-IS is the OSI standard equivalent to OSPF, and provides similar benefits. There are two new OSI standards for routing which fit the IS-IS category and compete with OSPF. End-System-to-Intermediate-System (ES-IS) protocol is used to route between end nodes and intermediate nodes, while the Intermediate-System-to-Intermediate-System (IS-IS) protocol is used to route between network nodes. An extension of IS-IS, Integrated IS-IS (also called Dual IS-IS and proposed by DEC), can support both OSI and TCP/IP networks simultaneously. This protocol helps TCP/IP users migrate to the OSI platform. DECnet V has standardized on supporting both versions. OSI IS-IS standards are popular in the international community, where OSI is the standard. IS-IS standards do not have a hop count limit. Eventually, a standard should emerge which combines the capabilities of both OSPF and IS-IS.

6.2.3 Point-to-Point Protocol (PPP)

Point-to-point protocol differs from the previously-mentioned router protocol in that it was developed for serial line communications between routers, often multiprotocol routers of different vendor origins. PPP is a data link layer protocol designed to encapsulate IP internetwork data. Created by the IETF in 1988, it has superseded the older asynchronous Serial Line Interface Protocol (SLIP) which was limited in both speed (56K bps maximum) and protocol support (TCP/IP only). PPP can support the configuration and management of links between multiple multiprotocol routers via a serial interface in both synchronous and asynchronous mode. HDLC could also be used, but only through encapsulation. PPP also supports addressing. PPP would not be used in a larger network design employing some other form of Internet protocol.

PPP can support communications between routers using many protocols in addition to TCP/IP, such as DECnet, OSI Internet protocols, and AppleTalk.   PPP has many advantages, such as providing addressing information exchange between routers having proprietary addressing schemes, employing data compression techniques, and offering levels of encryption. Its security features allow the network to check and confirm the identity of users attempting to establish a connection. PPP operates like many of the services and protocols which rely on clean, reliable transmission mediums; it discards any packets received in error, letting the higher-level protocols sort out the retransmission. PPP would be attractive for router network design when a small number of network routers were used without another Internet protocol, and with a higher-level protocol to account for discarded data. PPP does not have to tie up the entire transport circuit, it can be shared with other serial line protocols. PPP defines both a Link Control Protocol (LCP) for link establishment, configuration, and testing and an Internet Protocol Control Protocol (IPCP) for network control. PPP has been adopted by hub and router vendors alike.

6.2.4 Routing Protocols Comparison

Table 6.1 shows a comparison matrix between communications protocols and their native routing protocols.

TABLE 6.1 Communications and Routing Protocols

Communications Protocols	Routing Protocols
OSI	ES-IS, IS-IS, Integrated IS
TCP/IP	EGP, RIP, OSPF
XNS, Novell IPX	RIP
Banyan VINES	RTP
AppleTalk	RTMP
DECnet	DECnet
All	PPP

6.3   NETWORK and TRANSPORT LAYER PROTOCOLS

With this groundwork of the physical and data link layer protocols, we will delve into the midlayer protocols providing for the end-to-end network and transport of the data. Since many of the network and transport protocols work closely in unison, each architecture implementation will be examined.

6.3.1 TCP/IP

The Internet Protocol (IP) portion of TCP/IP was originally designed by the academic community using ARPANET. IP started as part of TCP/IP as a military standard, and has now become the "protocol of choice" for third-level networking. IP is not controlled by any single vendor and is the least proprietary internetworking protocol, as well as the most popular. IP routing is usually used in conjunction with TCP and UDP. IP has the benefits of a connectionless packet network, which routes packets through available paths around congestion and points of failure. It provides "best-effort" routing in a connectionless mode. IP does not provide end-to-end reliable delivery, error control, retransmission, or flow control. It instead relies on TCP to provide these functions. Also, IP can support a variety of higher-level protocols. The network designer must understand that this service is dynamic yet unreliable. Data delivery is not guaranteed, and when it is not delivered the network does not try to correct and retransmit unless TCP or some other higher-level protocol is used. TCP/IP implementations typically constitute a router, TCP/IP workstation and server software, and network management.

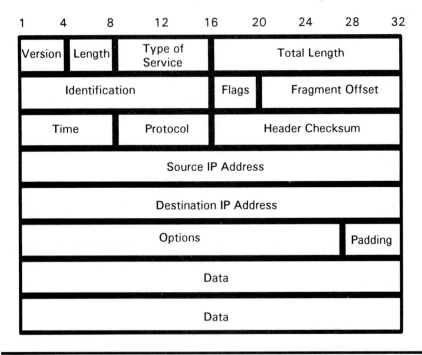

Figure 6.4 IP Datagram Format

The datagram format of the IP packet can be found in Figure 6.4. The version field specifies the IP protocol version used. The length field specifies the datagram header length. The type of service field is a future enhancement for OSPF routing. It specifies a three-bit priority "precedence" of 1-7, a one-bit "low delay" bit (when set), a one-bit "high-throughput" bit (when set), a one-bit "high-reliability request" bit, and a two-bit "unused" field. The type of service field could be a powerful enhancement for priority routing of higher-layer protocols such as SNA. The identification, flags, and fragment offset fields all work together to control fragmentation of a user datagram. The time to live field specifies how many devices the packet can pass through before it is declared "dead". The protocol field identifies the higher level protocol type (i.e., UDP). The header checksum acts as a 32-bit check. Source and destination IP addresses are added, and the user data is placed in the data field. There is also an option field which can specify routing and time-stamp options.

Figure 6.5 shows an example of IP routing of file transfer protocol (FTP) data. Application 1 (acting as an IP host) passes data (the File Transfer Protocol frame encapsulated with a TCP or UDP header) to a network (named Charlie network) utilizing an IP addressing (encapsulation) scheme. The data block contains addressing which specifies the destination of the data, in this example, file transfer protocol frame with header. The TCP or UDP header provides the end-to-end transmission confirmation. Charlie network then identifies the network protocol from the protocol ID type field, and the global IP address of the destination network, and adds its IP header to the data. The router routes IP packets based on the destination network address, not the destination host address. This IP header consists of a network identifier and a station identifier. The network then decides if the destination address is attached to the network, or if it is remote and must be accessed via a transport element (such as a router or bridge). If the device is local to the network, the data is routed, the IP address stripped off, and the data sent forward to the destination device. This is shown as Device A on the same network. If the device is not local, the network determines the address of the next network element and adds this subnetwork address to the existing IP packet. This is shown as Device C. Figure 6.6 steps the reader through the IP routing process after a router extracts the IP destination from the IP datagram.

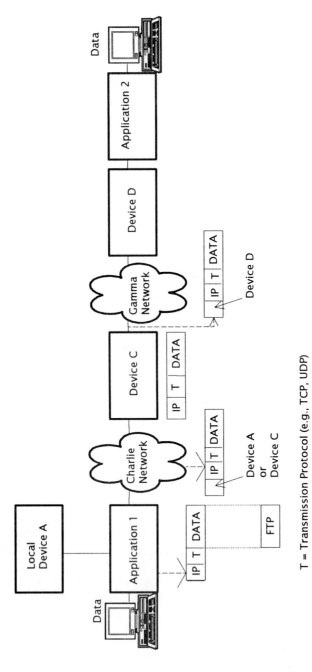

T = Transmission Protocol (e.g., TCP, UDP)

Figure 6.5   IP Routing of File Transfer Protocol Data

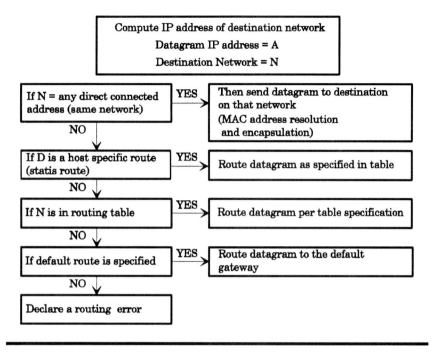

Figure 6.6 IP Routing Process

When Device C receives the packet, it strips off the header which designated its own address. At this point, the packet is back to the original data with the destination IP header. Device C then goes through the same process as the previous network did, finding whether the destination address is local to Device C, and if not, packaging the packet within another addressing header for Device D via Gamma Network. This process continues until the packet is routed to the destination address which, in this example, is Application 2 off Device D. Each protocol layer is then stripped off down to the original user data.

Figure 6.7 shows the TCP/IP protocol stack and its relation to the other transmission control protocols, such as User Datagram Protocol (UDP), Internet Control Message Protocol (ICMP), and routing control protocols. Routers send error and control messages to other routers using the ICMP. These include a Time to Live (TTL) telling how many routers the packet can pass through before it is declared "dead", a congestion control message called source quench, a notification of a parameter problem, and an echo request/echo reply "ping". UDP provides the router the ability to distinguish among multiple destinations within a single host. UDP provides a "doorway" to other application routing protocols, such as RIP (but not OSPF).

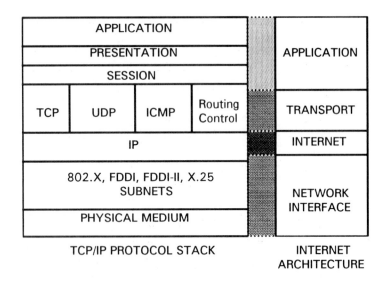

Figure 6.7 Internet Architecture Comparison to the TCP/IP Protocol Stack

Address Resolution Protocol (ARP) is used to connect Ethernet to TCP/IP. The workstation sends out the ARP to get a TCP/IP address. The router then builds up the ARP table and tells the user their MAC address. ARP then encapsulates a 32-bit IP address into an Ethernet frame which is delivered to Ethernet broadcast address to assure ubiquitous delivery to all devices. Accidental regeneration of these broadcasts can cause "broadcast storms" which can bring a network to its knees. This is particularly visible in bridged networks. The ARP acts as the IP to the MAC layer. The ICMP routes to the host the router information. The EGP is an autonomous system used to communicate between routers on different logical networks.

IP routing addresses can also be bought from Internet. There are Class A, B, and C licenses which provide four-, three-, and two-byte addresses for all users on the network. Class C are relatively easy to obtain, Class B are for large network service providers, and Class A are rarely given out except to extremely large IP network providers.

Transmission Control Protocol (TCP) is a connection-oriented protocol that maintains end-to-end reliability over a full-duplex connection. TCP works similar to the X.25 protocol, where a "sliding window" keeps track of acknowledged delivery of every byte that is received by the destination and sends more data after a given number of acknowledgments (window size) are received. The window size is the amount of data that remains unacknowledged. This window is either increased during periods of low network usage or decreased during network congestion.

Figure 6.8 illustrates the TCP frame format. The source and destination port fields correlate the TCP port numbers to a specific application program in the host. The sequence number field identifies the position of the senders byte stream of the data. The acknowledgment number field identifies the next octet to be received. The HLEN provides the length of the header. The code bits field determines the use of the segment contents (e.g., SYN for synchronize sequence numbers and RST for reset connection). The window field tells the amount of data the application is willing to accept. The urgent pointer field specifies the window position where the URG bit is set. The options field is software application specific. One field is reserved for future use.

TCP/IP protocols bridge UNIX, DOS, VM, and MVS environments. A majority of UNIX users employ TCP/IP for internetworking. Many Network Operating System (NOS) vendors are now integrating TCP/IP into their NOS platforms. Examples include Novell Netware and Banyan Vines. Some users access TCP/IP for terminal emulation, Network File System (NFS) for file storage and distributed file transfer for UNIX, File Transfer Protocol (FTP) for batch file transfers, Simple Mail Transfer Protocol (SMTP) for electronic mail, and Telenet for remote logon. There are no print services defined in TCP/IP. Application programming interfaces (APIs) which interface the user to TCP/IP are varied.

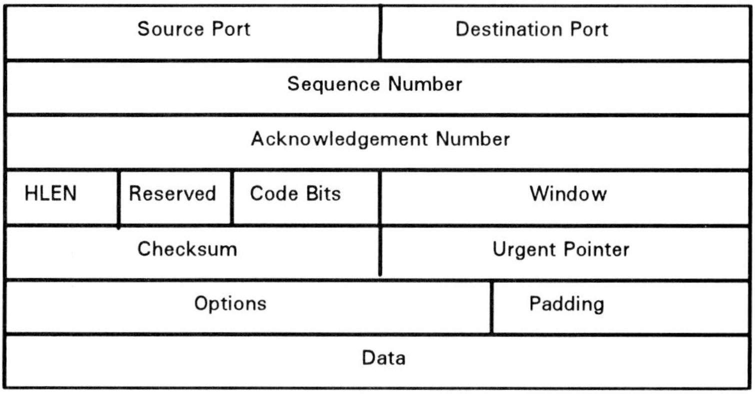

Source Port			Destination Port	
Sequence Number				
Acknowledgement Number				
HLEN	Reserved	Code Bits	Window	
Checksum			Urgent Pointer	
Options			Padding	
Data				

Figure 6.8 TCP Frame Format

The ISO standards for internetworking (ISO internetworking protocols) are less defined than those for TCP/IP. OSI standards have fewer products and less support than TCP/IP. Studies indicate that users of TCP/IP will eventually migrate to the OSI standard. There are currently many international TCP/IP interests, such as the commercial TCP/IP service offerings of Finland (Lanlink) and Sweden (Swipnet).

6.3.2 DECnet Phase IV and V

Digital Equipment Corporation's standards set is Digital Network Architecture (DNA), of which DECnet is a network protocol defining a hierarchical network architecture. Multiple areas are established across the wide area network, with each area supporting up to 1023 attached end nodes or hosts. The maximum number of areas is 63. There are two levels of routing within DECnet. Level one defines intraarea routing, and level two defines interarea routing. DECnet assigns its own MAC address to each station on the LAN.

The relation of DECnet to the OSI reference model can be found in Figure 6.9.

DECnet Protocol Stack	OSIRM
OSI Application Layer Protocol	APPLICATION
OSI Presentation Layer Protocol	PRESENTATION
OSI Connection Oriented Session Protocol (DNA Session Control)	SESSION
OSI Transport Protocol (LAT, NSC NetSvcControl)	TRANSPORT
OSI Internetwork Protocol (LAT, MOP, DECnet Routing Protocol)	NETWORK
IEEE 802.3, X.25/X.21, FDDI (Ethernet, Customer Interface, DDCMP, HDLC / SDLC)	DATA LINK
IEEE 802.3, X.25/X.21, FDDI (Ethernet, Customer Interface, DDCMP, HDLC / SDLC)	PHYSICAL

Figure 6.9 DECnet Protocol Stack Compared to the OSI Reference Model

DECnet has announced DECnet Phase V, also called advantage networks. Phase V contains enhancements to older versions and now uses the bottom four layers of the OSIRM, as well as being backward compatible to Phase IV. It also bundles TCP/IP functionality with the OSI protocol stack. DECnet V also supports both File Transfer and Access Management (FTAM), and provides a gateway function to the TCP/IP File Transfer Protocol (FTP). It can be seen that DECnet V supports all of the older protocols (in parenthesis) as well as their newer protocol stack which resembles the OSI reference model.

6.3.3 XNS, IPX, and AppleTalk

XNS, or Xerox Network Systems, was originally conceived for integrating Xerox office applications. XNS standards are defined in the Xerox Gray Book of standards in two levels. The layer 0 transmission media manages the interaction of data between the device and the network. The XNS synchronous point-to-point protocol could be used at this layer. The layer 1 Internet Datagram Protocol layer defines the data flow across the network via packet addressing via the Internetwork Datagram Protocol (IDP). The layer 2 transport layer defines end-to-end connectivity through echo protocol, error protocol, and Routing Information Protocol (RIP). The layer 3 control layer manages device resources, data structures, and data formatting for display devices. The layer 4 application layer manages the application data manipulation. Figure 6.10 shows the five layers of XNS in correspondence to the OSI reference model. In normal XNS operation, each packet is routed like a datagram over the best link available.

Internet Packet Exchange (IPX) protocol is a standard defined by Novell. IPX routing also uses echo protocol and error protocol, as well as Routing Information Protocol (RIP), and has evolved as a derivative of XNS. IPX protocol also works in conjunction with Service Advertising Protocol (SAP). The protocol stack is identical to XNS.

AppleTalk Phase II is another routed protocol defined by Apple Computer, Inc., supporting both data link access methods discussed. AppleTalk runs over the 802.2 LLC portion of the LAN Data Link Control layer. Some protocols defined under AppleTalk include AppleTalk Address Resolution Protocol (AARP), Datagram Delivery Protocol (DDP), Routing Table Maintenance Protocol (RTMP), AppleTalk Echo Protocol (AEP), Zone Information Protocol (ZIP), and Name Binding Protocol (NBP).

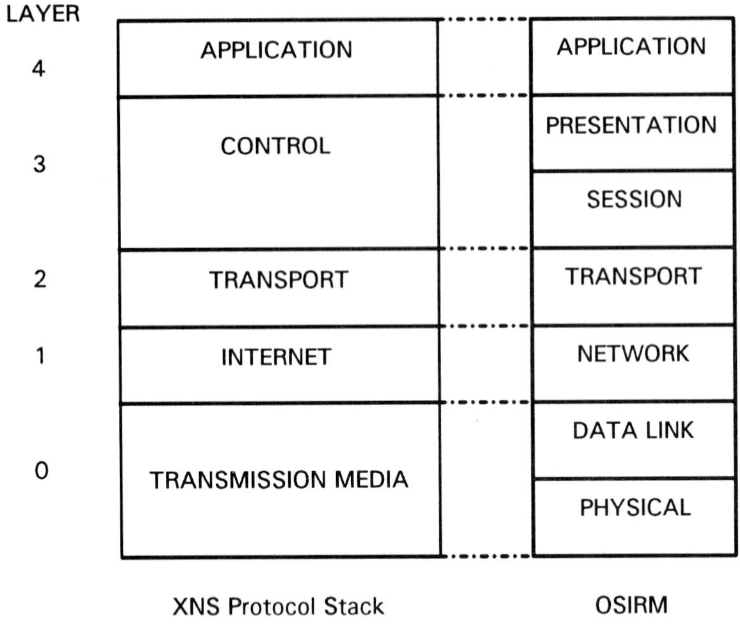

LAYER

4	APPLICATION	APPLICATION
3	CONTROL	PRESENTATION
		SESSION
2	TRANSPORT	TRANSPORT
1	INTERNET	NETWORK
0	TRANSMISSION MEDIA	DATA LINK
		PHYSICAL

XNS Protocol Stack   OSIRM

Figure 6.10 XNS Protocol Stack Compared to the OSI Reference Model

EtherTalk and TokenTalk are applications which run on Apple-Talk for Ethernet and Token Ring networks. They both use the IEEE 802.2 Logical Link Control (LLC) protocol and Subnetwork Access Point (SNAP) protocol. The protocol stack for AppleTalk is similar to the OSI Reference Model, with the exception of the Apple-Talk Presentation layer which combines the functionality of both the OSI presentation and application layers. AppleTalk can also be carried encapsulated within DECnet through the "tunneling technique" through the use of dedicated point-to-point VAXs.

6.3.4 SNA/APPC/NETBIOS/NETBEUI

SNA is still the predominant corporate mainframe architecture, accounting for over 60 percent of worldwide data communications networks. Figure 6.11 shows a typical SNA network depicting the layers of devices, Physical Units (PU), and the other naming conventions of SNA. The entire area controlled by one host is called the "domain".

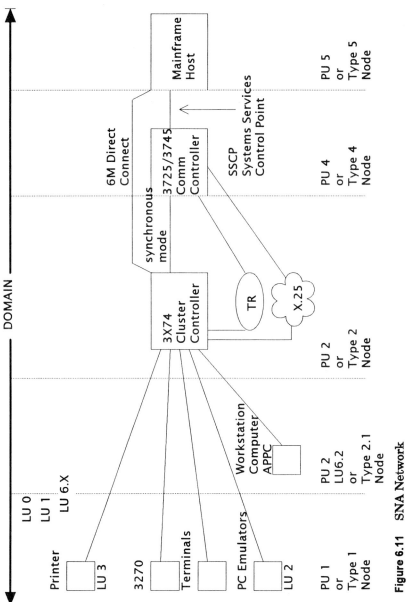

Figure 6.11 SNA Network

In the 1980s, IBM shifted its networking design away from mainframe controlled communications and more towards peer-to-peer communications. The hardware or software at each PU is called the Logical Unit (LU). LUs provide for the access and transfer of information across the domain and to other domains. The Systems Services Control Point (SSCP) defines the single point for domain control. A network device PU, LU, and SSCP is combined to form the network addressable unit (NAU), which forms the network address for a given device. Sessions are then established between NAUs. The SNA feature called SNA Network Interconnection (SNI) provides a front end processor with two 9600 bps circuits which are logically perceived as a single 19,200 bps line. These circuits, combined with multiple paths (one leased line and one dial-up) will provide the necessary diversity between FEPs. See Figure 6.12.

SNA routing is performed through the proprietary configuration of host VTAM and FEP Network Control Program (NCP). One method of *routing* point-to-point 3270 traffic from an IBM 3174 Cluster Controller is through SDLC "tunneling", also called synchronous pass through. Here the router encapsulates the SDLC traffic into an IP packet and routes through the network. Inter-LAN traffic is handled through either cluster controllers or front end processors acting as bridges, converting the LAN packet to an SDLC frame. The other method is through IBM source route bridges.

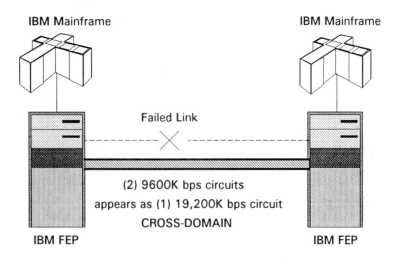

Figure 6.12 SNA Network Interconnection (SNI)

Some devices, such as the SNA Network Access Controller/Group Poll Concentrator (SNAC/GPC), manufactured by Synch Research, Inc., can aggregate multiple cluster controller links into a single SDLC point-to-point or multipoint circuit. This improves performance and reduces polling, saves bandwidth, and the circuit can now be placed over the WAN-routed network (through the router). A similar device, the SNAC/Token Ring Converter (SNAC/TRC) aggregates these cluster controller links onto a single Token Ring network access line. A similar model provides an alternative for older IBM FEPs to connect to Token Ring networks. Figure 6.13 shows an example of both the SNAC/GPC and the SNAC/TR.

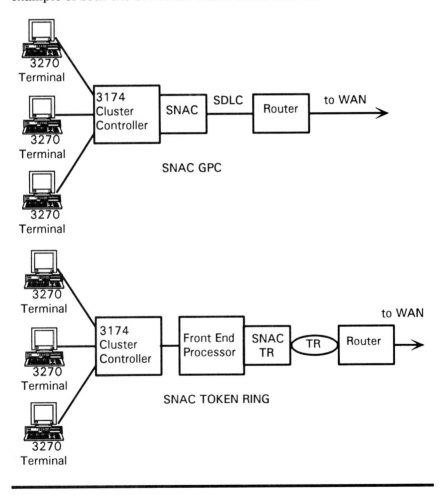

Figure 6.13 SNAC/GPC and SNAC/TR Examples

Advanced Program-to-Program Communication (APPC), also called LU 6.2, provides for peer-to-peer intelligent sessions between peripheral PU2.1 nodes. This constitutes an LU 6.2 device to LU 6.2 device session without involving the host using VTAM and the front-end processor using NCP. However, usage and management statistics can be gathered to be controlled by the host. In this configuration, the IBM host would run VTAM, the 3745 FEP Communications Controller would run NCP, and the workstations would run LU 6.2. APPC supports both dynamic and automatic routing between LU 6.2 devices, but it does not support multiple protocols nor mainframe to terminal traffic. The main limitations to APPC is the huge amount of memory (up to 500K) required to run on a workstation and the lack of software support. LU 6.2 can also act as an application gateway for a PC on a TCP/IP LAN to talk to the mainframe via a direct link. The LU 6.2 and APPC are not hardware specific, and allow many non-IBM architectures to communicate with SNA devices. Also, IBM has its own non-SNA devices which need help to be SNA compatible.

While NETBIOS is more of a user protocol, it is predominantly used as the PC LAN program network and transport protocol in Token Ring implementations. The IBM NETBIOS Extended User Interface (NETBEUI) allows NETBIOS to be transparently passed over the 802.2 LLC protocol and interface, accessing the token ring adapter at the MAC layer. NETBIOS can also be used in conjunction with TCP/IP, but NETBEUI will probably be replaced soon by more efficient LAN networking protocols, or even by the IBM APPC protocol.

As a final note, IBM now offers frame relay on the network side of the 3745 FEPs. This often offers the most efficient networking option at speeds of T1. They also now offer LAN gateway capabilities through TCP/IP for MVS software loads. One example is the 3172 Model 3 (shown in Figure 6.14). The 3172, through the use of an Intel 80486 microprocessor and Interconnect Controller Program (ICP) V3, can now interface to Token Ring, Ethernet, and FDDI LANs.

6.3.5 APPN

Advanced Peer-to-Peer networking (APPN) is an IBM protocol concept or architecture which allows routing LAN traffic independent of a front-end processor or a mainframe, providing users a move away from FEPs and mainframes and toward routers — the true platform for devices using APPN routing. APPN is a dynamic routing and control architecture part of both SNA and Systems Application Architecture (SAA). It can provide a migration path from SNA to SAA. The APPN protocol provides a peer-to-peer architecture designed for distributed processing and LANs, adding much needed independent operational flexibility. APPN allows network devices to share data equally independent of the host, but over fixed communications paths. This is peer-oriented communications and biased toward program-to-program cooperative processing applications. The old method of routing SNA traffic was through IP packet encapsulation. With APPN, users can actually translate SNA traffic and route it in native SNA mode (without the need for encapsulation).

APPN support is provided in IBM 3174 Cluster Controllers, OS/2 Extended Edition, VTAM, and NCP software. APPN can support SDLC, X.25, Token Ring LAN, IBM 3270 and synchronous connections, and will probably support frame relay and FDDI in the future. APPN can run on RS6000 workstations, servers, and IBM LU2 devices (like 3270 terminals).

3172 Interconnect Controller
with Interconnect Controller Program (ICP) Version 3

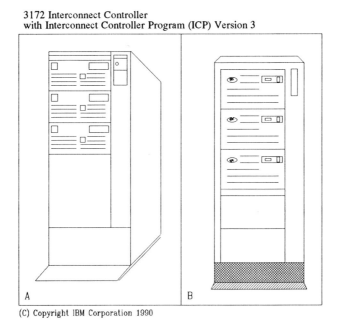

A   B

(C) Copyright IBM Corporation 1990

Figure 6.14 IBM 3172 Controller

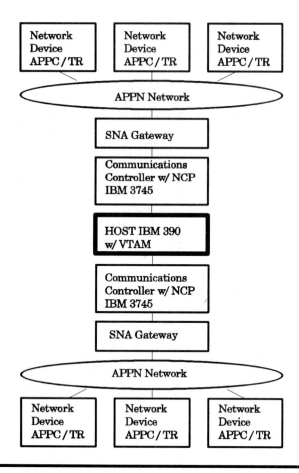

Figure 6.15 IBM Networking with APPN

Figure 6.15 shows an example of networking with the protocols discussed in this section. Two APPN networks with user network devices running APPC are accessing the IBM 3745 communications controllers. These front end processors are running NCP, which in turn is managed through VTAM in the IBM 3090 host computer.

6.3.6 XTP

Express Transfer Protocol (XTP) is a new high-speed network/transport protocol designed to operate over high bandwidth LANs such as FDDI. XTP carries a request for connection within the data packet, reducing overhead and message transfer time. XTP will operate on a special chip set to execute the protocol at speeds over 80M bps at the transport level.

6.4 REVIEW

In the last two chapters we have covered many of the common protocols spanning both telecommunications and data communications. Protocols from all seven layers of the OSI reference model have been discussed. This chapter centered on bridge, router, and higher-level protocols. Bridge and router protocols are the brains of computer protocol routing and bridging. Each router protocol was covered in detail, showing each with their own advantages and disadvantages. The predominant network and transport protocols such as TCP/IP and SNA were explained. Finally, the higher-level protocols were touched upon, but a detailed explanation is outside the scope of this text.

3

Switching
Technologies

This section is designed to provide the reader an in-depth study of the major circuit, packet, frame, and cell switching technologies which have begun to dominate data communications over the last decade and will continue to prevail through the end of the century. Starting with an overview of multiplexing, circuit switching, and packet switching, the reader will be equipped with the basics of the major switching technologies. A survey of packet switching shows why this has been the predominant data communications networking technology of the 1980s. It still remains the primary international data communications technology. Progressing to frame relay, a new paradigm is presented, defining frame relay as an interface, a protocol, and a service. Moving next to cell switching, the operation of the IEEE 802.6 Distributed Queue Dual Bus (DQDB) architecture provides the first service to emerge from this new fiber-based architecture — Switched Multimegabit Data Service (SMDS). Finally, Asynchronous Transfer Mode (ATM) technology and the Synchronous Optical NETwork (SONET) form the capstone of the whole pyramid of switching technologies. New technologies and evolving services form the core subject of this section. Future chapters assume a working knowledge of these basic technologies and services.

7

Multiplexing and Switching Technologies — An Overview

This chapter defines both multiplexer and switching technologies. Multiplexer techniques, like Frequency Division Multiplexing (FDM), Time Division Multiplexing (TDM), and Statistical Time Division Multiplexing (STDM), are covered. Each multiplexing technology will be explored in detail, focusing on the switching or lack thereof in each style of network. Circuit switching is discussed, as well as message switching and packet switching, primarily for background to the current modes of switched networks. Frame relay and cell switching will be discussed in detail in subsequent chapters. This chapter is a reference point and comparison of multiplexing to switching.

7.1 MULTIPLEXING TECHNIQUES

Multiplexing techniques take advantage of digital technology to share a physical medium between multiple users, all of which may require some or all of the bandwidth at any given time. This reduces the access cost for a user to get to the network, whether it be dedicated digital network comprised of private lines, switched services, or a genuine switched network. Some multiplexing techniques assign fixed bandwidth to each user, while others make more efficient use of the transmission facilities which interface to the network.

The terms "access" and "network" or "trunk" side have specific meanings. Figure 7.1 shows a common multiplexer or switch where the user access to the multiplexer is labeled as "access circuit", "access trunk" or "user access". The terms "network access", "trunk side" and "network side" refer to the trunks which interface the multiplexer to the data transport network and carry the aggregated multiplexed user traffic. Two forms of multiplexing, frequency and time division multiplexing, operate at the physical layer of the OSI reference model. Some fast packet multiplexers also perform operations within the data link and network layers.

7.1.1 Frequency Division Multiplexing (FDM)

Frequency Division Multiplexing (FDM) is an analog method of aggregating multiple voice channels into larger circuit groups for high-speed transport. The best example of FDM is the Plain Old Telephone Service (POTS) which is basically a telephone in every home and office, which frequency division multiplexes 12 voice grade, full-duplex channels into a single 48K hertz bandwidth group. These groups are then aggregated into larger groups (called supergroups) for transport across an analog network. In analog data channel communications, frequency division multiplexers are used to aggregate multiple low-speed analog data channels into a single full-duplex voice-grade line, typically with some form of line conditioning to ensure minimum data transmission error rates. FDM can also be used to aggregate low-speed data traffic into an aggregated data channel. Figure 7.2 shows an example of a frequency division multiplexer servicing twelve 2400 bps user data channels and providing a single 56K bps network trunk.

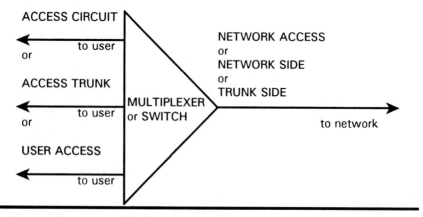

Figure 7.1 Common Multiplexer or Switch

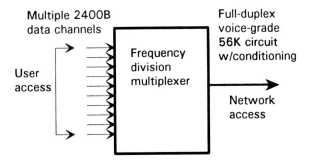

Figure 7.2 Frequency Division Multiplexer

7.1.2 Time Division Multiplexing (TDM)

Time Division Multiplexing (TDM) was originally developed in the public telephone network in the 1950s to eliminate FDM filtering and noise problems. In the early 1980s, TDM networks using smart multiplexers began to appear in some private data networks, forming the primary method to share costly data transmission facilities among users. In the last decade, TDMs have matured to form the basis of many corporate data transport networks.

TDM allows multiple users to share a digital transmission medium by using preallocated time slots. Figure 7.3 shows a standard time division multiplexer with eight low-speed users sharing a single high-speed transmission line to a remote multiplexer. TDM assigns a specific time slot to each low-speed channel — in this case; slots one through eight. These eight synchronous time slots are then aggregated to form a single high-speed synchronous channel. In this example, five users are accessing the network with 9600 bps synchronous data circuits and three are 2400 bps low-speed users.

All time slots comprise channels which are preallocated and occupy a predetermined bit layout of the combined transmitted signal. Time slots are dedicated to a single user, whether data is being transmitted or the user is idle. The same time slots are dedicated to the same user in the same order for every frame transmitted. Thus, time slot number 2 will always be dedicated to the same user, regardless of whether other users are transmitting data or not. Different time slots are dedicated to different channel sources, such as voice channels, data, or video. Multiplexer inputs typically include and can carry simultaneously asynchronous and synchronous data, digitized voice, and even video. After the T1 signal is transmitted across the network, it is received at the destination multiplexer node. Each channel is then demultiplexed at the receiving node. All transmissions through multiplexers are point-to-point.

A single T1 circuit can be configured for 24 to 196 allocated channels. The standard T1 1.544M bps data channel contains 8000 frames per second. Each frame is transmitted every 125 μS for a total of 24 multiplexed voice grade channels. Each frame contains eight bits for each of the 24 channels (8x24 = 192) plus one framing bit (1) for a total of 193 bits per frame. Each channel uses 64,000 bps. The total T1 transmission can be seen in Figure 7.4, along with the frame format.

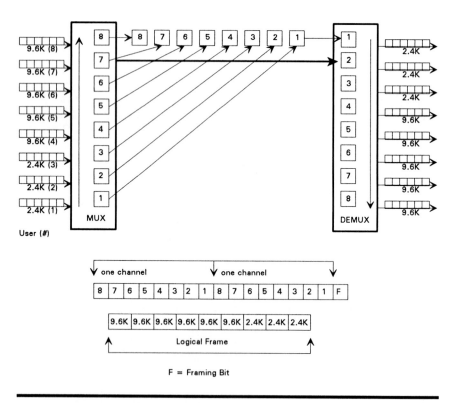

Figure 7.3 Time Division Multiplexer

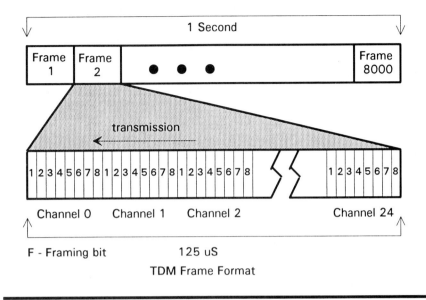

Figure 7.4 Tl Transmission Frame Format

International circuits often use the CCITT E1 Conference on European Post and Telegraph (CEPT) standard which supports a data transmission rate of 2.048M bps. Each E1 data channel contains frames with 30 or 32 multiplexed voice-grade channels. This allows 1.920M bps for voice channels and 128K bps for framing and synchronization. Multiplexers often use the DSX-1 standard for a physical interface with D4/ESF framing, while the CEPT/G.703 physical interface standard is commonly used for the E-1 interface. For a user requiring access speeds from DS0 to T1 access, the break-even point based upon tariff structures is six to seven DS0s or less.

7.1.3 Statistical Multiplexing and Concentrators

Statistical Multiplexing is also called Statistical Time Division Multiplexing (STDM) and operates similar to TDM, except it dynamically assigns time slots only to users who need data transmission at any given time slot. Efficiencies averaging up to 4:1 are gained by smart utilization of available time slots, rather than wasting them on users who are not transmitting data. The net effect is an increase in overall throughput for users since time slots are not "reserved" or dedicated to individual users — thus more dynamic allocation of bandwidth actually allows users more of the available throughput. Figure 7.5 shows a statistical multiplexer which mixes multiple low-speed asynchronous and synchronous user inputs for aggregation into a single 56K bps synchronous bit stream for transmission. The method used to multiplex the various-speed channels varies based upon the input and output protocols. These methods include bit, character-oriented, block-oriented, and message-oriented multiplexing, each requiring buffering and more overhead than basic time division multiplexer. It takes more "intelligence" to handle these refinements but the extra throughput often more than covers the extra cost to implement.

Figure 7.6 represents an excerpt from a statistical multiplexed data stream. It shows how a standard statistical multiplexed frame contains more overhead, control, and addressing information than frame formats for packet transmissions. Also, observe the synchronous data stream, where time slots are allocated to users who need the bandwidth. In this example, user inputs one, three, and seven have transmitted at this point in time, and each used a different amount of bandwidth. If the user input transmission requirements exceed the available circuit transmission speed out of the multiplexer, buffers begin to store the information until space on the transmission circuit becomes available.

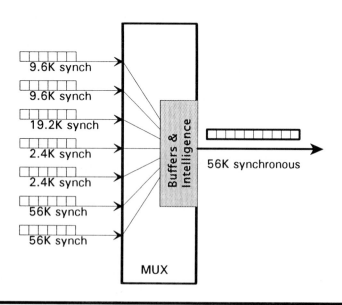

Figure 7.5 Statistical Multiplexer

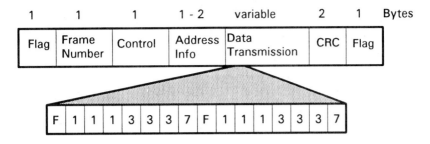

Example: Only 1, 3, and 7 are transmitting at given sample time, so they get as much bandwidth as they want.

Figure 7.6 Statistical Multiplexed Data Stream

One type of block-oriented multiplexer is the concentrator. Concentrators transmit blocks of information for each user as needed, adding an address to each block to identify the user and in most instances provide store and forward capabilities. This mode of transmission is similar to asynchronous block transmission, and multiplexers utilizing this technique are called Asynchronous Time Division Multiplexers (ATDMs). This form of statistical multiplexing is similar to packet switching but operates over a single dedicated circuit between two points, as opposed to packet switching which has multiple paths and destinations. The primary difference between concentration and multiplexing is that concentrators have additional intelligence to understand the contents of the data being passed and can route the information streams based upon the data within them. A study of asynchronous time division multiplexing will be made when we study Asynchronous Transfer Mode (ATM).

Another type of statistical multiplexing is statistical packet multiplexing (SPM). Statistical packet multiplexers combine the packet switching of X.25 with the statistical multiplexing of STDM. SPM operates similar to STDM in that it still cannot effectively transmit delay-sensitive information such as voice and video. And there is still the overhead delay of guaranteed delivery of packets, but efficiencies are gained in dynamic bandwidth allocation and sharing by assigning active bandwidth to the channels which need bandwidth at any given time. Each multiplexer groups the user data into packets passed through the network, multiplexer to multiplexer, similar to packet switching. Once the data is in packet form, certain value-added services such as compression, encryption, and error detection and correction can be performed by each multiplexer.

7.1.4 The Matrix Switch

Matrix switches provide a simplistic form of T1 multiplexing. They offer the capability to switch ports similar to a cross connect. They are composed of a high-speed bus for connection between ports. They are controlled and switched through a central network management center, and can manage the entire network from a single point. This allows for centralized control and diagnostics, as well as quick network restoration in case of failure. The major drawback is the possibility of a matrix switch failure which would bring down the entire network. Matrix switches can handle both DTE and DCE interfaces, and provide conversion from DTE and DCE to a 4-wire interface. Matrix switches can be accessed through gateways or interface units. Most interfaces are low-speed (9600 bps), but matrix switches also provide interfaces for T1 and LAN speeds. Matrix switches usually support in excess of 4096 ports. Satellite chassis form a method of distributing line interfaces. They aggregate many low and high-speed interfaces and transmit them to the matrix switch via copper or fiber. Figure 7.7 shows an example of a matrix switch.

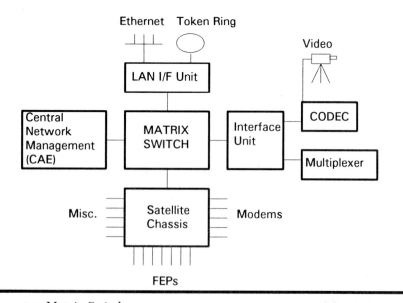

Figure 7.7 Matrix Switch

7.1.5 The Future of Multiplexing

The moving target of technology is providing new methods to route and distribute data. Multiplexers are coming upon hard times this decade. Bandwidth requirements are increasing. But, as they increase, so does the demand for more intelligence to manage that bandwidth in an efficient and user-controllable manner. Users no longer just want to aggregate their traffic into a single point-to-point circuit. Distributed computing and any-to-any communications needs are driving users to more intelligent devices, such as bridges and routers. Multiplexer vendors are trying to keep pace with offerings of interface cards, additional interface options, and cost reductions, but the intelligent switching and routing vendors are stealing much of the data transport market from the multiplexer vendors.

Many multiplexer vendors have begun to offer LAN and switching protocol access cards for their multiplexer products. Some are offering completely separate router hardware that is interoperable with their existing multiplexer line. These interface cards and routers provide an access interface of either single or multiple LAN circuits, frame relay channels, or even fast packet access. Most of these implementations are quick fixes and should be avoided in view of long-term connectivity problems for these types of interfaces and protocols. Much more efficient platforms to develop LAN and frame relay networks are proving themselves instead of multiplexers. With LAN/WAN interface speeds increasing in leaps and bounds (FDDI 100M bps and SMDS 155M bps), and router vendors selling cost effective CPE devices to perform frame relay, it doesn't make sense to purchase T1 and T3 multiplexers whose interfaces quickly becoming obsolete.

Many multiplexer vendors are forming alliances with LAN/WAN vendors to offer LAN/WAN and multiplex interconnectivity. Some examples include: Network Equipment Technologies and cisco Systems, Newbridge Networks and Wellfleet Communications, and Timeplex and RAD Network Devices. They offer bridge/routing capabilities either within or in conjunction with multiplexer product lines. But the primary driver for integrated LAN/WAN and multiplex equipment is WAN video and imaging at T3 speeds and above. The real test for the multiplex vendors is to offer this hardware at cost-effective prices compared to router and hub products.

While the multiplex vendors continue to modify existing product lines, or build alliances to further their multiplexer product base, there remain limitations to multiplexing as a technology for tomorrow's data transport. Multiplexers still have limitations compared to their routing and switching competition: slow bus speed architecture compared with router vendors, fixed circuit connections as opposed to dynamic switched connectivity, connection-oriented as opposed to connectionless services. A basic fact is that T1 multiplexers are not well suited for LAN/WAN traffic.

The user must clearly define short- and long-term requirements for varying types of traffic requiring transport before deciding upon a multiplexer solution. Interface and protocol options must be determined. This decision-making process will be covered later.

Two major factors exert pressure on multiplex vendors to modify their traditional support of low-speed asynchronous and synchronous traffic. The importance of public network interoperability in network standards and signaling, as well as the Digital Cross Connects (DXCs) discussed in the next section. Another important factor influencing multiplexer survival is the carrier pricing of switched services such as frame relay and SMDS. If these services are priced low enough, the TDM multiplexers will be forced into the role of local site concentrator and leave the wide area networking to routers and bridges. One example is the requirement of "network side" frame relay interfaces on multiplexers.

7.2 TYPES OF MULTIPLEXERS

Multiplexing is the process of aggregating multiple low-speed channels into a single high-data-rate channel. There are four major types of multiplexers for use in data network designs: the access multiplexer (or channel bank), network multiplexer, drop and insert multiplexer, and aggregator multiplexer. Capabilities and benefits of each type will be discussed. These multiplexers often also contain the capability for demultiplexing. Demultiplexing is the segmenting of a single high-rate channel into multiple low-speed channels.

Most forms of multiplexing are protocol-transparent and protocol-independent. The user interface for multiplexing is at the physical layer, and all layer two through layer seven protocols are transparent to the network. Figure 7.8 shows the protocol stack operation of a multiplexer network. This provides for fast end-to-end transmission times since no need to interpret the data during transport exists. All channels on a multiplexer are either configured to take up certain time slots in the TDM channel or dynamically allocated to the available bandwidth. Either way, the bandwidth between the source and destination node is a fixed transmission speed. Any changes to the bandwidth allocation must be performed to both the source and destination nodes. Frame relay eliminates these restrictions.

7.2.1 Access or Channel Bank Multiplexers

Access, or channel bank multiplexers, provide the first level of user access to the multiplexer network. These devices typically reside on the user or customer premises and are referred to as Customer Premises Equipment (CPE) devices. There, CPE devices are locally controlled and configured. Intelligent T1 access multiplexers are often used to combine voice and data signals into one high-speed data channel. These multiplexers can provide network access for a variety of user asynchronous and synchronous, low- and high-speed inputs including:

- terminal
- data telephone
- compressed voice and video
- front end processor (FEP)
- personal computer
- host computer
- remote peripheral
- analog and digital Private Branch Exchange (PBX)
- low-speed video
- imaging applications
- transaction-oriented device
- local area network (LAN)

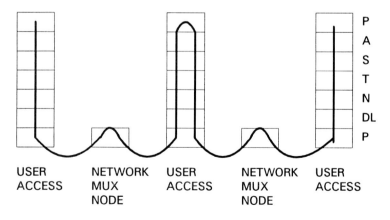

					P
					A
					S
					T
					N
					DL
					P

| USER ACCESS | NETWORK MUX NODE | USER ACCESS | NETWORK MUX NODE | USER ACCESS |

Figure 7.8 Protocol Stack Operation of the Multiplexer via the OSIRM

Access multiplexers usually provide one or more T1 trunks to the next class of larger multiplexers, the backbone multiplexer. Access multiplexers are less expensive and less growth oriented. They are best suited for static networks where there is little change, rather than dynamic networks where changes in network access are common. Access multiplexers can handle many of the circuit-switched interfaces (e.g., frame relay, X.25, SDLC/HDLC). The interface speeds of access multiplexers will vary and include:

- DS0
- Fractional T1
- T1
- E1

Two versions of the access multiplexer warrant further discussion. One version of the access multiplexer is the Fractional T1 Multiplexer. This multiplexer provides the capability for access speeds in fractions of a full T1. Fractional T1 is a private line service which provides the capacity of from one to twenty-three DS0 individual circuits in 56K bps or 64K bps increments. Some commonly used speeds include 56K, 64K, 128K, 256K, 384K, 512K, and 768K bps. Users which implement clear-channel B8ZS line coding can use contiguous bandwidth. Thus, the user accessing the fractional T1 multiplexer only incurs the cost of the bandwidth used. This was one of the first steps to bandwidth-on-demand, because the user only pays for a "fraction" of the T1, not the entire T1. Figure 7.9 shows an example of an access Fractional T1 multiplexer, with multiple fractions of T1 speeds for user access to a shared T1 1.544M bps network access circuit.

Another version of the access multiplexer is the Subrate Data Multiplexer (SRDM) shown in Figure 7.10. SRDMs provide multiplexing at the sub-DS0 level, aggregating multiple low-speed channels into a single DS0 channel. SRDM access speeds include 2400, 4800, 9600, and 19,200 bps. Again, the user accessing the SRDM multiplexer only incurs the cost of the bandwidth used. When we discuss Frame Relay access, a form of SRDM is used called the Frame Relay Access Device (FRAD). Both Fractional T1 and SRDM optimize the use of access trunks for multiple low-speed users who, without these devices, would have to purchase a full DS0 or DS1 even though using only a portion of its full bandwidth.

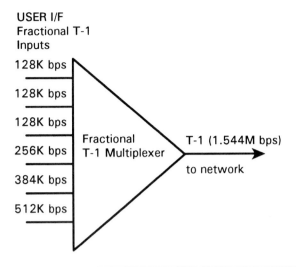

Figure 7.9 Fractional T1 Multiplexer

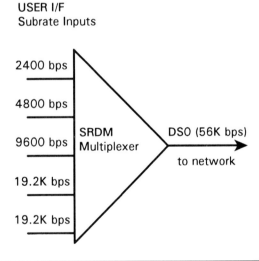

Figure 7.10 Subrate Data Multiplexer (SRDM)

7.2.2 Network Multiplexers

Network multiplexers accept the input data rates of access multiplexers and much higher, typically supporting T1 on the access side and T3 or higher on the network side. Their trunk capacity is also much larger than access multiplexers, ranging from a dozen user trunks to hundreds, and supporting many more network-side trunks at higher speeds. Network multiplexers provide the additional functionality of network management systems, local and remote configuration capability, and contain intelligent functions not found in less expensive access multiplexers. Private and public data transport network backbones are built using network multiplexers. Network multiplexers also better position the backbone portion of the network for port (increased access and backbone circuit capacity) growth and expansion.

Dynamic reroute capability is an important feature in network multiplexers. The multiplexer dynamic reroute capability can either be performed by predefined "routing tables" residing in software at each node or downloaded from the network management center, or through "routing algorithms" which update the network dynamically during changing network conditions. Algorithm control can also reside at either the individual nodes or at the network management center. Either way, it is important to fully understand these capabilities before deciding upon a multiplexer vendor.

Figure 7.11 shows an example of a network multiplexer and the many varied user input possibilities. Figure 7.12 shows a figure depicting Network Equipment Technologies' IDNX Model 90 network multiplexer. Figure 7.13 shows a figure depicting General Data-Comm's Megamux Transport Management System (TMS).

7.2.3 Aggregator Multiplexers

Aggregator multiplexers combine multiple T1 channels into higher-bandwidth pipes for transmission. These multiplexers are also sometimes called hubs (not to be confused with LAN hubs). Aggregator multiplexers are labeled based upon their aggregation and de-aggregation rates. These labels are:

M12 Multiplexers — aggregates four T1s to the rate of T2 (or vice versa demultiplex).

M23 Multiplexers — aggregates seven T2s to the rate of T3 (or vice versa demultiplex).

M13 Multiplexers — aggregates twenty-eight T1s to the rate of T3 (or demultiplex). This is shown in Figure 7.14.

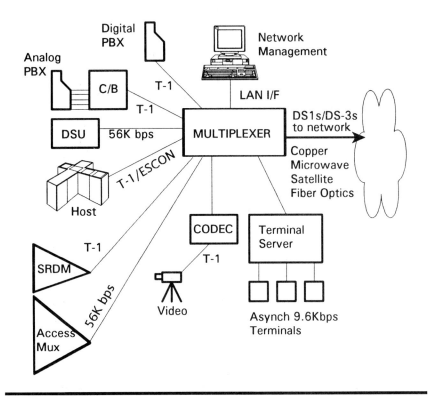

Figure 7.11 Network Multiplexer Example

M22 and M44 Multiplexers — provides configuration management and rerouting capability of 22 and 44 channels, respectively.

MX3 Multiplexers — aggregates different combinations of T1s and T2s to the rate of T3 (or vice versa demultiplex).

It is also important to note that synchronization of the aggregate circuits within many of these multiplexers (even as the individual DS1s within the DS3 M13 multiplexer) is not supported by many vendors. This will have a major impact when services depending heavily on channel synchronization (such as Synchronous Optical Network — SONET) are deployed and the multiplexer needs to be replaced.

Some aggregator multiplexers also offer the capability for switched N x 56/64K bps services. This provides the user multiple dial-up DS0 interfaces into a multiplexer, with the user impression of "bandwidth-on-demand" in 56/64K bps units. Carriers offer this service through the variety of multiplexer devices mentioned above.

This switched DS0 service is the most cost effective method of bandwidth on demand using a multiplexer, and is quite powerful when used by a statistical multiplexer. Switched 56 will be covered in detail when circuit switching is discussed. Aggregator multiplexers are slowly being replaced by the Drop and Insert multiplexers and Digital Access Cross Connects (DXC). Both types of hardware are covered in this chapter.

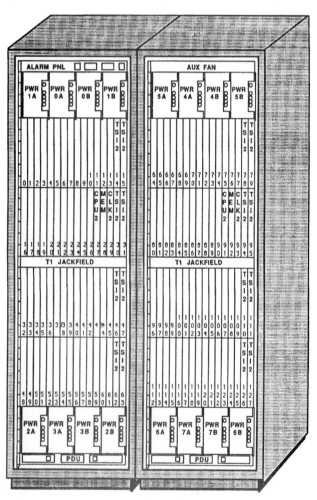

Figure 7.12 Network Equipment Technologies' IDNX Model 90 Network Multiplexer

Figure 7.13 General DataComm's Megamux Transport Management System (TMS) Network Multiplexer

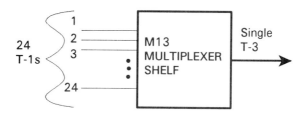

Figure 7.14 M13 Multiplexer

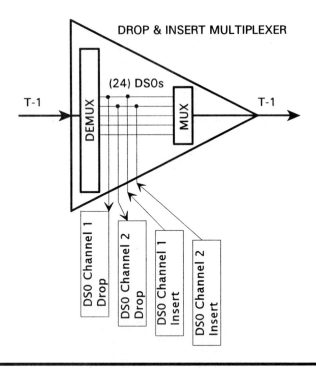

Figure 7.15 Drop and Insert Multiplexer Operation

7.2.4 Drop and Insert Multiplexers

Drop and insert multiplexers are special-purpose multiplexers designed to drop and insert low-speed channels in and out of a high-speed multiplexed channel like a T1. Channel speeds dropped and inserted are typically 56K bps or 64K bps. Each DS0 is demultiplexed and remultiplexed for transmission.

Figure 7.15 shows a drop and insert multiplexer operation, where two 56K bps channels (channels 1 and 3) are dropped out and replaced by two new user 56K bps channels before the total 24 56K bps channels are multiplexed and retransmitted. Some newer drop and insert multiplexers are adding circuit switching and network performance monitoring to their list of capabilities.

7.2.5 Selection of a Multiplexer

Since many options are available in multiplexers, each requirement must be analyzed to determine which type is the best fit for current and future applications. Some of the major decision criteria for all types of multiplexers are as follows:

- level of intelligence required
- speed of access (typically 9600 bps to T1)
- speed of egress (typically 56K bps to T3, and soon emerging SONET)
- capability to upgrade to LAN/WAN, T3, SONET speeds
- number of ports/cards per node
- maximum number of nodes
- public or private network access capability
- CPE compatibility
- virtual network partitioning capabilities
- protocol and interfaces supported (such as frame relay or LAN)
- nonproprietary architecture
- voice quantization schemes supported (PCM 64K bps, ADPCM 32K bps)
- network media interfaces (copper, fiber)
- physical interface standards supported (RS-232, V.35, ISDN, CCITT G.703)
- price versus functionality ($$$ versus feature function)
- warranty available
- topologies supported (point-to-point, drop and insert)
- types of framing (D4, ESF, B8ZS)
- network management and network diagnostic capabilities (terminal, graphical interface, proprietary or industry standard)
- clocking restoration
- fault isolation and reroute capabilities
- call reroute capabilities
- event reporting
- circuit bandwidth available after multiplexer proprietary overhead (1.344M bps of user data out of 1.544M bps)
- capability to select the primary alternate route
- degree which the multiplexer offers dynamic bandwidth allocation of all data speeds.

7.3 CIRCUIT SWITCHING

Circuit switching originated in the public telephone network. Let's look at the first telephone where each person had a dedicated circuit to every other person. This type of connectivity makes sense if you talk to very few people and very few people talk to you. Now let's move forward to modern day, where the typical person makes calls to hundreds of different destinations for friends and family, business, pleasure, etc. It is unrealistic to think that each of these call origination and destination points has its own dedicated circuit, since it would be much too expensive to provide a dedicated circuit to every other person to call. Yet a person picks up the phone in one city and calls a person in another city. When this call goes through, and both ends begin communicating, they are doing so over a temporary virtual circuit. That circuit is dedicated to the two people until they terminate the call. If they hang up and call back, another circuit is established in the same manner, but not necessarily over the exact same path as before. In this manner, common network resources (circuits) are shared between many users.

Figure 7.16 shows a simplified comparison of communications between eight users (e.g., LAN, MAN, PBX, Host) with (a) showing dedicated access circuits between each user and (b) showing circuit-switched access of a common pool of circuits between each user. In part (b), all users have access into the Local Exchange Carrier (LEC) via a dedicated tail circuit, but are circuit-switched as needed within the transport network. Data or voice is transmitted only via the physical layer. The data is not processed, but instead just passed across the network regardless of the content. The example shows User A talking to User H, and User D talking with User E.

Today, computers need to "talk" to each other in the same manner. Data calls which use circuit switching operate in the same manner, but the transmission is not restricted to voice. A point-to-point call is established as a virtual circuit and remains up until all data has been transmitted. Then the circuit is torn down. Each successive call is reestablished and torn down in the same manner. The bandwidth allocated to the call is dedicated until all data is transmitted and received. Circuit switching is the ideal technology for traffic which needs constant bandwidth and low call establishment and termination times.

When computers communicate over circuit-switched facilities, the originator first sets up the call with the destination. This is referred to as call establishment. Destination addressing information is transmitted from the user to the switch (in this example the LEC). Call message transfer is then executed after the circuit is established end-to-end and the originator receives a signal which confirms that the circuit is established. After the entire data message is transmitted, the call is terminated by the calling party. Figure 7.17 shows this process between two users, A and B. Delay is a major consideration with circuit switch design, but not shown in this figure. Reduction of delay to milliseconds can mean large revenue and service gains to the telephone network or data network providers, especially when user applications are delay-sensitive.

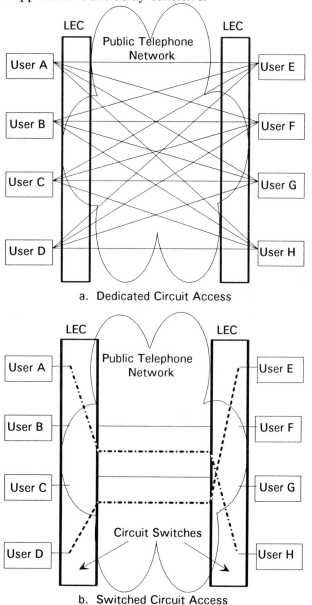

a. Dedicated Circuit Access

b. Switched Circuit Access

Figure 7.16 Dedicated versus Circuit Switched Access

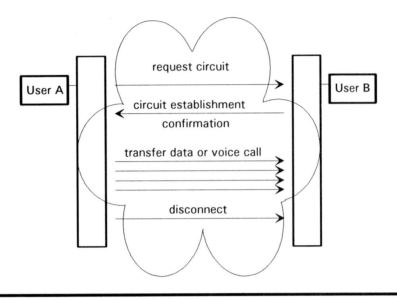

Figure 7.17 Call Establishment and Disconnect

Circuit switching still remains the most common type of Public Switched Data Service. Data circuit switching was much slower to emerge on the market than voice circuit switching, primarily because of the additional intelligence and digitization needed in data circuit switches to process high call-processing rates characteristic of data communications. Also, data communications networks need to provision much more network bandwidth resources to prevent blocking. Blocking occurs when all the available bandwidth pool between two points is being used and some calls must wait until circuit space (bandwidth) is available. Data communications equipment does not take well to busy signals or blocking, especially when the network is transporting information worth thousands of dollars per second!

Circuit switching has been used historically as backup for private line services, as we will discuss under Switched 56 services. It remains the most cost viable option for private network users, with most switched 56K bps data services selling at less than 25 cents per minute to as low as five cents per minute depending on time of day, usage, speed, error-free rate, distance to carrier, and other reliability factors. This price is getting very close to that of voice services. This pricing makes it a very cost-effective option to leased-line services. The higher the bandwidth, the more warranted switched data services should become. In fact, many IXCs are promoting services based on CPE-device support and lower-priced private line costs. The circuit switching service offered above is based on TDM techniques and provides protocol independent transport.

Users access a circuit-switched service by providing a dedicated channel into a circuit switch, which then communicates with other circuit switches within the switched network. Since voice networks were the first users of circuit switching, it stands to reason that voice traffic is not only compatible but also ideal for circuit-switched networks. The data communications user, however, needs three types of circuits for one call: the data circuit, a signaling facility, and a call management circuit. The calling procedures are the same for both voice and data calls. Circuit switching operates only at the first layer of the OSI reference model, providing intelligent methods of access to the physical medium (see Figure 7.18).

High-speed circuit switching of subrate T1, T1, and DS3 speeds is being offered by many of BOCs and IXCs. While some applications use high-speed circuit switching as an ideal solution, such as bulk data transport and applications requiring all the available bandwidth at predefined time periods, circuit switching seems to be the short term solution leading to more advanced packet- and cell-switching techniques. But until these services are available in mass coverage, circuit switching provides cost reductions and improves the quality of service in contrast to private lines, while improving the coverage of service.

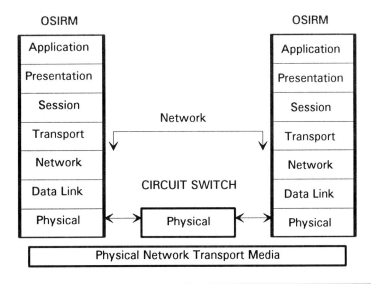

Figure 7.18 Circuit Switching via the OSI Reference Model

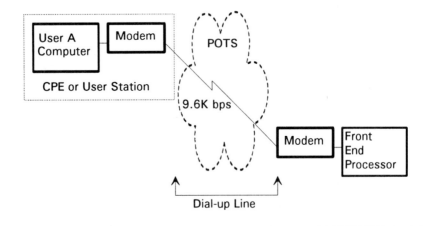

Figure 7.19 Remote Front-End Processor (FEP) Configuration

7.3.1 Dial-Up Lines

Dial-up lines represent low-speed dedicated point-to-point circuits. This type of circuit uses the Plain Old Telephone Service (POTS) for access, with access speeds typically at 9600 bps and below. Figure 7.19 shows a user who accesses a remote front-end processor via a 9600-bps dial-up line. A dial-up modem is used to access the telephone network. High-speed circuit switching can be compared to dial-up lines, where bandwidth is "dialed-up" as needed.

7.3.2 Switched N x 56/64K bps

This service was discussed earlier in multiplexing. Switched 56 is a service offered in both the private and public networking environments. Switched 56 Channel Service Unit/Data Service Unit (CSU/DSU) devices use dial-up lines to access multiplexers which offer the capability to the user of multiple dial-up DS0 interfaces into a multiplexer. This provides the user with the impression of bandwidth on demand in 56/64K bps units.

Figure 7.20 shows an example of switched Nx56/64K bps service. Switched Nx56/64K bps service provides the most cost-effective method of bandwidth on demand using a multiplexer, and is quite powerful and efficient when a statistical multiplexer offers the network interface. The interface for switched services is from the CPE directly to the Interexchange Carrier (IXC) Point of Presence (POP). This access can either be dedicated or switched. Figure 7.21 shows these concepts. Each type of access has its own merits and draw-backs, such as installation and usage cost, CPE costs, and reliability and availability of access trade-offs.

Many users implement switched 56/64K bps services as backup for dedicated circuits and to transport nonmission critical data traffic. Others use it for infrequent high data rate, constant-bandwidth data transfers. The typical traffic is short-term variable bandwidth data transfers, such as batch file transfer and LAN traffic.

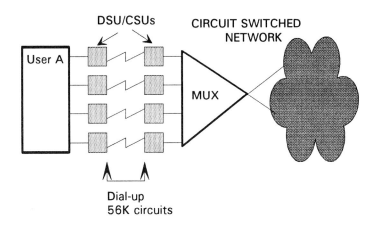

Figure 7.20 Switched Nx56/Nx64K bps Service

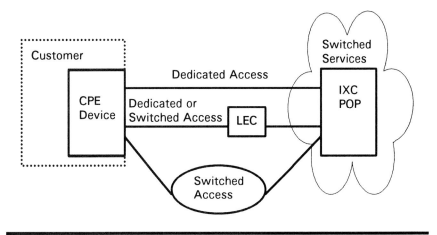

Figure 7.21 Switched Services Interfaces

Most switched services are at the 56/64K bps level, but new services offering higher-bandwidth multiples of 56K bps are now becoming popular such as switched 386K bps, T1, and T3. Some carriers offer noncontiguous and contiguous fractional T1 bandwidth-switched services. The noncontiguous services are offered in increments of 56K/64K and often use inverse multiplexing to divide an aggregate T1 or T3 into multiple fractional T1 or T3 network access circuits. Switched ISDN 64K services are also available. Videoconferencing is one example, where multiple 56K bps circuits are combined to form a single high-speed videoconference channel at speeds such as 112K bps or 224K bps. Imaging and CAD/CAM file transfer are good examples of high bandwidth switched traffic. Some examples of switched T3 service traffic includes video, imaging, and data center disaster recovery. MCI was the first carrier to offer all of these switched services, but all the major IXCs should be offering them by the time of publication.

7.3.3 Narrowband and Broadband ISDN (N-ISDN/B-ISDN)

Narrowband Integrated Services Digital Network (N-ISDN) has been offered as an alternative to the public telephone network Digital Data Service (DDS). N-ISDN offers additional services for data, including switched services and dual channels for simultaneous voice and data. N-ISDN is offered in two services: Basic Rate ISDN (BRI), or 2B+D, which offers the user two 64K bps bearer channels for data transfer and one 16K bps control and signaling channel, and Primary Rate ISDN (PRI), which offers the user a full T1 minus one 64K bps channel for control and signaling. Broadband ISDN (B-ISDN), which resembles packet switching more than circuit switching, will be discussed later.

7.3.4 Digital Access Cross Connects (DXC)

Digital access cross connects are central office devices capable of grooming multiple DS0 channels within a T1, and in some cases multiple DS1 channels within a T3. This provides a patch panel effect, where individual DS0s and DS1s can be mixed and matched between higher bandwidth aggregates. Digital access cross connects also perform reconfiguration, restoral and disaster recovery, testing, and provide monitoring capabilities. Some even provide the capability to switch and reroute circuits.

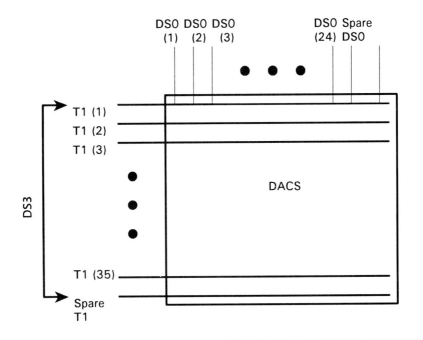

Figure 7.22 DXC 1/0

Figure 7.22 shows an example of a DXC 1/0, where 24 DS0 channels are groomed within one of the 35 T1s which terminate at the DXC. DXC also have the capability for network recovery and disaster recovery.

The major types of DXCs include:

DXC1/0 — support subrate DS0s reroute capability within a DS1.

DXC3/1 — support subrate DS1s reroute capability within a DS3.

DXC3/3 — support multiple DS3s reroute capability.

Figure 7.23 shows a network where DXC 3/1 and DXC 3/3s are deployed.

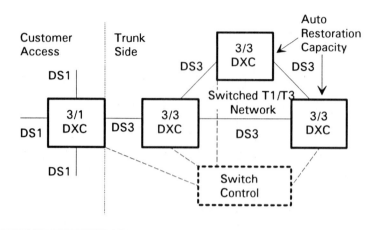

Figure 7.23 DXC 3/1 & DXC 3/3 Network Example

7.4 PACKET-SWITCHING TECHNOLOGIES

This section has been added to provide an overview of packet switching technology before delving into the detail in subsequent chapters. Packet switching is a broad term which began with X.25 Services. It is now used in one form or another to represent frame relay, fast packet, SMDS, and B-ISDN. We will attempt to clear up the confusion with associating these diverse technologies with packet switching.

Packet switching is much different from circuit switching. It allows multiple users to share the data network facilities and bandwidth, rather than providing specific amounts of dedicated bandwidth to each user. The traffic passed by packet switched networks is "bursty" in nature, and therefore can be aggregated statistically to maximize the use of on-demand bandwidth resources. While there is much more overhead associated with packet switching compared to circuit switching, this overhead guarantees error free delivery by the use of addressed packets which transit the network. Due to the connectionless characteristic of packet switching in contradistinction to circuit switching, the intelligence of the network nodes will route packets around failed links, where in circuit switching the entire circuit would need to be switched leading to service interruption. Much higher speeds can be achieved with circuit switching than with conventional X.25 packet switching (T1 and T3 as opposed to 56K bps which limits X.25), but new advances in packet switching and LAN technologies have led to speeds that match circuit switching.

7.4.1 X.25

X.25 was the first protocol issued defining packet switching. Access speeds range up to 56K bps. Trunks between network nodes are limited to 56K bps/64K bps (with the capability for fractional T1 speeds under proprietary implementations). X.25 contains the error detection and correction and flow control needed for the older analog transport networks of the 1980s. But much of this overhead and processor intensive operations are not needed in today's fiber optic networks. Packet switching is solely a connectionless service, and is good for time-insensitive data transmission but poor for connection-oriented and time sensitive voice and video. Packet switches pass data through the network, node to node, employing a queuing scheme for buffering and transmitting data. Data is received and passed if bandwidth is available. If it is not, data is stored in the queue until bandwidth is available (and to the extent of the memory buffer). The end nodes are responsible for error detection and correction and initiate error recovery. The newer fast packet services do not perform this queuing function, rather they drop the extra traffic which cannot be transmitted over a congested network. Congestion is one parameter which has not yet been solved in these new services.

X.25 allows numerous virtual circuits on the same physical path, and can transport packet sizes up to 4096K bytes. Both Permanent Virtual Circuits (PVCs) and Switched Virtual Circuits (SVCs) are supported in X.25, and the addressing scheme allows any user to send or receive data from any other user. Traffic can also be prioritized.

7.4.2 Frame Relay

While the primary use of frame relay is as an interface to a service, frame relay also defines an interface, a protocol, and a service. Frame relay is a connection-oriented service employing PVCs and SVCs similar to packet switching. Frames can vary in size and bandwidth is allocated on demand. Multiple sessions (up to 1000 PVCs) can take place over a single physical circuit, and these access circuits can range from DS0, through fractional T1 speeds, and up to a T1 (minus the overhead). Frame relay is only a transport service, and does not employ the packet processing of X.25, which guarantees end-to-end error and flow control. Instead, frame relay relies on the user to implement higher-level protocols in the upper layer of the OSIRM for flow control and error correction. Frame relay must transit reliable fiber optic transmission media with low bit error ratios, because errored or excess (blocked) data is simply discarded since error detection and correction is not built into frame relay. Chapters 8 and 9 provide a detailed explanation of frame relay. Later, frame relay design using frame relay circuits and routers is considered.

7.4.3 Fast Packet

Fast packet is a term which stands for a data transmission technique. It is also a term trademarked by Stratacom. It is not a defined standard, protocol, or service. Fast packet is a *backbone* technology (as opposed to frame relay *access* technology) which combines attributes of both circuit switching and packet switching. Fast packet can accommodate both (time) delay-sensitive traffic (e.g., voice and video) as well as data traffic not affected by variable delay. It also offers low network delay and high network resource efficiency. Fast packet resembles a circuit switch for constant-bandwidth traffic such as voice and video (isosynchronous) and a packet switch for "bursty" data traffic such as local and wide area network traffic, dynamically increasing the bandwidth for high bandwidth requirements and decreasing it for low bandwidth requirements.

Fast packet switching speeds up packet transfer and delivery through reduction of overhead since error detection and correction is not done in the intermediate nodes. The packets or cells passing through the network use the data link layer instead of the network layer. These packet or cell sizes can either be *fixed* (cell switching and ATM) or *variable* (frame relay). Fast packet can also provide protocol transparency. This causes minimal node delay, while providing addressing and routing capabilities. Fast packet technologies use advanced fiber optic transport media, such as T3 and SONET.

Fast packet multiplexing (FPM) is a general term for providing the capabilities of fast packet switching through multiplexing various types of traffic onto the transmission medium. FPM is characterized by a combination of both TDM and SPM, where packets of fixed or variable size, of fixed or variable delay-sensitive traffic, are statistically multiplexed over a network high-bandwidth circuit. Packets pass *through* network devices rather than into and out of them, thus providing minimal nodal delay. FPM can also dynamically reallocate bandwidth on any packet, regardless of whether it is within the multiplexer or partially in transit. This lends to variable or fixed packet or cell sizes based upon the transmission medium being used.

The most common form of FPM is the new Asynchronous Transfer Mode (ATM) technology, which uses a fixed cell switching form of fast packet multiplexing. We will discuss this technology in great detail in chapter 13.

The IEEE 802.6 Distributed Queue Dual Bus (DQDB) architecture provides the platform for a form of fast packet switching called Switched Multimegabit Data Service (SMDS). SMDS uses the IEEE 802.6 MAN technology discussed in chapters 11 and 12.

Broadband ISDN (B-ISDN) is a service which uses the ATM technology for transport. B-ISDN is defined by the CCITT as services which exceed the primary rate of N-ISDN. There are two types of B-ISDN: interactive for two-way transmission (similar to full duplex), and distribution for one-way traffic to one or multiple subscribers (similar to multidrop). Both will be discussed in Chapter 12.

7.4.4 Asynchronous Transfer Mode (ATM)

Asynchronous transfer mode (ATM) is another form of fast packet switching. ATM packetizes voice, data, and video and then "statistically multiplexes" the packets onto the same high-speed data channel. ATM provides two types of connection: virtual channel which provides logical packet connection between two users, and virtual path which defines a source to destination route for users.

ATM transmission services proposed include:

✤ full duplex 155.52 M bps
✤ asymmetrical transmission from subscriber to network at 155.52M bps in one direction and 622.08 M bps in the other
✤ full duplex 622.08M bps service

7.4.5 Integrated Circuit / Packet Switches

Switches are available which combine both circuit and packet switching capabilities within the same unit. These devices combine X.25 packet switching with T1 TDM techniques, as well as some DXC and channel bank functions. Thus, a large variety of traffic — from bursty LAN to dedicated T1 data streams — can be handled in the most efficient manner required, with the added capabilities offered by a DXC. X.25 and LAN routed traffic can be *packet*-switched, while LAN bridged, SNA, voice and video, and other delay-sensitive traffic can be *circuit*-switched. The primary advantage is seen when using this switch as a hub concentrator for many types of traffic through dynamic allocation of available bandwidth. It allows an integrated platform for both data transport and network management. Thus, the network designer can design one network, and not multiple networks for both circuit and packet switching, and consolidate equipment and operating costs of both. But one must also ask — can a Volkswagen beat a drag racer in a race for speed, or can a drag racer enter a miles per gallon race? The point is that this solution is not without the drawbacks of both technologies as we have discussed previously. One type of integrated circuit/packet switch is the Netrix #1-ISS integrated switching system. This transmission technique is similar to fast packet multiplexing, and can take advantage of the many IXC public network offerings available today.

7.5   REVIEW

The differences between multiplexing and switching are important. There are many methods of multiplexing, using FDM to TDM, STDM to FPM — each has multiple hardware devices to satisfy the various user needs for traffic aggregation. Looking at the future of multiplexing, switched services and LAN/MAN/WAN technologies are stealing away part of the multiplexer market. The list of features and functions of various multiplexer hardware helps decide what to use where for what need. Circuit switching services and technologies are quite valuable to constant data rate traffic, as well as valuable for dedicated circuit back-up. Packet switching technologies span from the older X.25 packet switching to newer techniques in fast packet and cell switching such as frame relay, DQDB (SMDS), and ATM. After understanding the advantages and disadvantages of each type of technology, it is apparent that many network designs require a "hybrid" of both dedicated and switched technologies. We will now discuss several of the emerging switched technologies for the 1990s.

8

Packet Switching

Since circuit switching has been covered, we will now turn our attention to the technology forming the basis of many advanced technologies. Packet-switched networks have been in existence for over 25 years. Packet switching provides the network environment needed to handle terminal to host bursty data traffic. Packet switching also serves many other user communities, especially in Europe, where it constitutes the majority of public and private data transport services. The CCITT X.25 packet-switching standard, along with a host of other "X dot" standards, was developed to provide a reliable system of data transport for computer communications over analog-grade transport mediums. Transport mediums have drastically improved since X.25 was developed, and now digital and fiber replaces much of the older analog communications facilities (this is less true in countries outside of the United States). Speeds have increased beyond 56K/64K bps, the maximum transmit rate of packet switches. Much of the overhead once associated with ensuring reliable data transfer over these poor mediums is not needed now. As a result, packet switching has taken a backseat to newer technologies which utilize less overhead and rely on clean transmission facilities. Yet we must understand the principles of packet switching, since packet-switching concepts form the basis of the newer technologies, and many international networks still use and will continue to use packet switching long after the world becomes fully digital and fiber predominates.

8.1   PACKET SWITCHING DEFINED

To understand the packet switching of today, we must go back to the original premise on which packet-switching technology was built. Packet switching was originally designed as a secure method for the transport of voice traffic.   Since then it has evolved into a data transport platform for statistical multiplexing of low-speed user traffic across large distances.   Elaborate transmission schemes are built into the protocols to accommodate for poor voice-grade transmission facilities.   With the advent of fiber optics, this superfluous overhead is no longer needed and is a hindrance rather than an aid.

8.1.1 Packet Switch Beginnings

Packet switching was first invented by Paul Baran and his research team for the RAND Corporation in the early 1960s and published in 1964 as a secure method of transmitting military *voice* communications. This was actually a project to enable the United States military communications system to survive after a nuclear attack.   By segmenting an entire message into many pieces of data and wrapping routing and protocol information around these pieces of data, "packets" were created. The routing and control information was to ensure the correct delivery and reassembly of the original message at the end-user destination.   These packets had a fixed maximum size assigned, typically 128 or 256 bytes long.   Through the use of multiple packets, the entire message could be transmitted over multiple paths and diverse facilities and reassembled in its original order at the destination.   In this manner of transmission, voice wire-tapping was virtually impossible because only a portion of the entire transmission could be tapped, and even that portion would be incomplete and garbled.

The next step in packet switch history was taken when the Advanced Research Projects Agency of the United States Department of Defense (ARPA DOD) implemented packet switching to handle computer communications requirements, thus forming the basis for the ARPANET.   This was also the first time layered protocols were used, as well as meshed backbone topologies.   Packet switching was chosen as the method to network wide area computer communications which consisted of large computing centers.   Soon after ARPANET, many commercial companies also developed packet-based networks.   These networks included AT&T Net 1000 (which never became a product offering), Western Union Safelink, GTE-Telenet, BT-Tymnet, ADPs Autonet, Telecom Canade Datapac, CNCP Telecommunications Infoswitch, and Infonet's packet network. Today, most of Europe is tied together with private and public packet-switched networks.

8.1.2 Packet Switching Today

Packet switching continues to play an important role in the '90s environment of distributed processing. Packet switches provide low-speed (56K bps and lower), efficient networking for bursty data transport applications. Many packet switches today can provide networking speeds up to T1 and E1, with packet sizes up to 4096 bytes. Packet-switched networks use the same architecture as switched networks defined in this text. The packet switching market in the United States and the rest of the world continues to grow, and is now a billion dollar equipment market alone. While the United States is now using predominantly fiber optics, the European and Pacific Rim countries' transport infrastructures are still growing, moving from copper and analog facilities to fiber and digital facilities. Packet switching remains a very popular technology, and will through the end of the decade.

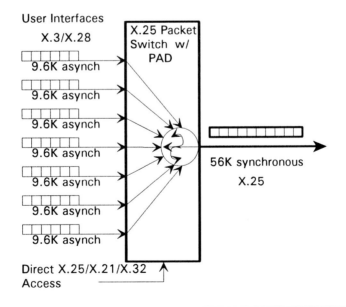

Figure 8.1 X.25 Packet Switch

Over 20% of United States public data networking is via packet-switched networks, where in Europe it is just the opposite — over 80% of public and private data networking is done through packet-switched networks. This is due to the high cost of leased and private lines and the poor quality of transmission facilities. What does packet switching provide to make it such a popular technology?

8.1.3 What Does Packet Switching Provide?

Packet switching provides logical user access and logical channel multiplexing onto a single line by interleaving low-speed user channels into a single high-speed data channel which then interfaces to the network. Figure 8.1 is a simplified version of this concept, showing ten 9600 bps X.3/X.28 circuits as user inputs to a packet switch and one direct X.25 access circuit. The single 56K bps circuit is the output trunk to the transport network. Each user logically thinks they have a full 9600 bps dedicated virtual circuit to their destination, when in effect they share the same network resources. Together, the total user bandwidth, if each user transmitted simultaneously, would be 96K bps, or twice the network transport bandwidth available. Packet switch networks are designed based upon the statistical chance that not all users will want to transmit at the same time, nor for the same duration. Large pauses in the transmission of data make this possible. In a circuit switched network, these pauses represent idle channel time and wasted bandwidth capacity. Thus, even though 56K bps is available in a circuit switched network, the entire bandwidth may rarely be used. When it is used the switch will queue the waiting packets until bandwidth becomes free. The packet switch includes buffers for those times when the inputs exceed the outputs and this queuing takes place. Buffering can be done and some delay by buffering tolerated, because data packet switching is non-time-sensitive.

Packet switching performs the same function as a statistical multiplexer, where the packet switch allows multiple logical users to share a single physical network access circuit, but also includes the additional intelligence of providing buffers which are used during times of overflow, queuing, and addressing of user ports. Packet switches allocate user bandwidth in virtual circuits (connectionless). This gives the user the perception of a dedicated circuit of full bandwidth when in actuality the network resources are shared over a virtual circuit. Virtual circuit bandwidth is only consumed when data is being transmitted, otherwise the bandwidth is not used. Oversubscribing users, by sharing of network resources in this manner, allows saving the cost of many low-speed communications channels, each of which could be vastly underutilized most of the time.

WHAT THE USER SEES - DEDICATED CIRCUITS

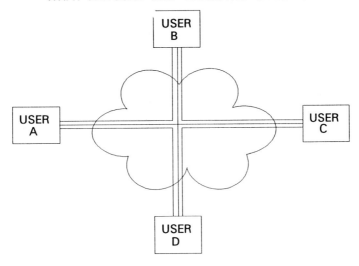

WHAT THE PACKET SWITCH NETWORK PROVIDES

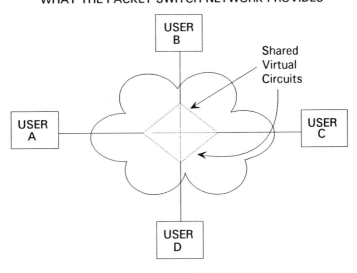

Figure 8.2 User's Perception of the Packet Switch Network

Figure 8.2 shows an example of the user's perception of the packet switch network (dedicated circuits) compared to what the network actually provides (virtual circuits). Packet switching retains the advantages of circuit switching, while providing much less delay because it does not require the time-consuming call setup and connect times associated with circuit switching. Packet switching also employs a queuing and nonblocking benefit through delay. The message may take longer to get there, but the chances of blockage are much less (depending on the buffer size) during periods of network congestion than with circuit switching.

Packet switching's first practical commercial application was designed for host computers to communicate through multiple nodes over a packet-switched network. Figure 8.3 shows this concept of DTE and DCE devices connected across a packet switch network. The X.25 protocol defines the protocols and procedures which manage the transfer of data between the DTE, typically a user terminal, and the DCE, typically the network node. The end user accesses the X.25 network through a communications controller, FEP, or host. A packet assembler/disassembler (PAD) could also be used as a DCE device for network access. We will discuss each of these in detail.

The CCITT X.25 standard defines services, facilities, and packet-switching options, as well as user interface standards for packet-switched networks. It provides a well developed standard interface for users and vendors, and provides additional end-to-end support for protocols higher than the HDLC and LAPB protocols.

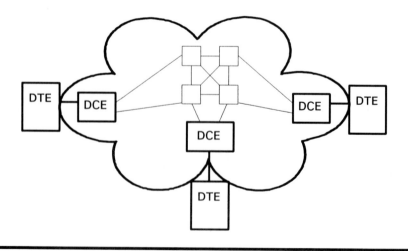

Figure 8.3 DTE and DCE Concepts

Advantages of packet switching include:

- format, code, speed conversions between unlike terminal devices
- all packets don't need to follow same path for network flexibility
- packetized long messages and short messages do not interfere with each other
- dial or leased line connections
- rapid exchange of short messages, consistent delay of long messages under steady load
- network functions transparent to users
- good for transport of small batch files
- asynchronous and synchronous interfaces
- transmission speeds 150 bps to 56K bps
- new networking speeds up to 256K, T1, and E1
- distributes risk among multiple switches
- reduce vulnerability to network failure (route around failures)
- better circuit utilization than circuit switching
- high network efficiency
- pay per bandwidth used, not based on distance
- international connectivity
- reduce cost of international circuits

8.2 THEORY OF OPERATION

Before we begin to understand packet-switching techniques, the types of traffic best suited for transport over a packet switch must be understood. We can then discuss the basic operation of switching packets between user devices, called DTEs. Finally, we can review the four types of traffic-affecting errors: errored packet, improper packet sequencing, loss of packets, and packet duplication. How does packet switching deal with each? The protocol operations which define each of these methods of error recovery is explained below.

8.2.1 Traffic Characteristics

Most data transmissions have a duration of less than five seconds, attributable to what is called "bursty traffic". This bursty traffic often travels predominantly in one direction, rather than equally bi-directional. When packet switch networks were first built, they provided an ideal transport for the typical terminal user who had many pauses between transmissions, thus causing the data to be bursty in nature. Packet switch networks also best accommodate traffic that is not delay-sensitive. Therefore, the two primary requirements which best suit packet networks are burstiness and delay-insensitive traffic. Because of the dynamic allocation of bandwidth techniques used by packet switches, bursty, delay-insensitive traffic is ideal for transport over a packet-switched network.

Figure 8.4 shows thirty low-speed terminal users at sites A, B and C communicating simultaneously with a single high-speed user host through a packet-switched network. The terminals could use low-speed X.3 circuits to request information downloads from the host, or possibly X.32 synchronous dial-up access through a modem. Terminal traffic *to* the host is very light compared to the data traffic received *from* the host. For instance, a small inquiry to the host can generate back a long response to the terminal. Also, there is a high probability that not all of the terminals will be requesting information at the same time. In fact, they will probably be transmitting and receiving less than 3% of the time each. The single 56K bps multiplexed X.25 line from the host into the packet-switched network shows the efficiencies of asymmetrical traffic patterns where the host will transmit to terminals much more traffic than it receives.

8.2.2 Basic Operation

Figure 8.5 shows a sample packet switch network with six nodes (1-6) and three users each (A, B, and C). Typically, there would be many more users on this network, but we have shown three for simplicity of the example. Each user device is acting as a DTE, and each network node is acting as a DCE. User A wants to transmit a message of size 1024Mb to user B. User A begins this process by transmitting the message to Node 1. Node 1 packetizes the message into four packets of 256K bps per packet (ignore the overhead associated with transmission until the section on protocols). These packets are then routed across the network toward Node 3 based upon routing tables predefined in the packet switches. Observe that the second packet, as well as both the third and fourth packets, were transmitted via a different route to Node 3. This alternate path routing could have been determined by increased traffic loads between Node 1, Node 5, and Node 3 (the original path which Packet #1 took) or a variety of other network conditions, such as failed circuits in either of those two paths. The packets are then received by Node 3, reassembled into the 1024Mb original message in their original order, and passed on to user B.

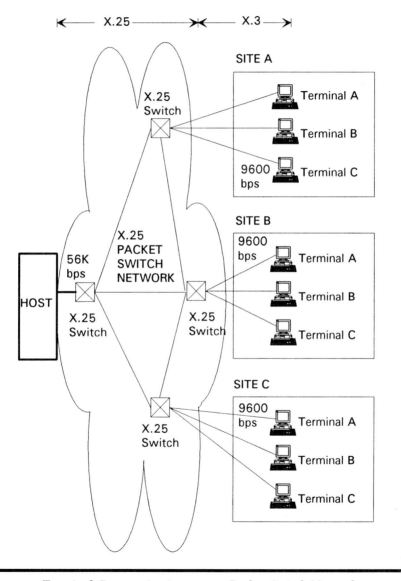

Figure 8.4 Terminal Communications over a Packet Switch Network

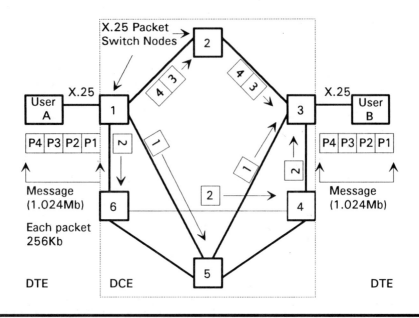

Figure 8.5 Sample Packet Switch Network

This example shows a user accessing the packet switch network through DTE devices. These DTE devices transmit data to the network via synchronous mode X.25 protocol through the use of packets. As packets flow through the network, each node checks the packets for errors and then retransmits if necessary. All operations are transparent to the user.

An elaborate acknowledgment and retransmission scheme is provided in packet switching, primarily because of the large amount of errors which could be experienced when using voice grade lines. This scheme provides flow control and error detection and correction, and limits the traffic flow by buffering to prevent congestion. Since packet switching was built on the premise of using voice grade lines for transport, much overhead, error correction, and buffering, were built into the protocol. Since the advent of fiber, many of these capabilities are no longer needed.

8.2.3 Acknowledgments

Under normal conditions, all packets are acknowledged by each receiving node as they pass through the network. The last example in Figure 8.6 demonstrates the acknowledgment function.

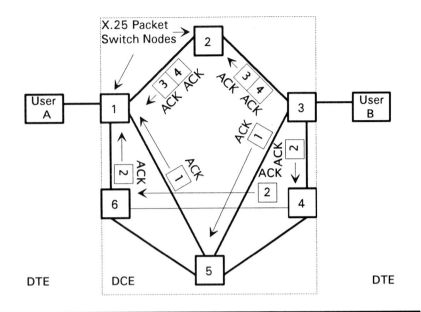

Figure 8.6 Packet Switch Acknowledgment Function

This example shows the acknowledgments which accompany each packet as it propagates through the network.

8.2.4 Packet Sequencing

Packet sequencing will be detailed later in the section on protocols. Basically, packet switching utilizes queuing and store-and-forward protocols with packet-sequencing information contained within each packet. Sequencing errors can be caused by delay incurred by message division (different packet delays) over simultaneous long and short hops.

8.2.5 Packet Error

In the last example, it was assumed that all four packets arrive at Node B. Figure 8.7 shows the result of an errored packet. If an error {E} occurs in packet number three before it arrives at Node 3 (a), Node 3 will detect the error and not send back an acknowledgment. If after a predetermined time (the time-out period) Node 2 does not receive an acknowledgment, it retransmits packet number three and packet number four (c). Node 3 then sends back an acknowledgment (ACK) when the correct packets are received (d). This one-way windowing can help or inhibit flow control. If the window size is seven (W=7), and packet number two is acknowledged, then packets three through seven have to be transmitted. Large window sizes can boost throughput on high-grade lines, or could cause long delays on noisy (error-prone) lines. A negative acknowledgment (NACK) can also be sent by the receiving node (in this case Node 3) telling Node 2 that the packet was in error. This NACK would reduce the time Node 2 waits before retransmitting the packet in error.

8.2.6 Lost Packets

Using the last example, if Node 6 receives packet two, sends back an acknowledgment to Node 1, and then fails, packet two will never be transmitted to User B. To alleviate this problem, each node only transmits acknowledgments to the previous node when it passes a packet on to the next destination node. In this case, Node 6 sends back an acknowledgment to Node 1 only after it passes the packet to Node 5. This scheme adds additional delay to the queuing process. If a certain level of packet loss can be tolerated, this scheme does not need to be implemented.

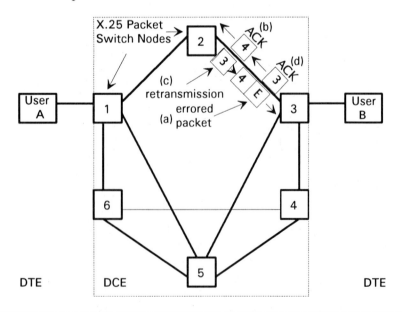

Figure 8.7 Packet Switch Errored Packet Example

8.2.7 Packet Duplication

The same packet also has the possibility to be duplicated throughout the network. This could be caused by a failure of a node to acknowledge a packet which continues through the network. The originating node, not receiving an acknowledgment, generates a retransmission to a different node, and thus the destination node receives duplicate packets. There are methods of using portions of the packet overhead to prevent this from happening.

8.3 X.25 INTERFACE PROTOCOL STRUCTURES

The X.25 interface protocols define two levels of control procedures. The subscriber link-across procedure defines the physical movement of data between two user locations. This is a local device protocol and is used to connect user devices to a network node. The packet layer DTE/DCE interface defines the transfer of information across the physical medium. This level contains user information in packets. It uses a version of HDLC which assures end-to-end error-free transmission. This level also provides virtual circuits to the user.

8.3.1 CCITT Recommendation X.25

CCITT recommendation X.25 defines the standard for packet switching. This standard is supported in the ISO OSI reference model (under ISO 7776 data link layer and ISO 8208 packet layer standards). It spans the physical, data link and network layer protocols. The X.25 protocol acts as the interface between Data Terminal Equipment (DTE) and Data Communications Equipment (DCE) for terminals and computers operating in the packet switching mode and connected to a public data packet-switched network by a dedicated circuit. Thus, the DTE devices communicate through DCE devices using X.25 protocols. The CCITT Recommendation X.25 covers the physical interfaces, link access procedures, packet layer interface, permanent virtual circuit procedures, packet formats, user facilities procedures, and many annexes for specialization of the protocol. Figure 8.8 shows the relation of DTE and DCE functions to the OSI reference model. This is an asymmetric relationship between the DTE and DCE.

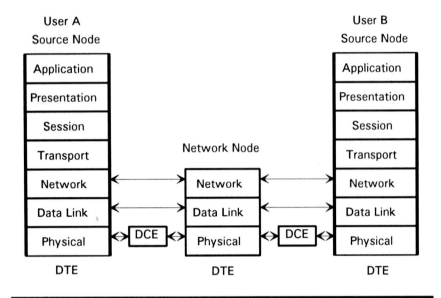

Figure 8.8 X.25 Packet Switching Compared to the OSI Reference Model

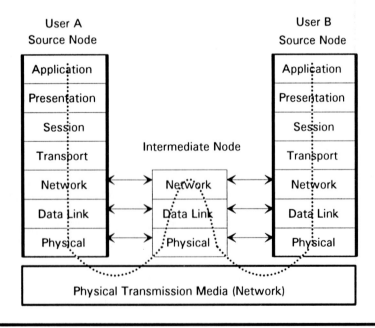

Figure 8.9 OSI Reference Model Layers Used in Packet Switching

8.3.2 X.25 and the OSI Reference Model

Figure 8.9 shows the OSI reference model and the layers used by packet switch protocols and specified by the CCITT Recommendation X.25. These include the physical layer protocol; X.21 or X.21 bis, the data link layer protocols; Link Access Procedures (LAP) and Link Access Procedures Balanced (LAP-B) (both for single circuits for DTE-to-DCE or LAP-B for multiple circuits), and the network layer protocol called the Packet Layer Protocol (PLP) which provides procedures for establishment of virtual calls and multiple simultaneous virtual calls over a single physical channel. Both the X.21 (including X.21bis) and LAP-B protocols are discussed in Chapter Five. X.25 also supports the V-Series of modem physical interfaces, as well as recommendations X.31 and X.32 semipermanent ISDN connections. Layers four through seven are either OSI compatible protocols or user specific. Figure 8.10 shows the packet-switch related protocols supported at each layer of the OSI reference model.

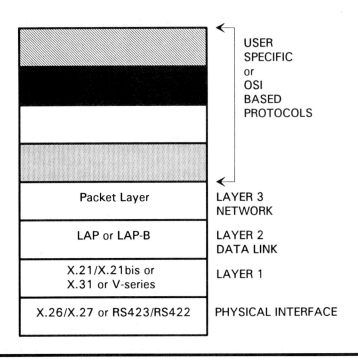

Figure 8.10 OSI Reference Model Specific Implementations

8.3.3 Data Link Frame Structure

X.25 network access can be accomplished over a single circuit or multiple circuits between the DTE and DCE. The single-link procedure (SLP) uses standard ISO HDLC framing and the multilink procedure (MLP) uses LAPB over multiple channels based on the ISO MLP standard. An example of the SLP frame (HDLC frame structure) is shown in Figure 8.11. A breakout of the MLP frame differences is also shown. Note that this is the same frame structure defined in the ANSI ADCCP specifications. All fields are used in the same manner as described for HDLC frames in previous chapters.

8.3.4 The Packet Layer DTE/DCE Interface

Each packet to be transferred across the DTE/DCE interface will be packaged within the data link layer HDLC frame information field. Figure 8.12 shows the SLP HDLC or LAP-B frame from the last example, illustrating the X.25 packet which takes the form of the information field. Note that the Level 3 packet (including packet header and packet data) forms the user data, or information, field of the Level 2 frame.

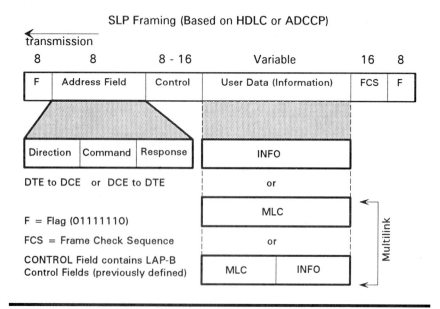

Figure 8.11 SLP Framing

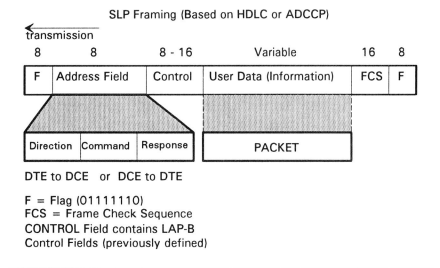

Figure 8.12 SLP HDLC or LAP-B Frame Showing Packet Payload

8.4   NETWORK LAYER FUNCTIONS

Now that DTE and DCE interactions are understood, the network perspective for the entire packet switch network is next.   Each physical interface can establish one or multiple virtual circuits to remote physical interfaces.   Each of these circuits is then assigned logical channel numbers similar to phone numbers but with local significance only.   Once these virtual circuits are established, end-to-end data communications is managed through the switching of X.25 packets.   This section describes the X.25 protocol as it defines the network layer features and functions.

8.4.1 Permanent Virtual Circuits and Virtual Calls

The network layer provides the user with up to 4095 logical channels designated either as Permanent Virtual Circuits (PVCs) or Virtual Calls (VCs) over one LAP-B physical channel.   PVCs are virtual circuits permanently established between a source and destination node.   This is similar to having a private or leased line dedicated at all times between two specific users.   These PVCs guarantee a connection between two points when demanded by the user.   The user always sees the virtual circuit as a dedicated circuit for his or her use only, whereas the network provides the same circuit as a shared resource to multiple users upon demand. Figure 8.13 shows the transfer of data between User A and User B. Notice that the virtual circuit remains established and only data (with the proper protocol information) is sent across the network between DTE devices. While the end points of the PVC or SVC remain fixed, the actual network path may vary.

Virtual Calls (VCs) or Switched Virtual Calls (SVCs) act like circuit-switched calls, with the characteristic of being connected and disconnected after the data has been sent between the source and destination node. Therefore, one source can connect to many destinations at different times, as opposed to always being connected to one destination. This is similar to the method of making a telephone call, with the phone off-hook, and talking during the duration of the call. Each time a call is made, regardless of the destination, it is re-established through the entire network. Figure 8.14 shows a standard packet sequence for the establishment of the virtual call, data transfer, and then call clearing. Note that the data transfer stage can last any amount of time.

SVCs can add even more delay than when using a PVC. SVCs, in addition to encountering connection blocking and connection delay, can encounter queuing and retransmission delays. These delay factors must be taken into account during the design.

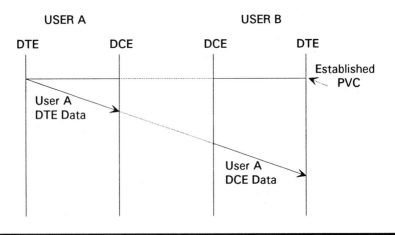

Figure 8.13 DTE to DTE Data Transfer Sequence

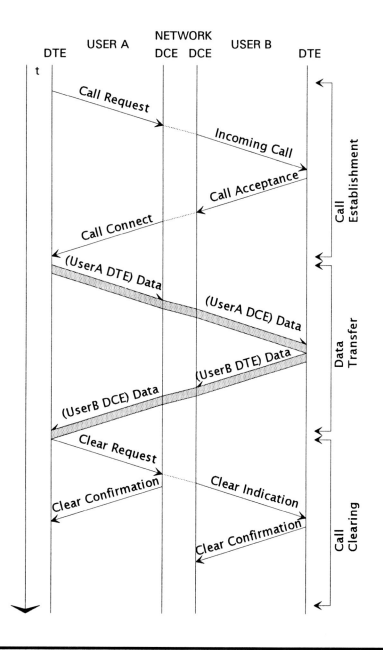

Figure 8.14 Standard X.25 Packet Transfer Sequence

8.4.2 Virtual Calls and Logical Channel Numbers

Logical channel numbers (LCNs) are assigned to each of the incoming and outgoing virtual calls for each DCE and DTE, respectively, as well as to all PVCs. Out of the 4095 logical channel numbers available per physical circuit, PVCs are assigned the lowest numbers, followed by one-way incoming virtual calls, then two-way incoming and outgoing calls, and the highest numbers are reserved for one-way outgoing virtual calls. Figure 8.15 illustrates an interpretation of logical channel administration as defined in the CCITT X.25 Annex A. These LCNs hold only local significance to that specific physical port, but must be mapped to a remote LCN for each virtual call. The packet network then uses search algorithms to assign these LCNs to each virtual call. These are also shown in Figure 8.15. Figure 8.16 shows the assignment of Logical Channel Numbers between two users. Note that the network layer of the OSIRM performs the relaying and routing of packets via the LCNs.

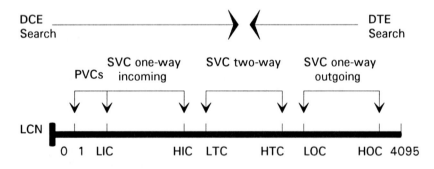

LCN - Logical Channel Number
LIC - Lowest Incoming Channel
HIC - Highest Incoming Channel
LTC - Lowest Two-way Channel
HTC - Highest Two-way Channel
LOC - Lowest Outgoing Channel
HOC - Highest Outgoing Channel

Figure 8.15 X.25 Logical Channel Definition

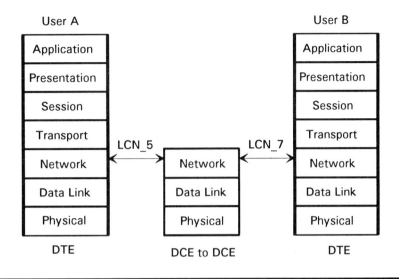

User A / User B

Application		Application
Presentation		Presentation
Session		Session
Transport		Transport
Network	Network	Network
Data Link	Data Link	Data Link
Physical	Physical	Physical

LCN_5   LCN_7

DTE   DCE to DCE   DTE

Figure 8.16 User Logical Channel Number Assignment

8.4.3 X.25 Control Packet Formats

Control packets are used for virtual call setup and termination. Figure 8.17 shows the format for a control packet. The general format identifier indicates the general format of the rest of the header, indicating whether the packet will be a call set-up, clearing, flow control, interrupt, reset, restart, registration, diagnostic, or data packet. The Logical Channel Group Number has local significance for each logical channel. We have already discussed Logical Channel Numbers. The Control Packet Type Identifier relates the packet type (indicated in the general format identifier) from DCE to DTE and from DTE to DCE. The fourth and any additional bytes contain information which is packet-type specific. Control packets perform many functions, including: call request and incoming call packets; call accepted and call connected; clear request and clear indication; DTE and DCE clear confirmation, data, interrupt, interrupt confirmation, receive ready (RR), receive not ready (RNR), reset confirmation, restart confirmation; reset request and reset indication; restart request and restart indication; diagnostic; DTE Reject (REJ) and registration request and registration confirmation. The example in Figure 8.17 shows the packet format for call request and incoming calls.

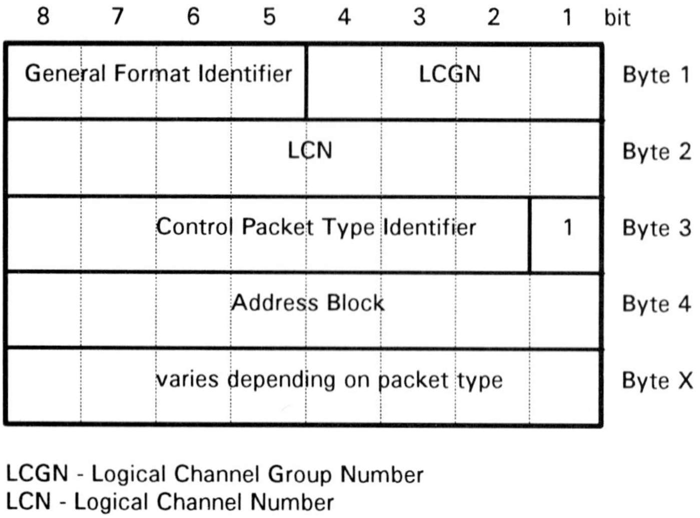

| 8 | 7 | 6 | 5 | 4 | 3 | 2 | 1 | bit |

General Format Identifier				LCGN				Byte 1
LCN								Byte 2
Control Packet Type Identifier							1	Byte 3
Address Block								Byte 4
varies depending on packet type								Byte X

LCGN - Logical Channel Group Number
LCN - Logical Channel Number

Figure 8.17 Packet Format for Call Request and Incoming Calls

A quick explanation of these control packets is in order. "Clear packets" are used to clear the user-to-user session (DTE-to-DTE). "Interrupt packets" are used when the user wants to bypass the normal flow control protocol. Interrupt packets are used for single-priority packet transmissions where the DTE must accept the packet. RR and RNR packets manage flow control and are initiated by user terminals. These packets work in a manner similar to their HDLC counterparts, and provide a level of flow control above the normal HDLC functions through the use of logical channel numbers. "Reset packets" will reset the entire PVC or SVC during data transfer. Restart packets will reset all of the PVCs and clear all SVCs on a specific physical interface port. "Diagnostic packets" are used as the catch-all identifier of all error conditions not covered by the other control packets. "Reject packets" reject a specific packet and retransmission occurs from the last received packet.

8.4.4 Normal Data Packet Formats

Normal data packets are transferred *after* call setup and *before* call termination. PVCs only require normal packet formats, as the virtual circuit is permanent and does not need to be set up or terminated. Figure 8.18 shows the format of a normal data packet. The Qualified Data (Q) bit distinguishes between user data and a user device control data stream. The D bit is set to 0 if the flow control and acknowledgments have local significance, and set to 1 to designate end-to-end significance. The LCGN and LCN together provide the 12 bits needed to form the virtual circuit number. The P(R) and P(S) fields designate the receive and send sequence count, respectively. The More Data (M) bit is set to 0 throughout the length of the message, and is set to 1 on the last packet of the message, indicating the end of the message. The M bit is used to chain packets together which serve as a single message, such as a 737-byte message which gets segmented into many 128-byte packets. The data field is prespecified to a maximum size (e.g., 16, 32, 64, 128, 256, 512, 1024, 2048, and 4096 bytes) and contains the actual user data (or padding).

8.4.5 Flow Control and Windowing

There are two methods of implementing flow control: through the control packets just defined and through packet sequencing. These methods are the same as those defined in the HDLC and LAPB standards.

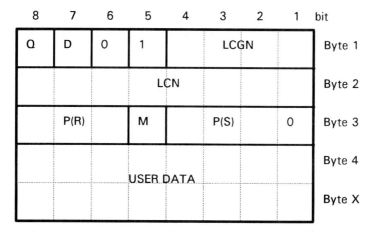

LCGN - Logical Channel Group Number
LCN - Logical Channel Number

NORMAL D - PACKET

Figure 8.18 Normal Data Packet Format

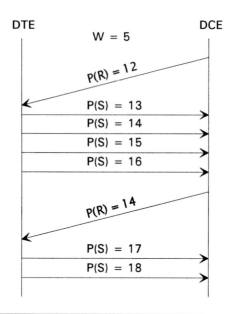

Figure 8.19 X.25 "Windowing"

Sequencing is used between the DTE and DCE device. This is also called sliding window flow control. Figure 8.19 shows an example of windowing using a windows size of five (W=5). The receive and send packet count, P(R) and P(S) respectively, keep track of the flow of packets between the DTE and the DCE devices. With a window size of five, the maximum number of packets outstanding on this logical channel is limited to five. The P(R) value represents the last packet received by the DCE. The DTE user must transmit packets lower than the window size (five) plus the receive packet number (P(R)). Thus, packets 13, 14, 15, and 16 can be sent, but not 17 (because 12+5=17 exceeds the limit). When the DTE receives the "packet receive" acknowledgment for packet 14, it now transmits up to packet number 18, then it again has to await for another P(R). Window size is directly proportional to traffic load on the logical channel, so resources should be used wisely, balancing the cost of providing more logical channels and bandwidth against maintaining performance. The default window setting is two.

While this method operates at both the data link and network levels, it allows the network to throttle individual logical channels rather than an entire physical circuit. Some protocols, such as TCP/IP, have the intelligence to reduce the window size during network congestion and increase the window size after the congestion has been relieved.

The control packet forms the basis of the flow control element of X.25. These packets operate between DTE and DCE and limit the rate of packet acceptance by updating the P(R). This flow control is negotiated separately in each direction, in the form of opening and closing windows. Receive Ready (RR) and Receive Not Ready (RNR) play an important role in postponing or closing and opening the DTE window during problem DCE conditions. Out-of-band data can also be used to control transmissions, using interrupt packets.

8.4.6 Datagram Function

Datagrams are single-packet messages which contain a destination address and are independently routed through the packet-switched network. All the information for a single transaction is contained within the datagram packet. These packets each select the best route through the network, do not require acknowledgments, are unsequenced, and thus are a form of connectionless service. Datagrams are ideal for when the user interface data transfer is for a very short period of time and where response time must be fast, or is not even needed, and where user control over the transmission is not required. Datagrams are also used when sending broadcast messages to many destinations. Some applications are more suited to datagrams, such as point of sale, credit card validation, or inquiry-response registration systems. Datagrams were dropped from the X.25 standard, but are compatible with the X.28 and X.29 standards. In fact, some internal network protocols, which are not controlled by the standard, still use datagrams. Datagrams are also used in the 802.X LAN standards.

8.5   USER CONNECTIVITY

User connectivity to the packet-switched network takes on many forms. The network user interface is accomplished through seven important standards: X.121, X.21, X.25, X.28, X.29, X.3, and X.32. We also need to understand the optional user facilities available, as well as the fast connect option.

8.5.1 User Network Interface

CCITT recommendations X.3, X.28, and X.29 have been defined as link-level protocols which provide asynchronous terminal DTE interface to X.25 networks. These standards are for interactive terminal DTE which do not have X.25 synchronous capability and can only communicate via low-speed asynchronous mode. CCITT recommendation X.25 provides synchronous user connectivity, and X.32 provides synchronous dial-up capability. Figure 8.20 shows three examples of Asynchronous and BSC DTE terminal connectivity options to an X.25 packet-switched network. Each of the interface protocols are defined as follows:

CCITT Recommendation X.121 — defines the international numbering plan for packet-switched networks.

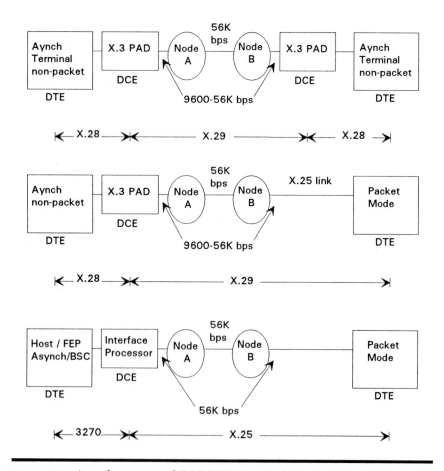

Figure 8.20 Asynchronous and BSC DTE Terminal Connectivity Options

CCITT Recommendation X.28 — defines the operational control of these functions between the (nonpacket mode) character mode terminal DTE device and the DCE PAD. Through the use of X.28 protocol, the DTE (through the PAD) establishes a virtual circuit, initializes the service, and exchanges the control and data packets. The X.28 protocol manages these controls and exchanges of data flows between the DTE and PAD through a terminal user command language.

CCITT Recommendation X.29 — defines the same controls but for the host computer destination (or origination). Information exchange can happen at any time over the virtual call. X.29 uses machine commands that identify packets between the host and the PAD.

CCITT Recommendation X.3 — defines a Packet Assembler/Disassembler (PAD) concentrator function for start-mode, or character mode, DTE devices. These terminal management functions include the bps rate, terminal specifics, flow control and escape sequences.

CCITT Recommendation X.32 — defines X.25 synchronous dial-up mode for DTE services.

There are two types of "user-to-network" interfaces using these protocols: packetized virtual circuits and packetized datagram. Each spans all three OSIRM levels as discussed. Virtual circuits assure sequencing of user data and an ordered flow of messages which require multiple packets. Packetized datagrams are messages unto themselves and do not follow the ordered packet flow of virtual circuits.

The other type of user access device is the packet assembler/disassembler (PAD), which takes asynchronous inputs from various terminals and assembles their individual transmissions into a single, synchronous X.25 circuit to the packet network. The PAD acts as a point-to-point statistical multiplexer, and PAD buffers are used to accumulate and accommodate the information streams not passed due to network congestion. The same device can also be used to disassemble the X.25 synchronous channel into individual asynchronous user interfaces. Many packet switches have the PAD functionality built into them, and provide for asynchronous user ports directly into the packet switch. Thus, the switch performs a function similar to the front-end processor and host. More common are the standalone PAD devices, including single-port cards for personal computers.

8.5.2 User Applications

The primary application used with packet switching is the terminal to host application access and vice versa. Packet switching was designed for terminals and PC interactions, and compressed voice can also be passed through HDLC over the packet-switched network. We have discussed the PAD as one of the methods of connecting remote IBM 327X and IBM 5251 terminal clusters to mainframes. There are three additional methods of connectivity:

- Software package and hardware protocol converter
- BSC 3270 uses both software and hardware protocol conversion
- Asynchronous dial-up to packet network

SNA hosts use the IBM X.25 frame format, called qualified logical link control, to format the SNA data within the X.25 protocol. Host can communicate with both packet capable and non-packet capable equipment. Terminal handlers implement the packet protocol for device interfaces not conforming to the X.25 standard. Some routers allow bridging and routing over the same X.25 network interface. SVCs encapsulating TCP/IP hosts through X.121 addressing can access other TCP/IP hosts on the router. It can also address DDN IP hosts and DDN addressing algorithm.

8.5.3 User-to-User Protocols

Users communicate with other users through byte streams, but these byte streams go through many changes as they are passed through a host or computer and then through the packet switch network. The protocols to control these user applications are transparent to X.25 and related protocols, as discussed.

8.5.4 Optional User Facilities (Section 6.0 of the Standards)

The CCITT X.25 standard defines procedures for optional user facilities which provide functions to end users at the packet layer. These facilities are specified in the control packet format. Some examples include:

On-line facility registration — permits the DTE to request registration of facilities from the DCE through use of the registration request packet. The DCE will in turn report back to the DTE what facilities are available.

Incoming and outgoing calls barred — prevents outgoing calls to the DCE or incoming calls to the DTE.

Additional options include extended packet sequence numbering, D-bit modification, packet retransmission, one-way logical channel outgoing and incoming, nonstandard default packet and window sizes, default throughput classes assignment, flow control parameter negotiation, and fast select, to name a few. More can be found in the X.25 standard under Section 6.0.

8.5.5 Fast Connect Option

There also exists a fast connect option for fast packet transactions. In fast connect mode, each packet has the call request format together with the data so the establishment of a virtual circuit is not required. This is similar to a datagram. There are two types of fast connect: fast select call and fast select with immediate clear.

Fast Select Call — The fast select packet from User A has both call request and user data (up to 128 bytes of data), and User B can respond with a call-accept packet which also contains user data from User B. The rest of the call connection and disconnect works the same as a switched virtual call. This operation can be seen in Figure 8.21.

Fast Select with Immediate Clear — This option is similar to the fast select, with the call request packet establishing the connection and the clear indication packet terminating the connection. Data flows for the fast select can be seen in Figure 8.22. This mode is similar to the datagram and is designed for single transaction based services, in addition to remote job entry (RJE) and bulk data transfer.

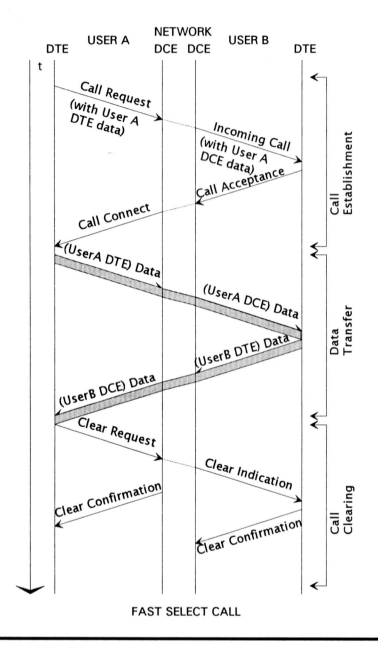

Figure 8.21 Fast Select Operation

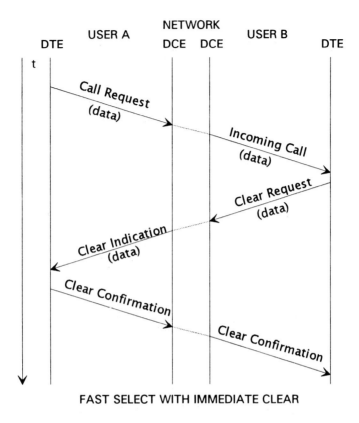

FAST SELECT WITH IMMEDIATE CLEAR

Figure 8.22 Fast Select Data Flows

8.6 PERFORMANCE AND DESIGN CONSIDERATIONS

Regardless of how fast the packets enter the network switch, they go into a queue before being switched through each node. If the queue is full, the node will buffer the packet until it can pass through the queue and bandwidth is available. The amount and length of queuing is directly related to the blocking delay through the network. Some packet switches degrade in throughput as their queues fill up and thus throughput decreases. The greater the packets per second passed the greater the delay or congestion. Errors in the network transport can also cause more queuing and delay. Overhead is also a major consideration. The overhead incurred per packet is anywhere from 64 to 256 bits per packet. Since typical messages are 256 to 1028 bits, overhead would account for 25% of the total transmission bandwidth. This directly reduces efficiency.

Performance in a packet switch network is measured in packets per second throughput in relation to the delay incurred from switch ingress to egress. These figures calculate to throughput and delay characteristic curves which should be supplied by any prospective vendor. Packet delay through a typical packet switch node is 50 to 200 milliseconds due to packet processing. Typical packet processing per node is anywhere from 300 packets per second low end up to 10,000 packets per second. Switches should be favored which have constant packet per second processing at all levels of traffic throughput. Also, the switch performance should be constant, irrespective of packet size. Some packet switches drastically degrade performance and packet processing as the packet size increases above 128K or 256K. These larger packets are characteristic of batch processing, which uses packet-switched networks because it cannot justify dedicated or leased line circuits. This type of traffic absolutely requires good performance and constant throughput, regardless of packet size.

Another consideration is protocol conversion. The greater the amount of protocol conversion which the switch or PAD has to perform, the lower the throughput and performance. Also, as the amount of PAD padding increases, the actual data throughput of the service degrades because of the increased processing power required.

8.7 X.75 INTERNETWORKING PROTOCOL

CCITT recommendation X.75 defines the protocol structure and procedures for internetworking multiple X.25 packet-switched private data networks (PSPDNs). Thus, the access of any user of an individual X.25 network is extended across multiple X.25 networks, providing the capability to share network resources and data across the international arena. X.75 can also be used to connect the larger backbone packet switches to one another. X.75 provides address translation in conjunction with the X.121 addressing protocol standard. X.75 functions reside in the network layer above X.25 functions. All functions of the X.75 protocol are similar to the X.25 protocol (e.g., LCGNs, LCNs, PVCs, SVCs). X.75 also supports the multilink procedures to support multiple links between STE. The STE is a Signaling Terminal Exchange, and acts as the internetwork interface point and performs both packet-transfer procedures and packet signaling.

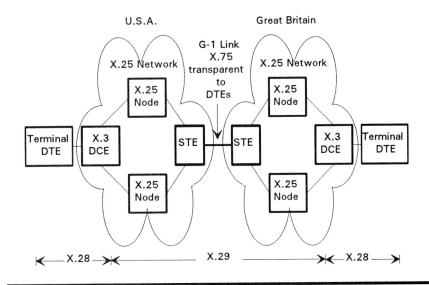

Figure 8.23 X.75 Packet Switch International Gateway

Figure 8.23 shows nonpacket DTEs from two separate X.25 networks in the United States and Great Britain communicating via the X.75 protocol. The actual communications still take place over the X.28 and X.29 protocols, and the X.75 network protocol is transparent to the users. The DCE in this case is part of network node functionality, and communicates direct X.25 into the X.25 network. This is an example of X.75 being used as an international packet-switched service. The links between these STEs can either be an A1 link, which is used between two adjacent DESs, or a G1 link, which is used between a source STE (United States) and a destination DSE (Great Britain). Many countries now have some form of private and public packet-switched networks and use X.75 to link these networks. Most vendors and X.25 networks implement different versions of the X.75 standard, changing or supporting one or more of the options available. One cannot assume that just because the network supports X.75 that it will be able to communicate with another X.25 network supporting X.75.

8.8 REVIEW

As packet switching emerged as a technology from the voice encryption requirements of the military, it laid the groundwork for many of the newer frame and cell-switching technologies. These will be discussed in the next few chapters. Packet-switched networks provide the best technology for geographically dispersed locations which require any-to-any connectivity, a mix of protocol and traffic types, and where end-to-end data integrity is of paramount importance. While packet switching has been the leading switched data technology for over thirty years, fiber optics and virtually error-free transmission facilities are quickly making it obsolete in the United States. As the digital fiber optic infrastructure is laid, so the robustness of the public network is allowing the interrelation of technology such as frame relay and cell switching with the error detection and correction only required at the ends of the transmission. In the global market, however, packet switching continues to dominate, and probably will until the end of the century. Packet switching requires physical and user interfaces, features, and functions, data link protocol, and network protocol operations, making X.25 packet switching both a private and public network protocol. User-to-user interface and transport protocols operate in conjunction with X.25, and with the X.75 internetworking protocol. Finally, performance and design considerations for packet networks show that performance relates to packets per second throughput compared to the delay in the packet switch by the packet. These principles for performance will apply to the upcoming packet technologies as well — frame and cell relay, ATM, and SONET.

9

Frame Relay, Part 1 — Theory

The theory of operation behind frame relay breaks down into three classifications: frame relay as an interface, frame relay as a signaling protocol, and frame relay as a network service. The benefits of each will be explored at a very high level. More detail will be given in later chapters. Standards have made frame relay a success. Frame relay has become an overnight standard, unusual in the standards arena, through the cooperation of the American National Standards Institute (ANSI) and the International Telegraph and Telephone Consultative Committee (CCITT), with a few select vendors supporting it. Through it an understanding of each standards committees' definition and each providers implementation of frame relay a user can interpret the multiple standards that define one common protocol, interface, and service. There are two types of frame relay offerings, private and public networking, defined by these standards bodies. Frame relay protocol structure is covered here in detail to explain how frame relay fits into the OSI reference model. Frame transmission theory rounds out this chapter and leads to a discussion of the operations and services in detail. The user's view of frame relay access, the architecture, and then the network provider view, ends this chapter. Performance of frame relay and a summary of available service offerings is made available for reference.

9.1   WHAT IS FRAME RELAY?

X.25 packet switching has predominated the data communications
marketplace over the last decade. Now a new transport technology is
needed to provide higher throughput, higher bandwidth, more cost
effective packet-style data transport and the ability to take advantage of new fiber optic transmission facilities. Frame relay is the
next step in packet technology transport. Frame relay provides an
upgrade to existing packet switch technology, by allowing user
transport speeds up to T1 (1.544M bps — with theoretical speeds up
to 45M bps) while switching frames of fixed or variable size over
Permanent Virtual Circuits (PVCs) and Switched Virtual Circuits
(SVCs). This switching and routing of frames across a frame relay
network is accomplished though the use of HDLC-based frames.
This upgrade is defined within the ISDN standards as a new packet
mode data service.

Figure 9.1 shows a typical frame relay network supporting a variety of user access, including T1/E1 multiplexer, bridge, router,
gateway, front-end processor, an X.25 packet switch, and a Frame
Relay Access Device or Assembler/Disassembler (FRAD).

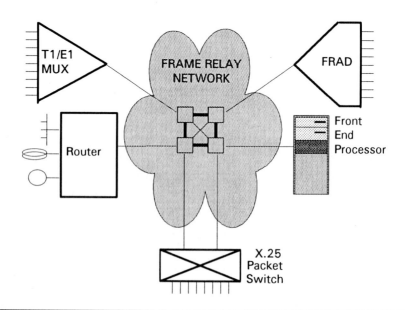

Figure 9.1 Typical Frame Relay Network

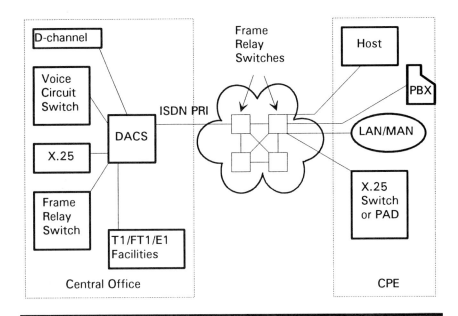

Figure 9.2 Frame Relay ISDN Access

Figure 9.2 shows similar network access elements, but also shows an ISDN feed into the network and how it would communicate with these other user access devices through a Digital Access Configuration System or Digital Cross Connect (DXC).

Frame relay fills the technology gap between X.25 packet services and SMDS and ATM broadband services. It also competes with these emerging broadband standards. The throughput and data rates are much higher than older packet-switching technology. Figure 9.3 again shows competing services, but with frame relay providing cost effective DS0 to T1 access for very bursty types of traffic. Also, notice the dashed line which extends frame relay out to T3 speeds, still under development. The proliferation of frame relay, and its competitive posture vis-a-vis broadband services, will depend primarily upon the expediency by which services such as SMDS and ATM cell switching become available, and how fast frame relay interfaces pervade the market. Frame relay is best suited for LAN interconnection because of its high speeds and low delay over physical lines sharing multiple virtual circuits. Frame relay bridges the gap between connection-oriented and connectionless protocols. Thanks to the cooperation between ANSI and CCITT standards, frame relay has become a true international standard. Either way, frame relay is a high bandwidth, cost-effective bandwidth on demand transport solution available now, and not just in the United States.

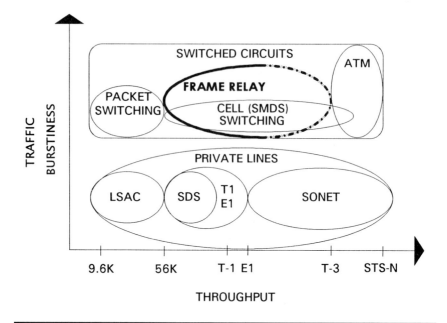

Figure 9.3 Frame Relay Compared to Competing Services

Frame relay should be approached through three viewpoints. Frame relay standards define three implementations of the technology: that of the interface, as a network signaling protocol, and as a network-provided service. It is important to note that frame relay is not a switching technique. After discussing these viewpoints and the advantage of each, a discussion of the standards which defined them will follow.

9.1.1 Frame Relay as an Interface

The first classification of frame relay is as an interface. For the network designer, frame relay defines both a packet access technique which provides bandwidth on demand and a data link OSIRM layer two interface. In one aspect, frame relay refers to a fast packet user interface to the frame relay network via the physical and data link layer. This interface transmits frames to a public or private network service which, in turn, transports them to a destination address. Frame relay CCITT standards are defined as a deviation of the ISDN bearer services, and frame relay has evolved from these ISDN interface standards. This evolution is evident in the comparison of PRI to the frame relay layer one interface. Frame relay specifications are also defined in the ANSI specifications. The adoption of frame relay as both an ANSI and CCITT interface make it a true international standard. In a practical sense it is, in effect, *only an interface*, operating over BRI and PRI ISDN, V-series, DDS and DDN, Fractional T1, X.21, and eventually T3 and, possibly, SONET interfaces, with the current preponderance using the non-ISDN interfaces.

Frame relay becomes a cost-effective solution for the transport of bursty data, such as LAN traffic, and its statistical multiplexing capability makes it an ideal choice for aggregation of multiple private lines up to T1 speeds. Multiple logical circuits are combined within a single physical circuit, thus better utilizing network bandwidth while lowering both CPE hardware and network service access costs. New advances in 50M bps HDLC framer chips may push frame relay transport rates up to T3 speeds. This will amplify the benefit of multiple users on a single access line, which in turn leads to reduced network access costs. Figure 9.4 shows a Frame Relay Access Device or Assembler/Disassembler (FRAD) at two customer premises. Each FRAD has mixed protocol subrate or T1 inputs ranging in this example from 2400 bps to 56,000 bps, which are aggregated into a single frame relay T1 trunk and transmitted to the frame relay network. This functionality will be explored in greater detail in the next chapter.

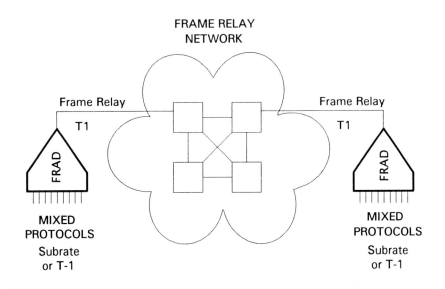

Figure 9.4 Frame Relay Access Device or Assembler/Disassembler (FRAD)

Bridge and router vendors have endorsed frame relay by making frame relay software for interfaces available. The standards which support frame relay as an interface have seen wide industry support from industry vendors who service the Data Terminal Equipment (DTE) and Data Communications Equipment (DCE) arenas, customer premises-style equipment (CPE), as well as Central Office (CO) interface support. Some of the benefits of using frame relay as an interface are:

* true international network interface standard
* multiple users per physical access line
* reduce network access hardware costs
* wide industry support
* support of both DTE and DCE industry vendors (CPE and CO vendors)
* cost effective for transport of bursty data traffic
* high speed of access due to low packet overhead
* provide higher throughput to high-speed applications
* meet throughput requirements of enhanced computing power applications

9.1.2 Frame Relay as a Signaling Protocol

Frame relay can also be construed as a signaling protocol. The OSIRM layer two (data link layer) is split by frame relay standards into two major areas: core services and user-defined (and selectable) services. Both services are defined by both the ANSI and the CCITT, and a few select vendors have written local management interfaces (LMIs) which either work in conjunction with the standards or have been replaced by them. This will be covered in detail later when frame relay is defined as circuit-mode ISDN through the CCITT Q.931 connection control procedures. Frame relay as a protocol will be discussed later when frame relay protocols are fitted into the OSIRM. Frame relay transport is also transparent to higher layer protocols. The many benefits of using frame relay as a protocol are:

• various size frame transport (and switching — depending on the provider)
• increased performance over older packet technologies
• reduced overhead for backbone networks
• reduce delays
• maximize link efficiency
• eliminates complexity of older packet protocols

- performs multiplexing functions
- discards incorrect and excess data
- encapsulates user data into variable length frames
- reduced nodal latency
- congestion causes discard of good and bad data indiscriminately
- eliminates congestion and error control overhead of X.25
- out-of-band call control
- transparent to level three and higher protocols
- transparency to link layer procedures
- improved bandwidth utilization

A few of these advantages can also become disadvantages. Discarded frames and lack of error correction can cause many user data retransmissions; and encapsulation can cause user data fragmentation. Bottom line, a private line or dial-up modem may still be the best solution. Now for a look at frame relay services.

9.1.3 Frame Relay as a Network Service

Frame relay is quickly becoming one of the primary LEC and interexchange carrier interfaces of the '90s, and is providing the interim solution to private and public network broadband data transport services. Users are demanding higher-speed and higher-performance switched data services to interconnect LANs and WANs. Client-server computing is on the rise and applications are becoming more bandwidth intensive. Network-attached devices are becoming more intelligent, offering to offload much of the network intelligence as well as handle the higher-level protocol functions with their increased processor capabilities. Frame relay service offers significant improvements to throughput and delay over traditional X.25 service. It also provides dynamic bandwidth allocation (bandwidth-on-demand) which cannot be achieved through private-line or circuit-switched networks. Frame relay does not provide the error correction or flow control attainable in X.25. The standards do offer some congestion notification (FECN and BECN), but a frame relay service relies on user applications to interpret it or actually provide this needed function. It also relies on higher-level user protocols, such as TCP/IP, to provide the retransmission and guarantee of delivery rather than duplicating it at the network layer. Frame relay does provide the concentration and statistical multiplexing of X.25 packet switching, while providing the short delay and high-speed switching of TDM multiplexers. Permanent Virtual Circuits (PVCs) and

Switched Virtual Circuits (SVCs) are established from one-to-one or many-to-one (multicast) end points, with a dynamic route through the "cloud". Frame relay becomes a major player in WAN-switched networks, where it provides a cost-effective virtual private network to a plethora of LAN users.

While frame relay service provides many of these improvements, it relies on a virtually error-free digital (fiber) transmission facility. This reliance on fiber optic transmission media is a major consideration in frame relay network design, and a major contributing factor to the popularity of frame relay services. While many LAN/WAN vendors are providing frame relay interfaces, it may prove to be the short-term solution for a long-term service. Frame relay service revenues are expected to outpace interface revenues by the middle of the decade, accounting for a multibillion dollar market, but by then frame relay may very well yield to broadband services such as SMDS or ATM.

In Figure 9.5 the differences of viewing frame relay as a network interface, as a method of switching via DLCIs via the data link layer (frame relay protocol), and as a network service can be seen. This figure shows a public frame relay network providing a service to the customer premises equipment through frame relay interfaces. The benefits of using frame relay as a service are:

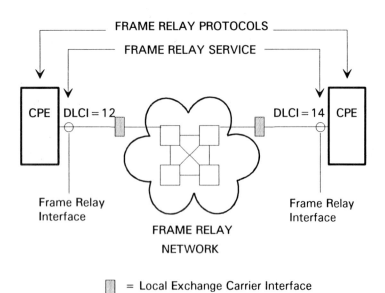

Figure 9.5 Frame Relay Interface, Switching, and Network Service

☞ ideal for bursty, protocol-oriented LAN traffic (fills the great need for LAN-to-LAN connectivity)

☞ cost effective — lower equipment costs than other high-bandwidth transport alternatives

☞ transport speeds of up to T1 (soon T3)

☞ utilizes all available bandwidth

☞ eliminate the WAN point-to-point bottleneck

☞ backwards compatible with older technologies

☞ public network access service offering

☞ fills gaps between X.25 and broadband services

☞ provides true bandwidth-on-demand

☞ STDM rather than TDM

☞ high speed and low delay

☞ takes advantage of fiber optics, error-free transmission

☞ next step in broadband services evolution

☞ multiple virtual sessions over single physical access to service

☞ intelligent workstations have replaced terminals — they can now handle high-layer protocols

☞ any-to-any connectivity

Again, some of these advantages could also turn out to be disadvantages.

9.2   MULTIPLE STANDARDS DEVELOP

The standards for frame relay have developed through the help of multiple standards committees, user groups, vendor consortiums, and a variety of other interest groups on an international scale. This broad-based cooperation has caused frame relay to become one of the fastest standards ever to develop. The standards about to be examined have been established worldwide in a very short time, driven by a common user-to-network interface and service. Since vendor implementations of frame relay interfaces, protocols, and services span numerous standards, it becomes extremely important for the network design engineer or manager to understand what the standards define and where to find specific details on their implementation. In the next few sections we will examine how the CCITT and the ANSI Telecommunications Committee standards define frame relay, as well as how each standard has developed. Much of the detail of each committees' implementation will be covered in following sections and chapters.

Frame relay originated in the CCITT in 1988 as a derivative of their Recommendation I.122. This derivative was actually from the Link Access Protocol-D (LAP-D) signaling portion of the new Integrated Services Digital Network (ISDN) signaling standard. Frame relay is now a protocol defined by both the ANSI and CCITT standards bodies. Frame relay is also a transport technique offered by many network providers though a service referred to as frame relay. In the context of providing a network service, frame relay is a connection-oriented Frame Mode Bearer Service as defined in the ISDN standards by both standards bodies. The ANSI T1S1.1 Committee, an architecture and services team under the Services, Architecture, and Signaling T1 Sub-committee, first standardized on a frame relay interface in T1S1/88-2242, which is a frame relay Bearer Service - Architectural Framework and Service Description. This standard does not include voice or video, and is compliant with the HDLC (High Level Data Link Control) protocol. This provided a common, cost effective interface to existing equipment on the market. The CCITT has standardized frame relay under the I-series recommendation I.122 Framework for Additional Packet Mode Bearer Services.

Both ANSI and CCITT ISDN standards use the LAP-D data link layer - Layer 2 of the OSIRM, and the ANSI standard also has provisions to supplement frame relay using the LMI extensions. The frame relay service per se is defined in CCITT I.122 and ANSI T1S1. The relation to ISDN is part of CCITT Q.921 and Q.931. Service Descriptions are defined in both ANSI T1.606 and CCITT I.2xy, while congestion management is defined in the ANSI T1.606 Addendum as well as I.3xz. The core aspects are defined in ANSI T1.618 as well as CCITT Q.922. And finally, the access signaling and framing is defined in both the ANSI T1S1/89-186 recommendation and CCITT Q.931. Each of these standards continues to evolve.

Frame relay is sometimes confused with the term "fast packet". Fast packet is a generic term used for many high-speed packet technologies, such as frame relay and cell relay, and has been used to represent multiplexer upgrades to faster X.25 packet switching. Fast Packet is also the trademarked name of StrataCom's Fast Packet Technology, which defines the transmission of voice and data using a packetized format. Frame relay and frame switching are synonymous only with the CCITT switching implementation of frame relay called Type II. A later chapter provides an overview of fast packet technologies.

Due to the disparity between the ANSI and CCITT standards for frame relay interfaces, protocols, and services, we will look at each implementation separately: the ANSI standards, the CCITT standards, and proprietary solutions. The LMI specifications, which span both standard organizations, will also be covered. It becomes apparent quickly that frame relay is more than just passing frames from a user to the network provider. In fact, it takes more than one standards body to produce a workable interface, network, and service for the complete end-to-end implementation of frame relay. Many of the ANSI and CCITT standards actually rely on one another to provide the complete interface, architecture, and services to embody the full complement of frame relay implementations.

9.3   CCITT STANDARDS

CCITT standards are first defined as part of a working group's recommendations. These recommendations are then formalized into draft recommendations and given numbers with letter prefixes. These draft recommendations, pending acceptance, are then published as CCITT recommendations. The "I" recommendations tend to provide the framework for services, protocols, and operations while the "Q" recommendations tend to define the detailed operations of subjects such as signaling, transport, and implementations. As noted, frame relay shares much of its architecture and protocol structures with the CCITT ISDN standards. This consequentiality to ISDN will be discussed first, followed by a description of each recommendation that plays a role in frame relay.

9.3.1 Definitions

Frame relay is referred to as an end-user service under the ISDN bearer services standards. This defines frame relay as an "interface" between the user and the network service. The relaying of HDLC frames was first defined in CCITT Recommendation I.122, which was a broad standard and could apply to many services including ISDN. CCITT Recommendation I.441/Q.921 defines the frame relay implementation of the LAP-D core protocol functions, and further defines the HDLC framing process. Here the addressing of frame relay is defined through Data Link Connection Identifiers (DLCIs), enabling multiple logical channels per physical user interface. CCITT Recommendation I.441 also calls for three additional functions to be implemented in the upper half of the layer two in the OSIRM, namely, link utilization, flow control, and error recovery. These are the main functions implemented by the older X.25 packet-switching protocol, and are overhead-intensive. Thus, they are rarely implemented when using these recommendations, and these functions are often performed by the user DTE equipment or other higher-level protocol implementations.

ISDN standards are the same standards used for frame relay. All frame relay standards rely on the ISDN standards which use the OSIRM layer 2 LAP-D data link layer standard, using the D-channel for signaling and the B- or D-channel for the transmission of information, depending on whether it is Type I or Type II. Both types of the standard will be considered. As discussed in the section on ANSI, the CCITT recommendation I.122 defines the frame relay bearer services under the ISDN recommendations. There are two main types of frame relay defined by this recommendation. Type I, Private or Virtual Private frame relay, and Type II, or Public frame relay. These types define the service descriptions and network architecture by which the services are offered. See Chapter 10. The Q.931 standard was developed by the CCITT to handle the access signaling portion of frame relay. This standard relates directly to the ANSI T1.617. The Q.922 standard was developed to handle the core aspects of frame relay. It is an enhancement of the Q.921 standard and relates to ANSI Standard T1.619. The Q.921 standard defines the frame format and corresponds directly with ANSI T1.602.

9.3.2 Where Do I Find Frame Relay CCITT Standard Information?

There are many CCITT standards which relate to frame relay which can be found in Appendix C of this book. Several of these CCITT recommendations may be read before attempting to design a frame relay network. These recommendations have been listed below with a summary of their topics of interest.

CCITT Recommendation I.122, Framework for Providing Additional Packet Mode Bearer Services, Blue Book, ITU, Geneva, 1988.

This recommendation details the architectural framework for packet mode bearer services, under which frame relay falls. Service aspects are highlighted, as well as user-to-network protocols. These aspects clearly make the distinction between the two types of frame relaying (Type 1 and Type 2) and frame switching.

CCITT Draft Recommendation I.3xx, Congestion Management for the Frame Relay Bearing Service.

I.320 defines the interaction between LAP-D and other protocol layers. Not included in this standard, but directly related, are recommendations I.430 and I.431, which define the user-to-network physical layer interface.

CCITT Recommendation Q.920, ISDN User - Network Interface Data Link Layer — General Aspects, Blue Book, ITU, 1988.

This recommendation defines the purpose and terms of the Link Access Procedure on the D-channel, LAP-D. LAP-D confers information across an ISDN network employing layer 3 entities via the D-channel. LAP-D requires a duplex, bit transparent D-channel. A good description of the service access points (SAPs), terminal endpoint identifiers (TEIs), and data link control identifiers (DLCIs) can be found within this standard.

CCITT Recommendation Q.921, ISDN User - Network Interface Data Link Layer Specification, Blue Book, ITU, 1988.

Q.921 provides the detailed specifications outlined in Q.920, including frame relay structure, procedure elements, field formatting procedures, and the detailed operations of the link access procedure on the D-channel (LAP-D).

CCITT Draft Recommendation Q.922, ISDN Data Link Layer Specifications for Frame Mode Bearer Services, April, 1991.

Frame structure, procedure elements, field formatting, and data link layer procedures designed to support the user plane defined in CCITT Recommendation I.233 are defined in this recommendation. Again, these link layer procedures are based upon an extension of the LAP-D protocols and procedures. Link Access Procedures to Frame Mode Bearer Services (LAP-F) statistical multiplexing techniques are also defined, along with their subset Data Link Core Protocol (DL-CORE) which is provided in CCITT Draft Recommendation Q.922, Annex A, (see below). Also, new address field formats are provided for extended control using DLCI or DL-CORE.

CCITT Draft Recommendation Q.922, Annex A, Core Aspects of Q.922 for Use with Frame Relaying Bearer Service, April, 1991.

Annex A of Recommendation Q.922 contains both LAP-F and DL-CORE specifications. NOTE: Appendix I of the standard contains some examples on how to control network congestion by the use of dynamic window size.

CCITT Recommendation Q.931, ISDN User - Network Interface Layer 3 Specifications for Basic Call Control, Blue Book, ITU, 1988.

Q.931 primarily defines connection-control procedures for ISDN services, as well as the general aspects of the user-to-network level 3 interface. Layer 3 protocols are defined in Q.930(I.450) and Q.931(I.451), as well as X.25.

9.4   ANSI STANDARDS

ANSI standards are first defined as part of a technical subcommittee recommendation. These recommendations are then formalized into drafts and given numbers with the prefix denoting the type of standard (for our study, we will focus upon the T1 standards). These drafts, pending acceptance, are then published as ANSI standards.

9.4.1 Definitions

Upon close examination, the reader will find that many of the ANSI standards defined below bear a close resemblance to the CCITT Recommendations defined in the last section. In fact, many of the ANSI standards are designed to complement the CCITT ISDN Recommendations. The Exchange Carrier Standards Association (ECSA) has been coordinating the efforts of the ANSI T1S1 Committee in the development of frame relay specifications. While many frame relay implementations comply to ANSI standards, it is the LMI specifications with extensions which have made early implementations of frame relay possible. These interface specifications are enhanced by the LMI extensions as defined by the "Gang of Four" (StrataCom, Inc.; Digital Equipment Corporation (DEC); cisco Systems, and Northern Telecom, Inc.). ANSI T1.617, T1.618, and T1.606 provide customer interface standards for access speeds of DS0, Nx56K bps, Nx64K bps, and DS1, primarily defining the user-to-network (UNI) and network-to-network interface (NNI). T1.607 concentrates on the frame relay circuit switching procedures. T1.617 defines frame relay call control procedures, and T1.618 defines the framing and transmission of frame relay (also includes new congestion control techniques). Presently, permanent virtual circuits (PVCs) is the only call connection service being offered via the ANSI standards, but switched virtual circuits (SVCs) will also be supported as standards develop.

9.4.2 Where Do I Find Frame Relay ANSI Standards Information?

The pertinent ANSI standards which relate to frame relay can be found in Appendix C of this book. A few of these standards should be read and understood before designing a frame relay network. The more relevant standards are listed below, along with a summary of their topics of interest.

ANSI T1.602, Telecommunications — ISDN — Data Link Layer Signaling Specifications for Application at the User-Network Interface, 1990.

This standard defines the basics if ISDN data link layer signaling at the user-network interface.

ANSI T1.606-1990 — ISDN — Architectural Framework and Service Description for Frame Relaying Bearer Service, 1990.

This standard outlines the architectural framework used in frame relay service. It provides more of an overview and guideline, and is followed up by many other detailed standards. The frame relay service defined in this standard directly corresponds to the CCITT I.122 type 1 frame relaying service. User-to-network and network-to-network interfaces are also discussed.

ANSI T1.607 (ANSI T1S1/90-011) — Digital Subscriber Signaling System No. 1 — Layer 3 Signaling Specifications for Switched Bearer Service, 1990.

This standard defines much of the layer three network signaling. It contains an Annex S which compares the ANSI T1.607 layer three specifications to those of CCITT Recommendation Q.931. It also defines circuit switching procedures for frame relay.

ANSI T1.617 (old T1.6fr, ANSI T1S1/90-213R1) — DSS1 — Signaling Specification for Frame Relay Service, December 1990. Updated and republished as:
ANSI T1.617-1991 (old T1S1/91-352) — ISDN, DSS1 — Signaling Specification for Frame Relay Bearer Service, 1991.

The signaling specifications outlined in these documents include message format and element coding and frame relay call control procedures. Our primary interest is in the two types of connection-control access techniques defined in ANSI T1.617. Case A is a two-step frame mode call establishment method which provides either a frame relay connection on demand or a semipermanent frame relay connection. Case B is an integrated access method which requires the ISDN call setup control mechanism. This provides a frame relay connection on demand only. Both of these access methods will be defined in the next chapter. ANSI T1.617 tends to supersede T1.607 but still refers back to it in sections where no changes have been made to the old standard.

ANSI T1.617 Annex D (ANSI PM91-1) — Additional Procedures for PVCs Using Unnumbered Information Frames, May, 1991.

This annex provides some additional options for permanent virtual circuits (PVCs) using unnumbered frames, primarily the notification and recovery from loss of PVCs and link integrity checks. This annex uses local in-channel signaling and defines methods of implementing call control with unnumbered frames.

ANSI T1.618 (old T1.6ca, ANSI T1S1/90-214R1), DSS1 — Core Aspects of Frame Relay Protocol for Use with Frame Relay Bearer Service, November, 1990. Updated and republished as:
ANSI T1.618-1991 — ISDN — Core Aspects of Frame Protocol for Use with Frame Relay Bearer Service, May, 1991.

This standard defines frame structure, formatting, new forms of DLCI addressing, the placement of the DL-CORE protocol into the ISDN protocol architecture, and new congestion and control methods. This standard and CCITT Recommendation I.122 are designed to complement, rather than compete, with the CCITT recommendation providing the definition and infrastructure for frame relay services and the ANSI standard providing the implementation of the service. The standard also provides new methods for handling congestion control.

ANSI T1S1/90-051R2, Carrier to Customer Metallic Interface — Digital Data at 64k bps and Subrates, May, 1991.

These standards are important for the user who will interface to the frame relay network. This particular standard specifies the normal operating signals and the maintenance signals on DS0 and lower physical-speed access.

9.5   LMI EXTENSIONS AND PROPRIETARY SOLUTIONS

When the CCITT and ANSI standards were under development, four vendors selling in the commercial/government marketplace did not wait for standards to evolve but decided to write their own specifications defining frame relay standards between the user and the network elements.   The four vendors were StrataCom, Inc.; Digital Equipment Corporation (DEC); cisco Systems, and Northern Telecom, Inc. (comprising the famous "Gang of Four").   As is often the case, these vendors saw the need to get a product to market, along with the projected revenue stream which could be achieved with an early release frame relay product.   These factors drove them to adopt an interim set of specifications, called the LMI extensions.   These extensions have features which, even though being a proprietary implementation, complement and supplement both the ANSI and CCITT standards, as well as represent the views of private and public network suppliers.   This common platform of the Gang of Four for interoperability has become the de facto standard in the industry for the interconnection of CPE equipment, via a frame relay access interface, to central office switches.   All four vendors have announced their intent to conform to the standards as they become defined, and are doing so now.   Other vendors have joined the original Gang of Four and have formed the frame relay Forum.

These proprietary LMI features handle the information exchange between the network and the user-attached devices, providing standards for such things as support of automatic reconfiguring of devices and fault detection.   LMI features also enhance service by providing the user with status and configuration information on the PVCs active at that time.   OSIRM level 1 connectivity is also addressed, along with customer network management functions.   Figure 9.6 depicts the user-to-network interface as defined by the LMI specifications.   Specifically, it shows where the LMI standards exert their influence.   The standard defines only permanent virtual circuits (PVCs) between data termination equipment (DTE) with frame relay networking equipment.   It does not define data communication equipment services, but enhancements are provided in the extensions for frame relay service.

There are two major types of LMI extensions, the standard set and optional extensions.   The standard extension set is used by almost every major CPE vendor providing a frame relay interface in their equipment. Many of the optional extensions are not used, and some remain under development. All LMI extensions should be used in conjunction with the ANSI standards as defined in Section 9.4. Also, many of the LMI extensions have already been implemented in recent ANSI standards. As the ANSI and CCITT standards evolve, they will also incorporate the same functionality as found in the LMI Extensions (primarily because of vendor influence in these standards committees). Some standards have even written annexes which contain sample implementations of extensions and options to the base standard.

This section includes some of the framing and addressing details which will be clearly defined later in the chapter when we discuss transmission theory. The frame structures for the LMI extensions will also be defined later in the chapter. The reader may want to re-read this section after completing the chapter.

9.5.1 Standard LMI Extensions

The LMI common extensions are based upon the message formats of the CCITT Recommendation Q.931, and define enhanced versions of these messages for configuration and maintenance. These standard LMI extensions perform the following functions:

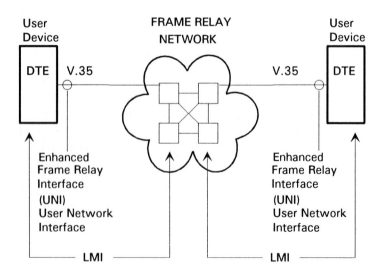

Figure 9.6 Frame Relay User-to-Network Interface

✽ notify user of PVC status (active and present DLCI)
✽ notify user of add/delete/change PVC (removal or failure of DLCI)
✽ notify user of physical link "keep alive" signal and logical link status

The standard extensions identify:

❉ maximum frame size of 8196 octets
❉ support for 1024 DLCI addresses
❉ common extensions
❉ setting of FECN/BECN bits, DE congestion bit
❉ support for multicasting
❉ global addressing
❉ now only PVCs and a maximum of 1024 logical connections.

All of these extensions are transmitted through the network on DLCI 1023. This is the logical channel which has been designated for LMI extensions. After we learn the standard frame formatting in Section 9.7 on transmission theory, we will discuss the many types of LMI frames and their use in the frame relay network.

9.5.2 Optional LMI Extensions

The optional LMI extensions define four key areas of user functionality additions. These four optional extensions include:

❉ Multicast Capabilities
❉ Flow Control
❉ Global Addressing Convention
❉ Asynchronous Status Updates

Multicast Capabilities. The multicast capabilities allow multiple LAN interconnected user devices to function with simpler address resolution. Multicast replaces the need to poll a LAN/WAN for a destination address for sending packets by simultaneously broadcasting to all routers in a predefined multicast group. Multiple DLCIs are received for each multicast group. The network device then replicates the transmitted frame into multiple frames of predefined multicast group of DLCIs.

Flow Control. Enhancements to congestion indication and user notification of such conditions are provided in this extension. This style of flow control is similar to XON-XOFF, but can be used only for unidirectional data flow. This mode of flow control should be used in conjunction with the ANSI standard for homogeneous implementation of frame relay standards. The primary users of this extension will be bridges and routers which cannot implement the ANSI standard congestion and control mechanisms.

Global Addressing Convention. This capability allows the network to provision DLCIs on a port-by-port basis; in effect, using one DLCI for each port or end device so that any user on the network can communicate and use the same DLCI for a given destination every time. This extension allows only a total of 1024 DLCIs for the entire network. Each DLCI is assigned to one specific network port, providing the same termination point for each given DLCI number. The termination point always remains the same, regardless of origination port. This is much different from the 1024 DLCIs per port as defined in the ANSI standard, and would be used only in small private network implementations due to its total network port limitation.

Asynchronous Status Updates. This enhancement allows the network to notify the user device of a change in logical channel DLCI status. This acts as an option to the ANSI standards for flow control and congestion control.

The optional functional extensions listed above will notify users of any changes to the multicast group, multicast virtual circuit availability or lack thereof, and the configured source DLCI for the broadcast endpoint. Other optional extensions provide an asynchronous update to DLCI changes, explicit notification of a deletion of any PVC, including Multicast, network buffer status, and minimum transmission bandwidth. There is also the capability of notifying the user device of the minimum bandwidth provisioned by the network for each permanent virtual circuit.

There are a few vendors who do not conform to either type of frame relay implementation. These vendors fall under the proprietary category and should be avoided. At best, they will provide only a short-term solution.

The LMI extension document mentioned above was published in late 1990. Since then no revisions have been issued. Each vendor has published specific implementation specifications for the standards at the back of the document in the appendices. Since then additional support has been added by all four vendors, and they would need to be contacted for this detail. The document is *Frame Relay Specifications with Extensions*, Document Number 001-208966, Rev. 1.0, September 18, 1990 (Digital Equipment Corporation, Northern Telecom, Inc., StrataCom, and cisco Systems).

9.6   PROTOCOL STRUCTURE

Frame relay interfaces and services span many protocols, from the level one physical layer to the layer three network layer of the OSI reference model. Frame relay protocols have taken the pre-defined ISDN functionality and redefined it to split the core and procedural services on the logical link layer. Frame relay also provides an addressing function through DLCIs similar to the Ethernet and cell-switch addressing. The following is an overview of the protocol structure whose implementation will be defined in more detail later.

9.6.1   OSIRM and Frame Relay Overview

Figure 9.7 shows the OSIRM layers and protocol structure of frame relay transport. Frame relay transport comprises only the first two layers of the OSI model, the physical and data link layers. The end-to-end protocol operations of all layers above the physical and data link are "transparent" to the frame relay network. The physical layer interface can range from a DS0, through fractional T1 and, finally, include a full T1. Signaling is similar to standard synchronous channel connection. Layer two utilizes the CCITT link access procedure (LAP-D) data link layer protocol. Layers three through seven are handled by the application program. This allows the actual packet processing to be accomplished by the application CPU, and frees up network resources to handle data transport. This forces the applications to perform the error correction, addressing, and sequencing. Some frame relay implementations can perform error correction and addressing functions in a limited fashion. These will be discussed later in the chapter.

Frames are transmitted between nodes at the OSI layer two data link level. These frames contain addressing, error detection and control for the frames themselves, but not the data contained within. The nodes establish permanent virtual circuits (PVCs) and route the data through this point-to-point serial connection. Frames are routed by destination addresses. If an error is encountered, the entire frame is discarded. No retransmissions are requested. Figure 9.8 shows and example of an end-to-end frame relay network.

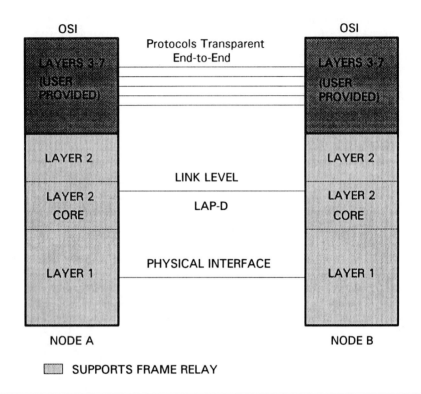

Figure 9.7 Frame Relay and the OSI Reference Model

Core functions of OSIRM layer two include frame error checking (on the address portion only — not the actual data) and retransmission in the event of errors. User physical connectivity and frame relay transport services are provided through the physical and data link level, while the network services touch upon the data link and network layers to provide the user services. Figure 9.9 shows the three layers of the OSIRM and how the standards relate to them. These concepts become clearer as each protocol function is explained in detail.

Virtual circuits are established between the ingress and egress points of the network. Figure 9.10 shows a public frame relay network providing access for a customer premises device through the Local Exchange Carrier (LEC) and into the Interexchange Carrier (IXC). Note that the router is providing the frame relay layer two protocol function, and the network (either the LEC, the IXC, or both) is providing the frame relay service which encompasses parts of level two (data link layer) and level three (network layer). Router access is through the local loop. This local loop is often the most expensive portion of the frame relay service network access.

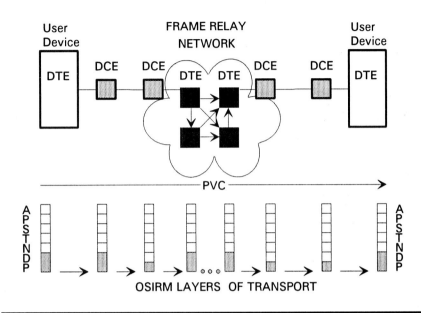

Figure 9.8 End-to-End Frame Relay Network

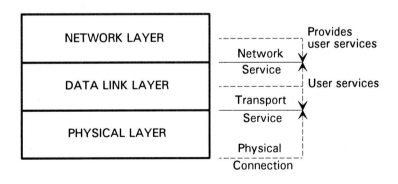

Figure 9.9 OSI Reference Model Interfaces to Frame Relay

9.6.2 Layer Two Protocol Structure

To understand frame relay as a service, we must first develop an understanding of the architectural framework on which this service is offered. An explanation of frame relay standards developed under the OSIRM gets us started. Figure 9.11 shows the seven-layered OSI reference model. We will examine part of layer one, all of layer two, and part of layer three of the OSIRM, which defines frame relay services, protocols, and interfaces.

Frame relay service, as it relates to ISDN packet mode bearer service, is concerned with two logically separate levels of the data link layer, defined as the control plane (C-plane) and the user plane (U-plane). Each provides different services, and we will discuss each in the context of how they relate to the protocol structure. The concept of user and control planes has been defined as a fundamental practice for ISDN protocols. The frame relay version of the ISDN standards has attempted to establish one set of protocols for all C-plane frame relay services, and separates the U-plane from the C-plane to allow user customization of frame relay services.

A further in-depth study of the original intent of the ISDN control plane standards is left to the reader.

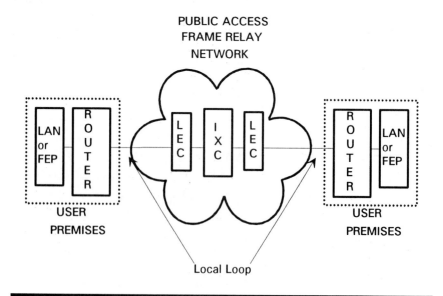

Figure 9.10 Public Access Frame Relay Network

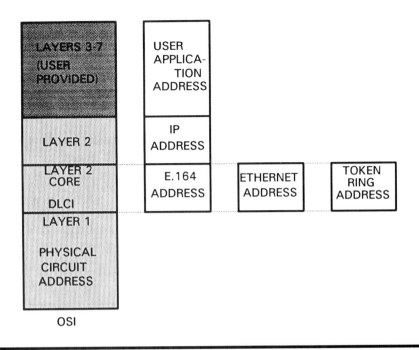

Figure 9.11 OSI Reference Model and LAN/MAN Comparisons

The C-plane can perform frame relaying by two methods: Virtual Calls (VCs) and Permanent Virtual Connections (PVCs). The virtual connection calls are defined by CCITT Recommendation Q.922 at the data link control layer (layer two) and Recommendation Q.931 at the network layer (layer three). Virtual calls are setup and released dynamically by these two standards. The second method of frame relaying is through PVCs administrated by the network provider and established at subscription time. The U-plane provides the data transport of the user data via the physical access line and through logical links.

There are two types of C-plane and U-plane interfaces: "user-to-network" and "network-to-network" element connectivity. Figure 9.12 shows the user interface to the network element, showing each relevant ANSI and CCITT standard, as well as the split C-plane and U-plane functions. The U-plane is split into core functions and user-selectable terminal functions. The C-plane is also split into layer two and layer three services, or the procedures necessary for signaling. Figure 9.13 shows the network-to-network element interfaces.

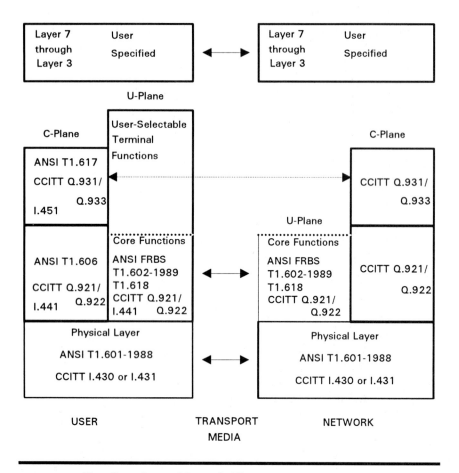

Figure 9.12 User Interface to Network Elements

Core Services and Procedural Services have been separated in the next two sections. This separation roughly corresponds to the U-plane and C-plane distinction, respectively. These services will be defined as they appear in the CCITT and ANSI standards.

9.6.3 Core Services

The core functions, defined by ANSI T1.602-1989 and CCITT Recommendation Q.921 with the user selectable terminal functions, make up the core services. The core services roughly correspond to the U-plane functionality, which defines user-selectable frame relaying services. Together they span the data link layer of the OSI reference model, and define the user's interface to the frame relay network. Some of the services provided by the U-plane are as follows:

* error detection
* congestion management
* bit-level transparency
* frame delimiting
* layer two address field frame multiplexing
* operating between each hop (user-to-network or network-to-network)

Figure 9.14 depicts the end-to-end transport via the U-plane. Notice that, again, all protocols above level two are transparent to the frame transport. These core services are discussed later in the next section on transmission theory.

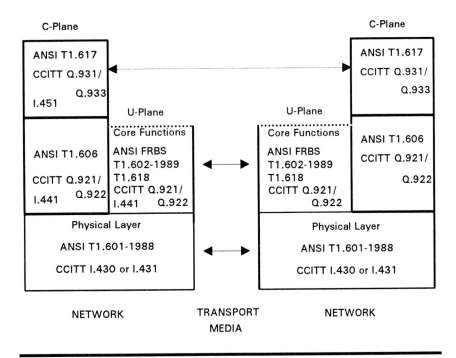

Figure 9.13 Network-to-Network Element Interfaces

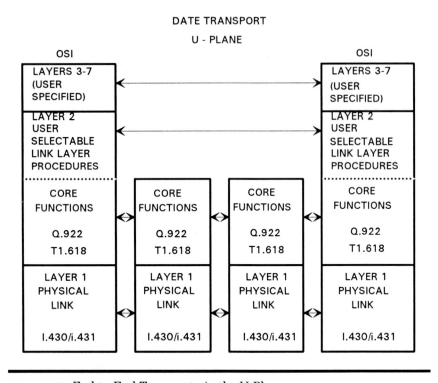

Figure 9.14 End-to-End Transport via the U-Plane

9.6.4 Procedural Sub-layer Services

The procedural sublayer provides procedures for data transport from the user device to the network and between network devices. It is different from core services in that it operates with the entire network in mind, providing end-to-end connectivity across the network, and bridging the OSI reference model layers two and three. This is where true signaling information is managed. In the standards, this relates roughly to the control plane (C-plane) and manages call control and the negotiation of network parameters. Some services provided by the C-plane are:

- ✤ error recovery
- ✤ flow control
- ✤ timer recovery
- ✤ mode setting
- ✤ acknowledgments

❖ XID exchange
❖ user-to-user, end-to-end service (the entire PVC)

These control services will be discussed later in the next section on transmission theory.

9.7   TRANSMISSION THEORY

With an understanding of the protocol structure of frame relay protocols, discussion can now center on how frames are assembled, the many types of frames and the data contained in each, addressing structures and schemes, and the detailed frame structure and operation of the LMI extensions. As mentioned previously, the reader may want to reference Section 9.5 after studying the operation of the LMI extensions covered here.

9.7.1 Overview

The high throughput of frame relay is achieved with the side effects of loosing error detection and correction, sequencing, and addressing overhead functions resident in traditional packet-switching technologies. This is the equivalent of removing the OSI layer three functions from the X.25 packet-switching protocol. One of the primary advantages gained is the relatively short delay incurred during data transport. Frame relay also provides fast reconnect and statistical multiplexing, consistent with what is expected of TDM and statistical multiplexers, respectively.

Highly reliable digital facilities with fiber optic transmission are preferred in the frame relay environment because virtual error-free transmission media is essential. All error detection and correction, as well as flow and congestion control, is handled in the customer premises rather than within the network. Thus, errored frames are discarded by the service and are left to the user for retransmission. Congestion control is also left to the user. While this eliminates overhead and delay (typically found in X.25 networks), it introduces an entirely new role for the user premises equipment. In some parts of the world where the transmission facilities are not fiber, such as Western Europe, traditional X.25 packet services will probably prevail for quite some time until the transmission infrastructure is upgraded to error-free fiber media.

High throughput and statistical multiplexing seem good to the user, but the price the user pays may be high. With the advent of frame relay and MAN services, the well-defined line between data transport services and LAN/WAN user services turns fuzzy. As we have discussed, these transport services leave the responsibility of end-to-end data delivery guarantees to the user. Many users are not ready to sign up to this responsibility, and public network providers offering frame relay service must provide some solutions to capture public switched service users.

Frame relay differs from multiplexer networks in that the statistical properties of frame relay only allocate bandwidth as needed. In a multiplexer, channels are allocated to specific users, and, in effect, bandwidth is wasted during idle times on some channels while another channel could have been using that extra unused bandwidth. This allocation works best for local area network connectivity, where sporadic and bursty traffic could demand the entire channel during a given time with little or no bandwidth requirements the rest of the time. When coupled with a frame relay public network service, using frame relay switching, the ultimate result is achieved: a switched public network bandwidth-on-demand service operating on a "pay as you use" scale. Statistical multiplexing can achieve this, either on-premises or via the network employing frame relay switching.

Figure 9.15 shows a comparison of frame relay to conventional T1 multiplexing. When using a multiplexer channel, the user pays for the entire channel and any empty preassigned slots not needed are wasted. When the frame relay user accesses the network, they do not pay for bandwidth not needed, but only for what is used.

TDM MULTIPLEXER vs. FRAME RELAY DEVICE

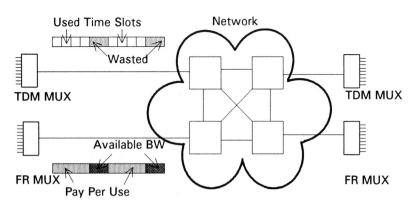

Figure 9.15 Frame Relay Compared to TDM Multiplexing

MULTIPLEXER
PRIVATE NETWORK

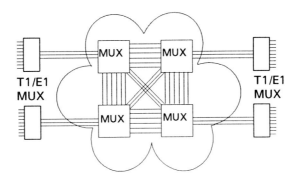

PUBLIC SWITCHED FRAME RELAY NETWORK

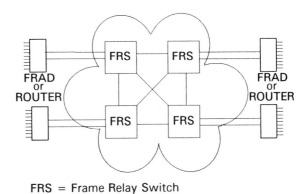

FRS = Frame Relay Switch

Figure 9.16 Frame Relay Network Compared to Multiplexer Network

Figure 9.16 shows a standard private multiplexer network compared to a public switched frame relay network. This figure shows the number of circuits required to support a fully meshed multiplexer network in comparison to the number of circuits needed for a frame relay network. Both approaches provide the same amount of user access, but fewer access lines are needed with frame relay.

Another major aspect of frame relay in comparison to older packet technologies is that frame relay switches merely passes frames along to the destination address. Thus, intermediate nodes do not perform any packet (or frame) disassembly or error detection/correction. DLCIs change at each node in the network. In relation to establishing circuits throughout the network, PVCs are established in the frame relay network. They are not changed other than through network operator intervention. Switched virtual frame relay circuits (SVCs) will allow users to modify their own network configurations and even enable dial-up connections to the frame relay network. To date, SVCs have not been well defined, nor implemented in frame relay services.

9.7.2 Basics of SAP and DLCI

Before discussing the recent frame relay standards implementations, we need to understand the basic foundations of the LAP-D and framing structures. First to review CCITT Recommendations Q.920 and Q.921 and then move to the newer Recommendation Q.922 and ANSI T1618 standards.

CCITT Q.920 defines the terms and basic concepts of DLCI data link addressing. The service access point (SAP) is the logical level data link interface from the user to the network. The SAP provides services to layer three protocols. There are multiple data link connection endpoints associated with each SAP and, at the link layer, these are referred to as data link connection identifiers or DLCIs. Figure 9.17 depicts the relationship between the SAP and the DLCI, showing how multiple DLCIs communicate at the data link layer from one SAP to another.

Two data link addressing values are assigned within the DLCI called the SAPI and the TEI. The SAPI, or Service Access Point Identifier, defines the service access point on the user or network side and how the data link layers relate to the layer three management entity through call control procedures, or packet mode communications of either Q.931 or X.25, or the layer two management features. The TEI, or Terminal Endpoint Identifier, defines the specific endpoint for the given SAPI, thus completing the virtual circuit. The selection of the SAPI and TEI values are detailed in CCITT Recommendation Q.921.

Figure 9.18 defines the address field structure of the CCITT Q.921 Recommendation. Notice the five-bit SAPI address and the seven-bit TEI address. One bit is reserved for a command response indicator and two bits for extended addressing (both covered later). While these basics are important, the true blend of CCITT and ANSI standards for the framing and address formats is in the CCITT Draft Recommendation Q.922 and ANSI T1618, as well as with the new LAP-F functionality.

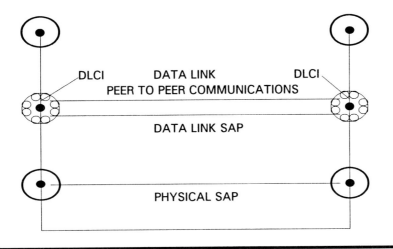

Figure 9.17 SAP to DLCI Relationship

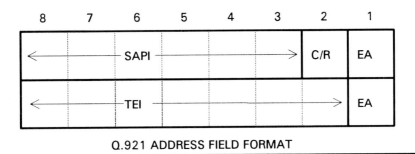

Q.921 ADDRESS FIELD FORMAT

Figure 9.18 CCITT Q.921 Address Field Structure

9.7.3 Frame Structure

The frame format used by frame relay services is a derivative of the ISDN Link Access Protocol D-channel (LAP-D) framing structure. Figure 9.19 illustrates the original LAP-D frame structure.

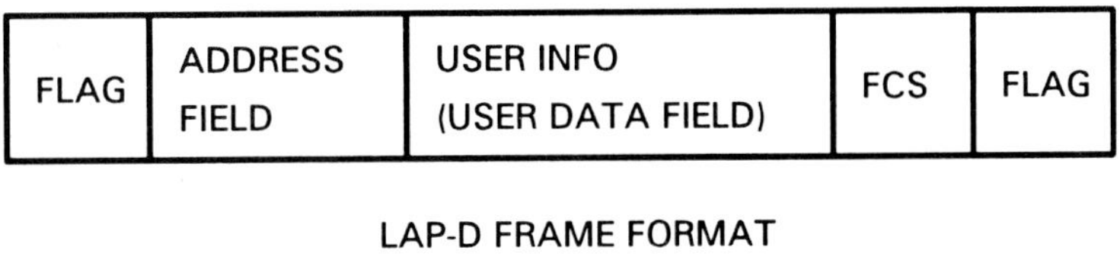

FLAG	ADDRESS FIELD	USER INFO (USER DATA FIELD)	FCS	FLAG

LAP-D FRAME FORMAT

Figure 9.19 LAP-D Frame Structure

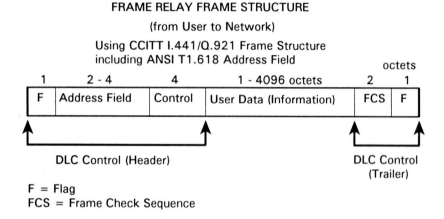

Figure 9.20 CCITT Q.921 Frame Structure

Figure 9.20 shows the standard frame relay frame structure based upon the CCITT Recommendation I.441/Q.921. This frame structure is part of the core services previously defined. Data is transmitted across the network using various implementations of this frame structure. The figure also shows the data link control header and trailer, to be detailed in a moment. The address field contained within the frame is based on the ANSI standard. The arrow depicts the flow from customer interface to the network. The first and last fields serve as flags, each being one octet long. The second field is the address field, taking up two octets. The third field is for user data. This field can be any number of integral octets up to 4096. The fourth and fifth fields are reserved for Frame Check Sequences (FCS), each one octet long, respectively. This provides a service with only six octets of overhead.

9.7.4 Address Field Structure

The address field resides within the frame. The original address field defined for frame relay was modeled after the LAP-D DLCI address frame, where the DLCI was composed of the SAPI and TEI. The structure has changed to accommodate congestion control signaling in the FECN, BECN, and DE bits. Figure 9.21 shows the address field breakout from the frame structure. There are two DLCI fields, which together form a 10-bit DLCI address which ranges from zero to 1023. This DLCI identifies the logical channel connection within the physical channel or port for a predetermined destination. The C/R bit is a command response indicator which is not used at this time (thus, it may be set to either value). The Forward Explicit Congestion Notification (FECN) bit is a toggle which tells the remote user that network congestion was encountered by the frame transmitted across the physical media, and that the user should take action to prevent data loss. The Backward Explicit Congestion Notification (BECN) bit works the same, but notifies the user of congestion in the data on the returning path. An increase in the frequency of FECN and BECN bits received is a good indication of the congestion through the network. The Discard Eligibility (DE) bit, when set (at 1), indicates that the frame should be discarded during congestion conditions as opposed to discarding other frames with a higher priority (those set at 0). This bit is used by carriers to determine what data they will discard when the total traffic exceeds the Committed Information Rate (CIR) and exceeds network bandwidth resources.

FRAME RELAY FRAME STRUCTURE

(from User to Network)
Using CCITT I.441/Q.921 Framecture Stru
including ANSI T1.618 Address Field

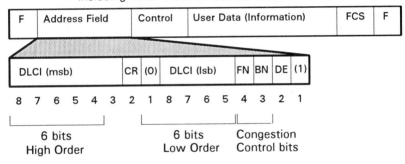

F = FLAG

FCS = FRAME CHECK SEQUENCE

DLCI = DATA LINK CONNECTION IDENTIFIER

CR = C/R = COMMAND RESPONSE INDICATOR

FN = FECN = FORWARD EXPLICIT CONGESTION NOTIFICATION

BN = BECN = BACKWARD EXPLICIT CONGESTION NOTIFICATION

(0) = EAB (0) = EXTENDED ADDRESS BIT ZERO

(1) = EAB (1) = EXTENDED ADDRESS BIT ONE

DE = DISCARD ELIGIBILITY BIT

Figure 9.21 Frame Relay Frame Structure Breakout

Unfortunately, the FECN, BECN, and DE bits are not often used, not because they do not send out an alert of a congestion problem, but because many CPE applications accessing the network cannot understand or make use of this information, and consequently ignore it. In this case, data is lost; either within the network or at the network access point. The Extended Address (EA) bits act as address field delimiters, set at 0 and 1 respectively. These bits have been used by ANSI to extend the DLCI addressing range.

Each user CPE device with multiple logical and physical ports must have a separate DLCI for each destination it wants to transmit to. These DLCIs are built into the switching/routing tables of each device on the network. The DLCI only has local significance because it establishes a virtual circuit to one of 1024 other ports in the network. Therefore, each network may have up to 1024 active DLCIs, or 1024 active virtual circuits. For small networks, this is not a limitation, but can become a problem for larger networks requiring many termination DLCIs. Some options are available for circumventing these limitations, such as combining the DLCI with a physical port and node number, NPA/NNX addressing scheme, or some other IP addressing scheme to allow a larger base of common addresses. Either way, the DLCI changes on multiple hops through the network, and only identifies a point-to-point virtual connection between two ports, whether they are defined as user or network ports.

Some vendors, such as StrataCom, use additional addressing schemes such as edge and node identifiers, while other router vendors translate DLCI addresses into Internet Protocol (IP) addressing schemes. When frame relay network design is discussed, the addressing options available for implementation will be reviewed. The ANSI specification provides the Extended Address (EA) bits with the capability to extend the DLCI address by two bits, thus breaking the barrier of the 1024 DLCI address limitation.

9.7.5 Proposed Address Structures of ANSI T1.618

ANSI Standard T1.618 defines two additional methods of extending the address field. The current version discussed, the default, is the use of two octets for the address field. There are options for identifying three and four octet address fields. A representation of the three octet extended addressing is seen in Figure 9.22. A representation of the four octet extended addressing is seen in Figure 9.23. The Data Link Control Indicator (D/C) indicates that the octet serves as an extended address DLCI.

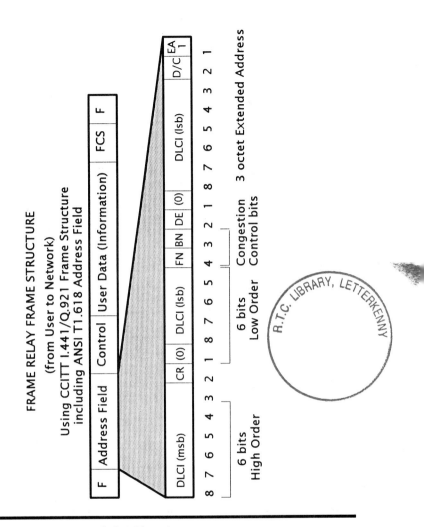

Figure 9.22 Three Octet Extended Addressing

ANSI T1.618 also defines the methods of assigning DLCI values based on the bearer channels (B-channel and N-channel) and on the D-channel. Both methods can use the 2 octet address scheme, but three and four octet extended addressing is defined only for the bearer channels. When using the D-channel, only DLCIs from 512 to 1007 can be assigned using frame relay connection procedures.

When using the B-channel or N-channel bearer channels, the DLCI values have a much stricter rule base. Table 9.1 shows the differences in using the two, three, and four octet DLCI addressing structures and how these DLCIs are assigned.

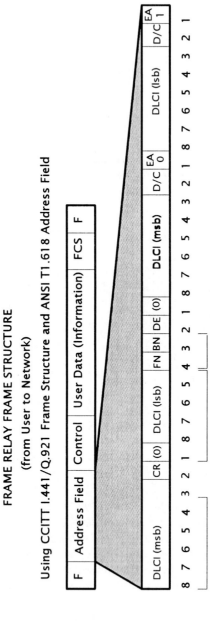

Figure 9.23 Four Octet Extended Addressing

TABLE 9.1 DLCI Addressing Structures and Assignments

2-octet DLCI value	3-octet DLCI value	4-octet DLCI value	Function
0	0	0	In-channel Signaling
1-15	1-1023	1-131071	Reserved
16-1007	1024-64511	131072-8257535	Assigned using FR procedures
1008-1022	64512-65534	8257536-8388606	Reserved
1023	65535	8388607	In-channel Layer Management

FRAME RELAY FRAME STRUCTURE

(from User to Network)
Using CCITT I.441/Q.921 Frame Structure
including Data Encapsulation Types

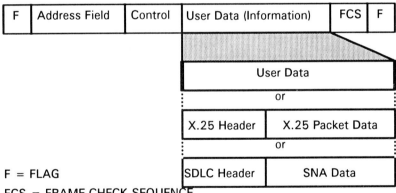

F = FLAG

FCS = FRAME CHECK SEQUENCE

DLCI = DATA LINK CONTROL IDENTIFIER

CR = C/R = COMMAND RESPONSE INDICATOR

FN = FECN = FORWARD EXPLICIT CONGESTION NOTIFICATION

BN = BECN = BACKWARD EXPLICIT CONGESTION NOTIFICATION

(0) = EAB (0) = EXTENDED ADDRESS BIT ZERO

(1) = EAB (1) = EXTENDED ADDRESS BIT ONE

DE = DISCARD ELIGIBILITY BIT

Figure 9.24 Frame Relay Frame Structure with X.25 or SDLC Encapsulated Packet

9.7.6 Data Field Structure

The data field can vary in size up to 4096 octets long. This size is limited by the integrity of the Frame Check Sequence (FCS). The data can be either pure data — when using a direct connection to a device which provides a frame relay interface — or it can be encapsulated packets of a different protocol. In the second case, an X.25 or SDLC packet, complete with packet header, subnetwork addressing, and user data, is encapsulated within the frame relay data field. This subnetwork packet is then unencapsulated at the distant end. Figure 9.24 shows the frame relay frame structure breakout with an encapsulated X.25 packet or encapsulated SDLC packet.

9.7.7 Frame Check Sequence

The frame check sequence assures the data integrity of the frame. If there is an error, the frame is discarded. The protocol does not correct the frame, so the packet will have to retransmitted. The FCS is defined in ANSI Standard T1.618 as a 16-bit Cyclic Redundancy Check (CRC-16).

FRAME RELAY FRAME STRUCTURE

(from User to Network)

Using the Local Management Interface (LMI) Extension Framing

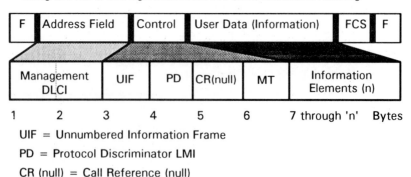

UIF = Unnumbered Information Frame

PD = Protocol Discriminator LMI

CR (null) = Call Reference (null)

MT = Message Type

Figure 9.25 LMI Extensions Frame Format

9.7.8 LMI Extension Framing

Figure 9.25 shows the LMI Extensions' data link layer frame format. It is similar to the Q.931 frame format, but uses the management DLCI 1023, the LAP-D unnumbered "frame byte" with poll bit set to zero, a protocol discriminator "byte set" to identify LMI, and a "null dummy" call reference. There is also the message-type byte, which defines the type of information elements in the information frame, where the details of the message reside.

There are two types of messages: STATUS_ENQUIRY and STATUS. STATUS_ENQUIRY is sent by the user device to request a status message from the network. The STATUS message is then sent from the network to the user device telling the status of PVCs in the network connected to that user device. Information elements are sent to the user device when this status is requested. Information elements can contain a KEEP_ALIVE_SEQUENCE which proves that both the user device and the network element are active, or a PVC_STATUS which gives the configuration and status of an existing PVC, or a REPORT_TYPE to indicate either the type of inquiry requested by the user device or the status message content. Figure 9.26 shows the same LMI extension frame but includes a sample STATUS message format. Notice that the message, REPORT_TYPE, KEEP_ALIVE_SEQUENCE, and PVC_STATUS have filled the information portion of the frame.

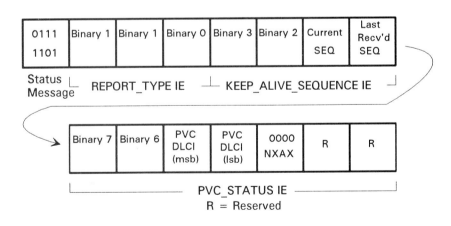

Figure 9.26 LMI Extension Frame Including Sample STATUS Message Format

The message and information elements discussed above are used to perform the "heartbeat process" consisting of heartbeat polling, keep alive sequencing, and adding new PVCs. These procedures ensure synchronization between the user and network devices, report new virtual circuits, report the deletion of old virtual circuits, and ensures link integrity between the user and network device. The LMI procedures also provide this information so that additional monitoring and maintenance can be accomplished by the user devices.

9.8 REVIEW

Frame relay defines and represents three areas: an interface, a protocol, and a network service. The standards which govern frame relay include CCITT, ANSI, and the LMI extensions written by the "Gang of Four" frame relay vendors. When frame relay is defined as an OSI Reference Model layer two protocol, it contains two separate functions: that of core services which provide the user-selectable services and that of the procedural sub-layer for frame relay transport and signaling. Finally, the transmission theory of frame relay covers the basic protocol structure of the SAP and DLCI, and the many types of frame and address and data field structures, as well as the LMI Extensions parameters for transmission. Now for the operations and service aspects of frame relay.

10

Frame Relay, Part 2 — Operation and Services

This chapter discusses the operation of frame relay and the services which can be provided. It defines both the American National Standards Institute (ANSI) and the International Telegraph and Telephone Consultative Committee (CCITT) implementations of user connection modes, which are the user interface options to the frame relay network. User access parameters will be discussed in detail such as the interface, applications, interface signaling, management of permanent virtual circuits, and methods of managing congestion and user bandwidth — all topics near and dear to the user's heart. Frame relay networks can be accessed in a variety of ways through a variety of hardware, and we will explore each option. The network view comes next, where options for network addressing, multiplexing and switching are explored. Frame relay will be compared to X.25 packet switching operation. Performance is the major design consideration in a frame relay network. Parameters affecting it, including buffers, bit error management, overhead, and delay, as well as the network management of queuing and statistical multiplexing, are explained. Frame relay services are emerging now in the marketplace. This chapter outlines what services will succeed from both a national and an international view. Tariffs, rate structures, and the committed information rate (CIR) will be studied, along with national and international frame relay services.

10.1 ISDN FRAME MODE BEARER SERVICES — A SHORT HISTORY

Most frame relay services are being offered through non-ISDN connectivity. But since frame relay interfaces and service standards primarily revolve around the ISDN frame mode bearer service, a few terms will prove valuable to know. User equipment is designated by TE1, or the ISDN frame mode terminal, or NT2, which defines network termination equipment conforming to the X.200 reference series. An ET is a network exchange termination, an FH denotes a frame handler and a RFH denotes a remote frame handler. All will be discussed in detail. The NT1 termination has been omitted for clarity and is not included in the present discussions or figures. The terms for these devices and their logical connectivity can be found in the X.31 Packet Mode Services standard.

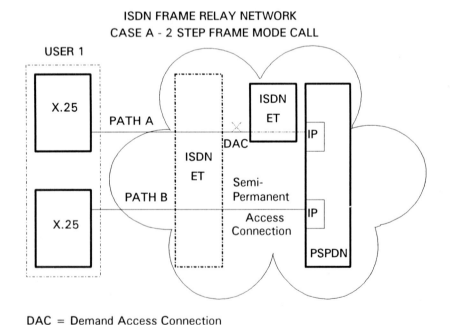

DAC = Demand Access Connection

Figure 10.1 X.31 PSPDN Access (Case A)

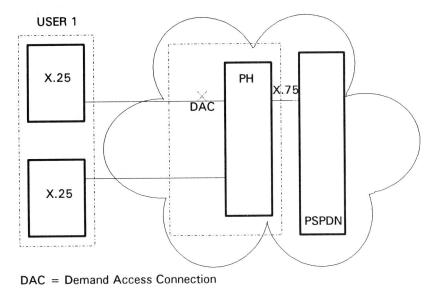

ISDN FRAME RELAY NETWORK

DAC = Demand Access Connection

Figure 10.2 ISDN Virtual Circuit Data Service

Figure 10.1 depicts the first of two types of access defined in X.31, which is the access to a packet switched public data service. The second type, shown in Figure 10.2, defines access to an ISDN virtual circuit data service, such as frame relay. The next section shows how these principles of access connectivity also apply to the ISDN frame mode bearer service frame relay implementation, but use a frame handler in place of the Packet Switched Public Data Network (PSPDN). Both PSPDNs will be replaced by the frame handler and remote frame handler as part of a frame relay implementation.

While the X.31 Recommendations were designed to accommodate in-band signaling procedures for X.25 packet mode services under ISDN standards, the X.31 Recommendations have been replaced by the CCITT I.122 additional ISDN frame mode bearer service standards. This recommendation uses out-of-band signaling, and defines four major types of services: frame relaying 1, frame relaying 2, frame switching, and X.25-based additional packet mode services.

The last chapter covered the protocol structure defining frame relay protocols. This included the physical layer functions and the logical link layer split into two planes: the Control Plane (C-plane) and the User plane (U-plane). The user plane was further divided into core functions and user-definable functions.

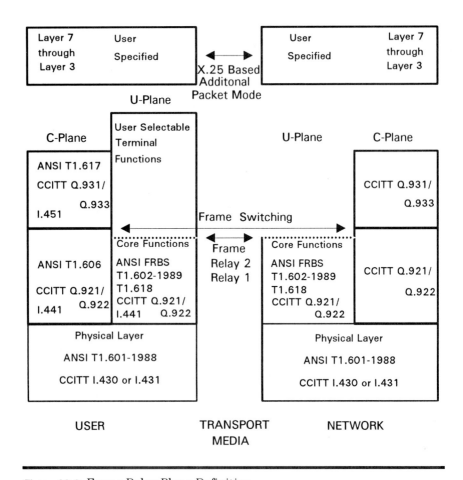

Figure 10.3 Frame Relay Plane Definition

Figure 10.3 shows these planes as defined, illustrating how the four services relate to these planes.

The four bearer services are defined by different functions defined within the CCITT Recommendation I.122, depending upon the U-plane function and whether they are "user" or "network" interfaces. Table 10.1 shows the four services and their applications to each bearer service. This table can be related to Figure 10.3.

Understanding the history behind the packet mode connection service definitions, and how they relate to this new ISDN bearer service called frame relay, the details of implementing these access techniques can now be explored.

TABLE 10.1 User Services and Functions

Bearer Service	User-Defined Function	Network Service
Frame Relaying 1	Core I.441 Functions	Core I.441 Functions
Frame Relaying 2	I.441	Core I.441 Function
Frame Switching	I.441	I.441
X.25-based additional packet mode	I.441 - X.25	I.441 - X.25

10.2 ANSI USER CONNECTION MODES

Two types of user connection modes are defined by ANSI T1.617 Section seven. These access types define the management of call control and access to frame-handling facilities. They are labeled as Case A and Case B. Our primary interest is in these two types of connection control access techniques as defined in ANSI T1.617. Case A is a two-step frame mode call establishment method to provide either a frame relay connection on demand or a semipermanent frame relay connection. Case B, on the other hand, is an integrated access method which needs the ISDN call-setup control mechanism. This provides a frame relay connection for "on demand" only. These call procedures may flow on either the DLCI=0 logical channel within the B-channel or the H-channel as in Case A, or in the D-channel only approach as in Case B. The ANSI T1.617 standard also provides extensions to Q.931 circuit-mode procedures, and provides new information elements. Examples of each case are covered in detail below.

10.2.1 Case A

Case A is referred to as two-step frame mode call establishment, because there are two steps involved in establishing a frame relay call to the network. This mode of operation provides either a frame relay connection "on demand" or a semipermanent frame relay connection. In Case A, frame relay connections are established through the B-channel or H-channel. This provides circuit-switched access to a remote frame handler. Either the user or the remote handler can initiate the connection. Figure 10.4 shows a user labeled as User 1 connecting to a frame relay (ISDN) network.

In the first step, the user (labeled TE1 or NT2) establishes a circuit-switched connection indicating that the channel is used for frame relay. This circuit mode bearer connection is established using ANSI T1.607 procedures. (This step is not required if a semipermanent connection already exists between the TE1 and the RFH over the B-channel or H-channel (in-channel signaling) and is equivalent to Path A in the last figure.) This call setup process begins by sending a SETUP message on the D-channel containing both called party address information and bearer capability information. This message also includes information transfer capability, circuit transfer mode notification, and bit rate information. It also sets the layer two protocol information to core aspects of frame relay to distinguish the type of service to be used.

Once this circuit is established, the user employees in-channel signaling to initialize the logical link through DLCI=0. Thus, the connection control message flow is on logical link DLCI=0, not the D-channel. This moves the call setup into the frame relay connection setup phase. The second step is when logical link DLCI=0 is used to pass connection control messages over the previously established channel. This step is equivalent to both Path A and B in the last figure, and is defined in both ANSI T1.607 and ANSI T1.617. Information flow and signaling are discussed in later sections.

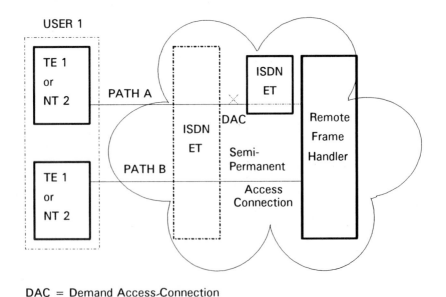

DAC = Demand Access Connection

Figure 10.4 Case A — 2-Step Frame Mode Call Establishment

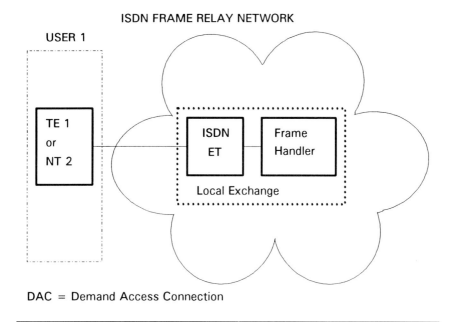

ISDN FRAME RELAY NETWORK

USER 1

TE 1
or
NT 2

ISDN
ET

Frame
Handler

Local Exchange

DAC = Demand Access Connection

Figure 10.5 Case B — Integrated Access

Thus, for frame relay connection "on demand", the TE1 communicates with the RFH to establish a frame relay connection to a target TE on the other side of the network. The control messages are passed via Q.93X in the DLCI=0 logical channel within the bearer channel. For a semipermanent connection, the TE1 communicates with the serving switch (ET) to establish circuit mode connections to the remote frame handler (RFH). Control messages are passed via the D-channel as defined in Q.931.

10.2.2 Case B

Case B is referred to as integrated access because call establishment is integrated with the D-channel signaling process. This provides a frame relay connection "on demand". Frame relay access is established via the frame relay virtual call service on the local ISDN (frame relay network). Because of this there is a need for ISDN call setup control. Frame relay data transport can occur over the B-channel, H-channel, or the D-channel. These channels can act as the physical connection for frame mode calls. Again, either the user or frame handler may initiate the connection. Figure 10.5 shows a user labeled as User 1 accessing the frame relay network local exchange via path C. The Local Exchange can either be a Local Exchange Carrier (LEC) or an Interexchange Carrier (IXC).

The user decides whether to use a specific bearer channel, allow the network to select that channel, or to use the D-channel for the frame relay call. For frame relay connection "on demand", the TE1 communicates with the FH at the Local Exchange to establish frame relay connection to a target TE on the other side of the network. The frame relay bearer connection is established via the D-channel using SAPI 0 indicating Frame Mode Bearer Services, per the ANSI T1.617 procedures (and Q.931). Annex B and D of ANSI Standard T1.617 defines additional procedures for permanent virtual connections which we will be defined in detail later.

10.2.3 Incoming and Outgoing Calls

When a user wishes to place a call, the service type (A or B) must first be determined. If the call is placed using type A service, the user must observe the two-step process listed above by first establishing the circuit-switched connection (or if this has been done already, skip to the next step). The frame relay connection setup is then established and information flow begins. When a user places a call using type B service, he will establish the virtual call using the signaling procedures (in SAPI 0) as described above. Figure 10.6 shows the various types of connection establishment methods for the frame relay services discussed above, and includes the pertinent standards for each.

All parameters of the demand access connections mentioned above are negotiated during the call setup time with the virtual circuit setup and release controlled by means of the enhanced ISDN access protocols and procedures. The parameters for the semipermanent connections are pre-arranged at the time of subscription and the permanent virtual circuits are released through administrative procedures by the network provider. In both services mentioned above, the connections, signaling, and exchange of information for incoming calls is similar to that of outgoing calls.

10.3 CCITT USER CONNECTION MODES

We have already discussed the beginnings of user connection modes defined in both CCITT Recommendations I.441 and X.31, as well as the four types of connection modes: frame relaying 1, frame relaying 2, frame switching, and the X.25 connection mode. Now to view these ISDN standards as they apply the HDLC framing to frame relay services. These ISDN standards are the same standards used for frame relay. All frame relay standards rely on ISDN standards using the OSIRM layer 2 LAP-D data link layer standard, with D-channel signaling and the B or D channel used for the transmission of information, depending on whether it is Type I or Type II. Both types of the standard will be covered.

CCITT Recommendation I.441/Q921 defines the frame relay implementation of the LAP-D core protocol function, and further defines the HDLC framing process. This is where the addressing of frame relay is defined through Data Link Connection Identifiers (DLCIs), enabling multiple users on a single user interface. CCITT Recommendation I.441 also calls for three additional functions to be implemented in the upper half of the OSIRM layer two (data link layer): link utilization, flow control, and error recovery. These are the main functions implemented by the traditional X.25 packet-switching protocol, which is overhead intensive. Thus, these three functions are rarely implemented when using this recommendation and are performed by the user DTE equipment or other higher-level protocol implementations.

	Access Connection Establishment	Frame Relay Connection Establishment
Demand	Case A: Demand T1.607 Circuit Mode Bearer Service Procedures	Case A: Demand T1.617 Procedures on DLCI 0 In the Bearer Channel
	Case B: Demand T1.617 Procedures	Case B: Demand T1.617 Procedures On D-Channel
Permanent Access Connection	Permanent	Case A: Demand T1.617 Procedures on DLCI 0 In the Bearer Channel
		Case B: Demand T1.617 Procedures On D-Channel
Permanent Frame Relay Connection	Permanent	Permanent

Figure 10.6 Access and Frame Relay Connection Establishment

The Q.931 standard was developed by the CCITT to handle the access signaling portion of frame relay. The Q.922 standard has been developed to handle the core aspects of frame relay. The Q.921 standard defines the frame format and corresponds with ANSI T1.602.

As discussed in the section on ANSI, the CCITT Recommendation I.122 defines the frame relay bearer services under the ISDN Recommendations. There are two main types of frame relay defined by this recommendation: Type I, private or virtual private frame relay, and Type II, or public frame relay. These two types define the service descriptions and network architecture by which these services are offered. Within the CCITT recommendation, there exists two distinct types of frame relay services, labeled Type I and Type II. Both methods are discussed below.

10.3.1 Type I

Type I frame relay, also called private or virtual private network frame relay, contains the functionality of the Link Access Protocol — D Channel (LAP-D) as defined in the CCITT Standard I.422 and Q.921. This standard does not include the D channel call processing capability, which distinguishes it from Type II. Type I is the simplest form of frame relay. It performs the statistical multiplexing and bandwidth-on-demand functions needed by LAN bridging and private line data communications, while concerning itself only with OSI level one and level two transport.

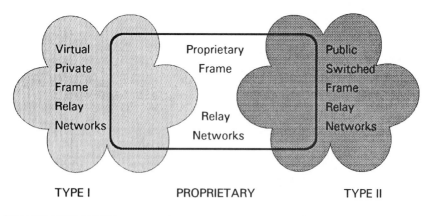

Figure 10.7 Frame Relay Implementation Types

Figure 10.7 shows frame relay implementation types and how they relate to one another. Type I is the Permanent Virtual Circuit (PVC) implementation. Typical applications for Type I frame relay include LAN bridging and routing and many of the initial vendor data transport interface offerings.

10.3.2 Type II and Q.931

Type II frame relay, also called the Q.931 implementation, defines the public switched frame relay implementation. This implementation not only follows the CCITT Standard I.422, but includes the D-channel for call management as well. This allows Type II services to interface directly to ISDN public networks. The ISDN primary rate interface Q.931 signaling standards are implemented using the D-channel. The Q.931 is a message-oriented control protocol used for call setup and tear down. Type II offers public network access and integrated data services (ISDN) capabilities. As public network frame relay services proliferate, Type II will eventually catch up to and outpace Type I services, providing a nonproprietary network solution which is no longer vendor specific. This is the Switched Virtual Circuit (SVC) service as defined in CCITT Q.921 and Q.931.

The D-channel signaling defined in Q.931 is an extension to ISDN signaling and provides one method to monitor the status of network devices. This standard maps to the ANSI T1.607 recommendation. This is needed, because it is not possible to remotely monitor the status of a single device through the physical interface when multiple virtual circuits are being passed over the same physical interface. D-channel signaling allows for the remote monitoring of multiple virtual channels over the same physical interface.

Virtual circuits established in this mode are always available as a network resource, and traffic for a given virtual circuit always uses the same (or alternate) path through the network. Some of the LAP-D-Core functions of this implementation are:

* transmission error detection
* frame delimiting
* frame alignment
* frame multiplexing/demultiplexing (using address field)
* data transparency
* frame inspection (assures that the frame consists of an integer number of octets prior to zero-bit insertion or after zero-bit extraction)

This implementation was the first frame relay standard developed. Typical applications for Type II frame relay include ISDN public network voice and data enhanced offerings and user access to the public switched network.

10.4 FRAME RELAY ACCESS — THE USER'S VIEW

Major applications, as users of frame relay networks, do so through the use of T1/E1 multiplexers, statistical multiplexers, packet switches, bridges and routers. These devices have traditionally operated independently and have rarely interfaced through a common transport media, other than perhaps a digital data network of dedicated point-to-point circuits. Frame relay provides a common transport platform for many of these devices.

Users want one-to-one, one-to-many, many-to-one, or many-to-many connectivity via various types of circuit interfaces. In previous chapters, these circuit interfaces were shown to communicate over either fixed or switched circuits. Frame relay offers the best of both types, with the capability to define fixed-virtual circuits over a switched network service efficiently and quickly via packet transmission techniques. Frame relay user service can give the impression that every device on the network is logically connected to every other device. The greater the mesh in the backbone network, the greater this truth.

Certain functions were defined last chapter in our protocol section. We showed that the OSIRM layer two (data link layer) was split by frame relay standards into two major areas: core services and user-defined (and selectable) services. We will now cover some of these user-definable services.

10.4.1 User Interface

By eliminating the need for multiple access lines into a switched backbone, whether a private or public frame relay network, users can reduce a significant portion of their entire networking costs. Since access charges constitute the majority of data networking costs, reductions due to frame relay can be significant enough to justify the purchase of additional hardware, services, and even support systems. The user can realize huge cost savings, but must also plan for supporting the control mechanisms not inherent in frame relay, such as congestion control and error correction. The actual physical user interface is typically an RS-449 or V.35 connection, although ISDN interfaces may soon become popular. With frame relay the user can reduce the number of interfaces to the network to multiple V.35 or four-wire if a CSU/DSU is needed.

Frame relay access will thus reduce the costs associated with network access. Part of this cost in access is when a user must go through a LEC via a "tail circuit". The user should perform an analysis of bandwidth requirements as defined later in this book, but as a general rule, try accessing the network with a circuit providing more bandwidth than required, and pay for only what is used. This method will cost close to the same in the short run, but will often provide the performance of the port access speed. It provides an opportunity to upgrade as soon as more bandwidth is needed, which will be very, very soon.

10.4.2 User Applications

Typical synchronous traffic might include long network connection times, excessive call setup and takedown times, long transmission sessions, nonbursty traffic patterns, and PVC connections. This is different from typical packet network characteristics, which may include short network connection times, short call setup and take-down times, short transmission sessions, and bursty traffic patterns. As mentioned previously, user applications which traverse frame relay networks and utilize frame relay services must have the intelligence to control both congestion control and error correction and retransmission of data due to these conditions.

10.4.3 Interface Signaling

Most of the methods of user-to-network interface signaling have already been covered, but additional elements in the channel may require additional signaling. Transmission equipment such as CSUs, DSUs, Extended SuperFrame Monitoring Units (ESFMUs), and other channel conditioning devices may require in-band or out-of-band signaling. This should be transparent to the frame relay transmission, while providing maximum throughput, line efficiency, and minimum response time degradation and delay.

ANSI has also defined some annex recommendations. Though not necessary for standards compliance, they can make user management much more efficient and thorough. These issues include PVC status, congestion recognition and signaling, fairness and guaranteed user throughput, and other future expansions and enhancements. Now for a look at a few of these enhancements.

10.4.4 PVC Management

When a Permanent Virtual Circuit (PVC) is established between two physical ports, and one or multiple DLCI addresses are established over this link, there is a need for both network elements (user and network device or network and network device) to manage the status of the link. PVC management is defined in both ANSI T1 specifications and the LMI Extensions. Both specifications define three main areas: PVC status, DLCI verification, and physical interface keepalive heartbeat.

The LMI extensions define DLCI address number 1023 as the Local Management Interface (LMI) address. The PVC status and configuration information features provided by the LMI include notification of PVC availability or unavailability, deletions, additions, and presence of PVCs, and a keep-alive heartbeat sequence which verifies that the physical interface is active.

Annex B of ANSI Standard T1.617 defines additional procedures for permanent virtual connections which apply to implementing both PVCs and SVCs on the same interface, Case B with D-channel signaling implementations, and Case A with in-channel signaling and PVC-only implementations. The primary benefit of this annex is the notification of and recovery from PVC outages. Annex D provides additional procedures for PVCs using unnumbered frames. It is for operational purposes in Case A in-channel signaling only.

10.4.5 Congestion Notification and Control

Congestion increases with the traffic load across the network. Network congestion occurs when the traffic attempting to be passed is greater than the available bandwidth (after overhead). Typically, network transport hardware will have some buffering capability, but when this capability is exceeded, a congestion condition occurs. In basic frame relay transport, as discussed last chapter, when the network reaches this congestion point it will begin to discard frames until there is no longer a congestion condition. When network congestion causes the network to discard frames, the user or network access devices must have the intelligence of higher level protocols to provide end-to-end transmission and retransmission.

Implicit. There are two types of congestion control used to manage frame relay data transport: implicit and explicit congestion notification. *Implicit* congestion notification infers the use of a Layer 4 transport protocol, such as the DOD Transmission Control Protocol (TCP), in either the network device or the user premises equipment. These protocols work similarly to the transmit and receive windowing in X.25 packet switching, but manage the end-to-end transmission of "frames" instead of "packets". TCP reduces the window size, or the number of frames transmitted, according to network delay or frame loss. This allows the end users (or the network access device) to accommodate network congestion and avoid discarding frames and retransmissions, but it also implies that the user will take the responsibility to manage congestion control. Many of the approaches discussed below can be used in conjunction with TCP implemented at the user device. Thus, flow control could be adjusted by the TCP upon receipt of congestion information from the frame relay network.

Explicit. The second type of congestion control is *explicit*. Explicit congestion notification comes in three flavors: Forward Explicit Congestion Notification/Backward Explicit Congestion Notification (FECN/BECN bits) and Consolidated Link Layer Management (CLLM) addressing. ANSI Standard T1.618 clearly defines congestion control with the FECN/BECN bits and manages message notification through the CLLM mechanism. Each method will be discussed in detail.

FECN/BECN Bits. Flow control is built into the frame relay address field in the form of FECN and BECN bits. As discussed previously, the FECN bit is a toggle which tells the remote user that network congestion was encountered by the frame as transmitted across the physical media, and that action should be taken to prevent data loss. The BECN bit works similarly, in that it notifies the user that there is congestion in the data on the returning path. An increase in the frequency of FECN and BECN bits received is a good indication of congestion in a network.

Figure 10.8 depicts a frame relay network connecting a host in Dallas with many remote user devices in Charleston. In the process of downloading massive files from the mainframe in Dallas to the users in Charleston via a PVC (shown as the dashed line), network congestion occurs at the Atlanta node. The Atlanta node sets the FECN bit to a "1" and notifies the Charleston node of impending congestion on those DLCIs in the defined PVC. The Atlanta node will also set to a "1" the BECN bit which is sent to Dallas, informing of the congestion condition and allowing Dallas to implement some type of throttle-back flow control until the congestion condition, and the BECN bit, changes. The FECN and BECN bit setting will continue throughout the network on all DLCIs until all devices which define circuits through Atlanta are notified of the congestion condition.

Unfortunately, the FECN/BECN bits are not often used, not because they do not function as designed and alert of a congestion problem, but because many CPE applications accessing the network can neither understand nor make use of this information. Thus it is ignored. If this is the case, data is lost; either in the network or at the network access point. Wellfleet was the first and, so far, the only vendor to read and use both FECN and BECN bits to address congestion issues.

Consolidated Link Layer Management. The second form of congestion management defined by ANSI is the Consolidated Link Layer Management (CLLM) function. CLLM reserves one of the DLCI addresses on a frame relay interface strictly for transmitting control messages to user devices during conditions, when there are no frames to transmit, yet a congestion notification still needs to be sent. This CLLM message is a contingency for notifying users of congestion activity outside the conventional framing structure, since there is no provision in the standards for "empty frames" which contain only congestion control information. The CLLM also contains a list of DLCIs that correspond to the congested frame relay bearer connections — all users affected are then notified of congestion. Multiple CLLMs can be transmitted, and will be needed in a network with many DLCIs addresses.

An example of CLLM in use is depicted in Figure 10.9. ANSI T1.618 defines the message based on the ISO 8885 definition of the use of XID frames for the transport of congestion management information. This figure shows the same network configuration as in the last example, but now Atlanta is notifying Dallas of network congestion through DLCI number 1023.

One further note on the use of CLLM: since both the ANSI-defined CLLM and the frame relay Specifications with Extensions which defines Local Management Interface both use DLCI 1023, they are mutually exclusive, and cannot be used simultaneously.

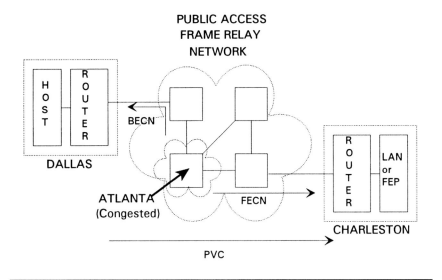

Figure 10.8 Congestion Control Example

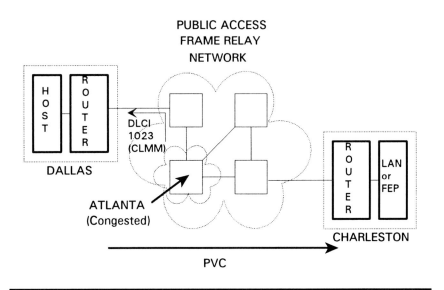

Figure 10.9 Consolidated Link Layer Management (CLLM) Function

10.4.6 Bandwidth Limitations

There are many ways to limit the amount of bandwidth provided to a user at any time. This may be a critical factor during congestion conditions, when many users of both high-bandwidth and low-bandwidth requirements are contending for limited resources. One method of delimiting the bandwidth to select users is through the intelligent and selective use of the Discard Eligibility (DE) bit discussed in the last chapter. This bit could also be used as a two-tier priority scheme. Other methods employ vendor proprietary implementations.

10.5 NETWORK ACCESS DEVICES

A variety of hardware devices exist on the market which make frame relay network access possible. These devices range in complexity and include bridges, brouters, routers, gateways, FRADs, multiplexers, and even voice switches. This is not a complete list, just the more popular devices. While many frame relay service providers offer a frame relay "cloud" which promises ubiquitous access, it befalls the user to employ intelligent premises devices to concentrate, aggregate, multiplex, and translate existing traffic and protocols into a frame relay interface to the network. Some users will even deploy a "hybrid" solution, using one or more of the above devices.

Various pictures of vendor hardware are shown throughout this book to give an idea of what these devices look like. It should be noted that they vary from the size of a personal computer to that of a central office switch, even though they provide the same interfaces and functionality. While processor and memory chip size, as well as pure processing power and the method by which it is accomplished, distinguish these vendor products, much of the protocol conversion, processing, and switching is done through software. This shows that the size of the hardware is influenced primarily by the number of access and egress ports on the device, and is often not representative of the functionality provided.

When purchasing frame relay access or switching hardware, ensure that the vendor is compatible with the standards discussed here. Support should be available for the DLCI and all frame relay header bits, FCS review, the use of FECN, BECN, and DE bits, congestion control conforming to the standards, and multiple protocol support.

10.5.1 Bridge/Router Access Devices

Many bridge and router vendors are providing frame relay interfaces in their products. Since the architecture of these devices lends itself to offering multiple multipurpose ports (as opposed to access and trunk side distinctions as traditional multiplexers and switches) they view frame relay as a software implementation on synchronous cards and provide it as they would any other protocol. Because of this flexibility, bridge and router vendors were one of the first genre of vendors to offer a frame relay interface. Some examples of vendors offering frame relay interfaces include cisco Systems, Inc., Wellfleet Communications Inc., and Proteon. Figure 10.10 shows the Wellfleet Concentrator Node (CN), Link Node (LN), and Feeder Node (FN) from large to small, respectively.

Figure 10.10 Wellfleet Routers — CN, LN, FN

10.5.2 Frame Relay Access Device (FRAD)

Figure 10.11 depicts the connectivity of a frame relay Access Device or Assembler/Disassembler (FRAD) to the private or public switched data network. Access speeds are via DS0, fractional T1, or full T1. A FRAD aggregates multiple data network access circuits (i.e., SNA/SDLC, Bisynchronous (BSC), Asynchronous, X.25, 2780/3780 RJE, DEC Hosts) from terminals, hosts, and other various network elements into a single frame relay access circuit, performing the frame packetizing function. FRADs also contain the powerful capability to carry SNA and LAN traffic over a single interface or network access circuit, and some even provide "spoofing" or some level of PU4/PU5 emulation.

FRADs can perform a level of congestion control outside the frame relay network through SDLC congestion control techniques as discussed in previous chapters. By generating and acknowledging the polls of each individual session, the FRAD can reduce or eliminate polling from being transmitted across the frame relay interface and thus drastically reduce the congestion caused by polling on each session. The drawback is the intense network management necessary to perform this function.

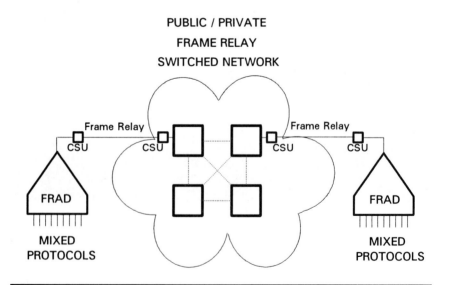

Figure 10.11 Frame Relay Access Device or Assembler/Disassembler Network Example

Most FRADs support frame relay transport with transparent transport of X.25 and SDLC protocols over frame relay PVCs and SVCs. Some FRADs also support the ISDN PRI and SMDS SNI interfaces. Figure 10.12 shows a FRAD providing access to many of the devices listed above. The network access side shows a frame relay access circuit with X.25 and SDLC transparent protocol passing. Typical interfaces are RS-232, RS-449, V.35 and DS1. Trunk side interfaces are typically DS0, FT1, T1, T-3, with ESF, SF, B8ZS formatting.

There are many advantages of using FRADs:

★ uses existing equipment and network hardware
★ minimizes new capital required to access a frame relay network
★ performs some FEP and cluster controller functions (possibly eliminating the need for them)
★ acts as a protocol concentration/translation device
★ facilitates multiprotocol communications

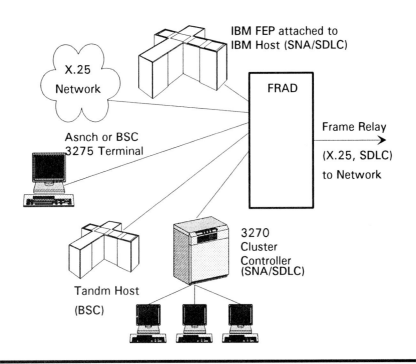

Figure 10.12 FRAD Connectivity Example

Another device used to transmit video over frame relay is the video coder/decoder (codec) — manufactured by Bolter Communications. This codec transmits "pixel updates" for only the portion of video which changes. Thus, data is only transmitted in bursts when movement occurs. Since the term "FRAD" has multiple meanings (e.g., frame relay access device, frame relay assembler/disassembler), many special-function FRADs will emerge to aggregate user application protocols for access to frame relay networks.

10.5.3 Voice Switch as Access Devices

Some voice switch vendors are providing frame relay access cards to aggregate a frame relay channel over existing dedicated trunks. One example of a vendor providing this functionality is Northern Telecom. Northern Telecom provides a frame relay interface for their DMS-100 central office and DMS-250 carrier tandem switches, accommodating both DS0 and T1 frame relay interfaces.

Also emerging on the market are a few frame relay switches. These devices integrate PAD, line concentration, and frame relay switching into a single product. These frame relay switches also act as remote protocol concentrators and feeders to the frame relay network.

10.5.4 Multiplexer as Access Devices

Frame relay has now become the obsession of many intelligent multiplexer vendors. An existing synchronous interface card can provide the frame relay protocol intelligence and it works as an access trunk to the network or as a user input port to the multiplexer. StrataCom's Integrated Packet Exchange (IPX) multiplexer now has a frame relay interface, as well as Newbridge's new 3600 Mainsteet multiplexer. Another classic multiplexer and packet switch vendor is Netrix, who was one of the first multiplexer vendors to implement frame relay (see Figure 10.13). T1 multiplexers have an added advantage in that they can exploit frame relay switching through either dedicated T1s into the network or as "partitioned" frame relay over existing trunks. Some reconfiguration difficulties occur in this scenario, but this does provide frame relay switching.

As noted in the last chapter when discussing fast packet switching, many multiplexer vendors are providing fast packet capabilities in their products by packetizing all originating traffic into a fixed-length packet for transmission across the network. While this is an efficiency upgrade to common TDM multiplexing, it still provides a proprietary implementation which must be replicated at the destination end. There are other disadvantages to this implementation. Multiple dedicated circuits are still required for access to the network. Also, packets are not built with multimedia from all users accessing the multiplexer, but instead are built for one user at a time. This implementation is not true statistical multiplexing, but mainly an implementation of multiplexing using packets for transport. The actual multiplexing is accomplished by the interleaving of packets from different access circuits and bandwidth is allocated by the number of frames, not bits.

Netrix Corporation's
#1 Integrated Switching System
Sample Application

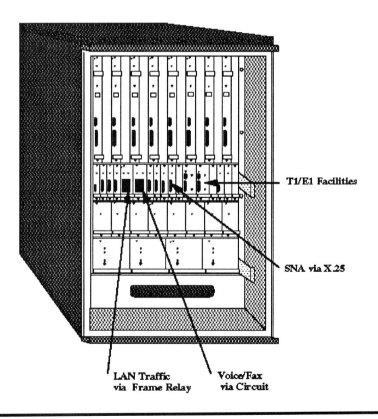

T1/E1 Facilities

SNA via X.25

LAN Traffic
via Frame Relay

Voice/Fax
via Circuit

Figure 10.13 Netrix Multiplexer with Frame Relay Interface

This multiplexing technique, while not providing true frame relay, still allows efficiencies. The primary difference between this style of fast packet and frame relay switching is that fast packet multiplexers still have dedicated paths for the data, while frame relay routes are based upon the DLCI in the address field. Also, multiplexers employing frame relay must allocate a portion of their trunks to frame relay, and this bandwidth is dedicated whether it is used or not. A true frame relay device, such as a router employing a frame relay T1 has the flexibility of "dynamically allocating" all available bandwidth to whatever application needs it at a given time. There are also cost efficiencies gained through the addition of only an interface card, rather than a new router.

There is another major drawback to the multiplexer frame relay implementation. Each multiplexer provides buffers for times of congestion control. This offers severe latency problems for users who ride SNA, or network operating systems which retransmit frames which may be sitting in a network node buffer — for which the user did not receive acknowledgments because they thought the packet was lost. This causes even greater network congestion. The router frame relay switching approach seems to be the best solution, where only the packets which are truly lost are retransmitted.

10.6 FRAME RELAY SERVICE — THE NETWORK VIEW

Understanding the theory of frame transmission, the protocols which manage this transmission, user access methods, and the hardware devices which interface to a frame relay network, an in-depth study of frame relay as a network service now can be done. To do this, the methods of addressing using DLCIs, logical channel multiplexing, and permanent and switched virtual circuit implementations must be understood.

10.6.1 Frame Relay Addressing

Each frame is sent from an originating DLCI to a destination DLCI, where the frame check sequence is verified. If the frame does not pass the FCS, it is simply discarded, with no indication to the network or user. If the frame does pass the FCS, the DLCI is located in a routing table. If the DLCI has been predefined for this PVC, it will be routed to its final destination. If the DLCI has not been defined for this PVC, it is discarded. If it is the destination node, it is passed through the logical and physical port to the user.

Figure 10.14 depicts a user CPE device residing in Dallas, who wishes to send exchange data with two other locations in Washington and Boca Raton. The Dallas CPE connects one T1 frame relay access line into the private virtual network cloud. The DLCI address for Washington is 85, and the DLCI for Boca Raton is 120. Washington and Boca Raton also want to send data to Dallas, and they use DLCI address 22 and 35, respectively.

All DLCI addresses are assigned using an extension of the CCITT Recommendation Q.931 call control procedures. These and other ANSI standards have already been discussed. ANSI Standard T1.618 proposes two additional methods of extending the address field. These were discussed last chapter.

Figure 10.15 shows a virtual frame relay network using DLCI addressing. It can be seen that user device A communicates with network device 1 through a single physical connection using multiple DLCIs for each virtual circuit. Each of these DLCIs have local significance, meaning that each set of DLCIs is established between each device on the network. Network device 1 then communicates over a different set of DLCIs to network device 2. Network device 2 then passes the frames on to user device B, the destination, and through yet another set of DLCIs again with local significance.

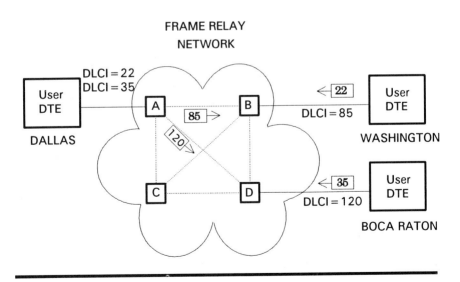

Figure 10.14 DLCI Addressing

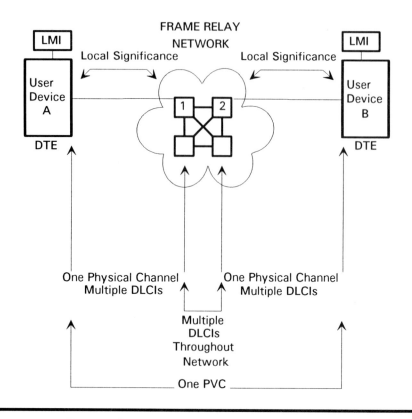

Figure 10.15 Frame Relay Network DLCI Addressing Example

The Frame Handlers (FH) and Remote Frame Handlers (RFH) perform much of the frame-relaying service between the physical ports and the mapping of the logical DLCIs between all ports in a group. It is the responsibility of the frame handler to identify the local physical channels which terminate on either the local exchange (Case B) or in the frame relay network (Case A). The DLCIs assigned to a frame handler for a group will be unique. The frame handler will also perform the following functions:

- ❖ maps in-bound to out-bound DLCI
- ❖ performs FCS and correct for retransmission
- ❖ discards corrupted frames
- ❖ writes out-bound DLCI value into the frame address field
- ❖ coordinates transport of frame out the physical channel

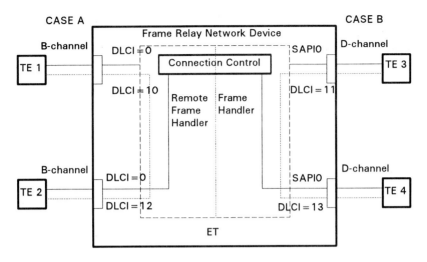

CASE A with DLCI = 0 in-channel signalling using B-channel
CASE B with SAPI0 integrated access signalling over D-channel
with data flows also over D-channel.

Figure 10.16 Remote Frame Handler Operation

Figure 10.16 shows the frame handler (or remote frame handler) operation. A frame relay network device is acting both as a frame handler for Case B integrated access and as a remote frame handler for Case A access. The Case B integrated access is using the D-channel for both signaling and data flow. TE3 is sending frames on DLCI=11 to TE4 DLCI=13. Call establishment is done via SAPI 0. The Case A access is via the B-channel for data transfer and call set-up, using DLCI=0 for call establishment. TE1 is sending frames on DLCI=10 to TE2 DLCI=12.

10.6.2 Logical Channel Multiplexing via Frames

Through the use of the DLCI addressing, multiple user logical data streams can be multiplexed and demultiplexed within the same physical data channel. Each physical channel can contain up to 1023 logical channels, each identified by a DLCI value (or more with extended addressing). Logical channel 1024 is reserved for CLLM. These multiplexed users are assembled into frames and transmitted across the network. These frames retain their order of transmission and reception. Each protocol is negotiated during the call establishment procedure.

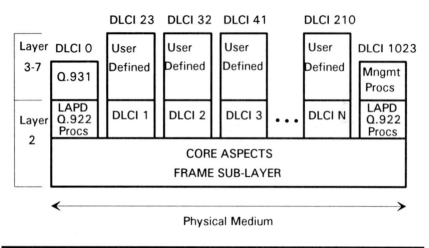

Figure 10.17 Frame Level Multiplexing

Figure 10.17 shows a standard frame representing the physical media connectivity, the data link layer core aspects connectivity as well as the LAP-D and Q.922 signaling and communications procedures. The figure also shows how the data link control procedures are multiplexing the various user DLCIs into each frame. In this example, there are various logical channels attempting to transmit. The figure shows DLCI 23, 32, 41, and 210, and assumes there are others between 41 and 210. These users are employing their own level three and higher user-defined protocols. We also see the frame establishment using the Q.931 procedures and DLCI 1023 being used for the link management procedures.

Each logical channel can operate with a different data link control procedure and a different network layer service. The many different types of data link control procedures which can be carried on a logical channel are:

* LAPB
* LAP-D
* LAP-D Extended
* IEEE 802.2
* SDLC

There are different types of network layer services which can ride over a logical channel. These protocols are transparent to data transport across the frame relay network. These protocols and services include:

- IP
- X.25/X.75
- SNA
- ASCII

These protocols and services are offered over various types of traffic and circuits including:

- priority traffic
- revenue bearing traffic
- time sensitive traffic
- single destination traffic
- multiple destination traffic
- permanent virtual circuits
- switched virtual circuits

10.6.3 Switched Virtual Services

Both ANSI and CCITT are publishing standards for Switched Virtual Circuit (SVC) frame relay operation. The standards (based on ISDN Digital Subscriber SSNo.1) define a signaling interface to build a switched virtual circuit. When these service offerings are implemented, they will provide an excellent platform for public frame relay offerings of switched services. They will also ease the management of virtual circuit addresses.

Switched virtual circuits will allow sending DTE to transmit the address of the receiving DTE along with the data at call setup time. When the first switch receives this address and data, it establishes the path to the receiver. This method eliminates the need for pre-configured PVCs. Some of the functions available on SVCs will include DLCI control and addressing scheme assignments, user channel negotiation, and service parameters negotiation (maximum frame size, throughput, transit delay), all on a switched service offering. Wide implementation of SVC services is slated for the mid-1993 timeframe. SVCs are outlined along with PVCs in CCITT Recommendation I.122.

10.6.4 Functionality Comparison to X.25

There are seven functions provided by frame relay also provided by X.25. These include address translation, discard of incorrect frames, fill of interframe time, FCS checking and generation, flag recognition and generation, recognition of invalid frames, and transparency. The concept of PVCs and SVCs is also similar. This together constitutes about one fourth the services provided by X.25, and frame relay can be viewed as a subset of X.25 packet-switching technology and functions. It can be seen that while the absence of services contributes to a less intelligent network, the benefit is apparent of a reduction in overhead to account for many services which are not needed if clean transmission mediums are used. It also provides much higher speeds of transmission. Also, people are familiar with packet switching in general, and can easily learn frame relay. Still, frame relay is not a replacement for X.25 packet switching if poor transmission facilities are inescapable and guaranteed delivery is a must.

10.7 PERFORMANCE

The first measure of performance should be built into the digital transmission network which carries frame relay services. Since frame relay does not provide the OSI layer three services found in X.25, the digital fiber optic transmission network should have performance characteristics which accounts for minimal or no errors.

The link layer core parameters define the controls which most affect performance within the network. Maximum frame size can affect performance and degrade throughput if it is set too low. Throughput is another parameter which can be set to allow for no more traffic than what can be handled by the network, thus putting the onus on the user to manage throughput class rather than poor network performance controlling it for him. Burst size settings which are too large can also effect performance if the network is congested and drops "DE" traffic. Proper implementation of the frame relay standards discussed in the last two chapters call for services to try to pass as much "DE" traffic as possible before discarding traffic, so some level of network overbuild is required to account for congestion conditions.

The relatively short delay incurred during data transport should be typically 5 to 10 milliseconds per node with routers, as opposed to the 30 to 50 milliseconds higher delay in other similar transport. Frame relay also provides fast reconnect and statistical multiplexing.

The magnitude of the statistical multiplexing can vary typically from 4:1 to 1:1 depending on traffic load patterns. This, coupled with the high transport speeds, makes frame relay the best alternative to packet switching that uses X.25 techniques.

Table 10.2 shows the performance characteristics of frame relay in comparison to circuit switching.

TABLE 10.2 Circuit Switching versus Frame Relay

CIRCUIT SWITCHING	FRAME RELAY
Voice, video, CAD/CAM	Bursty traffic
Protocol transparent	Circuit reconnect and reconfiguration
Switch delay	Statistical multiplexing
Polling	Connection-oriented service

Now to explore some of the performance considerations of frame relay networks in detail.

10.7.1 Queuing

Queuing in a frame relay network becomes a function of the network device buffers which process the frames. There is very little queuing as compared to X.25 packet switching. Remember, frame relay is designed to throw away frames when congestion conditions occur and warn the sending and receiving devices that a congestion condition has occurred. In this aspect, there is no queuing. Some vendors have implemented buffers which serve as queuing devices for a short period of time, but this is often limited to milliseconds. This can be extremely detrimental to session-oriented traffic (i.e. SNA). Any additional queuing needed should be performed by user devices such as a Packet Assembler/Disassembler (PAD) and Frame Relay Assembler/Dissembler or Access Device (FRAD). When the queuing capacity is exceeded, the frame relay device will generally begin to discard data, usually even before it enters the network. The order in which data is discarded can be managed by the DE bit, hardware configurations and flags, or proprietary implementations — only if the data passes from the access device, past any buffers, and into the network.

10.7.2 Buffering During Congestion

Two bottlenecks can exist — network congestion can occur when a network node receives more frames than it can process or it transmits more frames than the network can accept — all buffers being filled. When the information overflows the buffers, packets begin to be discarded. Congestion is typically handled at the network layer. This is again a reason for intelligent application transmission protocols. The network will rely on these higher-level protocols, such as TCP and OSI Transport Class 4, to inform the applications that they need to provide retransmission. Unfortunately, much of the traffic best suited for frame relay transport is bursty in nature, and prone to large bursts which will easily flood the network and overflow buffers. As congestion increases there is a direct negative effect upon throughput. Thus, implementations which allow large sustained bursts above CIR and discard based upon DE settings, coupled with large maximum frame sizes, are the least detrimental to throughput.

10.7.3 Delay

Since the frame size is variable so, therefore, is the total transmission delay. Both voice and compressed video do not tolerate delay, and therefore frame relay is not a good transport mechanism for these types of data. All applications using frame relay must be able to tolerate variable transmission delay, as well as the retransmission of data.

The primary performance measurement for frame relay is transit delay as defined in ANSI T1.606-1990. This defines the transit delay of a frame relay Protocol Data Unit (FPDU) from the first bit crossing the first of two boundaries (transmit) to the last bit crossing the last of two boundaries (receive). This is shown in Figure 10.18.

Transit delay is measured by:

$$t_{transmit} - t_{receive} = t_{transit\ delay}$$

This relatively short transit delay is typically 5 to 10 milliseconds per node with routers, as opposed to the 30 to 50 milliseconds higher delay in other similar transports. Typical one-way, cross-country transit delay for 128-Byte frames should be less than 300 milliseconds. The real delay is inserted during protocol and addressing conversions, as well as the data exchange. When a network offers a frame relay service, the only delay incurred other than the DLCI switching is the hardware, and high performance can be achieved. The protocol conversions usually take place in the user access device and are not part of the end-to-end transit delay through the network.

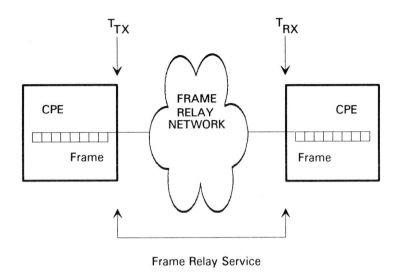

Frame Relay Service

Figure 10.18 Frame Relay Service Transit Delay

10.7.4 Error Rates and Lost Frames

Bit errors have the same effect as congestion. They cause the frames in error to be discarded. The network again relies on the higher level transmission protocols to perform network recovery. This demonstrates the disastrous effects which poor transmission facilities have upon a frame relay network, causing much retransmission of data. Transmission facilities, fiber optic cables, with error rates of 10^{-13} are desirable for best performance.

There are two performance measurements defined in ANSI T1.606-1990. Residual Error Rate (RER) is defined as the percentage of total transmitted FPDUs to the total correct FPDUs delivered between two boundaries. This is calculated by the formula:

$$1 - \frac{\text{total correct FPDUs delivered between boundaries}}{\text{total offered FPDUs between boundaries}} = \text{RER}$$

The second measurement is the lost frames incurred in a given period of time between two boundaries. This is usually measured in frame losses per second.

10.7.5 Statistical Multiplexing

Frame relay also provides fast reconnect of virtual circuits and performs statistical multiplexing. The magnitude of the statistical multiplexing can vary typically from 4:1 to 1:1 depending on traffic load patterns. This statistical multiplexing capability increases performance because multiple logical channels can be configured to be multiplexed over different physical channels. If one physical channel through the network goes down, the logical user channels can be rerouted and multiplexed into a different physical channel.

10.7.6 Overhead and Throughput

As mentioned before, there is very little overhead when using frame relay. This is due to the absence of the end-to-end services now provided by user high-level protocols. This also increases performance, since more information can be transmitted over the physical channel. The larger the frame size accepted by the network, the greater the throughput. Throughput can be calculated based upon the number of data bits successfully transferred from one boundary to another within a set period of time.

10.8 SERVICES

Frame relay has emerged full force on the public and private data network market. While most providers of frame relay hardware are CPE vendors, the high-end multiplex and switch vendors are now beginning to provide frame relay interfaces, protocol conversion, and switching service capabilities. The Interexchange Carriers (IXCs) seem to be the best positioned to provide frame relay services between LATAs. Figure 10.19 shows the connectivity provided by the IXCs across LATAs and outside the LATAs (alternate access scenario). Each of the IXCs (MCI, Sprint, and AT&T) and LECs have taken a different stance on offering frame relay services. In the near future, most if not all of these providers will offer frame relay services. The other providers of frame relay service are the traditional X.25 public packet switch network offerings, such as Tymenet, Telenet, and CompuServe.

Users must decide first by analyzing the benefits of accessing a public frame relay network or building a private frame relay network. If the decision is made for a public network, the user must decide between the IXC, independent information services network provider, and LEC frame relay services. This decision will be based on both intra- and interLATA access and pricing. Pricing is, in turn, influenced by a variety of bandwidth and access methods including the committed information rate (fixed), service access and transport distance, and amount of data successfully transmitted (usage). Another very important decision factor should be the method by which the service provider's access and backbone network is built. Where the tariff structure follows a short-term view of the service, the service provider architecture is the long-term insurance that costs and service quality will remain constant. How the IXCs frame relay services connect is also important, as well as international frame relay interconnectivity. With the expansions of international business, as well as the high cost of international circuits, many users see frame relay as optimizing costly international bandwidth. Each of these service options will be discussed in detail.

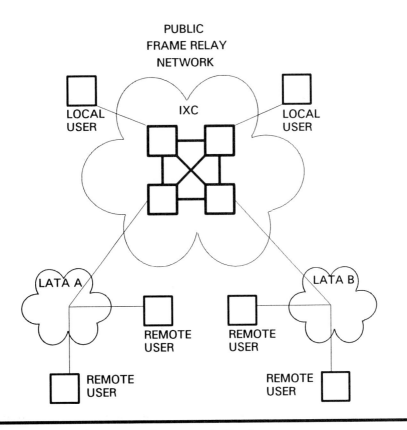

Figure 10.19 Public Frame Relay Network Topology

10.8.1 Public vs. Private

Now that the user is interested in frame relay as the transport technology of choice, does he or she build a private frame relay network, or subscribe to a public frame relay service? This decision is determined by cost efficiencies, operations and network management, alternate service and product offerings (standalone or bundled), and the corporate position on outsourcing. Some companies are retaining private networks to carry voice traffic and converting their data transport to frame and cell relay services until ATM/B-ISDN becomes available. Others are converting their private networks into data and partly to frame relay private networks.

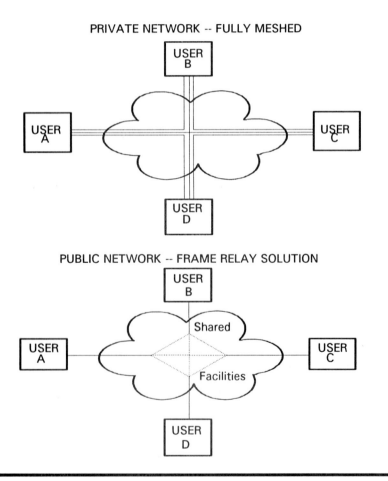

Figure 10.20 Frame Relay versus Private Network Solution

Public frame relay services can provide a network service to users to replace private networks built with private lines. Figure 10.20 shows the difference between a frame relay network service and the private network solution. This leads to a network with virtual circuit access to any other node on the network, a virtual fully meshed network without paying for many private lines, a reduction in facility costs in the number of network access ports, and true bandwidth-on-demand with respective usage billing. Obviously, the more nodes a service provider has in the network, the better the availability of the service will be.

Private frame relay network designs are based on three types of technologies:

1. T1 multiplexers employing fast-packet frame relay technology
2. Bridges or routers employing dedicated links for frame relay interfaces into a network
3. Fast packet switches employing dedicated T1s (expensive for large networks)

These types vary widely, and existing customer premises equipment may also influence the decision. The network design manager should consider all of the alternatives presented in this book before deciding on what type of technology to use to deploy frame relay services.

Figure 10.21 compares three types of frame relay access. The top diagram shows a circuit switch with frame relay riding through a dedicated T1 trunk. The entire T1 trunk is used and paid for, but only a portion of the bandwidth is being used at any one time. The second diagram shows a multiplexer employing fast-packet technology, where a dedicated circuit is again used, but this time the unused portion is dedicated to frame relay traffic. The most efficient and cost-effective access from the user's viewpoint is underscored in the last diagram. Here the user is accessing the network via a T1 circuit, but is using only bandwidth when needed. In this approach, the user pays only for the bandwidth being used, when he or she uses it. Each of the public frame relay service offerings use ones of these methods to provide access to customers.

10.8.2 Public Frame Relay Service Offerings

There are three main groups providing public frame relay services: the IXCs, the RBOCs, and the typical IP packet type service providers (e.g., Tymnet, Telenet). Most often the IXC (such as MCI Communications, WilTel, AT&T, and Sprint) provides the value-added services in addition to frame relay transport, additional voice service bundlings, flexible pricing schemes (i.e., fixed CIR and usage based) and higher and more flexible access speeds. The major IXC's frame relay service offerings are shown in Table 10.3. The major computer interconnect frame relay service offerings are found in Table 10.4. The RBOCs are also beginning to explore intraLATA frame relay offerings.

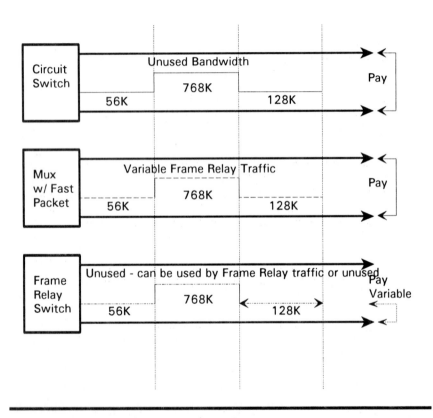

Figure 10.21 Frame Relay Access Types Comparison

TABLE 10.3 Major IXC Frame Relay Service Offerings

MCI Communications	Virtual Private Data Service 2.0
Sprint Data Group	Sprint frame relay Service
WilTel	WilPAK
AT&T Network Systems	InterSpan

TABLE 10.4 Major Computer Interconnect Frame Relay Service Offerings

CompuServe	FRAME-Net
BT North America	EXPRESSLane

IXC frame relay service offerings have the following additional advantages:

- little capital investment required
- backbone capacity distributed
- extensive IXC bandwidth capacity
- extensive IXC POP coverage
- flexibility of location and interface speeds
- multiple user access over a single dedicated physical port/circuit
- DS0, FT1, and T1 interfaces.
- centralized network management
- frame relay in combination with other switched services
- ubiquitous public network access
- can migrate users to other technologies as available
- usage based and flat rate billing (users take advantage of best one) with tariff structures
- carrier-provided CPE in many instances
- outsourcing packages through joint-marketing agreements
- control of network resources
- skilled service and equipment support structures
- inherent reliability, redundancy, and survivability
- universal access
- intercompany switched and permanent virtual sessions

Additional benefits of public networks over private networks have been discussed in earlier chapters such as: value-added services, network redundancy, and network management. Most, if not all, conform with ANSI T1.617, ANSI T1.618, ANSI T1.606, LMI Extensions and support the following capabilities:

- PVCs and SVCs
- extended addressing option
- access rates of DS0, Nx56K/64K, DS1, E1,
- PVC rates as low as 4K increments
- some analog access rates
- up to 1023 DLCI addresses
- 2, 3, or 4 octet DLCI addresses

- 4098 octets frame size
- FECN, BECN, DE bit support for network and CPE devices
- multicasting
- global addressing
- closed user groups of DLCI addresses
- PVC in-band local status
- interenterprise, intercarrier frame relay internetworking (i.e. NNI)
- end-to-end delay of service less than 300 milliseconds
- high availability and low error rate
- automatic NM and billing

Most offerings are based upon either the StrataCom Integrated Packet Exchange IPX-32, routers, the Netrix #1-ISS, or Siemens MAN/Wellfleet switch/router hardware. CompuServe and AT&T are offering frame relay service through the use of StrataCom IPX multiplexers. Sprint offers a frame relay service based on Telenet Processor/III 4900 packet switches and TP 7900 packet multiplexers. It is the author's opinion that the router/cell switch solution is the most versatile of the current offerings. This architecture allows the user a clear migration path to 802.6 and ATM. Many frame relay services are already full-blown products for carriers. Pricing for frame relay services ranges from flat rate CIR-defined to usage based. The correct selection of one or both remains different for each customer and application.

One of the IXCs, MCI Communications, has chosen to offer not only frame relay, but also cell-switched services based on the DQDB architecture and SMDS service. This provides the user the capability to preserve embedded hardware investments in frame relay technology when the transition comes to cell switch technology, also true ubiquitous access for any user. Carrier offerings such as this can alleviate the problem of having multiple T1 or T3 circuits between user locations. Its combines the access frame relay interface with a complete transport service. A public switched network frame relay service is ideal for users with a large number of sites over a large geographic area. Eventually, many of the major providers may move to a T3 backbone similar to the MCI-offered cell-based backbone.

10.8.3 Committed Information Rate (CIR)

The Committed Information Rate (CIR) has different meanings, but the most common use seems to be the amount of transport bandwidth guaranteed as available to the customer at any given time, even if the network is congested. This CIR states a maximum bandwidth which can be used within a given PVC between two predefined end-points. One example of a CIR for a T1 is 1.2M bps. Thus, at any one time the user can transmit up to 1.2M bps without any chance of blockage or lost data. Some carriers allow the user to exceed this bandwidth but do not guarantee delivery of data. Others simply discard data that exceeds the CIR. Some providers, such as MCI, provide the best of both worlds. MCI offers the CIR and allows the subscriber to try to use additional bandwidth over and above the CIR, marking all traffic above the CIR as "DE". In effect, this is a statistical approach that not all the customers will need bandwidth at all times. Since MCI overengineers capacity in the backbone network (as do many other providers), there is *statistically* a good chance that bursts exceeding the CIR will pass without appreciable data loss. Some carriers such as Sprint offer CIR and non-CIR services, thus allowing users to subscribe to CIR and use the DE bit to discard traffic in excess of the CIR during network congestion. Few providers offer usage and fixed (CIR) based pricing.

FECN and BECN are not being implemented by many frame relay services because there is no way to relate the network congestion conditions back to the user CPE devices. Even after a method is developed (there are proprietary schemes now being used), this information will have to be passed to higher-layer protocols (such as TCP) to effectively slow down the transmission of user data until the congestion condition is alleviated. FECN and BECN implementations can either work with the user device to establish flow control or slow down the amount of acknowledgments, which decreases the window size. Each customer is allowed a committed burst size to pass all traffic if the committed burst is not exceeded. If this CIR is exceeded by a given value, the first frames to be discarded are those marked for DE. Any transmissions whose frames exceed both the CIR *and* the excess burst rate will be discarded automatically until the level of congestion is brought down to the point where DE set frames are discarded and non-DE frames are passed. Be aware, however, of the danger of discarding frames marked with the DE bit during long periods of congestion. The applications will react by retransmitting lost frames and congestion will intensify. Window flow control protocols such as Transmission Control Protocol (TCP) can be used to scale down the volume of retransmissions and spread the load over a greater time period. This slows the effective offered load and decreases the throughput gracefully until congestion clears.

10.8.4 Oversubscription

Oversubscription is the capability to oversubscribe or "overbook" the CIRs coming into a single physical access port. For example, 200% oversubscription would allow eight 32K CIRs to be provisioned over a single 128K bps physical access port. 500% oversubscription would allow twenty 32K CIRs to be provisioned over a single 128K physical access port. The amount of oversubscription allowed by a frame relay service provider can be an important factor if there are many low speed user CIRs contending for a single physical network access port. High oversubscription rates can afford a greater throughput and equipment cost savings.

10.8.5 Rates and Tariffs

Rates and tariffs play an important role in the success or failure of a frame relay service offering. Currently, there is a need for a tariff-controlled public frame relay service competitive with other alternative services, and priced so that it is much cheaper than dedicated lines or switched-circuit data services. One question which a tariff must address is how much must the user pay to access the service from a network provider, as opposed to building a private frame relay network? These services must not only be price competitive to the frame relay provider competition, but also to competing technologies.

The pricing of frame relay services is based on many factors:

- ➤ access speed used per port
- ➤ total Frame Bandwidth Allocation (FBA) per port (combined virtual circuit speeds per port)
- ➤ Committed Information Rate (CIR)
- ➤ surcharge per port
- ➤ local access charge per access speed and distance
- ➤ fixed rate or usage based
- ➤ bulk rate or specialized pricing per customer
- ➤ points-of-presence (POPs) for access and corresponding tail circuits
- ➤ user distance to POP

Private line costs often limit users to low-speed dedicated circuits to ship LAN-to-LAN traffic, creating a wide area network bottleneck. If these circuits are analyzed, they are flooded during the busy hour and not utilized much during off hours (or nonbusiness hours). The lower the volume of traffic and trunk utilization, the better usage-based billing becomes. The eventual offering of SVC service will enable any user to talk to any user, but this brand of service will find heavy competition with SMDS services employing E.164 public network addressing schemes.

For many users, the initial frame relay deployment should be limited, allowing the user to learn his or her own traffic patterns. When these are known, the user can decide upon mass deployment of usage pricing, flat-rate pricing, or a hybrid of both. Initial users should also be subject to price caps so that the usage-based service does not become more expensive than the flat-rate fee or a leased line. Prices should not penalize users who are trying the service and are not familiar with their traffic patterns. When the peak traffic load is unknown, prices should be flexible and primarily usage-based. There will still be access charges, but a frame relay network using fewer dedicated access circuits in the local loop will decrease bottom line access costs. All in all, frame relay services must be priced to make the risks of these new services minimal and the rewards tempting and cost effective.

One question users should ask is: What is the minimum guaranteed line speed? Some lower-speed applications need only subrate access, and higher-speed access could be overkill. Users taking the conservative approach can buy into frame relay with only a FRAD or router, and grow access as needed. Equipment investments for frame relay access are minimal when compared to voice access.

Frame relay billing can be based upon:

- the number of packets transmitted in a given time period
- transmission facility charge
- per PVC per bandwidth on each
- additional CPE costs (e.g., DSUs, routers, FRADs)
- PVCs flat rate, distance sensitive from origin to destination
- SVCs variable rate based upon distance and usage per call
- number of frames received by destination point
- additional charge for PVCs and SVCs changes
- all of the above

WilTel offers frame relay service based upon "distance-insensitive" flat rates for bandwidth and number of PVCs used. Additional access costs vary depending on the hardware required and initial hardware investment fees. Sprint charges a link fee of $200, plus the PVC CIR rate. Cross country T1 dedicated access costs approximately $6000. 1M bps FR local loop access costs is about $2000. CompuServe service includes a flat rate $1050/month per location 56K access, and $1700/month per location for 256K access. MCI is the only service provider who is offering both usage-based and flat-rate billing options, but has not yet tariffed their offerings. Table 10.6 shows the more common service offerings and their rate structure by PVC access CIR speed (as of late 1992). These rates will obviously change with time and tariffs.

TABLE 10.5 Service Offerings By Rate Structure

PROVIDER	32K/56K PVC access	1024/512K PVC access
BT	$6k	No
CompuServe	$3.5k	$23k
MCI Communications	*nontariffed	*nontariffed
Sprint Std	$3k	$13k
Sprint Reserved	$7k	$36k
Sprint Hybrid	NA	$20k
WilTel	$4.5k	$26k

Note that MCI Communications is pricing both usage-based and flat-rate frame relay service on a nontariff, case-by-case basis. MCI also supports unidirectional assignment of these parameters (for each PVC) for greater traffic billing flexibility and planning, strong customer network management enterprise systems, and the capability for sustained bursting over the assigned CIR.

10.8.6 Carrier Frame Relay Interconnectivity

At present, frame relay interconnectivity between service providers is limited at best. The Frame Relay Forum has an intercarrier committee designated to define frame relay network-to-network standards for this type of interconnectivity. The Frame Relay Forum's Network-to-Network (NNI) interface interim standard will be the model for further ANSI and CCITT standardization. These efforts have been somewhat hampered by a lack of standardization in frame relay implementations for the following parameters:

- ❖ CIR usage
- ❖ congestion control implementation
- ❖ transfer of network management and billing information problems
- ❖ ability to pass LMI information
- ❖ FECN, BECN, DE support
- ❖ number of DLCIs per port / IP addressing

These problems need to be resolved for true international frame relay service between the RBOCs, IXCs, and PTTs.   Until this happens, and a solid NNI interface is deployed, users are limited to a single carrier.

10.8.7 International Frame Relay

Since frame relay is the first true international standard to succeed, it is worth while to take a look at how the European and Pacific Rim markets are handling implementation. While many of the communications hub cities are now conducting frame relay trials, the service may find a longer incubation period in many of the outlying regions. With the exception of a few countries, most of Europe's and the Pacific Rim's digital infrastructure will not provide sufficient transport media required by frame relay. As discussed previously, frame relay requires virtually error-free fiber optic transmission media. Much of Europe and the Pacific Rim remains on radio and copper transmission facilities.

Another factor slowing the implementation of frame relay in Europe is the price of digital access circuits, which are significantly higher than their American counterparts. While this drawback is much more easily remedied than the transmission media problem, it represents a major stumbling block, as frame relay will primarily be an access service offered by the PTTs and their competitors. The future for frame relay in Europe is not completely bleak, however. The cost of frame relay hardware and services may be minimal compared to the high cost of international access. Price and availability will be the driving forces in the success of international frame relay. Trans-Atlantic and Pacific Rim fiber optic cables will provide the transmission media needed, and the secondary expense of turning up frame relay access will be the hardware. These cables, along with completed CCITT frame relay standards, will move many European countries toward frame relay. The other major stumbling block is hardware and software compatibility between countries.

The introduction and migration toward ISDN in some countries may speed the resolution of these roadblocks, but if the implementation and widespread use of X.25 packet switching provides any indication, frame relay may not be as valuable an interim solution for bandwidth flexible services as SMDS or ATM is in the United States. However, the international community is far from implementing ATM technology, so frame relay may provide the needed interim solution. Many carriers in the United States and international PTTs have joined to offer international frame relay access. Examples include British Telecom North America, who announced a worldwide frame relay network as part of its schedule to roll out a B-ISDN network over the next three years. This service called ExpressLANE, will offer 56K/64K bps access in the United Kingdom, France, Netherlands, and Germany. Other providers such as MCI, Sprint, Graphnet, Infonet, CompuServe, and WilTel, and Cable & Wireless plan to offer international frame relay services, many up to E-1 access rates. Also, many frame relay vendors are making deals with international firms to provide frame relay equipment.

Initial pricing for international frame relay is based upon a flat monthly charge of about $3,000 to $4,000 for a DS0. This rate includes the dedicated port access, access line, and software for the CPE device.

10.8.8 Network Management

Network management is probably the most critical link in any frame relay network. Network management tools run the network. Network management adds the true value to any frame relay network — transforming it from an efficient protocol and interface into a true service. Local and remote hardware configurations, software modifications, protocol implementations, and control of both user access devices and network elements are but a few of the functions provided by a network management tool. Most vendors provide a proprietary network management device, often a terminal attached to an RS-232 port on the device. Others provide a feed to a standard network management element. While network management will be covered in detail in Chapter 21, it is important to note that the network management standard which all vendors should provide interfaces for is the Simple Network Management Protocol (SNMP). SNMP provides a common platform for multiple device reporting and network management. This is essential in a large network with multiple network and user devices supplied by different vendors.

OSI is now defining a Common Management Information Protocol (CMIP) which will become predominant in the network management arena by mid-decade. But these standards are still vaporware, and many vendors are only announcing their intent to support this standard when it becomes available.

10.9 REVIEW

This chapter started with a discussion of ISDN frame mode bearer services to give the reader a background into the standards that shaped both the ANSI and CCITT user connection modes and their implementations. Two methods of ANSI-defined connection modes define the management of call control and access to frame-handling facilities, and are labeled as Case A and Case B. The CCITT-defined connection modes were then discussed for frame relaying 1, frame relaying 2, frame switching, and the X.25 connection mode, and how the ISDN standards apply the HDLC framing to frame relay services. Type I and Type II services, which define the OSIRM layer two LAP-D data link layer standard for signaling, were also covered. The user-definable services, the other half of the core services, were defined next, and showed the many controls which can be exerted over the PVCs, as well as signaling protocols. Some methods of congestion notification and control were explained. A quick trip was taken through the various access devices used to interface to a frame relay network. Frame relay was then described as a service provided by the network, and various methods of addressing schemes and PVC management were shown. Performance in a frame relay network was discussed, along with ideas on how to maintain or improve it. The chapter closed out with an overview of service offerings with discussions on tariffs, CIR, oversubscription, choosing between a public and private frame relay network, public frame relay offerings by vendor, network management, and international frame relay services. It was also seen that a strong network management platform, based on SNMP, is required to mold frame relay from an interface and protocol into a complete service.

11

IEEE 802.6
DQDB MAN Theory

The Distributed Queue Dual Bus (DQDB) MAN standard defined in the IEEE 802.6 cell-switching proposal is fast becoming the next-generation data- and voice-switching architecture. This cell-switching architecture combines the best of two worlds — connectionless datagram public data transfer services similar to packet switching and speeds in excess of 155M bps. Central office switch vendors are the primary players for the first version of cell-switching to hit the telecommunications market: DQDB architecture which supports Switched Multimegabit Data Service (SMDS), where SMDS makes use of DQDB's Connectionless (CL) service. LECs, IXCs, and dominant carriers worldwide view cell switching as the technology base for public data networking.

Some experts say that DQDB IEEE 802.6 defined cell switching is only an interim solution until Asynchronous Transfer Mode (ATM) switching is implemented, but each version of cell switching has merits of its own. With the ever increasing call for more bandwidth to support such applications as image processing, high-resolution image transfer, and distributed processing, the existing DS3 speeds are found to be too slow. Cell switching eliminates bandwidth limitations across large geographic areas and brings data networking into a new era.

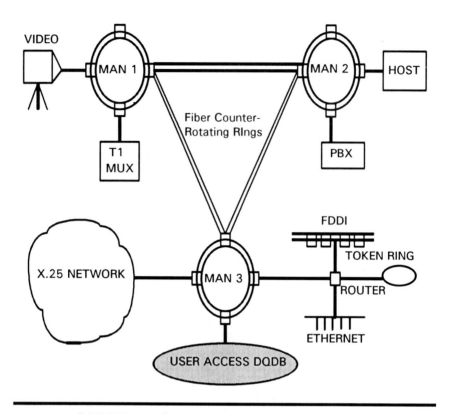

Figure 11.1 DQDB Metropolitan Area Network (MAN)

The term cell switching carries many meanings. In this chapter the first implementation of cell switching to hit the market is detailed: the IEEE 802.6 MAN technology. Cell switching should be defined in terms of standards, underlying architectures, initial services implementation (SMDS), and protocols. A detailed discussion of the DQDB bus architecture, the heart of metropolitan area networking, shows the clever design concept of DQDB which complements the new fiber transmission technologies. This chapter ends with a section on performance as it relates to the DQDB bus. An entire chapter will be devoted to ATM later.

11.1 TERMS DEFINED

The IEEE 802.6 DQDB standard defines the transmission technique used by Metropolitan Area Networks (MANs). It is defined as connectionless data transport service using cells as protocol data units providing switched integrated data, video, and voice services over a metropolitan geographic area (typically < 150km). Although connection-oriented isochronous services are also included in IEEE 802.6 standard, SMDS will initially support only connectionless data transport. Some vendors will provide their own proprietary version of isochronous transport. These multimedia data cell formats interface to DQDB subnetworks. The interconnection of these sub-networks forms the MAN.

The MAN provides a shared media for voice, data, and video transmissions over a local geographic area (typically less than 150km), as well as high-speed extension of each LAN and WAN attached. Cells are routed through the MAN wideband channels similar to packets in a packet-switched network, except that the bandwidth available is much higher (broadband transport speeds of 155M bps are available as opposed to 56K bps and 1.544M bps). All transmission is assumed to be via optical fiber. Figure 11.1 portrays an example of a metropolitan area network providing connectivity to a large variety of multimedia users.

Cell switching has taken two development paths: connectionless data transport in the form of IEEE 802.6 (DQDB), and connection (with connectionless planned) oriented in the form of Asynchronous Transfer Mode (ATM). SMDS services uses the IEEE 802.6 DQDB CL, while ATM offers an analogous Broadband ISDN (B-ISDN) service. While SMDS provides LAN/WAN interconnection, it is not limited to a geographical area (depending on the topology used). ATM and B-ISDN may well become the long term wide area networking technology, and are designed to use a SONET fiber network for data transport. Either technology will serve the purpose of eliminating long distance private lines in the near future. Now for an explanation of the IEEE 802.6 version of cell switching.

11.2 CELL SWITCHING STANDARDS

MANs and the DQDB saga began when the IEEE began work in 1982 on standards for transmission of voice, compressed video, LAN interconnectivity, and bulk data transfer. These ideas were first presented to the cable television (CATV) community, which in turn dropped the idea. Those same cable companies now have a wealth of bandwidth in cable TV and have still not used it. Burroughs, National, and Plessy initiated a second effort in 1985 with the slotted ring concept. This effort died when the leveraged buyout of Sperry Univac replaced required funding, and again MAN technology waited. The last effort began with a Bell Labs MAN standard proposal and was developed in parallel with the ex-Boroughs FDDI venture called MST (multiplexed slot and token). This new Bellcore MAN standard became the IEEE Project 802.

The IEEE 802.6 standard is based upon the distributed queue dual bus (DQDB) technology. The DQDB architecture (which resembles the Bell Technology Lab's dual coax with reservation MAN architecture called Fastnet) was invented at the University of Western Australia and hardware was first produced by QPSX LTD (a University of Western Australia and Telecom Australia spin-off). Multiple DQDB buses are configured to form a switched MAN. These buses provide a high-speed shared medium access protocol across point-to-point, bus, and looped bus topologies. Switched Multimegabit Data Service (SMDS) is a service offering for 802.6 as defined by the Bellcore standard, and uses the DQDB bus for transport. Metropolitan area networking is another term for the implementation of the DQDB under IEEE 802.6. The IEEE 802.6 Committee has chosen the DQDB bus as its MAN standard.

11.3 THE EVOLUTION OF MAN

Major technology trends are pushing multimegabit data communications to the desktop. As the costs of multi-MIP machines continue to drop, desktop personal computers and workstations are providing processing power equivalent to the mainframes of old. They continue to gain in processing power. In fact, an AST 286 PC now provides more processing power than the IBM 7090 used in the Apollo space project! Users are demanding higher-speed data transport with bandwidth-on-demand to support the bandwidth-intensive applications now emerging, such as imaging, three-dimensional motion graphics, and interactive computer simulations. True distributed processing has arrived in the form of the LAN. LANs provide large amounts of local capacity with good performance at a low cost. But with the advent of high-speed local networking among many users, the need emerges to connect these local environments over large distances in real time using large bandwidths. Thus enters the MAN. Figure 11.2 shows a comparison of LAN and MAN bandwidth capabilities.

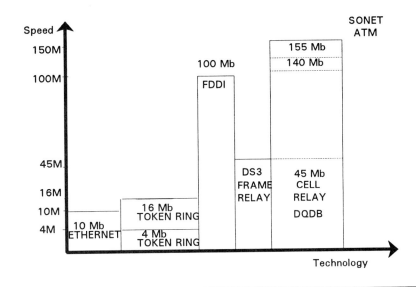

Figure 11.2 LAN & MAN Bandwidth Capabilities Comparison

However, the user still wants data transport to be transparent. The MAN is viewed by the user as a utility, with little or no data and addressing errors, lightning-speed response time, and pay as you use billing. The network resources accessed over the SMDS network must appear local. SMDS services do provide some error correction, congestion control, and quick response times, but much of the end-to-end data checking and congestion control is left to the users' application protocols.

MANs are the next step in the evolutionary process from local area networks to wide area networks. MANs offer high-speed public-switched services including data, voice, and video. MANs are also expected to pass several LAN/WAN protocols without protocol conversion (using, in effect, encapsulation techniques). While MAN technology was originally defined for LAN/WAN interconnection, it has become a major interim solution between packet technologies and ATM switching. A MAN's primary function is as a public network offering, but there are also provisions that port the capabilities to Customer Premise Equipment (CPE). Local Exchange Carriers (LECs) and Inter-exchange Carriers (IXCs) will provide MAN connectivity in the United States, while the European and Asian PTTs are also are finding a vested interest in this new technology.

MANs interconnect LANs and WANs, while providing switching, concentration, and high-speed data transport. The MAN operates on a shared DQDB bus. This bus operates like a LAN, where each station on the bus has equal access to all available bandwidth. Metropolitan area networks implementing the DQDB bus architecture to support SMDS will cut switched network costs. By eliminating numerous point-to-point circuits, similar to the functions provided by frame relay, SMDS networks perform statistical multiplexing and efficiently use bandwidth-on-demand.

Perhaps the predominant business case for these services is the enticement for users to eliminate numerous private lines in favor of fewer high-speed trunks utilizing the bandwidth-on-demand capability of SMDS. This drastically reduces the fixed cost of private lines, while achieving the benefit of paying for only the bandwidth used, when it is used. This will drive many corporations even further toward outsourcing. As mentioned in previous chapters, these advances seem good enough for the user, but they represent a nightmare for the service provider.

11.4 CELL SWITCHING PROTOCOL STRUCTURE

The IEEE 802.6 MAN protocols span both the physical layer and medium access control sub-layer of the OSI reference model. They also interface to the logical link control sublayer. Figure 11.3 depicts how the IEEE 802.X standards relate to the IEEE protocols and OSIRM layers.

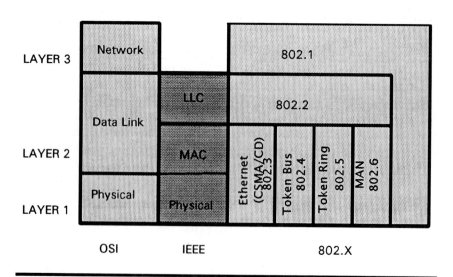

Figure 11.3 802.X OSI Comparison

11.5 DQDB THEORY OF OPERATION

The DQDB bus is a 45M/155M/622M bps dual bus which operates similarly to the token ring architecture mentioned previously. Fixed-length packets are placed within time slots which move from a time slot generator on one end of the bus to a terminator on the other end.

There are three implementations of the DQDB bus: the point-to-point bus (seen in Figure 11.4), the open dual bus, and the looped dual (folded) bus. These three bus structures can be combined to form many types of diverse architectures. The point-to-point bus requires one DS3 circuit, with a second DS3 circuit recommended for backup. This configuration is the most cost effective and least robust of the architectures, but it becomes very important for MAN Switching System-to-Switching System (SS-to-SS) meshed trunking topologies. The topology of the folded bus is designed to route around a link failure in the network. The open and folded bus architectures are discussed later in the section on DQDB bus operation.

The DQDB bus benefits are the efficiencies gained by the distributed reservation technique and its channel efficiency during normal operation. Transmission can occur over single and multimode fiber optics. Since SMDS networks are designed to accommodate a broad range of traffic, some overdesign is built into the network. As with frame relay traffic design, there are statistical methods for planning traffic patterns (explained in later chapters), but there may be large peak periods of traffic that could encounter some delay.

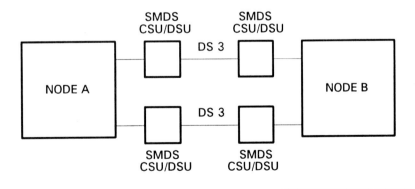

Figure 11.4 SMDS Point-to-Point Bus

11.5.1 Bus Defined

The DQDB architecture is based upon a unidirectional, full-duplex high-speed bus. The traffic flow in each direction is independent of the other. This transport medium of two fibers is shared between all users who interface to the bus in parallel.

The data transfer across the DQDB bus is formatted into 125-microsecond frames. Each frame contains a header and fixed-length slots. The frame is first generated at the node at the head of each bus, and then travels down the bus until the frame is filled with data. Each slot in the frame carries data between nodes. The slots can contain either asynchronous or synchronous data. SMDS implementation and segmentation of this frame is discussed under L3_PDU and L2_PDU later in this chapter.

The DQDB bus also has an automatic restoration feature which enables it to recover from a break anywhere along the length of the bus. This only occurs in a looped bus configuration.

11.5.2 Bus Architecture

The DQDB bus contains two layers of service: physical and DQDB bus (MAC). (See Figure 11.5.) The physical layer interfaces to the bus and the DQDB layer. The DQDB bus currently provides three types of feeds, with one service type still undefined. The DQDB provides a connectionless MAC sublayer to support a Logical Link Control (LLC) sublayer consistent with the IEEE 802.X LAN standards. This bus is queue arbitrated to provide three priority queues for MAC arbitration and fixed-length slots for data transfer. Variable-length MAC Service Data Units (MSDUs) are transported between entities using the LLC sublayer, and delivery is not guaranteed. There is no need to establish a connection between the entities for data transfer. This service is further enhanced by a MAC convergence function. There are three priority queues provided. The LLC layer will then provide the service of the OSIRM data link layer and manage communications between similar open systems architectures.

The DQDB bus provides a second type of interface to handle isochronous service between isochronous service users over a preestablished isochronous connection. This interface is nonarbitrated, and thus receives priority over the queue arbitrated bus in the form of preallocated isochronous time slots. These time slots are transmitted and received at constant time intervals. One example of isochronous service would be digitized voice, which would require constant transmission delays. Uneven delay would be readily noticeable by a listener using the service, and is undesirable.

The third type of DQDB bus interface provides for a connection-oriented data service. This supports a form of Permanent Virtual Circuit (PVC) but does not guarantee constant time arrival intervals for traffic and, in fact, is not necessarily permanent. This service accommodates asynchronous, bursty traffic. There is also a fourth service which remains undefined (to be defined at a later date).

11.5.3 Establishing a Service (Layer-to-Layer Communications)

The communications between layers is accurately described by the IEEE 802.6 Working Group DQDB standards as "state machine" notation. Layers communicate at mutually exclusive times with only one function active at any given time. Request and indication primitives are provided to and from the service provider.

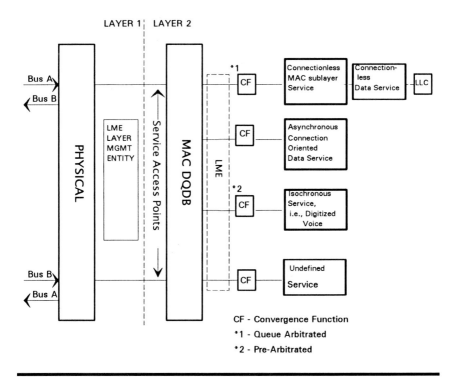

Figure 11.5 DQDB Bus Physical and MAC Protocol Layers

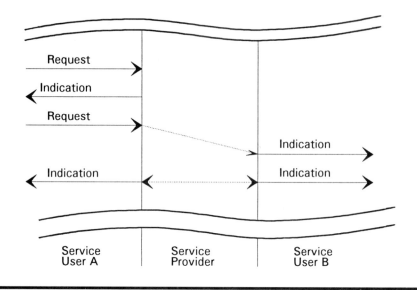

Figure 11.6 User Session Establishment

Figure 11.6 provides an example of two users establishing a connection across the DQDB network. User A submits a "request" to the service provider. The service provider returns the "indication" request to user A, identifying that the interface is available. User A then sends a "request" to the remote user via the service provider, and User B receives an "indication" that User A requests service. Both users are then provided an indication by the carrier that service is initiated and information flow begins.

Request and indication primitives only specify the service provided, not the means by which the service is provided. These primitives are transparent and independent of the network interface provided to the user. Services may have multiple primitives. When the information flow between primitives is shown in this manner, it is readily apparent that the DQDB service is a connectionless data service that uses multiple paths through the network.

11.5.4 Bus Operation

The DQDB operates similarly to that of a high-speed token ring network, with guaranteed speedy access at low bus utilization. There are two buses, A and B, called a bidirectional bus. Nodes can read and write onto either bus, and in either direction. Nodes are configured as stations.

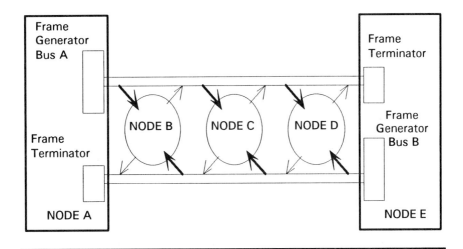

Figure 11.7 Five Node Open Dual DQDB Bus

Figure 11.7 depicts a five-node open dual DQDB bus. In the bus configuration, one station is designated as the head of the bus station and another station is designated as the end station. When a looped bus architecture is used, one station is designated as both the head and the end station.

Figure 11.8 depicts a looped dual DQDB bus architecture. It is the job of the head station to generate the frame synchronization for bus A. It is the job of the end station to generate the frame synchronization of bus B. Thus, HOB_A synchronizes Bus A and HOB_B may derive timing for Bus B.

It should be noted that the looped dual bus architecture does not perform like a ring. The station designated as both the head and the end station does not connect the A and B bus internally. The buses remain separate. Each bus begins with a head of bus (HOB) frame generator and ends with an end-of-bus frame terminator. Bus A and B never interface to one another. Although the buses do not connect, each node on the bus has both read and write capability with both A and B buses.

Another distinction which warrants discussion is that although the bus appears to pass *through* each node on the bus, it in fact only passes *by* each node. This provides for a highly reliable network, as a node failure will not affect the operation of the rest of the network. Neither bus operation would be interrupted. The looped architecture provides both a common point for timing into the network to ensure network synchronization, as well as a self-healing, fault isolation mechanism inherent to the architecture.

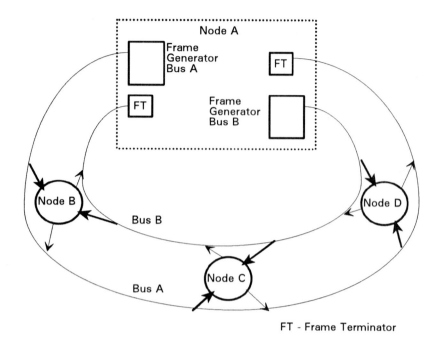

Figure 11.8 Looped Dual DQDB Bus Architecture

Figure 11.9 depicts a fully operational healthy network (a). When a link between two nodes fails, effectively cutting both buses in half (b), the network isolates the break and creates new frame generators at the nodes on each side of the break. The final healed network (c) will operate in this manner until the failed link is restored.

The frame, or cell, generated will transport 48 octets of data. This cell is also called a "slot" on the DQDB bus. This frame originates from the frame bus generator and travels along the bus until it is claimed by a node and filled with data.

When a node on the bus has data to send to another node, it starts its queue counter and gets into the bus queue. It does this by placing a request on bus A and sets its counter based on the number of nodes waiting in the bus queue ahead of it. After the node sets its counter, it will watch cells pass on the B bus, and decrements its counter each time a cell goes past. When the node's counter reads zero and it sees an empty cell pass on the B bus, it writes its data into that cell. The data is then routed according to the cell's header.

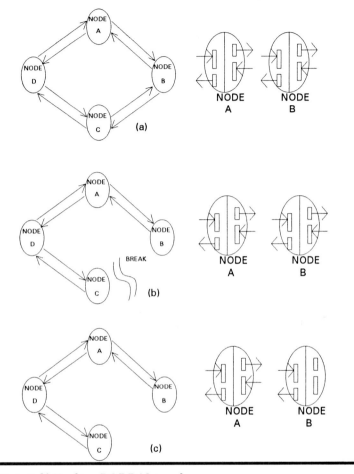

Figure 11.9 Self-Healing DQDB Network

11.5.5 Message Structure

The 53-octet message structure discussed earlier is divided into five bytes for header information and forty-eight bytes for information and error correction. Figure 11.10 shows a DQDB cell containing a DMPDU and showing the message structure. The header section contains a one byte packet identifier and a four-byte true header. The information section contains a two-byte label, a forty-four byte user data field, a one byte actual length identifier, and a one byte cyclic redundancy check. When SMDS service is used, this is called a Level 2 Protocol Data Unit (L2_PDU).

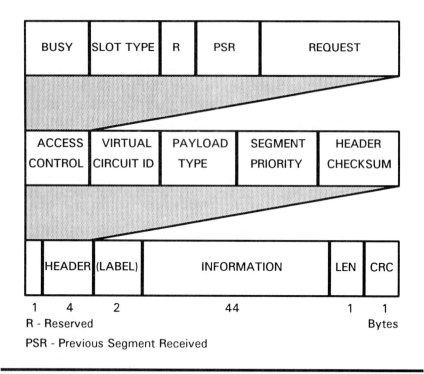

Figure 11.10 DQDB Cell Structure

11.5.6 DQDB MAC Protocol Interface

When the MAC service sends a MAC Service Data Unit (SDU or PDU) to the Logical Link Control (LLC) it performs some header and trailer encapsulation, as well as segmentation into fixed size cells of 53 octets in length. These cells are then re-assembled and stripped of their encapsulation at the receiving end.

Figure 11.11 shows the complete process. When the MAC layer receives the PDU off the DQDB bus, it formats and encapsulates it with three header and three trailer fields. The two header fields which form the Initial MAC Protocol Data Unit (IMPDU) include a 4-octet Common PDU header and a 20-octet MAC convergence protocol header (which is used only for connectionless service, such as SMDS). The third header field is a header extension. The three trailer fields include a variable length PAD, an optional 32-bit Cyclic Redundancy Check (CRC), and a common PDU trailer. Both the common header and trailer support the assembly and disassembly of PDUs for all types of 802.6 service.

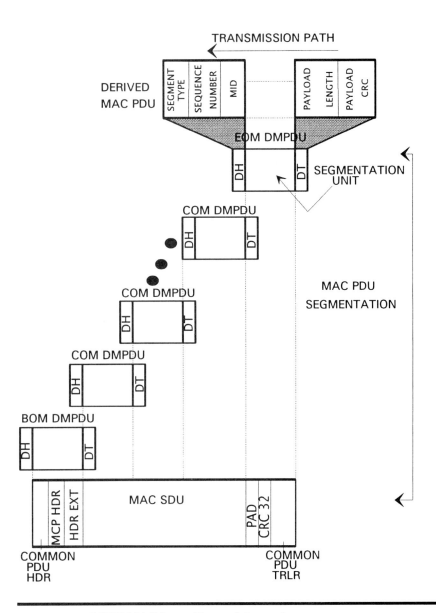

Figure 11.11 DQDB MAC Layer Protocol Operation

The initial MAC protocol data unit formed by the addition of the header and trailer fields to the MAC service data unit is then split into equal size cells. These derived MAC Protocol Data Units (DMPDUs) are then formatted and flow in the following order: beginning of message, "n" times continuation of messages, and, lastly, an end of message. These cells are then transmitted as shown in Figure 11.11, and reassembled in a similar manner.

11.5.7 Overhead and Throughput Degradation

When a DQDB network begins to experience an overload, it relies on three methods of flow control. However, this is not flow control in the traditional sense. The DQDB network does not store information as a packet network, rather it decides what type of data receives the priority of transmission.

Prearbitrated access enables the service provider to allocate bandwidth to isochronous connections with constrained jitter. This allows priority bandwidth to the services which are most affected by delay priority bandwidth — such as video and voice. This is the equivalent to a level one priority, and should be assigned with careful consideration. Video and voice are prime examples of data requiring primary access to bandwidth because of their intolerance for delay.

Queued arbitrated access would be level two in the bandwidth priority scheme. Queued arbitrated access has distance limitations. These are displayed in Table 11.1. These distances are derived from the distance of one 53-octet slot. The farther the distance between the stations, the less effective the priority queuing. This is usually allocated to lower-priority services, or services which can tolerate delay and retransmissions easier than video or voice. This service also assumes that bandwidth balancing is disabled. Bandwidth balancing will be discussed later.

For example, in Figure 11.12 there are three users on a DS3 DQDB bus. User A is located 150 meters from the MAN Switching System (SS), and should have no problem with employing queued arbitrated access with priority. User B is farther away but still within the two-kilometer distance limitation. User C may use arbitrated queuing, but is not guaranteed priority and may become the last in the priority queue behind User A and User B.

TABLE 11.1 Queued Arbitrated Access Distance Limitations

DISTANCE (meters)	SPEED (M bps)
2000	44.736
546	155.520
137	622.080

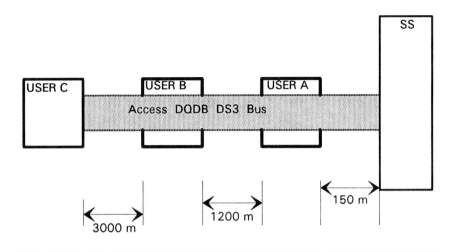

Figure 11.12 Queued Arbitrated Access with Priority Example

If User C does exceed the maximum distance limitation as in the previous example, and the service provider wants to provide equal access to all three users, bandwidth balancing could be applied. Bandwidth balancing would allow Users A, B, and C to receive an "equal share" of the available bandwidth by equally dividing bus bandwidth among all users on the bus.

The network designer should strive to design the subnetworks and CPE which interface with the DQDB bus in a manner to ensure that peak traffic conditions do not cause excessive delay and loss of traffic. This can be accomplished by the effective use of these techniques.

11.6 REVIEW

In this chapter the first implementation of cell switching via the 802.6 DQDB architecture was explained.   Metropolitan area networking concepts were discussed along with the evolutionary process experienced from local area networks to wide area networks, and now to metropolitan area networks.   The IEEE 802.6 standard was compared to well known IEEE protocols and OSIRM protocol layers. The DQDB bus, the basis for the IEEE 802.6 MAN technology, was discussed in detail.   Controlling bandwidth on the DQDB bus during peak traffic conditions was shown to be an important consideration. The next chapter will progress to a detailed discussion of Switched Multimegabit Data Service (SMDS) with relation to the IEEE 802.6 DQDB architecture.

12

SMDS and
Application of Theory

In the previous chapter, the DQDB bus was discussed in detail. The reader is now familiar with the operations of the DQDB architecture. Next, the Metropolitan Area Network (MAN) user interface will be explained. The major service offering for the DQDB is Switched Multimegabit Data Service (SMDS), and will be covered in detail. The service and architecture of SMDS and DQDB will be compared to other competing technologies such as FDDI and frame relay. This chapter ends with a discussion of design considerations when implementing IEEE 802.6-based networks using SMDS.

12.1 SWITCHED MULTIMEGABIT DATA SERVICE (SMDS)

Traditional switched, multiplexed data transport, and Local Area Networks (LANs) merge with the implementation of SMDS. As LANs continue to proliferate in offices worldwide, the need for networking high-speed services in the multimegabit range continues to grow. LAN/WAN administrators and managers are looking for alternatives to costly interpremises dedicated private lines. SMDS offers the ability to eliminate the geographic restrictions of distributed high-speed data processing though the use of the DQDB architecture. SMDS provides the subscriber with the capability to connect diverse LAN protocols and leased lines into a true switched public network solution. With the LECs and IXCs building public data SMDS networks, a corporation finds that outsourcing to the SMDS public networks may become an attractive solution and alternative.

12.1.1 SMDS as an 802.6 Service

SMDS is the first service offering of 802.6 as defined by Bellcore. SMDS is a service providing high-speed, connectionless public data service much like that of a packet-switched network. The transport layer of SMDS is provided by the DQDB MAN standards. SMDS operates over multiple MAN Switching Systems (SS) connected by Inter-Switching Systems Interfaces (ISSI). Figure 12.1 represents a typical MAN Public Data Network supporting multiple customer premises environments on multiple access DQDB.

Access DQDB refers to the use of the DQDB protocol as the basis for the SMDS interface protocol providing access to the MAN. The term MSS defines the complete set of components forming an individual switching system. Multiple SSs are interconnected by Inter-Switching System Interface (ISSI) to form a complete network. These terms will become clearer as each interface is defined in the network.

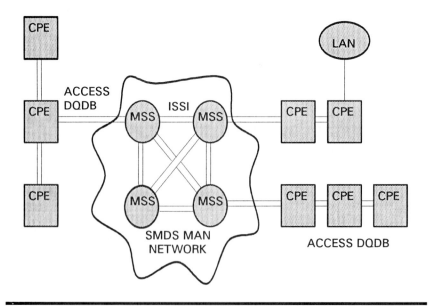

Figure 12.1 SMDS MAN Public Data Network

Users have many methods of interfacing applications to the SMDS network. Routers, bridges, brouters, and gateways typically provide four levels of protocol interfaces between the user and the SMDS network. Each of these are discussed.

It is important to note that Bellcore's SMDS service is defined only for connectionless data transfer operation. The service has not yet defined support of isochronous data operation. This may be a strategy by the Bell Operating Companies to keep market share in the voice market secured. Isochronous IEEE 802.6 services would make an attractive alternate offering to typical voice switches. At present, it seems as if half the RBOCs want isochronous service and half do not. Some problems with this include the need for a full T1 (rather than multiples of 64K bps — as in voice channels) and the rate adaptation required for different master clocks.

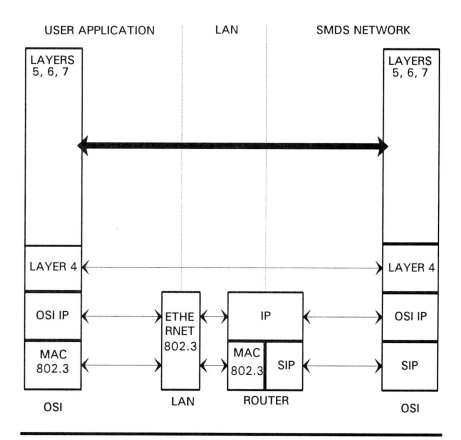

Figure 12.2 SMDS Protocol Stack Compared to the OSI Reference Model

12.1.2 SMDS and OSI

Figure 12.2 depicts the interconnectivity between a user application conforming to the ISO OSI standards and how that user communicates with the SMDS network. The physical layer has been omitted. An Ethernet LAN has been chosen as the layer two entity for this example. Notice that both the LAN and the router reside within the same geographical local area.

Layer two of the OSI reference model is represented by the Medium Access Control (MAC) protocol. The MAC will talk across the LAN to the MAC residing in the router. Here the router acts as a multiple-protocol device typically used for bridging and routing functions, as well as for SMDS network access. The router then provides the conversion from the MAC protocol to the SMDS Standard Interface Protocol (SIP). Using the SIP, the router now uses a DQDB bus providing SMDS to allow high-speed connectivity over large geographic areas. The OSI SMDS network interface also uses the SIP.

Layer three of the OSI is the Internetworking Protocol (IP). The OSI IP talks across the LAN to the OSI internetworking protocol resident in the router. The router then uses the same IP to route to the SMDS network OSI IP. IP addressing was discussed in previous chapters and will be covered in greater detail in Chapter 20. SNMP addressing is discussed later in this chapter. The OSI layer four is for all intents and purposes transparent to the router or switch, and ensures end-to-end transport of data between users. Layers five, six, and seven are also transparent and are handled between users by the application.

12.1.3 SMDS and TCP/IP

TCP/IP remains the predominant internetworking and transport protocol in the educational, scientific, and military environments. TCP/IP is a ARPA developed internetworking protocol. As OSI standards remain disputed and network engineers endure long standardization cycles, TCP/IP must be used to solve immediate protocol internetworking needs. TCP/IP spans multivendor networks, providing a common platform for end-to-end data management and error correction, as well as internetwork addressing. Many users find that TCP/IP will sufficiently meet LAN-to-LAN internetworking needs for the next five years.

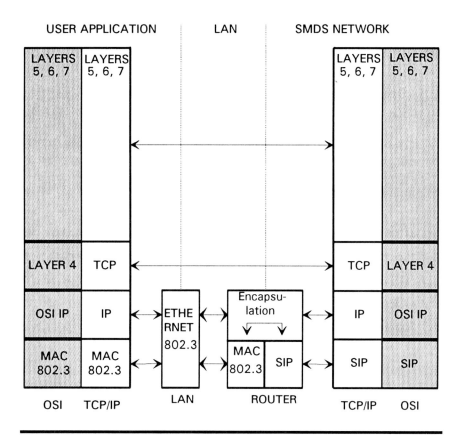

Figure 12.3 SMDS Protocol Stack Compared to TCP/IP Protocols

Figure 12.3 depicts a user employing TCP/IP to establish communication and data transport from the application through first the LAN, and then a router and, finally, to the SMDS network. The egress side of the network would usually resemble the ingress side displayed here. The OSI layers have been left on the figure for comparison but shaded out. The main difference between the OSI implementation and the TCP/IP is the use of the ARPA IP routing protocol as opposed to the OSI IP routing protocol. The transmission control protocol operates similarly to the OSI layer four transport protocol. The higher layers corresponding to the OSI layers five, six, and seven are not defined by the ARPA standard. RFC 1209 defines the specifics of implementing IP over SMDS, and is left to the reader for further study.

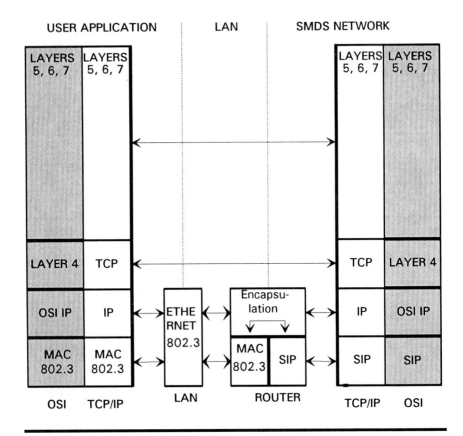

Figure 12.4 SMDS Bridging

Figure 12.4 depicts the same LAN connectivity as in the previous example but using a bridge device in place of the router. Many users have an existing product base of bridges, since bridges existed in the market before routers. Bridges are the simplest of networking devices, and are used normally to connect LANs of the same type. Figure 12.4 shows the main difference between the router and bridge. The bridge does not perform internetwork protocol functions, but instead only encapsulates data on one end and unencapsulates it at the distant end. The MAC interface is not translated into a SIP, but is instead encapsulated and transported between SIPs. All TCP and higher-level interfaces in this process remain the same (i.e., transparent to the bridge and SMDS network).

12.2 USER INTERFACE

The user interface to the SMDS network has three main components: the CPE device interface, the SMDS Subscriber Network Interface (SNI), and the SMDS interface protocol (SIP). Each of these levels is explained below.

12.2.1 The Customer Premises Environment

The candidate user environment, called Customer Premises (or Provided) Environment (or Equipment) — CPE — will likely contain multiple applications, using diverse protocols, riding multiple subnetworks. These subnetworks could be either collocated or geographically diverse. They may be composed of local area networks, wide area networks, leased or private lines, public multiplexer networks, or packet networks. The data speeds could range from low-speed asynchronous bursty traffic to high-speed bulk data transfers. Traffic patterns and quantities will vary along with intranetwork and internetwork protocols.

The customer's requirements can either be satisfied by interfaces directly into the SMDS network, or by concentration by a variety of devices. Typical CPE devices include modems, DSUs, CSUs, bridges, brouters, routers, gateways, pads, FRADs, and a variety of proprietary interconnection devices supporting one or more connectionless network layer protocols. Most vendors are also developing SMDS interface cards for existing equipment. Even the CRAY supercomputer has an SMDS terminal adapter built by Fujitsu. SMDS interface cards will soon be available in many products which now have frame relay interfaces, such as FEPs, minicomputers, and intelligent hubs. There is even a standard in development which interfaces SMDS bridges called 802.6A. This is a parallel effort to the 802.1g remote bridging standard still in development by the IEEE.

The internetwork and transport protocols should be identified, as they will play a major part in the network design. The user will likely have an existing equipment base which will need integration. The network element integration of non-SMDS devices is discussed in later design chapters. Major SMDS interface providers include Wellfleet Communications, 3Com Corp., Vitalink Communications, Advanced Computer Communications, and Ungermann-Bass. Other SMDS CPE vendors include Hewlett-Packard, Idacom, IBM, and Sun Microsystems. Since the SMDS CSU/DSU remains a proprietary device married to the CPE device, these major vendors are also presented: ACD Kentrox (proprietary with cisco Systems routers), Digital Link (uses standard DXI interface), and Verilink. These vendors support either the SMDS SNI interface, the DXI interface, or both. The Data Exchange Interface (DXI) was developed jointly by Wellfleet and Digital Link, and links router and SMDS CSU/DSU through a non-proprietary HDLC interface. Vendors of these interfaces should assure vendor interoperability over both RS-449 and V.35 physical interfaces, as well as support for DS-3 interfaces and HiSSI.

12.2.2 SMDS Subscriber Network Interface

The Subscriber Network Interface (SNI) is the subscriber's physical interface to the SMDS network. The SMDS Interface Protocol (SIP) operates across the SNI. The SIP provides for many customer premises environment devices to communicate over the Subscriber Network Interface using the DQDB protocol. This operation is called an "Access DQDB", which is distinguished as CPE-to-MAN Switching System access, as opposed to MSS to MSS access. The SMDS access DQDB is based on the open bus topology. Multiple CPE devices may be attached to one access DQDB via T1 or T3 circuits. These CPE devices may be a variety of devices, such as bridges, routers, gateways, or switches. If all CPE devices attached to a given access DQDB require autonomy, no other alien CPE may be attached to the same access DQDB. Thus, if there are multiple customers at a site, each customer must be provided a separate access DQDB into the SMDS network.

Figure 12.5 shows a typical multiple CPE arrangement. CPEs 1-3 are connected to one access DQDB. CPE 4 is connected to a separate access DQDB. The SIP within each CPE communicates with the SMDS network through the SNI. The MSS device provides access into the SMDS network for both access DQDB buses. The administration of CPE devices 1-3 is accomplished by a single subscriber labeled Subscriber A. Subscriber B represents an alien CPE which happens to be collocated with Subscriber A CPEs (not owned nor operated by the same user as Subscriber A). Thus, Subscriber B requires a separate access DQDB. Also notice that although Subscriber B has CPE 5 connected to CPE 4 via a common D4 framed T1, CPE 5 does not utilize a SIP, nor does it have a SNI to the SMDS network. Therefore, CPE 5 must transport its data to CPE 4, and it is the job of CPE 4 to provide access to the SMDS network.

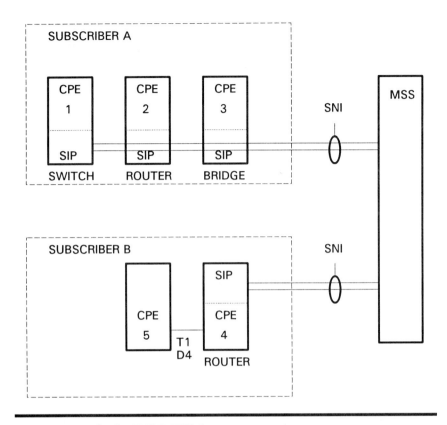

Figure 12.5 Multiple SMDS CPE Arrangement

As can be seen, it is common practice to feed a multitude of interfaces into the CPE device which supports the SIP. One example is to aggregate many synchronous and packet types of traffic, T1 and below speeds, into a router which has a DS3 DQDB interface into the SMDS network.

12.2.3 SMDS Interface Protocol (SIP)

The SMDS interface protocol (SIP) has been defined as the method by which the CPE communicates with the SMDS network across the SNI. The SIP is a connectionless protocol which communicates over the physical layer and MAC sublayer. There are three levels of the SIP: level 1 defines the physical interface, level 2 the segmentation and reassembling of the level 3 Protocol Data Unit (PDU), and level 3 defines the addressing information for the fixed data cell from the CPE. The PDU is variable in size up to 9188 octets. Figure 12.6

shows the encapsulation of the SDU into a level three PDU, labeled a L3_PDU. This L3_PDU is very similar to the DQDB IMPDU. It also shows the segmentation of the L3_PDU into level 2 PDUs, labeled as L2_PDU. Notice the transmission flow on this diagram is from right to left. The original SDU contains the user data.

12.3 USER APPLICATIONS

Many user applications are being developed for SMDS service and IEEE 802.6. The SMDS Interest Group is now defining services and products. One such service is video, which can be transferred as constant bit rate, uncompressed (as in NTSC 100M bps or HDTV at 500M bps), as lightly compressed video (45M bps), or as variable bit rate compressed (MPEG at 1-2M bps) or uncompressed (MPEG2 at 5M bps). Video CODECs are performing this compression, and some are even designed to transmit only the pixel updates as bursty data.

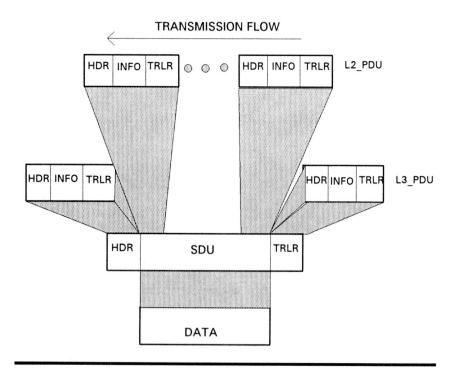

Figure 12.6 Service Data Unit (SDU) Encapsulation

QPSX Ltd. has proposed a standard for guaranteed bandwidth similar to a credit scheme. Users are guaranteed eligible to transmit M out of every N segments which add up to a fixed percentage of the total bandwidth. While multiple reservations may be outstanding, the bandwidth balancing modulus can be raised to guarantee that someone else "downstream" can use the slots unused by "upstream" users. These techniques can be used with both connectionless and connection-oriented operation.

One example of a user physical interface is that if a user interfaces to the SMDS network via a T1, he could use HDLC as a level 2 interface. If an interface speed of DS3 were available, he could use either HDLC or DQDB. Remember, SMDS provides the user with protocol transparent transportation of data.

12.4 FEATURES AND FUNCTIONS OF SMDS

The SMDS network provides a connectionless public network service similar to packet switching. Variable size data units containing user data of up to 9188 octets are segmented into "fixed"-size cells and transmitted across the SMDS network. Addresses are validated and high data transport rates are achieved, at the expense of data error detection and correction and link level control. In addition to these functions, the addressing, access classes, transport mechanisms, rates and tariffs, and network management aspects of SMDS must be understood.

12.4.1 SMDS Switching Systems (SS)

Bellcore standards define the SMDS Switching System (SS) as a collection of equipment that provides high-speed packet switching function in a network supporting SMDS.

Switching Systems (SSs) can be configured in a distributed architecture where multiple SSs would form the SMDS network. Figure 12.7 illustrates an example of the SMDS distributed architecture. Each LATA has CPE which communicates with their local SS via the System Network Interface (SNI). These switching systems are linked by the Inter-Switch System Interface (ISSI) and form the SMDS backbone network.

Switching systems can also take the form of a single switch in a centralized architecture, as shown in Figure 12.8. In this example all three CPE devices reside within the same LATA. There is one centralized switching system which provides the SMDS switching internally. There is no need for the ISSI. The planned locations for SS residences are central offices. This is where the main fiber junctions reside, as well as typically major network management hubs. The network design engineer needs to make the determination whether to support a single or multiple Switching Systems. Design strategies based upon traffic patterns will be discussed in later chapters. Switching systems will use all of the connection and connectionless oriented methods of the DQDB architecture for routing data between SNIs, as defined in Chapter 11.

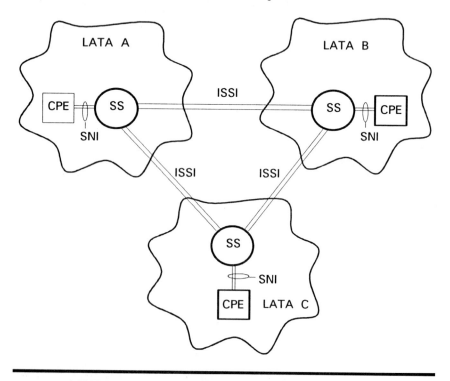

Figure 12.7 SMDS Distributed Architecture

12.4.2 Addressing

When a data unit is transmitted via the SNI to the SS, the source address is verified to assure that it was assigned to the SNI from which it originated. This will be performed for each data unit sent. The SMDS network also has the capability of authorization and address screening for both source and destination if private network service is required. It is the network designer's job to assure that logical private networks are built within the public data network, in effect provisioning a private virtual network, and that the subscriber receives every feature required by the SMDS network.

The addressing scheme used by the SMDS network is formatted with the same structure as the North American Numbering Plan (NANP) with ten-digit addresses. The SMDS service provider will assign each SNI one or multiple unique addresses. The subscriber will have full control over the use of each address, and may assign multiple SMDS addresses per CPE or the same SMDS address to multiple CPEs. No user in the network will have the same address. Many vendors will standardize on a maximum of sixteen addresses per SNI.

There are additional addressing functions available. The SS has the ability to reassign addresses to different SNIs. Group addressed data unit transport provides the CPE capability to transmit to up to 128 individual recipient addresses, similar to a broadcast. Each SS can support up to 1024 of these group addresses, and each address on the network could be assigned up to 32 group addresses.

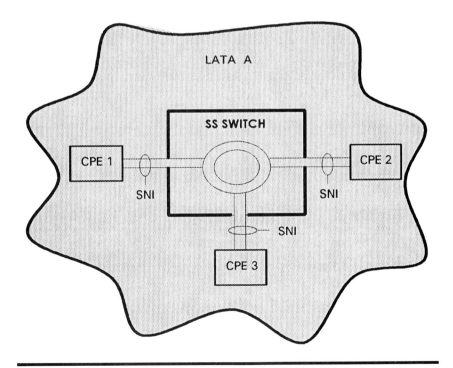

Figure 12.8 SMDS Centralized Architecture

TABLE 12.1 SMDS Access Classes

ACCESS CLASS	SIR (M bps)	LAN SUPPORT
1	4	4M bps Token Ring
2	10	10M bps Ethernet
3	16	16M bps Token Ring
4	25	16-34M bps Sub-rate
5	34	No Enforcement

12.4.3 Access Class

Access classes are assigned to SNIs based upon many factors, but primarily to control congestion conditions. Access classes are a method of providing bandwidth priorities for times when there is network congestion at the SNI. A network congestion occurs when there is an attempt by the network to transfer one or more SMDS data units without an interval of time between the units. This means that more data is attempting transfer than can be supported by the SIP across the access path. The access class places a limit per user on the rate of sustained information transfer used. When the user creates a burst of traffic across the shared SMDS access link, his access class determines the duration of time he controls the link.

There are five access classes defined in Table 12.1. This shows the class number assigned to the user, the maximum rate of transfer that the user can use over extended intervals, and the LAN traffic they were designed to support. It is apparent that T1 access does not warrant an access class, as the maximum amount of bandwidth achievable using SIP over a T1 SNI trunk is about 1.17M bps (due to overhead). Note that 34M bps is the maximum amount of bandwidth which can be achieved using SIP over a DS3 SNI trunk (due to overhead).

For example, suppose user A decides to aggregate one 16M bps Token Ring LAN, three 10M bps Ethernet LANs, and five T1 synchronous data channels through a router. The router aggregates these CPE into one DS3 SNI trunk into the SMDS MSS. Since the combined aggregate of these inputs (ignoring the inefficiencies of Ethernet and that it never achieves a 10M bps data transfer rate) would be approximately 53.5 M bps, the network provider assigns access classes to each CPE LAN. The T1 circuits are not assigned an access class, and are given 7.5M bps. This leaves 26.5M bps for the LANs. Since the 16M bps Token Ring is a mission critical LAN for the Stock Market, it is assigned an access class three, which is a 16M bps class. This leaves 10.5M bps for the three Ethernet LANs. Since we know that the maximum data an Ethernet LAN can transmit is 3.5M bps (due to collisions at low utilization), we assign an access class of one to each Ethernet LAN.

Thus, the total access class assignments will not exceed the available bandwidth of the DS3 trunk: 34M bps. Realistically, the router will perform some statistical multiplexing, and the characteristics of the user traffic on each LAN would dictate the bandwidth actually used by each CPE, but this analysis is good for defining the access class use. We could also approach the problem from the opposite direction and assign low access classes to each of the noncritical applications, and leave a full 16M bps of bandwidth for the Token Ring LAN without assigning it an access class. Either method will work. Egress classes are identical to the ingress classes shown in Table 12.1.

12.4.5 Transport

When discussing addressing, we found that access classes were assigned to CPE to provide for network congestion of multiple transfers of data units. There the user also has a choice on the number of data units which may be in transit between the CPE and SS at any given time. If he chooses only one, the SS will attempt to buffer the information when the maximum value of data units is exceeded. The user can also choose from two to sixteen at a given time, for which the SS would also perform buffering, with a greater chance of data loss. The user can also choose to not specify a limit, but this precludes the SS from buffering, and is not a wise choice for most users.

DQDB bus connectivity is established over 34M bps and 140M bps, 45M bps, and 155M bps links using CCITT G.703, ANSI DS3, and CCITT G.707-9 SDH standards, respectively. Fiber optic transmission media should be used to assure the best performance. Eventually, SONET will replace the existing LEC and IXC digital data networks.

The service offerings for SMDS will initially be T1 access speeds. These will then be increased to one of the SMDS access classes (4M, 10M, 16M, 25M, and 34M bps). Future offerings should include STS-N speeds.

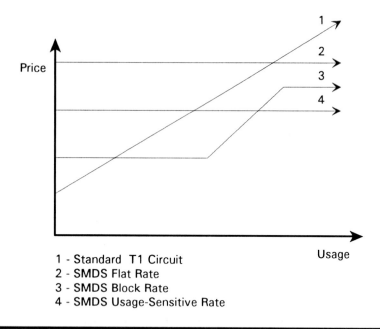

1 - Standard T1 Circuit
2 - SMDS Flat Rate
3 - SMDS Block Rate
4 - SMDS Usage-Sensitive Rate

Figure 12.9 Bell Atlantic Tariff Proposal

12.4.6 Rates and Tariffs

There are two major tariff structures proposed: the Bellcore and the Bell Atlantic. Bellcore offers an SMDS tariff structure based on the access classes (4, 10, 16, 25) as well as the full-line speed access (34M bps). The user is billed based on an average capacity used over a period of time. There are excess limits based on the bandwidth available, and excess data is discarded. On the other hand, the Bell Atlantic method is to bill based on a strict adherence to the access class, and the discard of data in excess of the access class, even though additional bandwidth is available (although they may offer a usage-sensitive tariff as well). Both services are either offered as a flat rate per access line or usage-based billing, which is much more amenable under the Bellcore method. Figure 12.9 shows an approximation of the Bell Atlantic T1 Tariff proposal for SMDS.

There are also fees for access hardware and the local loop, which could be substantial the farther service moves from the POP or LEC switch. It seems as if both the RBOCs and the IXCs will offer SMDS service, while European and Asian PTTs and dominant carriers are also taking a serious look at it. Regardless of what pricing scheme or access rate the carriers charge, the user will end up going with the lowest cost provider.

Bypass carriers such as Metropolitan Fiber Service (MFS) and Teleport are well positioned with over 30,000 miles of fiber to offer SMDS services, at the exclusion of the RBOCs. In fact, some IXCs will likely partner with bypass carriers to offer services such as SMDS.

12.4.7 Network Management

Network management for SMDS services comes in many flavors. Each vendor has a version of network management. Although some features are more necessary than others, the network design engineer must assure that they conform to the ISO/D15 7498-4 standard. This standard defines five categories of network management: fault, configuration, performance, security, and account management.

These systems must be tied together with an overall network management brain. One possibility is the newly emerging SNMP. In fact, TA-1062 specifies that the first customer network management application will be SNMP. Figure 12.10 shows a representation of network management connectivity. The ties to each element of network management are shown, with the overall element manager represented by SNMP feeds.

There are many other necessities to a good network management system. SMDS network management CPUs must have the capability for remote monitoring and software downloads, geographic diversity with real-time database updates, good statistical and performance reporting, many levels of user access security, a good vendor 7x24 support staff, and a friendly and graphical operator interface. The network designer will never know how important these requirements are until they have to administer and run the network management of an SMDS network. Network management will be covered extensively in Chapter 21.

12.4.8 Public vs. Private

SMDS is primarily a public data network offering, but could also be used in a private network. The SMDS will connect multiple nodes, referred to as Customer Access Networks (CANs). SMDS can provide transport for a variety of customer network access, including packet switched networks, synchronous data transport, ISDN, and LANs such as Ethernet and Token Ring.

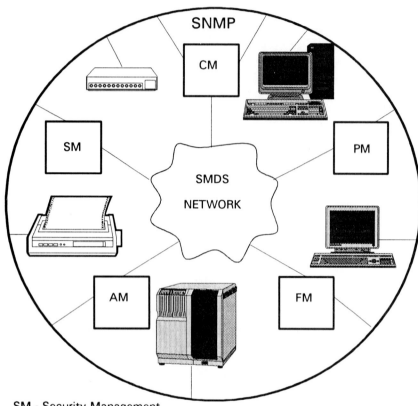

SM - Security Management
AM - Account Management
FM - Fault Management
PM - Performance Management
CM - Configuration Management

Figure 12.10 SMDS Network Management Elements & SNMP

Since SMDS is a public offering, it will be offered as a public data network service by the BOCs and IXCs. All of the RBOCs have announced SMDS offerings. Both will have to build and manage a large public data network based on 802.6 technology, which may eventually migrate to an Asynchronous Transfer Mode (ATM) network and very likely use transport over a SONET fiber backbone.

In the form of a public data network, SMDS provides access to the central office environment. In contrast to the central office, SMDS networks transport a variety of data, voice, and video in a high-speed switched environment through a public data offering. Figure 12.11 shows an example of SMDS networking.

RBOCs can either offer service between LATAs using the IXCs, or the user can bypass the RBOC through either the bypass carriers or direct connectivity to the IXCs through local loop dedicated circuits. RBOCs project that by year 2000 they will have over 30,000 consumer access lines installed.

National and international SMDS networks will be able to interface via the Exchange Access SMDS (XA-SMDS) specification defined by Bellcore. This standard will provide a seamless interface for multiple SMDS networks. The specification is defined in TR-TSV-001060 SMDS Generic Requirements for Exchange Access and Intercompany Serving Arrangements. This interface is similar to the DXI interface (Data Exchange Interface), which allows LAN bridges and routers to connect to SMDS through CSU/DSUs.

12.5 PERFORMANCE

The main focus on performance for SMDS networks is between the SNIs at a L3_PDU level. The CPE will measure the performance of the service. We have discussed how the SMDS network should be transparent to the user. Thus, the user CPE will measure performance based upon the availability of the service, the delay incurred between SNI-to-SNI communications, and the accuracy with which data is transferred. With these criteria in mind, and the fact that SMDS assumes error-free transmission, we, as designers, require that the transmission media be error free (or as close as possible to error free). We also require that the user have "smart applications" at each end to add data error detection, correction, and retransmissions, as well as end-to-end transmission congestion control and routing assurance.

When measuring the delay and accuracy of an SMDS service, the L3_PDU transmission is analyzed, because this is the true data unit being transmitted. All measurements are taken from SNI to SNI, as the user's interface to the network. Any performance criteria outside the SNI will be measured by the user; this is not part of the data transport network. When the service provider of the SMDS service starts providing CPE equipment such as routers, bridges, and other CPE gear typically managed by the user, the network performance becomes a true end-to-end function, and will require management of levels above the second layer.

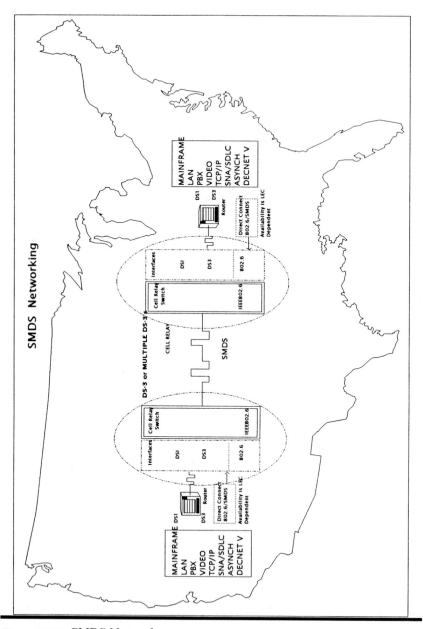

Figure 12.11 SMDS Network

12.5.1 Availability

Availability is measured by the ratio of actual service time to scheduled service time on a user-to-user basis (SNI to SNI). There are four methods of measuring the availability of the service: network availability should be at least 99.9%, scheduled service time should be 7x24, Mean Time To Repair (MTTR) should be less than two hours, and Mean Time Between Service Outages (MTBSO) should be greater than 1092 hours. Each of these availability measurements will change based upon user requirements, network configuration, equipment locations, and vendor specifications. A more detailed analysis of each calculation listed above can be found in later chapters.

12.5.2 Delay

End-to-end network delay is one of the most important considerations to users. This is especially true if they are used to certain expectations on their existing network, and SMDS service is being provided as the new service. Users will expect the same, if not faster, response times between SNIs. SMDS response times are similar to a hybrid of technologies: at low load — quick access similar to Ethernet, at high load — orderly access similar to Token Ring. SMDS does not have to take the tokens off of the ring as in FDDI (in FDDI, tokens must be removed or they would fill up the bus). The SMDS network response time portion of the overall network response time is measured in L3_PDUs. The delay will therefore be measured from inception of the L3_PDU from user A to the reception of the L3_PDU by User B.

Figure 12.12 shows an example of an interactive file transfer of medical images between a Token Ring LAN station and the IBM Host. The user contacts the host to request file transfer. The router acts as a CPE device in the building where the LAN resides. It takes the Token Ring PDU and translates (or encapsulates) it to the access DQDB bus protocol for transport over the SMDS network. The SMDS network provides the transport through MAN switching systems, and delivers the transaction directly into the front end processor, in this case an IBM 3745, which employs a direct access DQDB to the SMDS network. The FEP then interfaces to the host computer. The host application processes the request and transmits the file back to the requesting station. Obviously, there are many additional higher-level protocol functions that take place, but this illustration shows the "end-to-end" delay components.

In this example, we see that there are many areas where delay is added to the total transmission time. The Token Ring LAN access and transmission time, router protocol conversion or encapsulation and routing function, the SMDS network transmission delay, and the FEP and mainframe processing delay. Also, the transmission media in each of these steps adds delay. This is referred to as the total user delay, not the SMDS network delay. The SMDS network delay begins at the router SNI and ends at the Mainframe SNI, and is measured by the L3_PDUs between those interfaces. There is also delay inherent to the transmission medium used within the SMDS network.

The delay for the L3_PDUs are defined for individually addressed and group addressed L3_PDUs. With individually addressed L3_PDUs, the delay based on two CPE using DS3 access trunks should be less than 20mS. The delay based on two CPE using a DS3 access trunk at one SNI and a T1 access trunk at the other SNI should be less than 80mS. The delay based on two CPE who both use T1 access trunks should be less than 140mS. With group addressed L3_PDUs, the delay times should be less than 100mS, 160mS, and 220mS, respectively. The maximum lifetime allowed for a L3_PDU is 500mS. These delay figures are very important when analyzing the feasibility of an SMDS network over other types of technologies.

12.5.3 Accuracy

The accuracy of the network is achieved by the number of delivered or dropped L3_PDUs across the network. The ratio of errored L3_PDUs to the number of successful L3_PDUs is called the "errored L3_PDU ratio", and should not be more than 5 in 10^{13} L3_PDUs delivered. The ratio of misdelivered L3_PDUs should not be more than 5 in 10^8 L3_PDUs delivered. The ratio of L3_PDUs which are *not* delivered should not be more than 1 in 10^4 L3_PDUs. The last accuracy calculation which needs to be made is the ratio of duplicated L3_PDUs, which should be no more than 5 in 10^8 L3_PDUs delivered. These numbers would be adjusted based upon the same criteria as discussed in delay. For example, if the user were transmitting a majority of voice or video traffic, the misdelivery ratio would be of major concern.

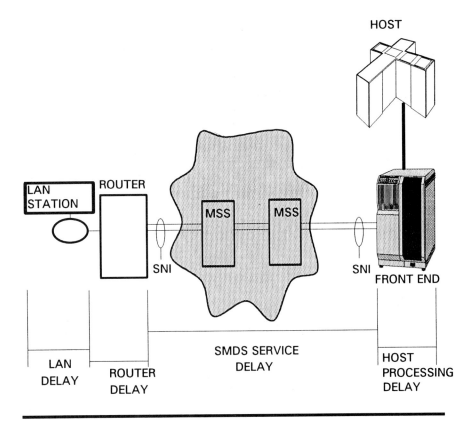

Figure 12.12 SMDS Interactive File Transfer Example

12.5.4 Reliance on Fiber

SMDS does not provide error correction nor control protocol functionality. It does do some level of error correction on the address, but this is minimal. When a user builds an SMDS network, he relies on digital fiber transport to provide them with an availability of 99.99% and low circuit error figures. The engineer should verify the availability and error ratio of every digital fiber circuit used. Digital radios are generally too noisy, and their errors too random, to be used as transport. The same goes for satellite shots. These standards are important to design into the communications system. Remember, the SMDS network does not perform flow control nor data error checking. It *assumes* the transmission paths are virtually error free! If the cell takes an error in the SDU (data portion), it will be passed on to the user. The concept is that a trade off exists between error correction for performance and high throughput.

12.5.5 "Smart Applications" Concept

Since SMDS networks provide only layer one and two protocol functionality, users of existing data services, such as packet switch networks, rely on the network to perform the error detection, correction, and end-to-end data transport control. The SMDS network does very little of these functions, and when it does, it does so for the addressing of PDUs, not the actual data within the PDU. What this means to the user is that smart protocols must be used at the CPE, such as TCP/IP and DECnet, to ensure end-to-end transmission.

Transmission systems may need the enhancement of Physical Layer Convergence Procedures (PLCPs) to enhance physical level interfaces.

12.6 BENEFITS AND COMPARISONS

A good understanding of high-speed data transport technologies is essential for the network design engineer to make a wise business decision. With SMDS, frame relay, and FDDI services vying for domination in the data market, these technologies must be compared. IEEE 802.6 and SMDS services provide the wide area network connectivity for LANs, and FDDI provides a WAN solution, so a comparison of the two is necessary. Since frame relay provides switched, statistical multiplexed services over long distances, we need to compare 802.6 SMDS with frame relay. Table 12.2 shows a comparison of 802.6 SMDS, FDDI, and frame relay technologies.

12.6.1 802.6 to FDDI

FDDI is similar to a Token Ring network, while 802.6 is similar to a packet switching network. With a token passing environment, users compete equally for the one token on the bus. Thus, distance limitations and long waits for a free token are inherent in the FDDI architecture. The DQDB bus eliminates the need to await the token through use of the reservation system.

Due to the limitations of FDDI when compared to 802.6 (more specifically the DQDB bus), 802.6 becomes a more versatile switching solution, and is suitable for isochronous data transport, including video and voice. FDDI will not support isochronous until FDDI II is introduced as a standard. SMDS, the main service offering of 802.6, does not support isochronous data transport as defined in the IEEE 802.6 standard. Perhaps the largest difference between the two technologies is the product market. FDDI is a private data network service marketed by private network equipment vendors. It is also sold as an interface by many central office vendors. In comparison, 802.6 is projected primarily for the public data network market. Equipment conforming to the 802.6 standards are predominantly built by central office vendors. This major distinction will be amplified when we discuss the vendor "time versus technology" issue later.

Some additional comparisons are that FDDI is available in the marketplace with a variety of vendor support. 802.6 is now coming into a bandwidth starved data market much in need of a high-speed switching protocol to replace packet switching. FDDI supports one speed of 100M bps, while 802.6 supports 35M bps, 45M bps, and 150M bps. With the use of a bandwidth aggregator such as a router, many low-speed interfaces can be combined into one FDDI or 802.6 network access.

SMDS also supports closed user groups, usage sensitive pricing, public worldwide access, and the 802.6 architecture will eventually support FDDI and IBM channel-to-channel connectivity.

TABLE 12.2 Comparison Matrix

CAPABILITY	802.6 & SMDS	FDDI	FRAME RELAY
Speed (M bps)	45 - 150	100	1.5 - 45
Statistical Multiplex-ing Capability	Yes	No	Yes
Data Transport Format	Fixed Data Cells	Token Passing	Variable Sized Frames
Distance Limitations	No	200 km	No
Voice/Video Capability	DQDB-Yes SMDS-No	No (FDDI-II can)	No
Service Offerings	Public Network	Private (MAN) Network	Private & Public Network
Vendor Support	Few Vendors	Many Vendors	Many Vendors
Service Orientation	SMDS-CL, IEEE 802.6-CO, CL, and Isochronous	Connection-less	Connection Oriented (PVCs & SVCs)
Futures	Migration to ATM	Higher Speeds (FDDI-II)	DS3 Speeds

12.6.2 802.6 (SMDS) to Frame Relay

Frame relay and 802.6 both offer a version of packet-switching technology. The main difference is that frame relay is connection-oriented and works on Permanent Virtual Circuits (PVCs), while SMDS is a connectionless service. Frame relay uses variable length frames as opposed to a fixed 53 byte cell size in SMDS. Frame relay does offer some congestion control, where SMDS is very limited.

Another major consideration is the vendor posture for each service. Frame relay is a technology sold by the computer market (i.e., primarily router vendors). SMDS is offered primarily by central office vendors, proving that the frame relay is a CPE-oriented service and SMDS a central-office-oriented service. This makes the transition to ATM and SONET much easier for the central office vendors, which might prove frame relay to be a short term access interface that will eventually disappear. Frame relay can coexist with SMDS and ATM, and some RBOCs may even make frame relay a service within SMDS.

Some issues which remain to be resolved include:

- interconnectivity between frame relay, SMDS, and ATM services will be difficult - emerging standards
- more industry support for frame relay than SMDS
- CPE hardware now supporting SMDS DQDB and DXI interface
- SMDS requires hardware upgrade, frame relay requires software change — thus harder migration to SMDS than to frame relay
- LEC still needs to invest in both
- billing more difficult for SMDS than frame relay
- both are international standards

12.7 DESIGN CONSIDERATIONS

SMDS design considerations are just now being determined. These include: limiting the distance between nodes and taking into account the nodes' position on the bus, network timing, network bandwidth capacity and overbuilding, and addressing constraints. Additional design considerations will be discussed later. Migration from 802.6 SMDS services to ATM B-ISDN services is also a consideration.

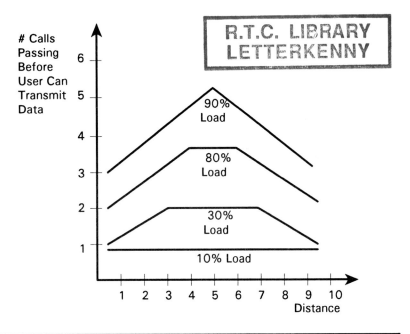

Figure 12.13 SMDS Delay

12.7.1 Distance Between Nodes

The overall distance between nodes and their position on the DQDB bus have a major influence on the amount of potential delay they may incur. Figure 12.13 shows that nodes on the center of the bus (from head and tail of buses) will incur greater delay than nodes at the head or tail of the bus. They will do so by having to wait for more cells to pass before they find an empty one (the head and tail nodes will be able to get the empty cells first). This has a drastic effect as the number of users and traffic on the ring increases.

12.7.1 Timing

Network timing can either be provided by an external source to the DQDB network (the network user) or provided by one and only one node per DQDB bus. If there is more than one DQDB ring, a master node will be chosen from which to draw network timing. External timing will always have priority over node timing. A stratum one clock source is desirable, but any source up to and including stratum three will suffice. Due to operating speeds, an accurate clock source is vital to operations.

12.7.2 Capacity

The most desirable way to build an SMDS DQDB MAN network is to over-build. The traffic base for a MAN is often many LANs. As seen in previous chapters, the traffic output from a LAN can vary up to 100 times the original projections. This is due to the very bursty nature of LAN-to-LAN information transfers. The design engineer must build enough capacity into the network for it to withstand these times of extremely high traffic volumes. This becomes a trade-off of cost avoidance and profit. Cost avoidance is planning enough capacity on the MAN to eliminate many private individual circuits, and profit is what will be gained or lost by doing so.

For example, it would not be good to eliminate forty-six T1 private lines for a router and one DS3, with the hope of their average utilization peak is only 50%. If their peak were to exceed the 50%, the DS3 would be out of capacity, and either the application would have to buffer the traffic, or the data will be lost. Now imagine if it is one-thirty PM and all the LANs attached to the network decided to perform massive file transfers. The utilization just went up 300%: there was not enough buffers and transport to handle even a third of the traffic: and the LAN's file transfers were delayed considerably. In a large network, this situation could bring the network to its knees. The designer needs to ensure a balance between the cost avoidance of extra circuits and equipment to plan for peak periods (and reduce delay during those times) and the profit or loss incurred if he or she effectively transports the user's traffic or not. This is why proper capacity planning is important.

If bandwidth hogging (bus fairness) is a problem, use a bandwidth balancing parameter of eliminating the use of 1 slot in N. At the same time, you may want to restrict the connectionless priority to 0 and add a higher priority to connection-oriented traffic.

12.7.3 Addressing Constraints

Bridging enhancements include the cf.802.1D defined 802.6-to-802.6 interface to a multiport bridge (forwarding on a cell basis) and the 802.X-to-802.6-to-802.X LAN-MAN-LAN bridging and 48-60-48 bit address conversion. As discussed previously, this can be done either through routers or through SNAP encapsulation. IEEE 802.6K will introduce 48-bit addressing.

12.7.4 Future PARs

Table 12.3 shows the current status (as of publish date) of the 802.6x extensions. The more prevalent ones have already been discussed in this chapter.

TABLE 12.3 802.6x Extensions

802.6x	Feature or Enhancement	Status
a	multiport bridge interface	in progress
b	DS3 premise extension	in progress
c	DS1 physical layer	finished
d	SONET physical layer	finished
e	eraser node	frozen
f	PICS pro forma	in ballot
g	layer management	in progress
h	isochronous	in progress
i	remote bridging	finished
j	connection oriented	needs work
k	additions to 802.1	TCCC ballot

Future directions also include speeds up to 622M bps, isochronous enhancements for end-to-end transparent cell transport (let the carriers sort it out), additional premises physical layer support, and efficient methods of guaranteeing bandwidth.

12.7.5 Migration Path from SMDS to B-ISDN

When Bellcore specified recommended standards for SMDS, they kept in mind that they would need to migrate to Broadband ISDN (BISDN) standards in the near future. The features and functions of SMDS, and the conformance to 53-octet DQDB and ATM cell size, were designed to provide a service which could migrate with a minimum loss of feature and function, quality of service, and performance, as well as minimizing the need to replace existing equipment. As BISDN services emerge, vendors supporting the new service will need to provide an incentive for users to migrate.

Major efforts to aid in the migration from SMDS to B-ISDN include:

- ➲ fiber oriented — based on cell-relay data format
- ➲ 802.6 and B-ISDN committees working together
- ➲ 802.6 method of dividing up bandwidth on B-ISDN to place 802.6 segments into B-ISDN
- ➲ ISO Sub-group SG6, Workgroup 6 working on ISO DQDB acceptance also with ETSI/NA5 MAN subgroup

Many of the vendors providing SMDS hardware, now primarily central office vendors, will be the same players providing B-ISDN hardware. Thus, it will be in their best interest to provide continuity in product line and protocol support for both services. Some similarities which should remain are the SMDS SIP and B-ISDN User Network Interface (UNI) with the flexibility of the SIP third level of protocol support. One example of this is Siemens-Stromberg-Carlson. They were the first central office vendor with a SMDS switch, and they have also announced plans to support ATM migration.

12.8 REVIEW

In this chapter the user interface to the DQDB MAN network was defined. More specifically, Switched Multimegabit Data Service (SMDS) was introduced. This is the first major service offering derived from the IEEE 802.6 MAN standard, and provides for connectionless public data service. SMDS direct interface to both the OSI and IEEE protocol layers was covered. We learned how to translate the 802.6 MAC protocol interface to the SMDS subscriber network interface, delving into the L2_PDU and L3_PDU frame structures which are the heart of the data transport service within SMDS. User applications were discussed briefly, including the HDLC interface. The user features and functions of SMDS were fully covered. The addressing scheme was discussed in detail, along with descriptions of how to use access class for congestion control. A section was also dedicated to describing transport speeds, along with how recent rates and tariffs affect SMDS offerings.

The designer was informed how important network management is. And many performance parameters were discussed in detail to ensure a virtually error-free network. Most importantly, considerations for designing an SMDS network were outlined, highlighting capacity planning as the critical key to success.

13

ATM and B-ISDN

Cell relay defines two major technology platforms: 802.6 Distributed Queue Dual Bus (DQDB) and Asynchronous Transfer Mode (ATM), over which Switched Multimegabit Data Service (SMDS) and Broadband ISDN (B-ISDN) services are offered, respectively. ATM technology, as well as the B-ISDN service implemented on an ATM network, will now be discussed. The success of both ATM and B-ISDN standards will rely upon an integrated deployment, as well as the use of a SONET infrastructure. This chapter provides the reader with a working knowledge of both ATM and B-ISDN. It will soon become apparent to the reader that the development of the CCITT standards has blended together many of the protocol and operations functions of both ATM and B-ISDN. The terms and standards which have created the framework of ATM and B-ISDN will be defined in this chapter. The theory of operation through cell structure and asynchronous multiplexing techniques will be discussed. All layers of the ATM/B-ISDN) protocol stack will be defined, along with the many methods of user and network connectivity. The many types of services B-ISDN supports will also be explained in detail. Next, focus will be brought to bear on performance, benefits, and design considerations. In the next chapter, the Synchronous Optical NETwork (SONET) transport infrastructure that these technologies and services transit will be detailed.

13.1 TERMS DEFINED

Asynchronous Transfer Mode (ATM) refers to a switching technique for broadband signals. It is often labeled as a standard for cell relay or fast packet. In the purest sense, ATM switching resembles the true definition of fast packet because it handles all traffic types through a fast-packet switching technique which reduces the processing of protocols and utilizes statistical multiplexing. ATM is primarily a connection-oriented technique which can transport both connection and connectionless services at both a constant bit rate (CBR) and variable bit rate (VBR). ATM uses virtual calls to transmit user information. This information is transported in the closest approximation to true bandwidth-on-demand service we have seen. ATM is an asynchronous technology because the cells transmitted by the user are not necessarily periodic.

ATM switching also provides the capability to standardize on one network architecture and platform, where ATM provides the switching platform and SONET provides the digital infrastructure and physical transport. Thus, the entire network operates on one switching and multiplexing principle for transmission of multimedia services. ATM is a WAN switching technique, as compared to the DQDB MAN switching technique. DQDB MAN networks provide a clear migration path to ATM WAN networks in the near future.

ATM is designed to provide the transfer mode for B-ISDN services. ATM is the first technology to provide true bandwidth-on-demand because users can vary bandwidth within a given call, while the B-ISDN service allows multiple users to optimize network resources by sharing the bandwidth efficiently. It is the first packet standard to support the full gamut of multimedia at circuit speeds higher than T1 and T3, to handle:

  ★ voice
  ★ framed data
  ★ video
  ★ imaging

while providing an alternative for low-speed X.25 packet-switched services as well as high-speed dedicated leased lines. ATM provides multiple switched virtual circuit connections to other users through a single access to the network. This makes ATM the first technology to combine the benefits of both circuit switching and packet or cell switching.

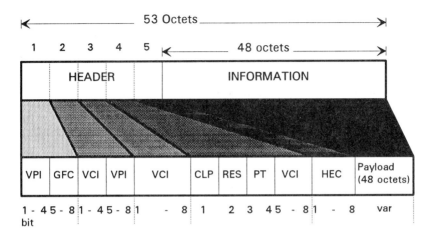

Figure 13.1 ATM User-Network Interface (UNI) Cell Structure

Broadband integrated-services digital network (B-ISDN) is the switched service that uses ATM switching techniques. Both ATM and B-ISDN were developed and standardized by the CCITT, and in 1988 the CCITT standardized on B-ISDN as the service to transit ATM technology. B-ISDN was developed to handle voice, video, and data applications on the same network, and within the same transmission. As we go through this chapter, it will become increasingly clear that both ATM and B-ISDN are intricately tied together, co-dependent for survival as standards in the data market.

13.2 STANDARDS

In 1988, CCITT Recommendations I.113 and I.121 defined the first two standards for B-ISDN: Vocabulary and Terms for Broadband Aspects of ISDN and Broadband Aspects of ISDN, respectively. These recommendations were revised in May of 1990. At the same time, eleven more Draft Recommendations: I.150, I.211, I.311, I.321, I.327, I.361, I.362, I.363, I.413, I.432, and I.610, were published detailing the functions, service aspects, protocol layer functions,

OAM&P, and user-to-network and network-to-network interfaces. Each standard was defined as a B-ISDN standard to operate over the ATM architecture. A complete list of these standards can be found in Appendix C. Most, if not all of these draft standards, will be found in the CCITT 1992 White Books.

13.3 THEORY OF OPERATION

ATM offers a wide variety of switched services to the subscriber through a single-network entry point. This subscriber information can be either delay-sensitive or delay insensitive data, constant or variable bit rate, requiring or not requiring timing, and connection-oriented or connectionless traffic. Either way, the user information field is transmitted transparently through the ATM network through the use of cells. While the cell structure is fixed, the header can vary based upon the type of user or network interface.

13.3.1 Cell Structure

ATM forms of packet called a "cell". The format of the cell is shown in Figure 13.1. Cells are 53 octets in fixed length. Each cell contains a 5-octet header and a 48-octet information field. All information is transmitted transparently through the network in these fixed-length packets. Thus, the information field is passed through the network intact, with no error checking or correction performed on the information field. With fixed-size packets, longer packets cannot delay shorter packets as in X.25 implementations. There is also a packet header which identifies virtual channels and paths.

There are two types of user interfaces: user-network interface (UNI) and network node interface (NNI). While Figure 13.1 shows an example of the UNI cell, Figure 13.2 illustrates an example of the NNI cell. Both are defined in the cell header.

13.3.2 Cell Header

The header information tells the switch which virtual channel and virtual path the cells are riding on. Each of these channel identifications are local to the multiplexer and can be changed at each switch. The combination of both the Virtual Path Identifier (VPI) and the Virtual Channel Identifier (VCI) establishes a node-to-node communications channel. Switch routing is then accomplished based on the

VCI and VPI. The ATM switch requires a connection to be established between the incoming and outgoing virtual channels before information can be routed through the switch. The ATM switch will then switch and route each individual cell from the incoming multiplexed cell stream to the outgoing multiplexed cell stream based upon the virtual channels identified within the cell header. In this context, ATM is truly seen as a connection-oriented technology. The cell sequence is maintained by the ATM switch, and each cell is switched at the cell rate, not the channel rate, to accommodate for variable bit rate transmissions.

Call Loss Priority Field (CLP) — There are two levels of semantic priority which allows users or network providers to choose which cells to discard during periods of network congestion. Both types are defined by a 1 or 0 in the CLP bit field within the cell. This is similar to the discard eligibility bit in Frame Relay.

The Payload Type (PT) field discriminates between a cell carrying user information or service information in the Information field.

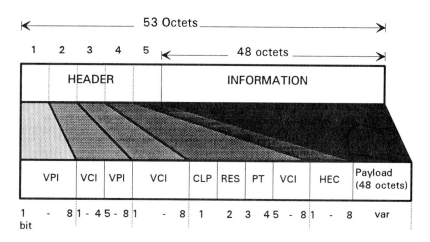

Figure 13.2 ATM Network Node Interface (NNI) Cell Structure

The Generic Flow Control (GFC) field is designed to provide shared public network access similar to the functionality of a MAN. GFC is used when there is a single S_B user access point servicing multiple terminal interfaces, such as those found in a LAN environment. Each terminal must receive equal access to the network facilities, and the GFC ensures that each terminal will get equal access to the shared network bandwidth. The GFC will manage the various LAN topologies and architectures.

The Header Error Control (HEC) field provides error checking of the header, and the Reserved (RES) field is reserved for header functionality enhancement.

13.3.3 Asynchronous Multiplexing

When the cells are sent to the ATM switch, they are placed into a queue until they can be multiplexed asynchronously with other cells for transmission. This multiplexing is statistical in nature, since the ATM switch removes idle traffic from the user traffic before transmission. A time slot will only be allocated to channels which have data for transmission, so that an ATM switch can achieve efficiency gains of typically up to 4:1 input to output. The switch will adapt the incoming bit rate to the transmit channel rate, inserting dummy cells when needed to achieve the aggregate bit stream rate (e.g., 155.52M bps). The remote switch will then delineate the good cells from the dummy cells based on header information. Thus, true bandwidth on demand is achieved when the incoming data traffic can use the entire channel when needed. This also achieves the maximum throughput and performance possible. This operation is shown in Figure 13.3.

13.4 ATM AND B-ISDN PROTOCOL STRUCTURES

ATM protocol functions roughly correspond to the first two layers of the OSI reference model. B-ISDN protocol functions correspond to levels two and three of the OSIRM. Figure 13.4 illustrates the ATM protocol stack with a breakout showing correlation to the ATM cell structure. The ATM protocol stack is entirely comprised of both the Physical and ATM layers, and the B-ISDN rides on top of these layers. The physical layer is either SONET based or cell based. The ATM layer is comprised of the virtual channel and virtual path sublayers. It will soon be seen that both the ATM and B-ISDN protocol model have the same physical and ATM layer. The ATM Adaptation Layer (AAL) is comprised of the convergence sublayer (CS) and the segmentation and reassembly (SAR) sublayer. This layer works with B-ISDN protocols to provide services. The relationship to the network and higher layers of the OSIRM in relation to the ATM layers will be explained. Each layer will be discussed in detail in this section.

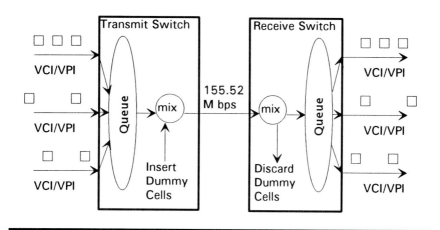

Figure 13.3 ATM Transmission

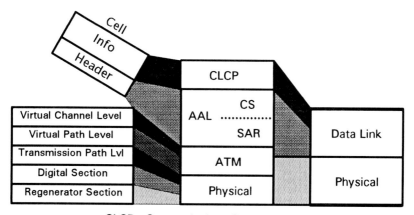

CLCP - Connectionless Convergence Protocol
AAL - ATM Asynchronous Transfer Layer

Figure 13.4 ATM Protocol Stack

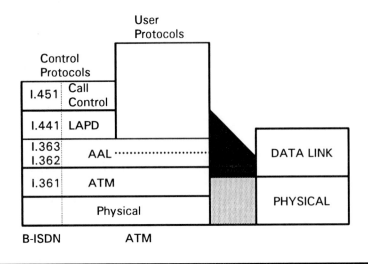

Figure 13.5 ATM and B-ISDN Protocol Stacks

Figure 13.5 shows the data link layer protocols and each of the ATM and B-ISDN protocol stacks as they relate to control and user protocols. Each of the B-ISDN network layers are portrayed as to their relation to the physical, ATM, and higher layers of protocols. All other layer protocol functions conform to those of N-ISDN as defined in CCITT Recommendation I.320.

13.4.1 Physical Layer

There are two major physical interface types available to ATM: SONET based and cell-based interfaces. When ATM uses a SONET-based interface, the combined transmission of multiplexed cells is inserted into the virtual channel and network monitoring and management functions are handled outside the virtual channel. These channels correspond to the SONET path, line, and section levels. The ATM/B-ISDN user-to-network interface access rate is fixed at either an S_B rate of 155.52M bps (SONET OC3, STM-1 rate) or a T_B rate of 622.08M bps (OC12 rate). Thus, ATM can easily be integrated with the SDH. While there are obvious advantages to embedding ATM onto a SONET infrastructure, there now comes a need for strict clocking which is not required with a cell-based interface (ATM-only architecture).

CCITT Recommendation I.432 defines a method of mapping ATM cells into an STS-1 SONET frame conforming to the G.707, G.708, and G.709 specifications. The ATM cell is first mapped into the C-4, which is then packed into a VC-4 channel. Many of the specifics are still in draft form.

The other method of interface is through a cell-based interface. With this type of interface, the combined transmission of multiplexed cells fills the entire access bit rate and network monitoring and management functions are handled in specialized nonuser information bearing cells. This transport layer is comprised of a transmission path level, digital section level, and regenerator section level.

As indicated, the user-to-network interface layer could either be SONET, or physical layer of FDDI (or even FDDI-II), or some other cell-based technology. The industry seems to be moving toward the SONET interface, but both options will exist for a long time.

13.4.2 ATM LAYER

CCITT recommendation I.361 defines both the cell structure and the protocol procedures for the ATM layer. This portion of the protocol stack is connection-oriented. The ATM layer builds the cell and provides for call transfer by placing the required header information around the user information. The transfer of the ATM cell is done independently of the data contained within the information field and of the applications using this field. ATM networks do not process any of the information within layers higher than the ATM layer. The ATM layer can also be thought of as part of the physical layer of the OSIRM. Also, the data contained within the cell cannot be directly related to the applications information being transferred, and there is no relationship between the network timing and the application transferring the data.

The transport functions of the ATM level are divided into the Virtual Channel (VC) and Virtual Path (VP) sublevels. The identification of the virtual channel for each cell is contained in the cell header, which contains both a Virtual Channel Identifier (VCI) and multiple Virtual Path Identifiers (VPIs). These VCIs and VPIs are then aggregated over a single Transmission Path (TP). Figure 13.6 shows this ATM layer VCI, VPI, and TP aggregation.

VC sublevel. This defines each individual unidirectional virtual channel transmission of cells from a specific user. Each VC is mapped to a VP and TP. VCs are then concatenated to form Virtual Channel Connections (VCCs), which define either point-to-point or multipoint connections across the network.

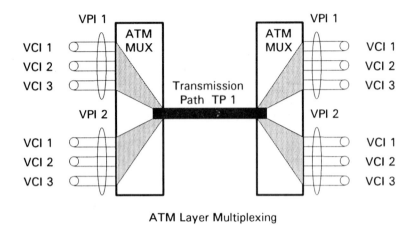

ATM Layer Multiplexing

Figure 13.6 ATM Layer VCI, VPI, and TP Aggregation

VP sublevel. This defines an aggregate bundle of VCs with the same user end point. VPs are concatenated to form Virtual Path Connections (VCPs), which define either point-to-point or multipoint connections across the network.

   Both VCCs and VPCs provide user-user, user-network, and network-network transfer. Network switching can occur at either the TP, VP, or VC level.

13.4.3 ATM Adaptation Layer (AAL)

CCITT recommendations I.362 and I.363 define the next layer of the ATM/B-ISDN protocol stack, the ATM Adaptation Layer (AAL). The AAL is split into the Convergence Sublayer (CS) and the Segmentation and Reassembly (SAR) sublayer. The convergence function supports connections between ATM interfaces and nonATM interfaces. Examples include the N-ISDN BRI and PRI interfaces and B-ISDN interfaces that do not conform to ATM standards. The SAR sublayer either segments large messages or assembles small messages into the information field of the ATM cell on the network side of the ATM switch, or performs the opposite at the egress side of the ATM switch.

SAR. This segments the higher-protocol-level data into the information field of a cell and vice versa.

CS. This provides a service-dependent AAL function at the defined service access point (SAP).

The AAL layer enhances and maps the ATM layer services to the upper-layer protocols, providing a bridge of user, control, and management functions from the ATM to upper layers. It does this by mapping the Protocol Data Units (PDUs) of the higher-layer protocols into the ATM cell and passing them to the ATM layer. It also extracts the PDUs from incoming cells and sends them to the higher-layer protocols. All AAL functions are service-dependent.

The AAL is utilized by either ATM switches that are application-dependent or it resides in the user CPE equipment. This layer helps to delineate the individual application messages which use the common network interface channel. All segmentation and reassembly is performed at the AAL layer for each type of service. The network transport remains the same for any service, and the application specifics remain outside the network, and are characterized by four classes: A, B, C, and D. This scheme minimizes the number of protocols AAL must support while defining the timing relationships, bit rate (constant or variable), and connection mode type (connection or connectionless) between source and destination. Information transparency is retained through the AAL layer. There is also some error detection at this layer. Figure 13.7 describes each of these classes and their service attributes.

ATM Service Class Attributes	Class A	Class B	Frame Relay Class C	SMDS Class D
Timing Relation Between Source and Destination	Required		Not Required	
Bit Rate	Constant		Variable	
Connection Mode	Connection Oriented			Connectionless

Figure 13.7 ATM Service Classes

While the Class A through D defines the classification of service, Recommendation I.363 defines the protocols which use these four types of services. Each protocol type is defined below, but we are primarily interested in Type 3 and Type 4. Protocol Type 1 through Type 4 roughly correspond to Class A through D.

Type 1 — constant-source bit rate

Type 2 — variable-source bit rate

Type 3 — connection-oriented data transfer, as well as user-to-network signaling transfer (e.g., message mode and streaming mode services)

Type 4 — connectionless data transfer (e.g., message and streaming mode services).

The primary focus of the standards bodies at this time is toward the Type 4 service, or connectionless data transfer. The format of this service is compatible with the IEEE 802.6 connectionless mode that was previously discussed.

13.4.4 Connectionless Convergence Protocol (CLCP) Layer

The connectionless convergence protocol (CLCP) layer corresponds to the equivalent layer of the same name in SMDS. This allows the SMDS service to be offered over ATM switching. This layer manages the addressing, quality of service selection, and carrier selection. Another service is under the review of ETSI: the connectionless broadband data service (CBDS). CBDS is similar to and compatible with SMDS, and encompasses both the MAN and B-ISDN standards.

13.4.5 Summary of Protocol Structure

Figure 13.8 shows a summary of the three layers of the ATM/B-ISDN protocol structure.

13.5 USER CONNECTIVITY

B-ISDN will support the following connections:
    * switched

* permanent
* semipermanent
* point-to-point
* point-to-multipoint

and supports the following services:

* demand
* reserved
* permanent

in both connection-oriented and connectionless modes.

As mentioned before, the user interfaces through either a VCC or VPC connection. VCCs and VPCs offer the user quality of service parameters (e.g., cell loss priority), switched and semipermanent VCCs, call sequence integrity, and traffic parameter negotiation and usage monitoring. Many of these parameters are still under development.

UPPER LAYERS	UPPER LAYERS PROTOCOL FUNCTIONS
CS	Convergence
SAR	Segmentation Reassembly
ATM	Generic Flow Control Call Header Generation/ Extraction Cell VCI/VPI Translation Cell Mux/Demux
TC	Cell Rate Decoupling Cell Delineation Transmission Frame Adaptation Transmission Frame Generation/ Recovery
Physical	
PM	Bit Timing Physical Medium

Figure 13.8 ATM and B-ISDN Protocol Structure Summarized

User to Network Interface (UNI)

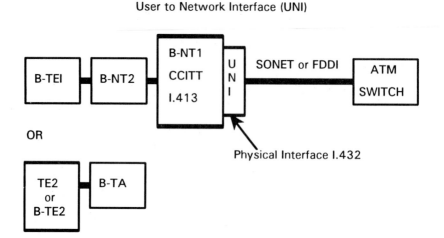

Figure 13.9 B-ISDN User-to-Network Interface

13.5.1 User Interface

CCITT Recommendation I.413 (B-ISDN access reference model) defines the CPE interface to an ATM network. B-TE1 refers to the B-ISDN Termination Equipment at the CPE, and serves as the interface point to the ATM network. Figure 13.9 shows these interfaces. Physical interfaces use the standards as defined in section 13.2.

In a B-ISDN network, the user device is called the Terminal Equipment (TE). This device interfaces either directly to another user or to the network. Each element in the network has a connection point, and the segments between connection points are called connection elements (CE). There are also user-network and IXC signaling elements. Figure 13.10 shows an example of these elements.

Two types of interfaces have been discussed: the user network interface (UNI) to the network and the network node interface (NNI) for node connectivity within a network. From these various types of connectivity, it becomes apparent that users can easily migrate from Frame Relay, SMDS, and N-ISDN to ATM cell-based services over B-ISDN.

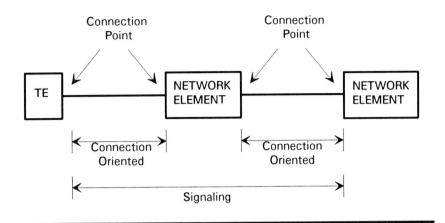

Figure 13.10 User-to-Network B-ISDN Signaling Elements

GFC may eliminate the need of a router as the network interface. This would take the place of 802.3 Ethernet LAN or FDDI, because it can support both asynchronous and isochronous services. There is also the option of using S_B interface terminals rather than LAN or FDDI interfaces to access the ATM network.

13.5.2 The Connectionless Server

When subscribers to B-ISDN want connectionless service they access the ATM network through a Connectionless Server (CS). Each user provides a peak bit rate ATM connection (operating at an average bit rate) into the connectionless server, which in turn is connected into the ATM switch. The connectionless server performs the routing and switching based on the first cell received from the user. A fast response time is achieved by the connectionless server's capability to transmit the packets across the ATM network as soon as they are received. This is the common method for LAN/MAN 802.X and FDDI internetworking over the ATM switch fabric.

13.5.3 Connectionless Mode

LANs and MANs are connectionless by nature (e.g., SMDS), and have traffic patterns that yield to bursty data transmissions through connectionless modes. The opposite is true for WANs, which are better using connection-oriented transmissions to effectively utilize the often point-to-point bandwidth and higher bit rates. What is needed is a method for LAN high-speed connectionless services to support Ethernet and FDDI over the WAN environment. ATM will provide the access for a WAN of this nature, but connectionless servers would manage and administer the network. Figure 13.11 shows the many types of user interfaces to an ATM network which uses a SONET transport infrastructure.

13.6 ARCHITECTURE

As ATM technology proliferates, the need to evolve away from traditional TDM architecture becomes apparent. The new architecture using ATM technology is based on ATM switches and cross connects. The current drive is toward building campus backbones out of ATM switch/hubs. These devices are also forming some of the first ATM WANs. Devices using the ATM protocol can either operate as a switch or a cross-connect. If both the VPI and VCI are used, true ATM switching is implied. If only the VPI is used, an ATM cross-connect operation is implied.

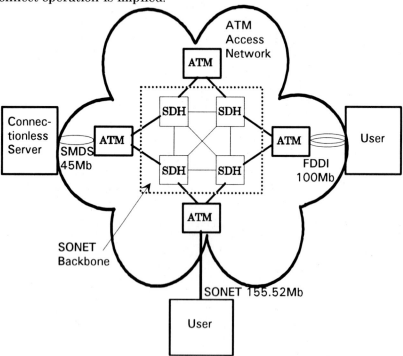

Figure 13.11 Methods of Interfacing to an ATM Network using SONET Transport

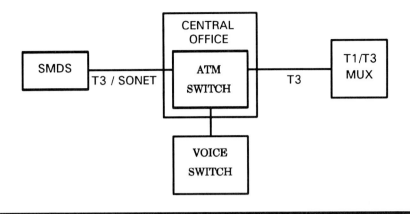

Figure 13.12 ATM Switch/DXC Functionality

If existing T1 and T3 circuit based services could be combined with ATM switches in the central office environment, the ATM switch would be deployed as a high-powered DXC. Bandwidth could be easily reconfigured to the sites that need it, while supporting the four service classes of B-ISDN. Figure 13.12 shows an example of an ATM switch serving the function of a DXC for an SMDS switch, a voice switch, and a T1/T3 Multiplexer.

13.7 SERVICES

CCITT I.211 defines B-ISDN services and the network capabilities required to support them. There are two types of services: interactive and distribution. Interactive (bidirectional transmission) services are defined by three classes of service: conversational (e.g., data, sound, video & sound imaging), messaging (e.g., document and video multimedia electronic mail), and retrieval (e.g., text, data, sound, still- and full-motion pictures and video).

Distribution services are divided into services with individual presentation control (e.g., data, text, graphics, still images, video with sound such as multimedia (TV) distribution) and without user individual presentation control (e.g., text, graphics, sound, still images such as remote news retrieval and advertising, respectively). The services *not* requiring user control are typically broadcast from a central point much like cable TV. The services requiring user control provide users with the capability to interact with the broadcast service, for example, touching the TV screen to select a video-marketed product. The standard also provides an in-depth look at video encoding techniques and methods for B-ISDN usage. Table 13.1 shows these classes of services.

TABLE 13.1 B-ISDN Classes Of Service

SERVICE	CLASSES OF SERVICE
Interactive	Conversational Services
Interactive	Messaging Services
Interactive	Retrieval Services
Distribution	Without user individual presentation control
Distribution	With user individual presentation control

As B-ISDN services develop, data services will begin to predominate over voice services. There are many standards bodies (previously discussed) such as the CCITT, the IEEE, and ETSI who are working with international vendors and service providers (e.g., MCI Communications, France and British Telecom, NTT) to develop both data and voice services. These vendors and service providers will, in turn, offer B-ISDN services through ATM-based public network B-ISDN services. But the users who will gain most from an ATM/B-ISDN network are those who deploy it for point-to-point multimedia services. Single-medium services can be transported over this type of network, but less expensive technologies and services are available which provide better alternatives.

Some of the first ATM/B-ISDN service offerings might include LAN/MAN interconnect via CBDS service over the ATM network for a maximum transmit bit rate of 100M bps or lease line ATM cross connects up to 600M bps speeds, with the first services probably PVC-based. Many carriers have already announced plans to offer B-ISDN services over an ATM and SONET infrastructure. They plan on capitalizing on the fact that bandwidth can be allocated in an almost continuous range from 55.52M bps on up, and changes to bandwidth allocation can also occur per call, or even within the call. This is a strong selling point for the first true bandwidth-on-demand services. Either way, all new services must fit under the four classes of service defined previously, and all vendor-provided ATM interfaces should conform to CCITT I.121, I.150, and I.361, as well as ANSI T1S1.5 standards.

Some vendors that have ATM switches that are SONET compatible include DSC Communication's MegaHub iBSS, Fujitsu (FETEX150), NEC America (NEAX 61E SMDS Service Node), and Siemens Stromberg-Carlson (SSC ATM switch).

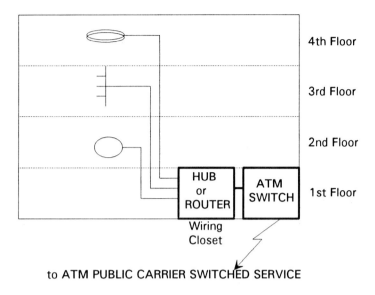

4th Floor

3rd Floor

2nd Floor

HUB or ROUTER   ATM SWITCH   1st Floor

Wiring Closet

to ATM PUBLIC CARRIER SWITCHED SERVICE

Figure 13.13 Intelligent Hub Connectivity to Public-Switched ATM Network

Many router vendors are also building large intelligent LAN hubs that convert LAN protocols to ATM cell formats. This will allow users to transmit multimedia applications from the desktop through the use of an ATM interface card in the workstation. Existing twisted-pair wiring would bring the multimedia cell down to the intelligent LAN hub, which would then transmit the signal to another workstation or to the ATM-switched network. Figure 13.13 shows an intelligent hub connecting to the public-switched ATM network via a 155.52M bps user interface.

Some of the equipment vendors who are developing ATM switch products are shown in Table 13.2.

TABLE 13.2 ATM Switch Vendor Products

VENDOR	SWITCH
AT&T	BNS-2000
Digital Link	DL3000 ATM switch
DSC Communications	MegaHub iBSS
Fore Systems	ASX-120, ASX-125 ATM
Fujitsu	FETEX150
NEC America	NEAX 613E SMDS w/ATM switch
Siemens-Stromberg Carlson	SSC ATM switch
StrataCom IPX	ATM Trunk Module for switch

13.8 PERFORMANCE

ATM switches can also quickly route around network failures without affecting calls in progress. This reroute capability can happen in less than a second, which is quite valuable when user traffic is session-oriented and cannot stand an outage of a few seconds. Thus, normal ATM network performance is excellent. There is also a capability through the use of the "payload type" field for a user to create a subchannel in parallel to the user information channel to monitor network performance.

Some measures of performance would be taken at the physical (SONET) level, TC level, VPC level, and VCC level. Statistics can be calculated for:

- errored ATM cell headers
- discarded cells due to header bit errors
- discarded cells due to header content errors
- cell loss ratio
- cell header error ratio
- end-to-end VPC performance
- end-to-end VCC performance
- quality of service,

all of which are defined by the standards listed above. Any user should also closely measure voice performance when supporting large amounts of bursty data and imaging traffic. Predominant operation of a single type of traffic could have adverse effects on the remaining network traffic.

13.9 BENEFITS AND COMPARISONS

ATM is a backbone technology, as opposed to frame relay which is an access technology. ATM handles both delay-insensitive and delay-sensitive (e.g., video and voice) traffic over the same channel. It can combine high-rate speeds of circuit switching with the flexibility of packet switching. B-ISDN benefits users of multimedia, but will be too expensive for single-media or small-bandwidth users.

ATM/B-ISDN also offers the following benefits:

- ☺ high performance for bursty traffic
- ☺ bandwidth efficiency

☺   service availability
☺   adaptable quality of service
☺   public network technology
☺   suitability for delay sensitive traffic
☺   processing of cells faster due to fixed length

13.9.1 B-ISDN Additional Advantages for ATM

The advantages which B-ISDN adds to ATM include:

+   can use various physical layer interfaces for transport
+   dynamic bandwidth allocation
+   network flexibility
+   flexible OAM&P

13.9.2 Drawbacks of ATM/B-ISDN

The drawbacks of using ATM/B-ISDN at this stage of development include:

☹   standards just now being defined
☹   congestion control needs work
☹   all new hardware
☹   does not support wireless communications or broadcast

Figure 13.14 is again shown, this time illustrating where ATM/B-ISDN services fit compared to frame relay and SMDS services.

13.9.3 Comparison to N-ISDN

When Narrowband ISDN (N-ISDN) was covered, it was pointed out that its standards were founded on voice network principles of connection oriented point-to-point services. B-ISDN over ATM will offer similar connectivity through connection oriented point-to-point virtual circuit services. There are many other major differences between B-ISDN and N-ISDN besides the obvious transmission speed increases. These are listed below.

1. B-ISDN is all packet switching, where N-ISDN is only packet switching on the D-channel (signaling only) — otherwise N-ISDN is primarily a circuit switching technology.

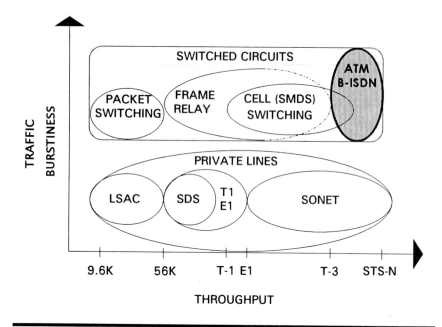

Figure 13.14 ATM/B-ISDN Services Compared to Other Switched and Dedicated Services

2. B-ISDN requires fiber optic cable, where N-ISDN can use existing cabling and even twisted pair. Thus, B-ISDN gains performance advantages through less error checking and more transfer of information given the same amount of bandwidth.

3. B-ISDN uses virtual channels with variable bit rates, where N-ISDN uses fixed channel bit rates.

4. B-ISDN is primarily connection oriented over virtual circuits — connectionless oriented service, broadcast mode, and polled and multipoint services require special arrangements of service class.

Also, N-ISDN services can be aggregated for transport over B-ISDN services.

13.9.4 CHALLENGES

The primary challenge of ATM networks is the implementation of congestion control management. Call service priorities are the easiest method for congestion control at this time. Section 4.0 of

CCITT I.311 defines four traffic control parameters: connection admission control, usage parameter control, priority control, and congestion control. These define call establishment, traffic volume and cell routing validity monitoring and control (defining parameters such as average rate, peak rate, burstiness, and peak duration for cells), user traffic flow and prioritization of that traffic control, and negotiating quality of service through methods of congestion control. While these implementations can achieve some measure of congestion control in ATM networks, there is still much development needed before they are widely used by B-ISDN services.

On a higher level, successful worldwide ATM implementation is contingent upon the dominant carriers and service providers, now offering DQDB/SMDS and FDDI services, to focus on delivering ATM based B-ISDN services. This demands a significant capital investment, but will pay off in a global standard that, together with SONET transport, can provide a strong technology base for entry into the next century. This is supported by market projections in the billions for the mid to late 90s.

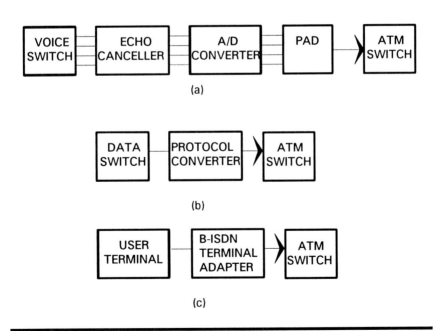

Figure 13.15 Three ATM Internetworking Examples

13.10 DESIGN CONSIDERATIONS (INTERNETWORKING WITH ATM/B-ISDN)

Many existing CPE devices, as well as voice and data networks, will need to interface with the ATM infrastructure. Figure 13.15 shows three examples of internetworking with an ATM network. The first (a) shows a voice switch interfacing with an ATM switch through an echo canceller, an analog-to-digital converter, and a packet assembler/disassembler. The second (b) shows a data switch (which could be an X.25 packet switch) interfacing with the ATM switch through some form of protocol converter. This is a prime example of interfacing an existing data network with an ATM network. The third example (c) shows an N-ISDN user terminal interfacing with the ATM switch through a terminal adapter. All three methods transform the original traffic (3kHz analog and 64K bps digital, 56K bps X.25 digital, and BRI N-ISDN) into the ATM 53 octet cell format, and the reverse process is needed at the remote end.

Most design considerations will be on the access side. This chapter has focused primarily on access side design aspects. Most ATM/B-ISDN networks will be offered by the IXCs and LECs, because of the high bandwidth multimedia service capabilities of the technology. It seems as if initial implementations of ATM/B-ISDN networks will offer SMDS and 802.6 access side connectivity into an ATM backbone.

13.11 REVIEW

B-ISDN services offered over an ATM technology seem to be the preferred future method of offering high-bandwidth multimedia services over an intelligent switching platform. Even though ATM operates through fixed cell structures which are statistically multiplexed and switched across a connection-oriented network, it also offers both variable and constant bit rate services, as well as both connection-oriented and connectionless modes of data transport. This is accomplished through the use of both physical and data link layer protocols (and network in the case of B-ISDN).

ATM can accommodate many types of user access and physical media transport, from traditional circuits to SONET. Protocols and classes of service offerings were then defined in the B-ISDN standards, which included: from voice to video, fax to data, and many types of multimedia user interactive services to provide broadcast services with user interaction. Performance considerations were covered, as was a summary of the benefits and drawbacks of using B-ISDN. N-ISDN was also compared to B-ISDN. Finally, the challenges and design considerations one must face when interfacing or building an ATM network has been clearly shown.

14

SONET

Without debate, fiber optics has thrown a whole new light on the future of communications. The costs of fiber optic transmission facilities are decreasing as fast as new methods of using fiber optic cables are growing. These new superhighways of fiber optic cable now ring most major cities and provide terabits worth of bandwidth across the United States. Fiber optics also span the globe. By the year 2000, fiber will become the predominant transmission medium. The advent of these fiber highways comes with concomitant technologies to effectively utilize this seemingly limitless bandwidth. The market demands of technologies such as DQDB, FDDI, and ATM, as well as services such as SMDS and B-ISDN, are driving the need for ultrahigh speed transport.

Synchronous Optic NETwork, or SONET, is the harbinger of this new technology wave. SONET is also referred to as the Synchronous Digital Hierarchy (SDH). This chapter explores the many facets of SONET, from its protocol structure, standards, and operation through the interfaces and hardware which allows SONET to be used as the terabit transport for such services as SMDS and B-ISDN and other services not yet dreamed of.

14.1 TERMS DEFINED

Synchronous Optical NETwork (SONET) is a Bellcore term for the Synchronous Digital Hierarchy (SDH) standardized by the CCITT in Europe and Asia. SONET was conceived as a method of providing a high-speed international fiber optic, multiplexed transmission standard for interface between the PTTs, IXCs, and the LECs. For all standard purposes, SONET is a transport interface and method of transmission only (i.e., it is not a network in itself). SONET is planned to eliminate the different transmission schemes and rates of countries such as the United States, Japan, and Europe through a common international rate structure. It will also provide a technology which allows the major IXCs and PTTs to standardize and control broadband network transport media.

The SONET standards continue to evolve through the guidance of Bellcore, the RBHCs, and MCI Communications. These groups have a vested interest in making SONET work with standards such as IEEE 802.6, SMDS, ATM, and B-ISDN. These vendors and service providers consider SONET a common interconnectivity medium for direct fiber services such as SMDS and B-ISDN. SONET will also lower transmission costs over the long run for everyone. As these services grow, so will the deployment of SONET transmission facilities. Again, certain of the IXCs are partnering with bypass carriers who are deploying SONET-based gear to offer direct-connect capabilities. SONET will be a true international standard for fiber transport interconnectivity.

SONET uses a transfer mode which defines switching and multiplexing aspects of a digital transmission protocol. Two types of transfer modes comprise switching technologies: synchronous and asynchronous. Synchronous transfer mode (STM) defines circuit switching technology, while asynchronous transfer mode defines cell relay technology. SONET uses both synchronous and asynchronous transfer modes through the use of a fixed data transfer frame format including user data, management, maintenance, and overhead.

Major terms associated with the SONET world are Path-Terminating Equipment (PTE), Line-Terminating Equipment (LTE), and Section-Terminating Equipment (STE). The primary user interface at the CPE is the PTE. The next level of access device is the LTE. Both can be switching, multiplex, or cross-connect devices. STE is primarily regeneration equipment.

14.2 STANDARDS

In 1984 MCI Communications requested the Exchange Carriers Standards Association (ECSA) to develop a standard for a "mid-span fiber meet" for IXC interfaces to LECs. Bellcore also requested a method of connecting add/drop multiplexers with DXCs. Both of these requirements, including that of a hierarchical network bandwidth scheme, were formalized in ANSI T1.1051988 and ANSI T1.1061988 (Phase I), as well as CCITT G.707, G.708, and G.709. These standards define the topics covered in this chapter.

SONET standards have been introduced in three phases. Each phase presents additional levels of control and operations, administration, maintenance, and provisioning (OAM&P).

14.2.1 Phase I

Phase I, as approved by the CCITT in 1988, defines transmission rates and characteristics, formats, and optical interfaces. This phase primarily defines the hardware specifications for point-to-point data transport. Phase I supports the initial requirement of an Optical Carrier-N (OC-N) midspan meet at payload level only. It also defines the standard data communications channels (DCC) with basic functions, as well as the basics of the framing and interfaces defined in this chapter.

14.2.2 Phase II

Phase II builds upon the mid-span meet for multiple-vendor connectivity and management. Phase II defines:

- OAM&P procedures
- synchronization
- SONET to B-ISDN interconnectivity
- pointer adjustments for wander and jitter
- CO electrical interfaces and network advantages
- imbedded operation channels
- Common Management Information Service Elements (CMISE)
- point-to-point, add/drop multiplexer capabilities,

and further defines DCC protocols.

Phase II also defines the intraoffice optical interface (IAO), which allows equipment to be interconnected at the central office. This interface is limited from 20m to 20km and operates at OC-1, OC-3, and OC-12 rates.

14.2.3 Phase III

Phase III builds upon Phase II by providing all of the OAM&P required for a midspan meet. Additional network management, performance monitoring, and control functions are added, as are DCC standard message sets and addressing schemes for identifying and interconnecting SONET network elements. This allows the passing of DCC information between various vendor implementations of SONET. Phase III also provides for ring and nested protection switching.

14.3 SONET STRUCTURE

The SONET Optical Carrier (OC-N) structure follows a strict structure and hierarchy, which maps the electrical hierarchies of many nations. These OC-N levels are then multiplexed to form higher-speed transport circuits which range into the gigabits range and provide an alternative for multiple T1 and T3 transmission facilities. SONET also solves many of the network management problems associated with T3 transmissions. The SONET architecture also maps somewhat to the OSI reference model, as do the messaging and file transfer protocols, and offers the only international standard for digital to fiber optic communications.

14.3.1 Basic Structure

Primary structure of SONET is built around Synchronous Transport Signal Level-1 (STS-1) transport through an Optical Carrier (OC-N) signal over fiber optics. An aggregate 51.48M bps STS-1 bit stream, when converted from electrical to fiber optic is called Optical Carrier-1 (OC-1), and is comprised of a transmission of 810-byte frames sent at a rate of 8,000 times per second. Here the electrical DS3 signal is converted to an OC-1 optical signal, and vice versa. SONET speeds range from 51.840M bps (OC-1) to 4.976G bps (OC-96), and true fractional speeds are achievable, providing virtually any division of bandwidth required. Any subrate signals below OC-1 are multiplexed to form a single OC-1 channel.

14.3.2 Hierarchy

Table 14.1 shows the SONET speed hierarchy by OC-level, illustrating the number of DS0s, DS1s, and DS3s equivalents.

TABLE 14.1 SONET OC-N Speed Hierarchy

OC level	Bit Rate (M bps)	DS0s	DS1s	DS3s
1	51.84	672	28	1
3	155.52	2,016	84	3
6	311.04	4,032	168	6
9	466.56	6,048	252	9
12	622.08	8,064	336	12
18	933.12	12,096	504	18
24	1,244.16	16,128	672	24
36	1,866.24	24,192	1008	36
48	2,488.32	32,256	1344	48
96	4,976.00	64,512	2688	96
255*	13,219.20	171,360	7140	255

(* = theoretical maximum speed)

Table 14.2 shows the closest SONET equivalent compared with the digital hierarchy existing today between countries. All values are in M bps.

TABLE 14.2 SONET Equivalent to Digital Hierarchy

SONET	T-Carrier	USA	Europe	Japan
VT1.5	DS1	1.544	2.048	1.544
VT6	DS2	6.312	8.448	6.312
OC-1	DS3	44.736	34.368	32.064
OC-3	-		139.264	97.728

14.3.3 Multiplexing

SONET provides direct multiplexing of both SONET speeds and current asynchronous and synchronous services into the STS-N payload. For example, STS-1 supports direct multiplexing of DS1, DS2, and DS3 clear channel into each STS-1, which is called a tributary.

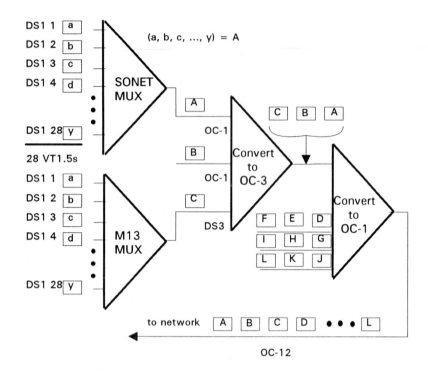

Figure 14.1 SONET Multiplexing

Figure 14.1 shows the method of multiplexing, where users "a" through "y" represent individual DS1 Virtual Tributary (VT) user access circuits which are multiplexed first into an OC-1, then into an OC-3, and finally into an OC-12 trunk. This figure also shows an M13 multiplexing 28 DS1s into a single DS3, which the SONET multiplexer converts to an OC-1 format. This M13 ends up bit stuffing 5M bps worth of overhead into the signal to boost up the contents and meet the STS-1 rate of 51.84M bps. Notice also that the synchronous order in which the subrate channels are transmitted remains constant.

The other major advantage of SONET is that each individual signal down to the DS1 level can be accessed without the need to demultiplex and remultiplex the entire OC-N level signal. This is commonly accomplished through a SONET Digital Cross Connect (DXC) similar to those discussed in Chapter 7.

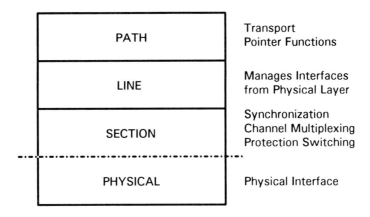

Figure 14.2 Four Layered SONET Architecture

14.3.4 SONET Architecture Layers

There are four layers to the SONET architecture: physical (or photonic), section, line, and path. These can be seen in Figure 14.2. Three of these layers, section, line, and path, roughly correspond to the layers of overhead present in the SONET frame. First, the physical layer defines the physical fiber type, path, and characteristics. The more common examples of the physical interface include 1550-nm dispersion-shifted fiber and 1310-nm conventional glass fiber. The physical layer also includes the many electrical interfaces which become virtual channels within the STS-1 frame.

Secondly, the section layer builds the SONET frames from either lower SONET interfaces or electrical interfaces. Thirdly, the line layer provides synchronization, channel multiplexing, and protection switching. This layer provides the communications channel between line terminating equipment at the point of STS-1 multiplexing to higher rates, as well as providing management between LTE and STE. The path layer manages the actual data transport across the SONET network, as well as the pointer function. Each layer will be discussed in detail in the section on overhead. Figure 14.3 shows an entire SONET network with each of these interface points.

14.3.5 SONET Protocol Stack

Figure 14.4 shows the protocol stack used by SONET. This stack parallels the seven layer OSI reference model. These protocols constitute the Operating Systems (OS) interface to SONET. Phase 1 provides a subset of this stack for file transfer, but uses the RS-232-C, LAPB and X.25 for file transfer and messaging. Phase II changes the lower levels to 802.3, 802.2 and TCP/IP or ISO 8473. Phase III establishes the standard shown in this figure.

14.3.5 OC-N Midspan Meet

The OC-N midspan meet allows CPE, LEC, and IXC hardware from different vendors to interface with each other via SONET. This interface covers from basic payload-only connections to payload transport, operations and maintenance information exchange, and automatic protection switching. It provides a single platform base for access from the central office to the CPE. Figure 14.5 illustrates the concept of a midspan meet for both the typical asynchronous interface as well as the new SONET interface.

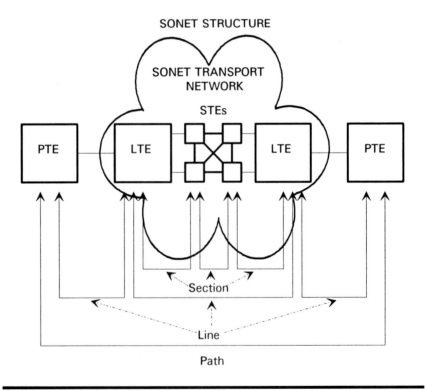

Figure 14.3 SONET Network Structure

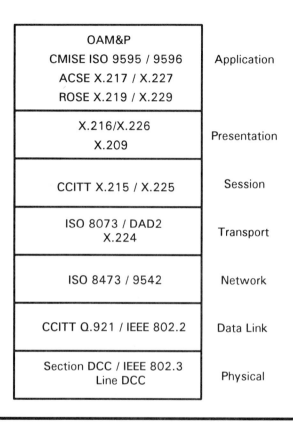

OAM&P CMISE ISO 9595 / 9596 ACSE X.217 / X.227 ROSE X.219 / X.229	Application
X.216/X.226 X.209	Presentation
CCITT X.215 / X.225	Session
ISO 8073 / DAD2 X.224	Transport
ISO 8473 / 9542	Network
CCITT Q.921 / IEEE 802.2	Data Link
Section DCC / IEEE 802.3 Line DCC	Physical

Figure 14.4 SONET Protocol Stack

14.3.6 Data Communications Channels (DCC)

SONET transmissions also contain communications channels which transmit network management information between network elements. This information includes alarm, control, maintenance, and general monitoring status. Each SONET terminal and regenerator (repeater) uses a 192K bps channel, and each optical line between terminal multiplexers use a 576K bps channel for the DCC. Initial implementations are using X.25 to perform the DCC functionality, which will eventually be replaced with the local communications network (LCN) interface.

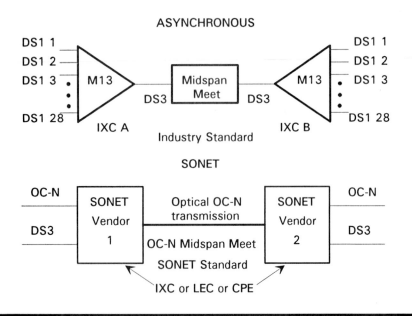

Figure 14.5 SONET Midspan Meet

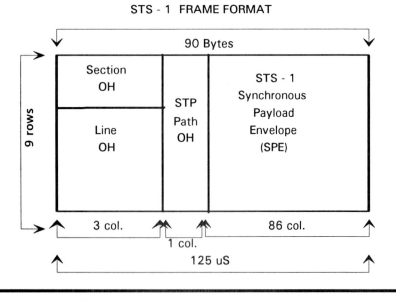

Figure 14.6 SONET Frame Format

14.4   FRAME FORMAT

SONET uses a unique framing format for which timing is the most critical element. SONET frame payloads are not synchronized by a common clock even though SONET is a synchronous technology. While there are many complexities to the standard SONET frame still under development, the two major pieces, payload and overhead, are functional in Phase I implementations.

STS - FRAME FORMAT

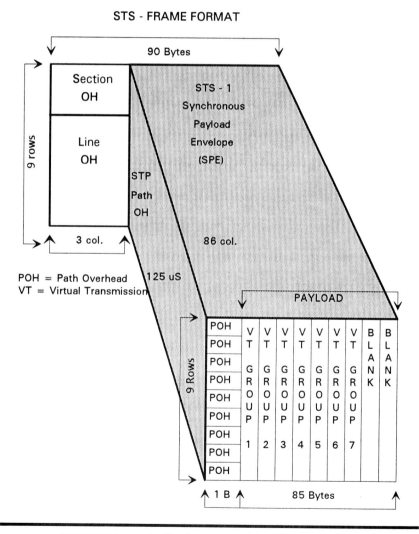

Figure 14.7 SONET Frame Payload Section Breakout

STS - Nc FRAME FORMAT

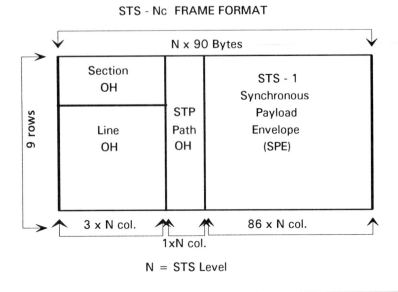

Figure 14.8 STS-Nc Basic Frame Format

14.4.1 Basic Frame Format

The SONET frame format can be found in Figure 14.6. Notice that the frame is comprised of multiple overhead elements (section, line, and path) and a Synchronous Payload Envelope (SPE). The frame size for an STS-1 is 9 rows by 90 columns (one byte per column) for a total of a 783-byte frame (excluding section and line overhead) transmitted at 125 microseconds. Figure 14.7 shows the breakout of the payload section of the SONET frame. For frames larger than the STS-1 level, each column is multiplied by N depending on the size of the STS (STS-N). Figure 14.8 shows the basic frame format of the STS-Nc frame. Figure 14.9 shows the frame format for a STS-N frame of undetermined size. Each shows the payload size in a breakout.

14.4.2 Payload

The payload is defined as the actual data to be transported across the SONET path. Payloads can vary depending on the OC speed of transport. Payloads can take many forms, such as typical T-carrier channels (e.g., DS3), FDDI or SMDS, or virtual tributaries of various sizes. The payload envelope of the frame can vary in size in 774-byte

(9 rows x 86 columns) increments, and the term used for the envelope is Synchronous Payload Envelope (SPE). The STS-1 SPE is comprised of the Path OH bytes and the STS payload capacity (N x 9 rows x 85 columns), and is not aligned to the STS frame.

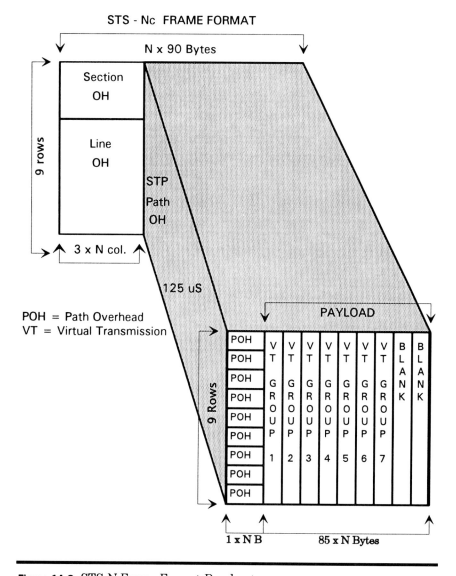

Figure 14.9 STS-N Frame Format Breakout

When a SONET frame is first created, it is easy to identify the beginning of the SPE data frame. When a frame is transmitted across the network, the line and overhead of each frame can be changed based upon timing changes throughout the network. Frame starts and stops begin to slip, and even the best synchronization schemes cannot keep the frame transmissions in perfect synchronization. Buffers should not be used to correct this problem of frame slip because this would introduce unwanted delay (125 microseconds at each transit node) which would also accumulate over time. So SONET uses pointers to provide synchronization.

14.4.3 Virtual Tributary

Virtual tributaries are the building blocks of the SPE. The label VTxx designates virtual tributaries of xxM bps. These virtual tributaries are labeled as VT1.5 for DS1, VT3 for DS1C (3M bps), VT2 for CEPT E1, and VT6 for DS2 (6M bps). Table 14.3 shows these VTs and their respective equivalents.

TABLE 14.3 Virtual Tributary Classes with Equivalents

VTxx	Data Rate (M bps)	Electrical Channel Equivalent
VT1.5	1.544	DS1
VT2	2.048	E1
VT3	3.152	DS1C
VT4	Open	Open
VT5	Open	Open
VT6	6.312	DS2

VTs are combined to form VT groups. These VT groups consist exclusively of three VT1.5s, four VT2s, two VT3s, or one VT6 within a 9-row by 12-column portion of the SPE. A complete SPE is then comprised of seven VT groups, two bit stuffed unused columns, and the path overhead column. There are seven VT groups of 6M bps groups per STS-1. These VT groups can be mixed to make up an STS-1, which can either contain multiple VCs or a single asynchronous DS3. VTs can either operate in "locked mode", which fixes the VT structure within a STS-1 and is designed for channelized operation, or "floating mode" which allows these values to be changed by cross-connects and switches and is designed for unchannelized operation. STS-3s can handle even larger payloads, such as ATM, FDDI, and SMDS payloads. STS-12 payloads will often be extremely large bandwidth-intensive applications, such as medical imaging or HDTV.

The common tributary is VT1.5, which supports a virtual tributary of 1.5M bps through a DS1 transport envelope. VTs run from the VT1.5 through a VT6 (DS2). Table 14.4 shows the many types of traffic payloads possible.

TABLE 14.4 Virtual Tributary Payload Type with Applications

PAYLOAD TYPE	APPLICATIONS
VT1.5 through VT6	Asynchronous DS1, Asynchronous DS1C, Asynchronous DS2, Asynchronous 2.048, Synchronous DS1, Synchronous 2.048
STS-1	Asynchronous DS3, Syntran
STS-3	DS4NA, DQDB/SMDS, FDDI, FDDI-II, NTSC Video, ATM/B-ISDN
STS-12	Imaging, HDTV

14.4.4 Synchronization and Pointers

The SONET answer to frame synchronization is through the use of pointers. Pointers are used by SONET devices to easily identify sub-channels down to the DS0 level within a SONET transmission. These pointers are located within the line overhead portion of each frame. The Synchronous Payload Envelope (SPE), is allowed to "float" anywhere within the SPE-allocated portion frame, and will often overlap multiple frames. The pointer number (pointers H1, H2, and H3) indicate the start of the SPE frame. If the frame experiences jitter or wander, the pointer will shift within the frame parallel to the SPE shift, thus maintaining its pointer integrity. The H1 and H2 pointers are then updated at each terminal across the network.

Figure 14.10 illustrates an example of an SPE which spans two STS-1 frames. In this figure, the H1, H2, and H3 pointers in the line overhead identify the start of the SPE-3 frame. The next STS-1 frame H1 and H2 identify the beginning of another STE, but the H3 octet holds the value of the last byte in STS-3. This continues on through the transmission. Also note that it is common for STEs to span two frames. Pointers are also used to identify virtual tributaries (VTs) within an SPE.

14.4.5 Overhead

The SONET overhead structure parallels the existing telephone network, with three layers to match section, line, and path segments. Each layer assures the correct transmission of the layer below it. Figure 14.11 shows path terminating equipment (PTE), line terminating equipment (LTE), and section terminating equipment (STE), and the overhead structure by which they are managed.

The section layer provides management of network segments between regenerators, which includes both switching and nonswitching elements. This function is similar to that of the electrical repeater discussed in Chapter 4. The section layer contains overhead for multiplex equipment framing information (A1 and A2), span performance monitoring BIP-8 parity counts (B1), STS identification (STS-ID) number (C1), 64K bps local maintenance communications channel (E1), 64K bps end-user channel (F1), and data communications channel (192K bps section operations channel) for remote monitoring and control (D1, D2, D3). Each STE regenerator in the network performs the section overhead function. Figure 14.12 shows the breakout of the section, line, and path overhead sections of the SONET frame.

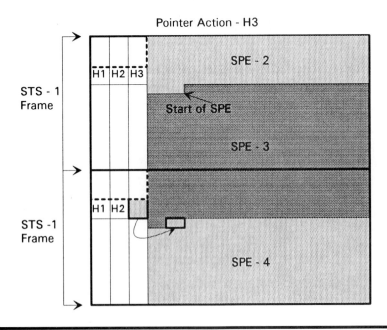

Figure 14.10 Synchronous Payload Envelope Operation

SONET OVERHEAD STRUCTURE

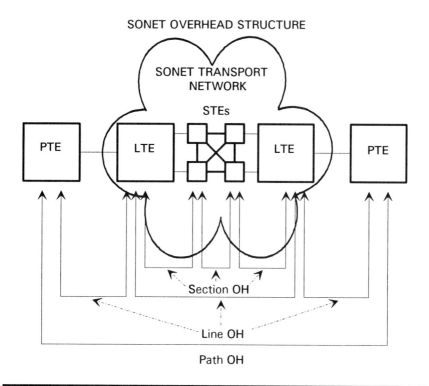

Figure 14.11 SONET Overhead Structure

The line overhead provides for reliable transport of payload data through parity checking (error monitoring) between elements. These elements perform switching functions through a BIP-8 check (B2), interoffice data communications channel (576K bps embedded operations channel) (D4 through D12), 64K bps order-wire channel (E2), 128K bps automatic line protection switching to backup circuits (K1 and K2), and payload pointers to show where the payload section begins (H1 and H3) and to identify the pointer action byte (H3) for control of synchronization. Growth or B-ISDN functionality (Z1 and Z2) is also included. The line and section overhead make up the transport overhead of the STS-N frame.

The path overhead provides the end-to-end performance monitoring similar to a network layer protocol. The path overhead operates between PTE devices, providing a BIP-8 (B3), STS mapping (C2), 64K bps user channel (F2), path error statistic conveyor (G1), multi-frame identifier (H4), trace from receiver back to transmitter (J1), and growth bytes (Z3, Z4, Z5). The path also maps the DS3 format to the OC-3 format and is associated with the STS payload capacity. It is an essential part of the SONET add/drop multiplexer (SADM). Path overhead is assembled at each PTE and rides through the network until is disassembled at the receiving PTE.

14.4.6 Bit Interleave Parity Check (BIP-8)

Parity is provided through a one byte bit interleave parity (BIP-8) code at each section, line, and path segment of the frame. The section BIP assures error-free transport between regenerators. The line BIP assures error-free transport between terminating devices, and the path BIP error-free transport between line termination equipment.

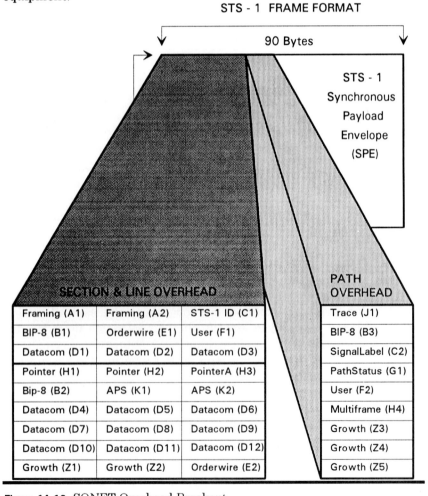

STS - 1 FRAME FORMAT

90 Bytes

STS - 1
Synchronous
Payload
Envelope
(SPE)

SECTION & LINE OVERHEAD

PATH OVERHEAD

Framing (A1)	Framing (A2)	STS-1 ID (C1)	Trace (J1)
BIP-8 (B1)	Orderwire (E1)	User (F1)	BIP-8 (B3)
Datacom (D1)	Datacom (D2)	Datacom (D3)	SignalLabel (C2)
Pointer (H1)	Pointer (H2)	PointerA (H3)	PathStatus (G1)
Bip-8 (B2)	APS (K1)	APS (K2)	User (F2)
Datacom (D4)	Datacom (D5)	Datacom (D6)	Multiframe (H4)
Datacom (D7)	Datacom (D8)	Datacom (D9)	Growth (Z3)
Datacom (D10)	Datacom (D11)	Datacom (D12)	Growth (Z4)
Growth (Z1)	Growth (Z2)	Orderwire (E2)	Growth (Z5)

Figure 14.12 SONET Overhead Breakout

14.4.7 Bit Stuffing

When the incoming tributary data rates cannot fully meet the STS-N rate, the SONET device performs bit stuffing to achieve the desired bandwidth. This is as simple as inserting extra bits into the data stream which are then stripped off at the destination SONET device.

Bit stuffing is also used for frame synchronization. This technique is used when the access hardware and network hardware are using different timing sources having clock frequency differences. This method locks the user into a point-to-point transmission in that any bits stuffed at the transmitting multiplexer must be demultiplexed and unstuffed in the same order. This may prove to be a major drawback to multiplexed SONET signals that need intermediate demultiplexing.

14.5 INTERFACES

The Network Node Interface (NNI) specifies the link between existing in-place digital transmission facilities and the SONET network node, as well as the process for converting the electrical signal into optical pulses for network transmission. This is the primary interface from the electronic world into the optical world.

14.5.1 Interface Options

There are three major SONET interface options:

① direct CPE or CO hardware interface
② gateway device to convert to OC levels
③ conversion within the SONET switch itself

All three interface options will be supported by the major IXC and LEC deployments of SONET in major metropolitan centers across the country. Most access to the SONET network is through the LTE, commonly a SONET add/drop multiplexer (SADM).

An M13 multiplexer can also be used for access to the SONET network. M13s can aggregate multiple DS level inputs into a single DS3 for access. They can also be used for synchronization across the SONET network through a single DS1 within the DS3, but this can cause data slips which, in turn, could cause lengthy delays and degradation of data rate. Figure 14.13 shows an example of the M13 multiplexer concentration via DS3 access.

14.5.2 User Access

Many vendors providing SONET access hardware are following the TR-08 SONET access standard. This seems to be the predominant standard. Typical SONET access interfaces include:

- SONET STS-N
- SONET OC-N
- SONET OVTG (fiber)
- European or Asian SDH
- Channel Bank
- LEC Telephone Service device
- DXC 3/1, 3/3
- M13 multiplexer
- ISDN
- DS0 through DS3 (incl. Fractional T1/T3, E1)
- LAN/MAN

14.5.3 Services Support

Typical services to ride over SONET networks include:

- Digital Data Service (DDS)
- N-ISDN
- B-ISDN
- X.25/X.75

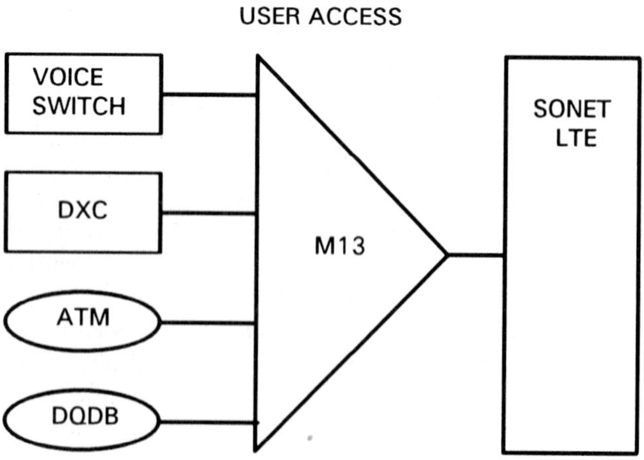

Figure 14.13 M13 Multiplexer Access to SONET LTE

- Frame Relay
- FDDI/FDDI-II
- 802.6/SMDS

With SONET, users will be able to dial up whatever bandwidth increments are needed. This achieves true bandwidth-on-demand, at up into the gigabit speeds! The standards bodies that work high-speed data services are now considering SONET to provide the much needed backbone infrastructure on which to base these services. An example of this prospect is the work currently under way by the FDDI Committee to provide an FDDI to SONET interface, with a DQDB to SONET interface not far behind. There are parallel efforts between the ANSI SONET T1X1.5 standards committee and the B-ISDN standards body (T1S1.1) for compatibility of B-ISDN services over SONET.

14.6   SONET HARDWARE

SONET hardware distinctions are possibly the most difficult aspect of SONET to understand. Many devices produce the same functionality, and even the vendors are split on the choice of hardware markets and the future of SONET equipment types. Some hardware both extends the distance of the CPE into the SONET network and allows both the user and provider to monitor and control the network in the same manner, rather than through proprietary T1 and T3 systems. We will also study the many types of switches, DXCs, and regenerators constituting the core SONET network.

14.6.1 SONET Terminating Multiplexers

Terminating multiplexers provide user access to the SONET network. Terminating multiplexers operate similar to the M13 multiplexer, and provide a public network SONET access to the user. Terminating multiplexers, also called PTE, turn electrical interfaces into optical signals by multiplexing multiple DS1, DS1C, DS2, DS3, or E1 VTs into the STS-N signals required for OC-N transport. These devices are configured in point-to-point configurations. Figure 14.14 shows end user LTE terminals which interface up to 48 DS1s into a single OC-3 interface for SONET transmission.

14.6.2 SONET Concentrators

SONET concentrators operate the same way as electrical concentrators and hubs, concentrating LEC OC-3 and OC-12 interfaces into higher transmission rates.

14.6.3 SONET Add/Drop Multiplexers

SONET add/drop multiplexers work in a similar manner to their electrical equivalents, but allow the provider to drop and add not only the lower SONET rates, but also electrical interface rates down to the DS1 level. The optical signal is converted to electrical, and these functions are performed electrically.

The current system of electrical transport and multiplexing is asynchronous at DS3 and lower speeds. This requires a huge investment in asynchronous equipment (multiplexers, DXCs). This method not only involves extensive overhead, but also the need for large numbers of multiplexers and DXCs. Multiple back-to-back M13 multiplexers and patch panels are used to break out low-speed channels from the aggregate DS3 signal. SONET eliminates these multiple equipment implementations and requirements through the use of SONET Add/Drop multiplexers (SADM).

Figure 14.15 shows a comparison of the asynchronous and SONET methods of add/drop multiplexing. Notice that the asynchronous digital method requires the M13 multiplexer to break down a DS3 into 28 DS1s, which are then placed into a cross connect or patch panel for drop and insert capability. The SDXC allows direct drop and insert of any DS3 through DS1 VT within the OC-N signal (in this case an OC-12). Also notice the reduction of equipment necessary to provide additional functionality.

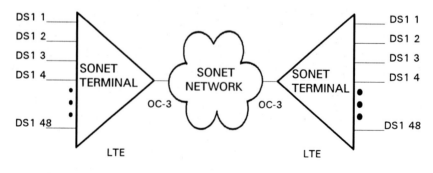

Figure 14.14 SONET Line Terminating Equipment (LTE)

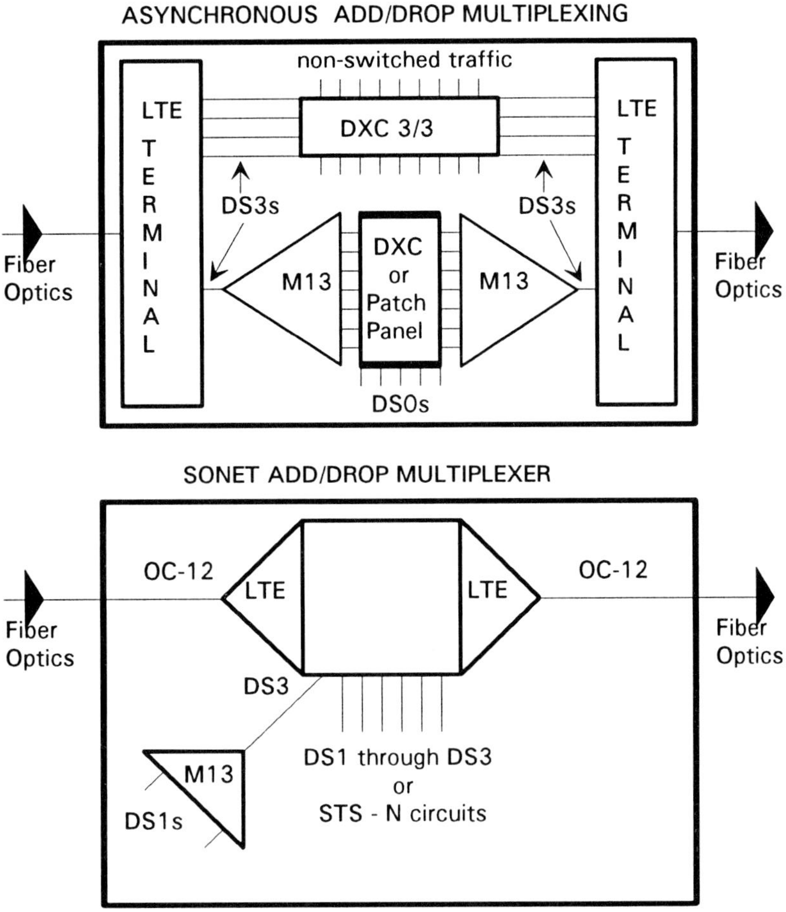

ASYNCHRONOUS ADD/DROP MULTIPLEXING

Figure 14.15 SONET and Asynchronous Add/Drop Multiplexing Comparison

SADMs are generally used for distributed point-to-point network connectivity. They are central office devices forming the building blocks of the SONET network. SADMs allow ease of expansion and are often used in SONET ring architectures. They generally operate at the higher transmission speeds of OC-3 through OC-48.

Some additional capabilities of SADMs include:

- operation and protection channels
- optical hubbing
- ring and time slot capabilities
- dynamic bandwidth allocation

14.6.4 SONET Digital Loop Carrier Systems (DLC)

Digital loop carrier systems (DLC) are used to concentrate multiple DS0 traffic from remote terminals into a single OC-3 signal. These devices are typically situated at the LEC and handle both voice and data traffic, providing an interface for non-SONET CPE, LEC, and CO switches to the SONET public network. DLCs also have many of the capabilities typically found in the LEC voice systems, such as operational and maintenance capabilities, and can handle access for many of the data services such as N-ISDN and B-ISDN. These devices are also used when a remote terminal cannot transmit an OC-N signal to the IXC CO switch and needs to go through the LEC.

14.6.5 SONET Digital Cross Connects (SDXC)

SONET Digital Cross Connects (SDXC) operate similar to standard digital DXCs, in that they allow switching and circuit grooming across all levels of the transmission down to the DS1 level, including those that interface to the SDXC without being on the incoming or outgoing transmission. SDXCs provide SONET OC-N level cross-connect provisioning capabilities and can also act as a SONET hub to provide both asynchronous and synchronous user or network access. SDXCs will help migrate LEC and IXC asynchronous networks in place today over to SONET. Most grooming and routing of SONET circuits is done through SDXCs. Figure 14.16 shows an example of a SONET DXC (SDXC).

SONET DXCs use pointers rather than traditional DXC slip buffers to mark the beginning of a DS1 frame and allow insert/extraction with minimal delay. The additional SDXC features include:

- network monitoring and testing
- network provisioning
- maintenance
- network restoral

SDXCs come in two flavors: broadband (BDXC) and wideband (WDXC). The lower-speed device is the WDXC, which provides cross-connect capability for floating VTs within a DS3 or STS-N. WDXCs can also provide a transparent connection between multiple DS1 interfaces and DS3 or OC-N termination. The higher speed device is the BDXC. The BDXC can both cross connect at DS3 (asynchronous or synchronous) and STS-1 signals and provide concatenation of multiple STS-1 signals to STS-N levels. At present, the only types of digital DXC that can be easily upgraded to SONET capability are the DXC 3/2 and DXC 3/3. SONET DXCs and ATM DXCs are similar and can coexist in the same network, but SONET DXCs can provide restoration for ATM switches that do not have restoration features. In this case, the ATM switch would provide the network access and conversion to ATM over optical OC-N channels, and the SDXC would serve as an optical DXC. SDXCs can also act as international SONET gateways.

14.6.6 SONET Broadband Switches

SONET broadband switches have taken a backseat to the broadband and wideband SDXCs. These devices provide the switching capability found in major voice switches, but operate at the higher OC-N levels. It seems that many SONET switch vendors will also include SDXC capabilities in their switches. A need also exists for this functionality in the CPE environment interfacing many LAN and MAN technologies. This device will also require the capability to allocate bandwidth on demand, and provide both digital data network interfaces as well as SONET interfaces. Figure 14.17 shows an example of a SONET broadband switch which interfaces many CPE broadband services.

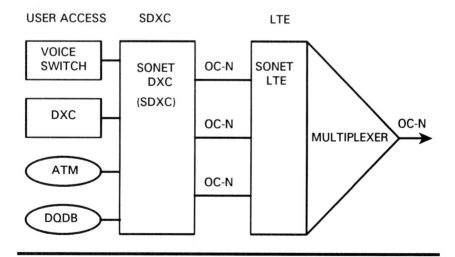

Figure 14.16 SONET DXC (SDXC)

Eventually, SONET switches will be replaced by a photonic switch which will provide both terabit transmissions and the capability to switch at any subrate down to DS0. This is the inevitable evolution of many of the devices covered in this chapter.

14.6.7 SONET Regenerators

SONET regenerators work in the same manner as the signal repeaters covered in Chapter 4, but they work through optical signal regeneration over fiber optics. Regenerators will reshape and boost signals incurring dispersion or attenuation over long transmission distances.

14.7 SONET Equipment Vendors

SONET hardware vendors can be separated into three camps. In the first are the vendors pushing the SONET DXC. In the second are those who offer drop and insert multiplexer products. In the third camp, vendors push integrated SONET message switches, such as Northern Telecom's FiberWorld SONET product line. Their view is that the new SONET switches now emerging will have all the capabilities of the SONET DXCs. One Northern Telecom product, the OC 48 S/DMS TransportNode regenerator, will support both switching and some DXC capabilities. Table 14.5 shows the major vendors and the type of equipment they market at present.

A shift is taking place among the central office vendors who provide high bandwidth transport devices over 1.8G bps toward the manufacture of SONET-compliant equipment. These vendors are offering hardware that complies with Phase I of the standard, with the intention of providing software modifications for Phase II and III compliance. Most RBOCs and IXCs are placing purchasing caps on their asynchronous hardware in favor of new SONET hardware.

The other major consideration for these providers is the size of their existing DXC base and the upgrade capability to SONET. Some vendors, such as MCI Communications, are well positioned to upgrade, having a majority of DXC 3/3s which will be provided SONET interface upgrades. MCI Communications is so far the only carrier who has begun to build a SONET network operating at the OC-48 level, preparing to interface with other SONET public networks on an international level. MCI Communications was also the first IXC in the United States to deploy a SONET switch in a move which will eventually futurize the very fiber of their switching platform.

In the international arena, many PTTs plan to use SONET primarily for transport and cross-connects and use technologies such as DQDB and ATM as the switching architecture. This may also become the norm domestically.

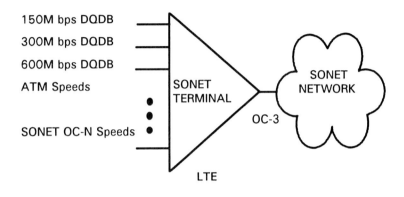

Figure 14.17 SONET Broadband Switch Interfaces

TABLE 14.5 SONET Equipment by Vendor

Vendor	Product	Type
AT&T Network Systems	ServiceNet 2000 product line (DDM-2000)	DXC
Alcatel	RDX-33, RDX-31	DXC
NEC America Inc.	IDC-B, IDC-W	DXC
Tellabs	TITAN 5500	DXC
Rockwell		DXC
DSC Communications	ECS1, ECS3, iMTN	DXC
Alcatel		D/I Mux
Fujitsu	FLM-150, FLM-600, FLM-2400	D/I Mux
Telco Systems		Access Mux
Ascom Timeplex	Synchrony STS300	Access Mux
Northern Telecom	FiberWorld	Switch
NEC America	Litespan2000	All

14.8 ADVANTAGES AND DRAWBACKS

The advantages to deploying a SONET-based network far outweigh the disadvantages. Some of the many advantages provided by SONET include:

- abundant support provided for broadband services
- true fractional DS3
- lower transmission costs
- easy long-range planning for network providers
- efficient management of bandwidth at the physical layer
- aggregation of low-speed data transport channels into common high-speed backbone trunk transport
- standardization of global transmission networks
- long life-span technology
- internetworking simplicity
- standard optical interface and format specification
- service provider's equipment quantity and costs reduction
- increased reliability and restoration
- increased bandwidth management with smart network management features
- vendor and service provider interoperability
- economic method of access to subrate multiplexed signals
- resources for managing the transmission network carried at different levels
- add/drop channels without the need to demultiplex or remultiplex
- reduced overall network transport delay
- monitoring, maintenance, and provisioning uniformity
- ATM, FDDI, FDDI-II technologies supported
- B-ISDN, SMDS, and other high BW services supported
- ability to build private SONET networks

Some of the drawbacks of SONET deployment include:

- strict synchronization schemes
- early deployment of *asynchronous* fiber will eventually be replaced by *synchronous* fiber to mesh today's synchronous networks with SONET.
- lack of control systems
- lack of bandwidth management
- need new hardware for add drop multiplexers
- near complete software dependence

❧ must retrain hardware maintenance staff for software access only

❧ much development on the OAM&P standards may require significant software changes

14.9 PERFORMANCE AND DESIGN CONSIDERATIONS

SONET provides the means for end-to-end in-service traffic performance monitoring through the three types of overhead discussed. SONET also uses pointers to eliminate the slip buffer delay of asynchronous transport. This significantly reduces transport delay from 193 bits down to 24 bits, almost an 8:1 decrease! It is interesting to note that with standard 56K bps or T1 transport, a majority of the delay was incurred just getting data into the pipe. With SONET transport, the delay limit is primarily dictated by propagation delay — at this time an unsolved limiter. Within the SONET standard, there are many performance parameters defined which can be reported by SONET-compliant equipment. These include:

- Severely Errored Framing Sections (SEF)
- Errored Seconds (ES)
- Severely Errored Seconds (SES)
- Coding Violations (CV)
- Line Depredated Minutes (DM)

While SONET provides the many advantages listed above, there remain many performance issues which either have been solved and require refinement or still need to be addressed. These issues include:

- error and jitter performance
- protection switching
- alarm/status indication
- provisioning
- performance monitoring
- pointer monitoring

In addition, SONET requires a minimum of a Stratum 1 clock source to minimize slips and maintain transmission integrity.

A note on SONET design: the SONET standards have been designed so that SONET can exist in both metropolitan rings and wide area network meshed networks. The metropolitan rings will provide higher transport and access speeds, while the WAN-meshed connectivity will provide large areas of coverage and remote access to the SONET network. Service providers are still debating the method of deploying SONET rings and meshed configurations. Some are building small, high-density ring networks which turn out to be more survivable than meshed networks. Others are building SONET rings with add/drop multiplexers and DXCs providing the wide area network connectivity over point-to-point circuits. Much of the network restoration in these configurations will be through the broadband DXCs which employ both on-line and spare dedicated circuits.

14.10 REVIEW

In this chapter the SONET standards were reviewed, from Phase I defining the basic operation and frame structure, through Phase II defining additional functions, OAM&P, and the IAO interface, and Phase III defining SONET vendor interfaces and additional OAM&P for the midspan meet. Next, the frame structure and multiplexing hierarchy of SONET was discussed, first understanding the highly structured STS-N and OC-N hierarchy, and how the messaging and file transfer protocols fit into SONET, and then the actual composition of the SONET frame. User interface options, user access to a SONET network through various interfaces, and the services that can use SONET were explored. The full line of SONET hardware available was analyzed to provide the reader with the capability of objectively choosing the switch, DXC, or regeneration device for any given application. SONET vendors providing this hardware were highlighted. Finally, an analysis of the advantages and drawbacks of SONET networks was undertaken, covering some performance and design considerations inherent in SONET network designs.

4

Requirements, Planning, and Choosing the Technology

Part 4 covers the aspects of defining user requirements, conducting an analysis of the traffic requiring transport, establishing a capacity plan for the next three to five years, and comparisons of circuit versus packet versus frame versus cell technologies. One of the most difficult aspects of a network design is fully understanding user requirements. The user's view of the network is often like that of someone who simply picks up the phone and dials, and, if there is a problem, he calls the phone company. If you are the phone company, and a user wants to connect more than just a telephone, this user must relate his requirements to you and, in turn, you will teach him telephony and how his specific phone works. Data network design is not much different, except there are multitudes of variables to the average user data connection. Chapter 15 helps us outline the right questions of the users to determine their exact communications needs. Chapter 16 provides methods to analyze user traffic to provide a preliminary location and configuration design. This will feed into the chapter on access network design, as well as how to walk through a data network capacity plan. Chapter 17 presents a comparison of technologies discussed thus far. The next few chapters leave many of the detailed calculations to more in-depth traffic analysis books, but provides the important calculations for user satisfaction and hardware and software acceptance.

15

Requirements
Definition

The network design is the product, and the user is the customer. The network is built for the customer, and it must conform to his needs, wants, and desires. The requirements analysis is the first and most important step in the network design. No single aspect of the design is more important than fully understanding the user's needs, for they ultimately dictate the technology, protocols, hardware, software, and resources devoted to both access and backbone design. The two major views of requirements are those of the user and those of the designer. The user looks at the network from the outside in, and the designer looks at it from the inside out, thus creating two myopic views which must merge to provide a comprehensive complementary analysis beyond simple network ingress and egress. Many questions must be asked before beginning the design to allow both parties to "get to know each other" and begin the marriage of user and designer. This working relationship is essential to the success of the network. Changes and inaccuracies in the user requirements can have devastating effects on both the access and backbone network design. While user requirements change, the designer needs to set a time when the addition of new requirements is frozen and the current ones analyzed. This chapter will examine how to look at the requirements from both the user's and the network design engineer or manager's perspective. But, first, to explore the many aspects of user requirements to prepare the stage for traffic analysis and the capacity planning phase.

15.1 USER REQUIREMENTS — THE USER'S VIEWPOINT

How often do users proclaim "I want it all, now." Users want as much network as they can get. Bandwidth requirements are exploding as LANs and MANs proliferate and the distributed computing environment grows. But as this environment grows, so does the tangle of protocols, architectures, support systems, and multitudes of other factors influencing and controlling the flow of data between users. This is an exciting age, when users have taken more control of their data transport. Corporate communication data networks are playing catch-up to the users. As discussed in the beginning of the book, the user has become smarter and more sophisticated, users no longer stand for the smoke and mirrors approach that corporate MIS has used in the past. In the past, the MIS application backlog averaged 1 to 2 years. If the user's true requirements are not satisfied, the user will go elsewhere. Thus begins the education process of both user and designer. First, however, we must clarify the level of user expectations and satisfaction.

15.1.1 User Expectations and Satisfaction

User expectations must be properly managed to ensure satisfaction not only at initial turn-up but through continued service of the network. The best way to ensure this is by meeting the initial levels of customer satisfaction. Clearly understand what is expected on both sides and set expectation levels accordingly. If this is a new network, make sure users know that initial service is in its test stages. Establish graduated levels of new technology introduction: technology trials, then alpha test, then beta test, and finally full service. These levels of user access are defined as:

Technology Trial. A test environment where both the user and network designer are learning from each other. Network downtime and reconfiguration are expected often. This phase is primarily for both parties to learn about the new technologies. The network services are limited in scope and functionality.

Alpha Test. A prerelease version of the network service where the customer is still learning from the network provider, but the provider now has the network design completed and hardware is fully operational. Downtime is minimal, and very few users are on the network. This is a small-scale version of the "real thing", often trialed internally, and expected to be "buggy".

Beta Test. This is the final test phase before commercial or public service availability. Most, if not all, of the bugs should be out of the system by the beta test and users should experience little to no downtime. This is close to the actual network offering, but since its new, everyone expects some instabilities.

Production. The network goes officially live! All of the bugs (hopefully) have been eliminated and the network is running with full functionality. There is no downtime (unless scheduled).

The final version of the production network should be transparent to the average user. Each user should feel like he has a full time point-to-point dedicated circuit to every destination imaginable. Let's look at the user "dream sheet" of expectations:

- ✷ no delays
- ✷ no cost to user
- ✷ no protocol or functional restrictions
- ✷ no physical or logical constraints
- ✷ no network errors (or very low error rate)
- ✷ ubiquitous access
- ✷ no circuit downtime
- ✷ network performs all protocol conversion
- ✷ internetwork interconnectivity
- ✷ broadcast capability
- ✷ confirmation of receipt (fax, interactive SVCs)

These are among a few of the epitome expectations and network perceptions of users. After a short time, the network is viewed as a utility, something which always works — not a luxury but a necessity. Back to our telephone example: the user picks up the phone and just expects it to work. Above all, remember that the user is the customer and the network designer is the provider. The primary responsibility of the network manager or designer is to provide the network as close as possible to the user's expectations. Keep a professional, friendly attitude toward users, treat them as peers, and they will treat you the same way.

The problem with many of the new technologies is that they require the user to understand these mid- to high-layer protocols more than ever before. It is the job of network engineers and managers to make as much of this environment as transparent as possible to the user, while still providing the feature functionality in a user-friendly manner to the desktop. The network should look like a transparent cloud as shown in Figure 15.1. Sit now in the designer's seat and understand user requirements from the network side or viewpoint.

15.1.2 User Involvement in Technology

Never before have users known as much about networking technologies as they do now. Users are forming consortiums, forums, user's groups, and other organizations to actually drive the standards defined in this text. The user becomes more informed every day. In some cases the user is more experienced than the network designer or manager. Do not overlook this source for network planning and design. Instead draw upon it. The user has a wealth of knowledge concerning his applications to tap into, so make sure the requirements gathering process is an educational process for both. The network process is a partnership where the sharing of information is bidirectional. Now that the pep talk is over, the user and designer are partners, and the next step is to get down to business and define requirements for the data network.

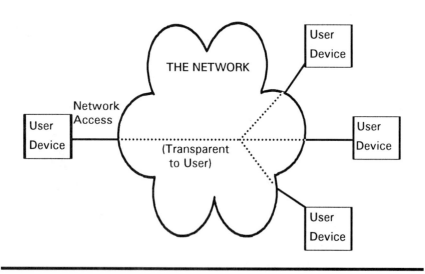

Figure 15.1 The Network

15.2 USER REQUIREMENTS — THE DESIGNER'S VIEWPOINT

The rest of this chapter discusses user requirements from the designer's viewpoint. Obviously, the network designer has to understand much more than what is provided by the user. While scanning the following sections, the reader should comprise a working list of each user's requirements. Refer back to earlier chapters for clarification on technologies, hardware, and services to meet these requirements. The reader should also note both user and network designer constraints. Once the user requirements are identified, a comprehensive traffic analysis for purposes of capacity planning will be explained and provided in the next chapter. The designer will never receive all of the user requirements, and even if you do the requirements will most likely change throughout the history of the design and life of the network.

15.3 TRAFFIC CHARACTERISTICS

Traffic comes in many shapes and sizes, conforms to many protocols and formats, travels in many types of patterns, and requires special methods of processing and handling. This section covers these aspects and more. The format of this section, and the approach of this chapter, is provided in a questioning tone. The network designer should ask the questions identified here for each application planned for the network. These values should also be compared to existing transport means, such as speed of the private lines now being used.

15.3.1 Message or Data Size

The two common formats for data traffic are the units of measure and the packaging of these units. Units of measure progressively explained include: bit, byte, octet, message, block, and file. These units can be packaged into packets, frames, cells, or left as plain units for transfer across a physical medium using transmission packaging (e.g., DS1). Another name for a data package is a protocol data unit (PDU) which is how the SMDS and ATM standards use the term.

TABLE 15.1 Data Transmissions Comparison

DATA SENT	DATA SIZE (bps)	TRANSMISSION SPEED (bps)	TIME TO SEND DATA
Basic text file	30K	56K	0.5 seconds
(10 pages)		T1	.02 seconds
Basic	250K	56K	4.5 seconds
spreadsheet		T1	.16 seconds
Single page	1M	56K	18 seconds
graphics		T1	.65 seconds
Average	3M	56K	54 seconds
FAX		T1	2 seconds
CAD/CAM	10M	56K	3 minutes
graphics file		T1	6.5 seconds
Compressed	100M	56K	30 minutes
video		T1	1 minute

Data is measured by the number of units or packages transmitted per unit of time. Some examples include packets/frames/cells per second, bytes per second, messages per hour, or even by the number of transactions per second (using a fixed or variable unit of measure for each transaction). These values translate into channel transmission rates, and the transmission rate determines the amount of time it takes to transmit the unit of data. Table 15.1 shows examples of data size and transmission times based on network bandwidth available (these times assume no overhead — and could be considerably less). In network design, average packet or message size is often used as the measurement.

The actual data size of each package is the size of the data for each user compared with overhead and framing in the packaging. Many technologies set limits of minimum and maximum packaging size, such as a maximum X.25 packet size of 4096 Bytes or a maximum frame relay frame size of 8096 Bytes. Also, determine if the packet/frame sizes need to be identical at each end of transmission.

Many protocols provide windowing for acknowledgments of data sent and received. Users of these protocols need to provide the window size required; minimum or maximum. But frequently, users require your expert recommendation, because they do not have the foggiest idea. Many of the technologies discussed tend to segment data as it transits the network. Confirm if the segmentation of user data can be tolerated by the user applications, or if the user applications require segmentation.

15.3.2 Sessions and Usage Patterns

What is the relationship between network devices? What are the characteristics of the user's sessions with the network? Some characteristics include:

- ✓ number of sessions
- ✓ number of call setups
- ✓ number of times it turns on
- ✓ number of calls
- ✓ polling intervals
- ✓ reporting intervals

What are the time patterns of user sessions? Time patterns occur in:

- ✦ peak work/traffic hours (Figure 15.2a)
- ✦ specific hours of bulk data transfer (Figure 15.2b)
- ✦ random times but predictable averages (Figure 15.2c)

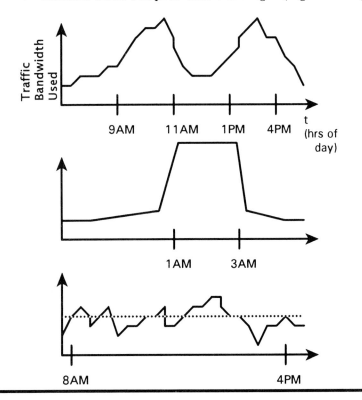

Figure 15.2 User Traffic Patterns

What are the traffic patterns during the peak usage times?

? peak burst of traffic (not same as peak traffic condition)
? sustained peak condition
? constant data flows
? change of protocols

One way to get a handle on existing traffic flows is to evaluate the specific amount of traffic on the network at time slots throughout the day, say at ten minute intervals. This sampling technique, when graphed on an X-Y axis, will resemble those figures seen above.

Another example is comparing calls per second versus data packets per second. Figure 15.3 shows an average response graph for displaying calls per second versus the total data packets per second (calls vs. PPS).

15.3.3 Call Setups per Second

An example of call setups per second can be generated based upon total network response time. Take for example a point-of-sale credit card verification machine. A retailer "zips" the credit card through the machine and it transmits a request for verification from a remote host computer. From the time the machine transmits the request to the time it receives a confirmation (or rejection) is the total network response time. If the machine is one of ten accessing a local packet switch node (through a PAD), the number of requests (calls) the node establishes in a given second is called "call setups per second". Figure 15.4 shows this example.

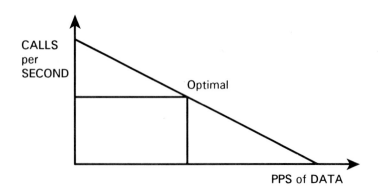

Figure 15.3 Average User Response

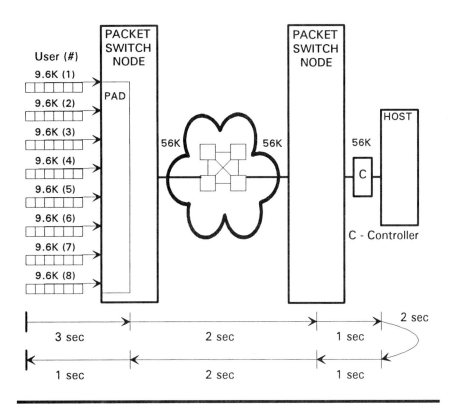

Figure 15.4 Total Network Response Time

By this example, it takes 3 seconds to dial the local packet switch node and transmit the request, 2 seconds across the packet switch network, 1 second from the remote packet switch node to the host, 2 seconds for host processing (verification and database lookup), 1 second from the host to the packet switch, 2 seconds back across the network, and 1 second to relay the information back across the modems (connection already established) to the automatic teller machine (ATM). Total network response time would be 10 seconds, with an additional 2 second delay for host processing for a total response time delay of 12 seconds. This may not seem like a long time, but compare it to a credit card transaction at a busy department store, which uses the same method of verification, and it will seem much longer.

Networks of this type are designed based upon standard Poisson distribution arrival rates, where the network is designed to accommodate overload periods of traffic for a short period of time at a certain grade of service with the understanding that there will be underutilized periods when queued data can be processed at even better grades of service.

Notice that there is little to no room for delay in this type of application. In fact, delays have a way of escalating in packet switch systems, because calls store up in the queue and new calls cannot even get into the queue until old ones are out. In frame relay and cell switching, excess calls are not queued or buffered, they are simply "dropped".

15.3.4 Call Routing

Another important factor is call routing. What route does a call take from origin to destination, and what are all of the possible routes it may take? For circuit-based technologies, the origin and destination are fixed. Even circuit switching services can provide planned origin and destinations. But with switched services such as frame relay, there are logical addressing schemes where any user can talk to any other user. How do you plan for the call routing in this scenario? The best method is to calculate "best guess" based on common user groups, known traffic patterns, and projections provided by the users.

15.3.5 Minimum Error and Data Accuracy

Some applications require a minimum amount of error or a stated data accuracy to certify a transmission. Excessive delay could cause the user to lose the session and need to reestablish, and experience time-outs, or worse yet, loss of data. Minimum error levels are stated as 1×10^{-X}, where "X" ranges from a value of 3 in poorer transmission networks to a value of 12 to 15 in fiber optic networks. The higher value of the negative "X" exponent, the fewer errors are incurred.

What costs will be incurred if the transmission is received late, out of sequence, or not at all? Relating these implications to dollars is one of the best methods of deciding what level of data accuracy to provide. Is error and flow control required, or can data simply be discarded (as in frame relay) assuming that the application will take care of error detection and recovery? Some technologies simply discard data during error and congestion conditions. If the user cannot tolerate this, is he willing to provide the high-level protocols needed for end-to-end flow and error control, or is he relying on the network or CPE devices to do this?

A good method of approaching the problem of data accuracy is through a risk analysis. A risk analysis can allow the user to relate data risks to financial loss numbers. For example, the loss of a T1 amount of data from a banking institution for 10 seconds could cause the bank to lose up to $200K in lost transactions. An even worse case would be the stock market for international trading, where millions could be lost in seconds, or brokerages, where large penalties are paid for network-induced late trades. Other risks that could relate directly to costs include: downtime of users or computers, routing of data to the wrong destination (possibly a competitor), and loss of customers due to downtime. These figures all add up to a bottom-line value of the assurance that data is delivered error-free and to the correct destination.

15.3.6 Prioritization

Users with multiple applications, traffic flows, and hardware devices must prioritize each. Find out what the user views as his priority scheme for each application, between applications, and even among multiple users. Find out which network elements carry the *mission critical* traffic, as well as those that have redundant systems for periods of failure. *Mission critical* traffic is defined as data containing the day-to-day critical business information which would cause severe financial or operational loss if corrupted or not delivered. Examples include: billing, command and control, money transfers, and account information. Prioritization is important for services such as frame relay, where the user who transmits over their committed information rate (CIR) will need to set the discard eligibility (DE) bit based on a traffic priority scheme.

15.3.7 Capacity and Throughput

Capacity and throughput are similar, but not the same. *Capacity* is the actual amount of resources available across a given path. *Throughput* is a measure of how much *data* can be passed across a medium in a stated period of time. For example, the capacity across a T1 synchronous line is 1.544M bps. If a TDM multiplexer using D4 framing used this line, the actual throughput would be 1.536M bps (1.544M bps minus the 8K bps framing overhead). Some multiplexers use additional available capacity for in-band signaling, network management, and other proprietary functions, thus reducing the throughput even more. In the *circuit* world, throughput is commonly measured in bits and bytes (or even in peak periods). In the *packet* world, throughput is measured in packets/frames/cells per second. For example, on a 56K bps capacity line a typical packet switch using 128 Byte packets could pass up to 55 packets per second (PPS):

56K bps * (1 Byte/8 bits) * (1 packet/128 Bytes) = 55 PPS

Thus, the theoretical maximum for 128-Byte packet sizes over a 56K bps circuit is 55 PPS, but in actuality due to overhead packets this number is usually lower than 50 PPS. The same formula for packet switching can be used for frames and cells per second throughput.

Some technologies utilize the available bandwidth more efficiently than others, providing for higher throughput. One example is Token Ring, which more efficiently utilizes the available bandwidth during the transmission of larger file transfers than Ethernet due to its token reservation technique. Another example is FDDI, which eliminates the need for token reservation. Either way, a data network should be designed with good throughput and extra capacity for the future. These steps are part of the capacity planning process discussed in detail in the next chapter.

15.3.8 Packet per Second Processing, Forwarding, and Filtering

Network hardware can measure performance through packet per second processing, forwarding, and filtering rates. Packet per second processing is how fast the switch interprets the packet header and does what it needs to do with the packet (either interprets, terminates, or switches it). Typical access switches operate at the 2,000 to 5,000 PPS rates with backbone switches exceeding 50,000 PPS processing. Packet forwarding rates are provided by bridges and routers to represent how fast they interpret packets and forward them to their MAC destination. Typical rates are 3,000 to 4,000 PPS forwarded, with many vendors now exceeding these rates. Packet filtering rates show how fast a bridge can determine if a packet is addressed to a point inside the LAN (filtered) or whether it should be passed on to the next bridge. Typical filtering rates are from 9,000 to 14,000 PPS.

These rates are commonly displayed as bridging throughput, with Wellfleet Communications leading the pack at 480,000 PPS. bridging throughput. There are also routing throughput rates in excess of 20,000 PPS, and some vendors such as Wellfleet Communications support rates much higher. Many of these high rates are based upon small packets (e.g., 64 byte), many of the rates go down as the size of the packet increases, and some are based on packets with no data in them. Source route bridging protocol for example usually offers the highest PPS. bridging rate, and the source transparent bridging protocol for Ethernet throughput is much better than that for Token Ring. FDDI packet forwarding rates are usually 75,000 PPS (for 22-Byte packets). Obviously, as more protocol conversion and routing takes place the PPS rate is driven down proportionally.

15.4 PROTOCOLS

The full suite of user protocols must be defined for both existing and planned applications. This includes all protocols for each architecture supported, such as all seven layers of the OSI reference model, which may include everything from the physical media requirements to the presentation software and operating systems. Multiple protocols per layer are often present. Previous chapters have identified and explained approximately 90% of the user protocols which will be encountered.

Next, define what protocols are to be used on both the application side and the network side. What functions are specific to applications that warrant special consideration? Also define how users will internetwork these protocols. The bridging, routing, or switching required will depend heavily upon the protocols. For example, some protocols such as NetBEUI and DEC LAT are not Internet-layer addressable protocols and must be bridged, while others such as FTP can be routed through the use of TCP/IP. Will the user require protocol encapsulation, conversion, translation, or just bridging? Are higher-level directory services required (e.g., X.400, X.500)? Also know if there are any special or proprietary implementations of these protocols and how they operate.

15.4.1 Connection/Session Oriented

Is connection-oriented or connectionless service required? Connection-oriented services could include channel-to-channel (host-to-host) or synchronous dedicated circuit (multiplexer), and connectionless services could be LAN/MAN (bridged or routed) traffic — standard LAN protocols are all connectionless.

Is polling/selection or nonpolling used? Examples of polling or selection services include selective or group polling (E-mail), stop-and-wait polling, sliding windows for packets or frames (e.g., BSC,

HDLC, SDLC)), or transaction processing (e.g., order entry). These are typically based on a client/server architecture. Nonpolling services and applications include Request to Send (RTS)/Clear to Send (CTS), Xon/Xoff, multiplexed, token-based reservation and non-reservation (priority and nonpriority), and CSMA and CSMA/CD random access with sensing (priority and nonpriority). Ethernet is the only common CSMA/CD protocol and it is always nonpriority. Also determine whether traffic is centrally processed, or if the processing is distributed over multiple locations.

15.4.2 Protocol Routing

Routing accomplished by a protocol can either be fixed or dynamic. This depends on many factors often outside the user's control and is a function of the network technologies and protocols. An example of fixed-protocol routing is PVCs in frame relay, where a user-assigned PVC will always route from one point to another. An example of dynamic protocol routing is X.25 packet switching (including datagram services), where the packets will route dynamically the "best way", including around a failure. Dynamic protocol routing can either occur in a true dynamic nature through the network and transparent to the user, or it can be initiated by user commands (as in high-speed circuit switching). The capability to dynamically reroute around network failures is often a requirement, so determine if the dynamic rerouting is serviced on a packet level or a circuit level.

15.4.3 Application Architecture

Applications will generally follow a specific application architecture which establishes the standard user and programmer interface. When the developers use an application architecture, they set the standard method of protocols for programming, file transfer, and data access methods for consistency across the network. Some examples of application architectures include:

* TCP/IP / Internet Architecture
* IBM Systems Application Architecture (SAA)
* DEC Network Applications Support (NAS)
* HP NewWave Computing

These application architectures tend to be found in more distributed applications. While many network designs are transparent to this level of protocol, many of the architectures listed above also define protocol levels affected by the network. The network designer and manager must understand protocols down to the application program interfaces (API), as problems with an application architecture could have repercussions down to the physical media.

15.4.4 Addressing and Naming Schemes

What are the user addressing and naming schemes currently being used? Are they flat addressing conventions or are they hierarchical in nature? Are they permanent, if not, can changes, modifications, and new adaptations be made easily? Can they fit in with a global addressing scheme? Comprehensive addressing questions can be asked after the reader has completed Chapter 19 and 20 on address design, but these are good starting questions.

15.5 TIME AND DELAY CONSIDERATIONS

Probably one of the most important aspects of a data communications network is time consideration. Decreasing transmission time and network delay is a major responsibility for the design engineer, and can relate to significant cost savings. Response times can be measured across the application, the network access, and the total network.

Delays in transmission times affect throughput. Sometimes, short delays are tolerable. Some technologies such as packet switching use buffering to queue excess data until it can be transmitted, thus incurring some delay in the storage. Other technologies such as frame relay simply block data or discard excess data. Interrupts and resets are additional forms of delay.

While timing plays an important part in the accuracy of data transport across the network, especially with new synchronous technologies like SONET, it becomes increasingly important to have accurate timing sources.

15.5.1 Access Time

Figure 15.5 shows a comparison of LAN/MAN access times for the IEEE 802.X and FDDI technologies. These curves assume that there are 100 stations on the LAN generating the same amounts of traffic with an average packet size of 256 Bytes.

For circuit-switched and multiplexed services not having to contend for a token or access to the common bus, the access time should be almost instantaneous (except during blocking). Access time differs from response time in that it only provides a measure for when the user can place data on the medium. Response time takes into account the round-trip delay from source to destination.

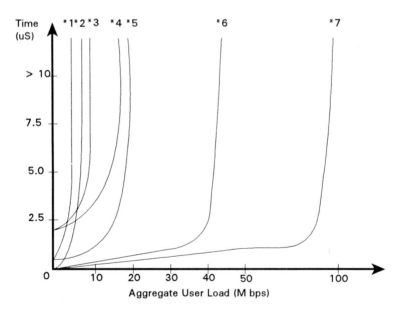

*1 - 4M bps Token Ring
*2 - 10M bps Ethernet
*3 - 10M bps Token Bus
*4 - 16M bps Token Ring
*5 - 20M bps Token Bus
*6 - 45M bps DQDB
*7 - 100M bps FDDI-I

Figure 15.5 LAN/MAN Access Times Comparison

15.5.2 Response Time

Is delay in response time a major factor? What are the response time requirements and what are the trade-offs versus costs of delay and blocking? Response time is a two-edged sword. Normally, a user will be quite aware of their response time (probably to the nearest second). If a new network service is provided that decreases response time, they will, of course, be happy. Let us take the example of a new LAN service such as FDDI, where they are the only test user during the first three months of service. Their response time decreases from 6 seconds to 2 seconds. They are ecstatic, until the FDDI ring adds more users, and their response time starts to creep back up toward 3 seconds. They still have half the response time they originally enjoyed before the FDDI network was installed, but they still complain that their response time has been degraded (and will continue to degrade as more users are added to the ring).

The lesson here is to manage the users response time and expectations accordingly. Delays may even have to be artificially imposed when a service is first installed to provide a constant delay to the user.

15.5.3 Delay vs. Throughput

How much delay can be tolerated when accessing the network? What amount can be tolerated during the end-to-end transport of data? What is the round trip delay? Usually, the more delay that can be tolerated the greater the throughput. What is the cost of X time delay in dollars? Does the network devices chosen to provide transport add appreciable delay (such as routers vs. bridges)?

Users want minimum delay and maximum throughput. Thus, a good network design should maximize utilization of resources while keeping delay in line with the user's expectations. When a new data service is offered, and few users are on the network, it may be necessary to introduce artificial delay. This is only temporary until more users start using the service. This provides the first users to experience a constant delay over time, rather than a decrease in performance.

One example of an application expensive for bandwidth is a card swipe reader. This device uses an entire call for call setup, data transfer of two packets, and call teardown. One alternative for this is to use the packet switch option of fast select. This will allow the entire process to occur in a single transaction.

15.5.4 Blocking vs. Storage vs. Queuing

Invariably, every network will eventually encounter some network congestion. It is important to understand how the user applications will handle these conditions. Congestion can have three major effects on user traffic. When a technology uses blocking, the user traffic flow is blocked until sufficient bandwidth is available to continue the transmission. This is the most drastic method of dealing with congestion, yet it is the method employed by many of the new technologies discussed in this text. Other technologies just "drop the information out onto the floor" during congestion conditions, assuming that the user has employed some form of higher-level protocol at the CPE transport through application level to stop the flow of information until congestion clears. Others notify the user and simply block the transmission flow until the congestion clears or the user slows down transmission of data. The determination to use or not use blocking is very important, as many delay-sensitive applications cannot handle blocking.

Storage and queuing are similar and often used together. Queuing is the method of allowing data to be stored in buffers in the order it was received until there is sufficient bandwidth to transmit it to its destination, when it will be sent out in the same order it was received. The primary example of queuing discussed already was the packet switch, which queues packets until transmission bandwidth is available. Storage is used in store-and-forward networks, as typically signifies the use of memory buffers in a FIFO arrangement. Both queuing and storage increase delay through the network.

15.5.5 Interrupts and Resets

The use of interrupts is rare but still should be taken into account. Interrupts occur when the user carries software-defined interrupts such as break keys. Ask if the user plans on using interrupts, and if so what interrupts will they use. Also ask if the interrupts need to be sent ahead of data. A second form of interrupt is the reset, which some applications will perform. Both interrupts and resets should be transparent to the transport network.

15.6 CONNECTIVITY

With a handle on the types and characteristics of traffic the user requires for transport, the types of user and network connectivity required to pass this information should now be explained.

15.6.1 User-Network and Network-Network Connectivity

The user would like a single graphical interface to the entire computing and communications environment. This user-network interface allows local or remote access to other high-level user protocols, with the lower-lever protocols and the transport in-between the local and remote application transparent. Symbols and icons of the graphical interface are preferred to text mode command prompts, forming a requirement for a true *human* rather than *machine* interface to the bit/byte/packet/cell world.

Other questions must be asked about the user-network interface. Some have been covered before. What is speed (data access and data transfer rate) of connectivity? What types of devices will be connected? What is the distance involved between user equipment and data communications equipment? What vendors and models are now used? What are the vendor model numbers? What mode of addressing is required for the access and backbone portions? What are physical, data link, etc., protocols required for basic connectivity? These questions are covered in this chapter, but they are by no means all inclusive.

Once the user-network interface has been defined, any network-network interface requirements for internetworking new and existing networks to the data transport network must also be defined. To what extent is ubiquitous connectivity required? Any-to-any connectivity can take on many forms and layers of protocols. Will the connectivity community be homogeneous or heterogeneous? Will private, public, or a hybrid network access and transport be required? Also, the question of what form of technology will be used for network interfaces and internetworking arises. Examples include:

- ✣ Point-to-point
- ✣ Multiplexing or concentration
- ✣ Switched (type: circuit, packet, frame, cell)
- ✣ Bridged or routed
- ✣ Hybrid

15.6.2 Geographical Requirements

Where do the applications reside? Are there specific geographical location restrictions based on topology, transport media availability or quality, user communities, existing facilities, or any other limitations?

15.6.3 Structure

What type of structure is required? Since most of the technologies described in this book lean toward distributed processing, this is often the answer. But a centralized structure such as the mainframe environment is also an option. Also, is the design hierarchical in nature? Hierarchical LAN/MAN designs was the first step in building WANs because of the importance of segmentation of data user groups.

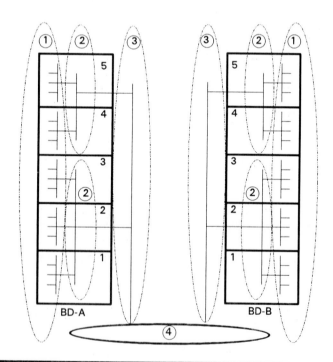

Figure 15.6 Hierarchical LAN/MAN Design

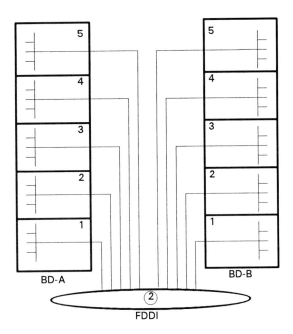

Figure 15.7 Nonhierarchical LAN/MAN Design

Figure 15.6 shows an example of a hierarchical LAN/MAN configuration. Here, Ethernet LANs are built in a three-tier hierarchy. Tier 1 is the local floor LAN segment, tier 2 is the departmental LAN segment (department 1 is floors 1, 2, and 3 of building A, department 2 is floors 4 and 5 of building A, and so on), and tier 3 is the building LAN segment. The fourth tier is the FDDI MAN connecting users in both buildings. This allows segmentation of user traffic and prevents "flooding" the FDDI MAN with information transfers it never needs to see. Figure 15.7 next shows an example of how to design a LAN/MAN without this hierarchical design.

15.6.4 Current Infrastructure

What is the current infrastructure being used? Are user communities heterogeneous or homogeneous? What impact will implementing a new platform or architecture have on existing facilities? The designer must understand the existing voice, data, and other media structures to effectively utilize all existing resources in the new network. Migration from an existing architecture to new platforms is often an arduous and difficult process. The designer must consider the use of temporary gateways as the first step at internetworking the current infrastructure with a new architecture.

15.7 AVAILABILITY/RELIABILITY/MAINTAINABILITY

A user is concerned with three major measures of network quality: availability, reliability, and maintainability. While availability and reliability calculations are defined thoroughly in the next chapter, a user also looks at availability related to available network capacity. A user always wants to have enough capacity to transmit his message (regardless of the message size) in a reasonable amount of time. The network allocates capacity across the network, rather than to a specific user. Some services have methods of allocating specific capacity to users, such as SMDS and ATM.

Maintainability is defined as the measure of effort a user or network provider must put into the network to maintain its operating efficiency. It also defines how the designer will expand the network as requirements increase.

Does the protocol require fault tolerance, and what is the level of error tolerability? How much sharing of resources and capacity is tolerable? Does the user protocol being used require guaranteed delivery? These are some of the questions that should be asked of the user regarding the reliability of existing systems.

15.8 AMOUNT OF USER CONTROL

To what degree does the user need control of network resources? Many public network service providers are struggling with this question. Today, many users want more and more control over their traffic than in the past. They require the capability of intelligent network management and control stations which they can use to reconfigure network resources. The designer must draw the line as to which resources the user can access and which he cannot. One option is to provide the user network status for resources but which cannot be directly modified. An interesting aspect of services such as frame relay is that they provide the user with network status for conditions such as network congestion which he may or may not be able to control. Remember, too much information can be dangerous, so carefully evaluate what the user can see of the network. The best solution is to provide a true level of transparency while providing optimal user control of network resources.

Control can take on either a centralized or distributed nature. Centralized control is often the case for networks managed by local or regional network management centers. Distributed control is often the case where many LANs are communicating over a WAN. The newer technologies require a little of both. The question must then be broached as to how much the user will actually use the control information provided by the network protocols.

15.9 EXPANDABILITY, SCALABILITY, AND EVOLUTION

The network must be able to react to change. Therefore it must be designed with the capability to expand, evolve, and have the option to be scaled either up or down. The network plan should define an evolutionary cycle, not a revolutionary change. It should also provide the capability to change configuration rapidly to meet changes in a dynamic user environment. This change could involve hardware and software configurations, addition or deletion of services, constant technology updates and upgrades, and protocol flexibility. Expansions and migrations should be planned in advance, based on user projections for expanding current applications or adding new ones. Dates should be provided even in the broadest form as available (e.g., 2H94).

One method of ensuring that the network can handle future growth is to overbuild the network with capacity, throughput, and processing power. This can become expensive, and can prove to be misguided if the excess allocations are in the wrong location, the technology changes, or the network requirements were overestimated in the first place. Effective network planning is the answer for this problem, along with education of the users on how to plan for the technologies covered in this book.

15.10 SERVICE ASPECTS

Service aspects such as network management, billing, security, user support, and disaster recovery are important topics with a broad range of standard and proprietary solutions. The best defense is for the network designer to understand the services offered by each technology and protocol and make the best decision based on each.

15.10.1 Network Management

Most users require end-to-end fault identification and isolation. This can be accomplished through a flexible network control and monitoring system which provides configuration capabilities and dynamic capacity allocation through a user-friendly network management interface.

15.10.2 Billing

For public data networks, billing is based on either tariffs or an individual case basis. Does the user require special billing arrangements? What are the departmental, companywide, or industry standard billing mechanisms, and does the customer agree with these methods? Is itemized billing required, and to what detail does each user analyze network costs?

For strictly internal company networks, how are the network and traffic transport costs charged back to the user departments? If it is strictly an internal enterprise network, can a chargeback scheme be implemented? What method of expense allocation accounting is preferred?

15.10.3 Disaster Recovery

What level of disaster recovery is required? The highest level of disaster recovery is one fully redundant system for every live system. Is the user prepared to incur the high cost of such redundancy, or is less required? Is outsourcing of the disaster recovery to an IXC circuit-switched service an option?

15.10.4 Security

What level of security is required? What levels of passwords and access are required? What resources require what levels of access, and how are they accessed? Does the user require end-to-end security? How sensitive is the information being passed?

15.10.5 User Support

User support should be tailored to the user's needs. Users with highly complex routing protocols, proprietary protocol implementations, and users at the executive level will require much more resources than a user of a circuit switched service, mostly because of the complexity of the problems that will occur. User expectations should be managed from the day the network is first announced through the day the user gains actual access to the network. The levels of user support should be agreed upon before the design is complete.

It is sometimes difficult to understand what the user defines as excellent service and quality service. These values vary based on the user's past experience with the network. One gauge of user support is to monitor existing systems support and the level of dedication and time required. Ask what services does the user currently operate, or access which will be required with the new network? It is important to establish lines of communications with users so that they will alert the network designer to future requirements when they are known. Start the long-term planning process in the initial design.

15.11 BUDGET CONSTRAINTS

While users try to spend as little money as possible on the network, they want as much network resources as possible. Either way, the designer must design the network within a given budget, accommodating the largest amount of users and applications with the available budget resources. This becomes a balance between cost, capabilities, and connectivity. How far can you push the economics and still provide functionality and connectivity. As seen, budget allocations for data related services are now larger than those for voice (on the average). But how do we justify these large data network expenditures?

A network becomes cost-justified if it improves operating efficiency and/or saves capital, provides a potential revenue increase, or avoids larger expenditures. Additional criteria include improving customer service and creating new services for the user. Ultimately, the question must be asked — can the network expenses be related back to the user and will the network produce revenue? If it does, how will the user or designer maximize that revenue?

There is also the concern of who controls the network budget, the network manager or the users. Most often, it is the network manager. This is heavily influenced by how much control the user has over network expenditures. The rule is, the more critical the application, the closer it gets to the revenue stream, the more control the user will have over what is purchased. When users have budgetary control, more care is exercised about what is placed on the network. When the network manager has control and the user is a customer, the user tries to place everything possible over the network. This is the point where departmental chargeback schemes are implemented. The manager can also allocate a communications budget to each department based on the requirements discussed in this chapter. Either way, network costs must be controlled. Customers and users may have an existing hardware base, so the cost of additional equipment, or savings thereof, may not be the major consideration. What happens when the budget cannot meet the requirements? Outsourcing then becomes a viable option.

15.12 POLITICS

Politics can cause a perfectly good network plan to go awry. Politics are part of the everyday business, and it is also part of the requirements process. The user requirements should be scanned for hidden political agendas and parameters based upon political rather than technical or economical factors. This is not to say that we should have the McCarthy purges of the 50s, looking for communist bits in every data stream. The network designer should be aware, and be prepared to dig a little deeper into the requirement, should it appear that executive politics, vendor or application favoritism, turf wars, or any other politically based bias is negatively influencing the network design decision.

15.13 REVIEW

There are two viewpoints to requirements: that of the user and the designer. Obviously, the designer viewpoint is much more extensive, and has warranted a detailed discussion. The many characteristics of traffic affects the network design. The characteristics of protocols and their routing, architecture, and the importance of addressing schemes plays it part. Control over time and delay is desirable as a requirement for many types of traffic. Methods of connectivity were also discussed, and the calculations to help the designer meet availability, reliability, and maintainability of users was outlined. A brief look was taken at the amount of user control provided by the service aspects of each network design. Two issues which plague every network designer were discussed: budget constraints and politics. The network designer, therefore, must compile all of the user requirements listed above and account for their influence in the traffic analysis phase.

16

Traffic Analysis
and Capacity Planning

Advanced protocols, traffic patterns and characteristics, and inter-networking has changed capacity planning into more heuristic guesswork than actual concrete calculations. Capacity planning used to be fairly straightforward for voice networks and private line data networks. Traffic growth and volume figures were calculated, and the appropriate number of dedicated circuits were added or subtracted each year. If capacity exceeded projections, more band-width was ordered. With the advent of distributed computing and communications environments, the number of capacity contingencies has grown at an exponential rate. The traffic matrix is no longer a two dimensional spreadsheet, but a multidimensional matrix includ-ing variables such as protocol types, multiple traffic flow patterns, multiple technologies, circuit options, and more. These new tech-nologies and traffic types are causing capacity planners to throw away the old traffic design books and simply over-engineer the network. In fact, even many LECs and IXCs, which are just now turning up these advanced services such as frame relay, SMDS, and B-ISDN, are also struggling with having to over-engineer networks until all of the unknowns are sorted out. In this chapter some new views on traffic analysis are presented — a new method for creating a traffic matrix, enhanced calculations applied to traffic patterns for both old and new technologies, and other measurements that may hopefully get the designer closer to accurately predicting the re-quired network capacity.

16.1 CREATING THE TRAFFIC MATRIX

Now that the user requirements have been defined, this chapter will compile and sort these requirements into meaningful data to perform a network design. The first step is to take a high-level view of the traffic requiring transport and categorize it into distribution patterns. The next step is to create the traffic matrix. The matrix will then be analyzed, taking the results and drawing comparisons and synergy between traffic patterns where appropriate.

16.1.1 Asymmetric vs. Symmetric Distribution

Is the distribution of traffic by direction asymmetric or symmetric? Asymmetric traffic lacks directional symmetry through imbalanced flows, speeds, and a variety of other characteristics. It originates from large sites to small, or vice versa, and does not follow a normal distribution. Access devices vary in quantity, design, engineering, and loading. On the other hand, symmetric traffic often originates from communities of similar interest, such as specific geographic regions, and is uniformly spread across these sites within each region, and the bidirectional quantity is similar. In symmetrical networks, many of the access devices are similar in quantity, design, engineering, and loading. Distributed networks often resemble asymmetrical rather than symmetrical traffic distributions. Figure 16.1 illustrates symmetrical and asymmetrical traffic distributions.

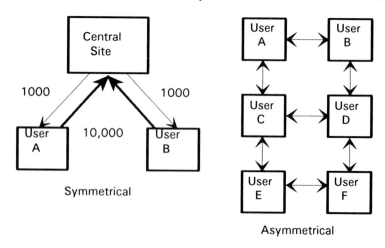

Figure 16.1 Symmetrical and Asymmetrical Traffic Distributions

i \ j	A	B	C	D	E	F
A	20	0	2	1	20	12
B	2	3	3	*	6	5
C	2	0	12	2	8	3
D	8	*	1	10	22!	5
E	12	4	0	31!	6	3
F	22	8	2	1	2	2

TO LOCATION (columns), FROM LOCATION (rows)

Traffic Flow Units (e.g., packets, bytes, calls per second)

Figure 16.2 Traffic Matrix

16.1.2 Creating the Traffic Matrix

The traffic matrix is an essential part of the access node design, and illustrates traffic flows not readily apparent otherwise. Traffic interfaces to the network through an "access node" or "access point". Access nodes could take the form of a concentrator (e.g., PAD or FRAD), access multiplexer, access switch, or any other device concentrating user inputs but not operating as a switching backbone. The traffic matrix helps define details about the access nodes required, such as location, size, operation, protocol support, performance characteristics, and device type.

The traffic matrix is, in effect, a spreadsheet mapping traffic flows from origin to destination. The "FROM" nodes are listed vertically down the left side of the graph, and the "TO" nodes are listed horizontally across the top. Figure 16.2 shows a sample traffic matrix, showing simple data units (e.g., bytes, calls, packets) per second for each site. Each box represents a traffic flow from the "FROM" node (i) to the "TO" node (j) (this is sometimes referred to as the "ij matrix"). All traffic which remains local to a node will be found in the "A-A" box, "B-B" box, and so forth. Notice that Node A has 20 intranodal traffic units per second, while Node B has only three. Notice also that Node B sends two traffic units per second to Node A.

It is good practice to start with a local geographic area as node A and work out from there, so that letters which are lexically close represent nodes in a geographic region. Keeping this scheme as much as possible simplifies the process. Traffic flows, or profiles, will be represented by any of the previously discussed formats, including packets, bytes, calls, or even protocols, priorities, or other dependencies. Traffic measurements used should have a similar message (packet) length and should be measured in mean busy hour throughput (in bps) or in the peak to mean ratio. Both measurements will be discussed later in the chapter. Traffic measurements should be measured from either the network or the application point of view. Thus, a matrix can be built which shows specific application traffic flows (such as FTP or Telenet file transfer protocol traffic) or generic packet flows, such as Ethernet packets from a LAN interface. Also mark off any invalid routes or preferred routes. In Figure 16.3, invalid routes are shown with a *, while preferred routes are shown with an !

16.1.3 Interpreting the Matrix

The matrix shows the connectivity required between each site. Let M be the binary matrix, with a 1 indicating connectivity and a 0 indicating no connectivity. The following are then true:

M = connectivity for one node (i.e., direct connectivity)
M x M = connectivity for a 2-hop path
M x M x M = connectivity for a 3-hop path
$M + M^2 + M^3$ = all paths up to path length of 3 hops

Note that this method can include "cycles" or loops (paths which traverse the same node more than once) which should be eliminated.

All local traffic identified as remaining local to a given node would be placed into the same access node. Circles are then drawn around boxes which represent a small geographic area which could be served by a single concentrator. This is done by grouping nodal traffic distributions in a local geographical area together to form larger access nodes.

i \ j	A	B	C	D	E	F
A	20	0	2	1	20	12
B	2	3	3	*	6	5
C	2	0	12	2	8	3
D	8	*	1	10	22!	5
E	12	4	0	31!	6	3
F	22	8	2	1	2	2

TO LOCATION (columns), FROM LOCATION (rows)

Traffic Flow Units (e.g., packets, bytes, calls per second)

Figure 16.3 Traffic Matrix

Figure 16.3 shows the same matrix from the last example. Node A has 20 units of traffic which remains local and Node B only has three. Also note that Node A sends no traffic to Node B, and Node B sends only two units of traffic to Node A. Since in this example Node A and Node B are in the same geographic region, the traffic from both nodes is combined at Node A where a single access node is placed. The small amount of traffic originating at Node B is "backhauled" to Node A. This also assumes that no other large amounts of traffic are generated or sent to Node B from other nodes.

This process continues until the number of access nodes required to begin the design is established. Ideally, one would like to place a separate concentrator node at each site, but often economics do not justify this, so the network design usually settles for smaller nodes at less concentrated regions and larger ones in more dense regions.

Network designs for multimedia and multiprotocol networks are much more complicated. These complex designs often require many traffic matrices such as a multidimensional matrix (such as a Z-axis forming a three-dimensional matrix to accommodate for interdependencies such as priority of information, budget trade-offs, or protocols), or in large networks design tools to perform these calculations. Ideally, this analysis would be performed by a design tool.

16.2 CALCULATIONS FOR PACKETS, FRAMES, AND CELLS

Reviewing some basic calculations will provide a foundation before progressing to the access node design. Each user interfaces to the access node via an access circuit. This access circuit is engineered for capacity based upon the traffic which passes over it. The common unit of measurement is packets, frames, or cells per second. These figures can be applied not only to transmission, but also to processing, filtering, and forwarding capability of the node. The overhead also affects the total throughput of an access line, as well as the technology used and the number of encapsulations incurred before actual transmission of the user data takes place.

16.2.1 Packets, Frames, Cells per Second

The following terms will be used throughout these calculations. The maximum achievable Packet Per Second (PPS) rate can be calculated as follows:

$$\text{PPS} = P_K \text{ (in bits per second)} * \frac{1 \text{ Byte}}{8 \text{ bits}} * \frac{(1) \text{ packet}}{\chi \text{ Bytes}}$$

For example, a DS0 circuit ($P_K = 56,000$) has a maximum PPS transmission with 128 Byte ($\chi = 128$) packets of:

$(56,000)(1/8)(1/128) = 55$ PPS.

The same calculations are carried forward for frames and cells per second, adjusting for the speed of the medium and the overhead involved.

16.2.2 Effects of Overhead

The number of PPSs calculated in the last example is actually somewhat misleading. It does not account for the overhead incurred in the switching and protocol handshaking operations. For example, take a frame relay fractional T1 access line at 512K bps. The frame size of 1024 Bytes with overhead of 12 Bytes (using extended addressing) per frame is used. The total frame size would be 1024 + 12 = 1036 Bytes. The actual frames per second throughput would be:

(512K bps)(1/8)(1/1036) = 61.78 frames per second

Consider the same example using a frame size of 56 Bytes. The same amount of overhead is 13 Bytes, for a total frame size of 56 + 13 = 69 Bytes. The actual frames per second throughput would be:

(512K bps)(1/8)(1/69) = 928 frames per second.

At first blush, it appears that throughput has been improved, but, actually, it has degraded drastically. The overhead in the first example was only 1.25%, whereas in the second example it jumped to 18.84%! These calculations are shown below. This illustrates that the larger frame sizes are more efficient and provide higher line throughput than the smaller ones, but only up to a certain point.

Overhead (Example 1) = 13/1037 = 1.25%
Overhead (Example 2) = 13/69 = 18.84%

In packet switching, the larger the packet size, the higher the probability of taking an error. For any given Bit Error Rate (BER), the probability of an error increases with packet (or frame) size. If BER is high enough, and packet/frame size is large enough, you are almost guaranteed an error for every packet/frame. This is why 128 Bytes was chosen as the best compromise for packet switching. For noisy lines, throughput can be increased by decreasing packet size to 64 Bytes. The added overhead is offset by reducing retransmissions. Figure 16.4 illustrates the theoretical range for packet/frame sizes versus delay incurred. Cell relay technologies such as DQDB, SMDS, and ATM have a fixed overhead per cell, effectively reducing throughput by approximately 17% (by DQDB standards, 9 bytes overhead out of the 53 total).

16.3 CALCULATING TRAFFIC PATTERNS

With this knowledge of the way packet/cell modes of traffic are measured, this user traffic can be used to calculate the patterns provided to the access network so that a model of the access node can be simulated. Also, the statistical characteristics of the traffic directly affects the access design.

16.3.1 Statistical Behavior of User Traffic

Traffic patterns can be calculated in many ways. User information arrives at the network node based on statistical arrival rates. Not all users need to talk to all other users at the exact same time. Therefore, statistical approximations can be used to model these traffic patterns. These calculations can either be based on mathematical calculations which generalize the messages or packets, their arrival rates, and then calculates the required bandwidth to support the total traffic. Or, they can be performed through packet-level modeling where each type of packet transmission is calculated, and bandwidth is sized from the aggregate. This chapter presents aspects of these two methods.

The primary parameters to be concerned about are the call arrival rate (λ), usually in calls per hour, and the average hold time or duration of message (τ) per call. With these numbers demand on the access node can be predicted in erlangs. Erlangs can be used in voice network modeling as well as circuit-switched data network designs.

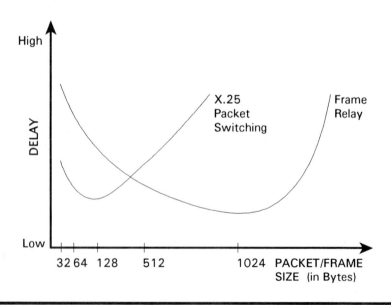

Figure 16.4 Theoretical Delay Curves For Packet/Frame Transmissions

16.3.2 Voice Traffic Modeling (Erlang Analysis)

The standard for statistically calculating an estimate of user demand based on random call arrival and holding was modeled by the Swede Mr. A. K. Erlang in the 1920s. This unit of measure, called the erlang, was calculated as:

$E = \lambda\tau$
where
λ = the call arrival rate in calls/hour
τ = average holding time in hours
E = one erlang

This measurement was developed for analog voice communications, where calls would arrive at a voice switch or circuit switch (see Figure 16.5) at λ and whose total transmission time would be τ hours. In this example, there are five users at Switch A. If each user calls every 12 seconds and talks for 6 seconds before hanging up, the total utilization of the transmission line is .5 erlang. (1/12)(6) = .5 It is obviously possible to offer 1 erlang for every user, but the laws of statistics shows that much lower numbers of circuits can be maintained since all users will not try to place a call at the same time. The actual utilization on an access line is normally much less than erlang. Erlangs also constitute 36 Call Century Seconds (CCSs). CCS's were first standardized by Bell Labs. A camera took a picture of the call peg counters on the electro-mechanical switches every 100 seconds — thus the term called century seconds.

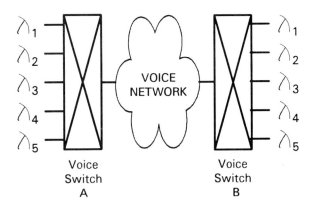

Figure 16.5 Voice Network

The number and duration of calls vary, and therefore so must the erlang calculation. The formula for calculating erlangs with multiple call durations is:

$$\sum_{n=1}^{k} \tau_n = \text{average Erlangs}$$

where k is the total number of calls completed in an hour and τ_n is the length of the call in hours.

Thus, if 100 calls of 150-seconds duration, 200 calls of 100-seconds duration, and 300 calls of 50-seconds took place over a one hour period, the number of erlangs would be 13.89.

Since all blocked calls are "cleared", blocking should be calculated from the erlang calculations above. The formula to calculate the probability of blocking is:

$$B = \frac{E^N / N!}{\sum_{k=0}^{k=N} E^k / k!}$$

where B = B(N, E) = percent of blocking as a function of the number of lines available and the number of affected Erlangs and
$N! = N(N-1)(N-2)...(3)(2)(1)$

For example, 20 calls per hour at 3 minutes per call would equal 1 erlang (or 36 CCS). The number of trunks required for 2% blocking would be four. There are Erlang B (lost calls cleared) tables in most voice and packet switch design books which graphically demonstrate blocking versus erlangs for various volumes of traffic. As the call arrival rate increases, the probability that a new call will be blocked or dropped increases.

16.3.3 Characteristics of Packet and Frame Traffic Arrivals

While erlangs work well predicting voice network and circuit-switched traffic rates, they do not work well with packet-switched networks. In packet-switched networks some level of queuing is employed so that packets are queued in buffers and transmitted when congestion ceases, rather than being immediately blocked. Also, packet-switched networks provide a mix of protocol and traffic types, whereas voice and circuit switched networks provide point-to-point, transparent homogeneous transport of information. Therefore, packet switching demands a different analysis of traffic handling.

16.3.4 Queued Data and Packet-Switched Traffic Modeling

There are three major queuing formulas for dealing with voice, TDM, and packet (queuing) models. Each of these formulas follow the same format:

AD / SD / #s / #w / P, where

  AD = Arrival Distribution
  SD = Server Distribution
  #s = Number of Servers
  #w = Number Waiting
  P = Population

The three major modeling formulas include:

Erlang-B : M / G / s / s,
which is used for voice-blocked calls cleared and TDM modeling.

Erlang-C : M / G / s / k,
which is used when k > s (the waiting room is greater than the number of servers) for voice with blocked-calls held or for operator services.

Packet : M / G / 1,
which is used for one server and infinite waiting and population, commonly used to model packet, frame, and cell networks.

The third formula is of primary concern: packet modeling with M/M/1, where M designates a Markovian process (or Poisson arrival distribution — G designated Gaussian arrival distribution previously). Thus, formulas are based on a single server with Poisson arrivals or equivalently negative exponential service times, and first-in-first-out (FIFO) service. Figure 16.6 shows this relationship between arrival rates, service times, and a single server. Buffers and buffer overflow will also be discussed.

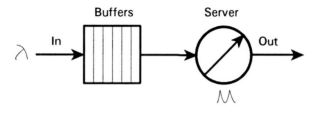

Figure 16.6 M/M/1 Single Server Model

16.3.5 Utilization and Capacity

Continuing with the packet analysis, the fractional utilization, ρ, can then be found with the formula:

$$\rho = \lambda b / C$$

where b = the average number of bits per packet,
C = the line capacity of the trunk in bps and
λ = the packet arrival rate in units of packets/second.
The service rate is defined as $\mu = c/b$ in units of packets per second.

The utilization calculated determines the average amount of time to serve each message (τ):

$$\tau = b / C.$$

When dealing with M/M/1 queues, as in packet switching, there is an average of N users in the queue:

$$N = \rho / (1 - \rho)$$

where ρ is the probability that the queue is not empty and $(1 - \rho)$ is the probability that the queue is empty. Figure 16.7 shows how the increase in utilization (ρ) causes an increase in the average queue size (N).

Their average wait time is:

$$w = N\tau,$$

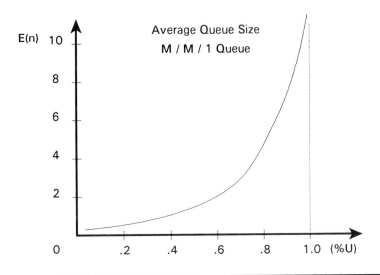

Figure 16.7 Relationship Between Utilization and Queue Size

then the average delay (d_{avg}) equates to the sum of the waiting time and the service time during stable queue conditions:

$$d_{avg} = w + \tau = \tau / (1 - \rho).$$

For example, if a packet switch has 5 users, each transmitting 10 messages per second at 1024 bits per message, with the packet switch operating over a 56K bps trunk, the following applies:

$\lambda b = (5)(10)(1024) = 51,200$ bps
$\rho = 51,200$ bps $/ 56,000$ bps $= 91.4$ % utilization
$\tau = 1024 / 56,000 = .0183$ Seconds
$N = (0.914) / (1 - 0.914) = 10.63$ users in queue
$w = (10.63)(.0183) = .195$ seconds average wait time
$d_{avg} = .0183 + .195 = .213$ seconds average delay

Note that the queuing delay is an order of magnitude much greater than the transmission delay. Now, one portion of overall network delay is known, that within the packet switch queue. Based on the queuing theory shown, the utilization of a trunk is directly influenced by the delay in queuing. Figure 16.8 shows how as system utilization (ρ) increases, so does the delay through the system (d_{avg}). This shows that the greater the utilization the longer the delay. The cause and effects are simple; if delay in queuing becomes a problem, either delete some users or add more trunks.

16.3.6 Buffer Overflow

While it is important to understand the limits of the queue and delay imposed, it is also important to calculate when a buffer will overload and data will be lost. The probability that there are k packets in the M/M/1 queue is approximated by:

$$P_k = \rho^k(1\text{-}\rho)$$

With this value, we can calculate the probability of overflowing a buffer capable of holding B packets of variable length (Pr[u>=B]). This value can be approximated as:

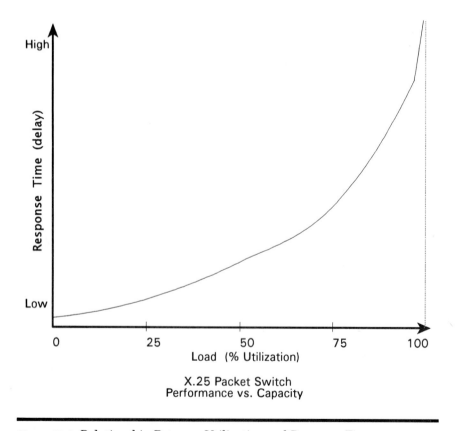

X.25 Packet Switch
Performance vs. Capacity

Figure 16.8 Relationship Between Utilization and Response Time

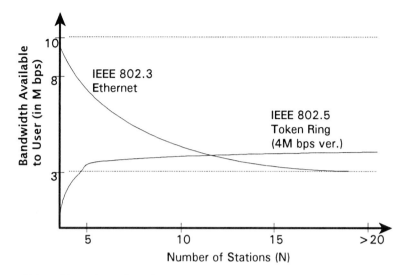

Token Ring -throughput increases when N increases
because less time spent token passing
Ethernet - throughput decreases when N increases
because inceased chance of collision

Figure 16.9 LAN Throughput Comparison

$$Pr[u{\ge}B] = \sum_{k=B}^{\infty} P_k = \rho^B$$

16.3.6 LAN Traffic Modeling

LAN traffic modeling is a difficult process and is subject to further
study, although some of the techniques discussed in the section on
DQDB MAN traffic modeling are applicable.   Many LAN traffic
characteristics, such as performance and throughput based on
number of users, have been thoroughly studied and are provided by
LAN equipment vendors.   Figure 16.9 shows a comparison of Token
Ring 16M bps LANs to Ethernet 10M bps LANs.   Notice that the
Token Ring throughput increases when the number of users increases because less time is spent token passing, whereas on the
Ethernet (CSMA/CD) LAN the throughput decreases as the number
of users increases due to the increased likelihood (and number) of
collisions.   FDDI and 802.6 MAN response curves resemble that of
the Token Ring, yet are more efficient.   More detailed calculations

can be obtained by the LAN traffic analyzers on the market which are increasingly becoming more sophisticated.

Table 16.1 shows some of the tools on the market for analyzing and modeling LANs.

TABLE 16.1 LAN Modeling and Analysis Tools

Manufacturer	Device
Comdisco	BONES
HP	4972A
IBM	Trace & Performance Analysis Tool (TAP)
Network General	Sniffer, Sniffer 7000, LANalyzer
VANCE	ATS1000

LAN bridge designs are concerned primarily with frames forwarded per second and frames filtered per second. Any design should also take into account packet and frame forwarding and filtering buffers, as well as the depth of LAN address table memory. Proprietary bridge and router overhead, as well as routing algorithm traffic (especially the distance-vector routing protocol), can generate substantial traffic in addition to user traffic.

16.3.7 MAN IEEE 802.6 DQDB and ATM Traffic Modeling

The DQDB bus operates like a LAN, but handles calls similar to the Erlang method, where messages contending for the bus have to wait until they can reserve a space on the bus. Call arrival rates are modeled differently. This new method will be explored.

The traffic characteristics of a LAN (bursty traffic) attached to a MAN are shown in Figure 16.10. The variables of concern are:

B = Bandwidth consumed during burst (peak rate) in bps
t_s = Average time slot used for transmission in seconds
n = Number of time slots between starts of bursts
t_b = Time of each burst (average) in seconds $(\mu^{-1}) = 8(D_b)/B$
D_b = Burst duration (Bytes)

Note that a burst may contain one or more packets. For more than one packet per burst, use $(t_b)(x) \Rightarrow t_b$, where "x" equals the number of packets per burst.

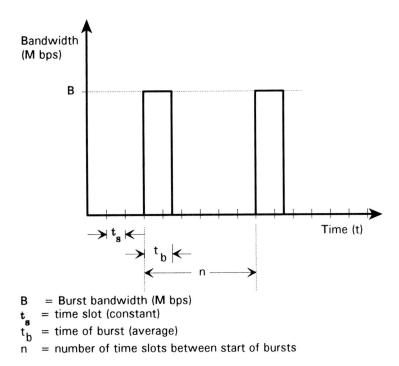

B = Burst bandwidth (M bps)
t_s = time slot (constant)
t_b = time of burst (average)
n = number of time slots between start of bursts

Figure 16.10 Characteristics of LANs Attached to MANs

Previous calculations can be taken one step further to model an IEEE 802.6 DQDB MAN where the traffic characteristics of the individual LANs attached are known. The required capacity of a DQDB MAN to handle all user traffic is calculated with the sum of the λs (packets per second) of the local, remote traffic from and to the MAN, and the pass-through traffic. The formula reads as:

$$\sum \lambda_l + \sum \lambda_{rf} + \sum \lambda_{rt} = \lambda'$$

where $\sum \lambda_l$ = local traffic,

$\sum \lambda_{rf}$ = traffic *from* remote MAN,

$\sum \lambda_{rt}$ = traffic *to* remote MAN,

such that $\lambda' = 1 / (n)(t_s)$,

where all λs are the sum of the users in that category and λ' represents the minimum required capacity of the local MAN. The capacity of the MAN would obviously be higher than the capacity required.

Since MANs often provide high-bandwidth connectivity to a small number of users, the traffic approximations just discussed become valid (where aggregations tend to have Poisson distributions). Huge bursts on the MAN can dwarf the normal large packet transmissions normally seen on the LAN.

Now for a look at a LAN/MAN traffic model which allows for the approximation of the number of LANs, based on traffic characteristics, that can be attached to a MAN.

The probability that a LAN time slot will be busy (ρ_b) is calculated as:

$$\rho_b = \text{Avg. \# of slots consumed by burst} \ / \ n = (t_b/t_s)/n = 8(D_b)/[(B)(n)(t_s)] = \lambda' \ / \ \mu$$

Suppose a DS3 DQDB network has the following characteristics:
    B = Class of Service (4, 10, 16, 25, and 34M bps) = 16M bps
    t_s = 53 octet time slot of 125 μsec
    n = 5
    D_b = 1024

The probability that a slot will be busy = $[(8)(1024)] \ / \ [(16\text{M bps})(5)(125\mu\text{sec})/13] = .8192 = 81.92\%$.

and the probability that a LAN will transmit onto a particular MAN slot, in theory, is:

$$\rho_m = (\rho_b)(\rho_{\text{Inter-LAN}})$$

where $\rho_{\text{Inter-LAN}}$ represents the fraction of inter-LAN bursts.

If there are N LANs connected to the MAN, the probability that k out of N LANs will be transmitting bursts onto the MAN at any given time is represented by the binomial distribution:

$$\binom{N}{k} \rho_m^k (1-\rho_m)^{N-k} = \rho_N(k)$$

Assume that the MAN is "y" times faster than each LAN (e.g., y = 4 for a DS3 SMDS MAN connecting to multiple 10M bps Ethernets).

If the number of LANs currently sending bursts exceeds y, then some or all of the LAN bursts will have to be queued or discarded. Approximating the DQDB MAN as an M/M/1 queue with average utilization (or throughput) of:

$$\rho_{MAN} = (N)(\rho_m) / y,$$

then the average M/M/1 delay is proportional to $1 / (1 - \rho_{MAN})$.

We can optimize the ratio of throughput to delay, usually called the queuing power (P), as follows:

$$P = \text{Throughput/Delay} = \mu (1-\mu) = [(N)(\rho m) / y] \cdot [(N)(\rho m) / y]^2$$

The queuing power versus network utilization is plotted in Figure 16.11.

If $\partial P / \partial N$ is set to zero ($\partial P / \partial N = 0$), the optimum number of LANs (N_{opt}) can be solved for which can be connected to the MAN:

$$N_{opt} = y / 2\rho_m$$

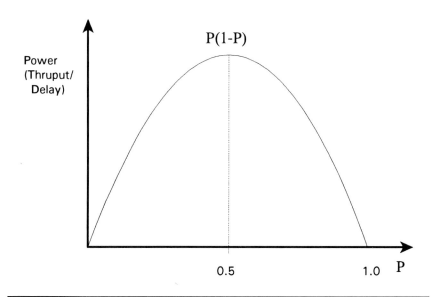

Figure 16.11 Queuing Power versus Network Utilization

16.3.9 Modeling the Unknown

When all approaches are exhausted, use the method most carriers and service providers are now using: over-build the capacity of the network, watch for degradation, blocking, and data loss, and add more (or less) capacity as required.

16.4 PERFORMANCE

Many parameters affect the performance of a switched data network. As a general rule, all conversion between protocols, speeds, switching techniques, and connectivity types can lead to performance degradation.

16.4.1 General Performance Issues of Packet Networks

Regardless of how fast the packets enter the network switch, they still go into a queue before being switched through each node. If the queue is not empty, the node will buffer the packet until it can pass through the queue and secure available bandwidth. The amount and length of queuing is directly related to the delay through the network. Some packet switches degrade in throughput as their queues fill up and overflow, and this degradation causes a loss of data due to congestion. Figure 16.12 shows the optimum operational point between throughput and response time based on an increase in load (utilization).

Errors in network transport can also cause more queuing and delay. Overhead is a major consideration as well. The overhead incurred per packet can be anywhere from 64 to 256 bits per packet. Since typical messages are 256 to 1028 bits, overhead could account for 25% of the total transmission bandwidth, and this would reduce network efficiency.

Performance in a packet switch network is measured by throughput in relation to the delay incurred from switch ingress to egress. The previous Figure 16.12 illustrates the throughput and delay characteristic curves which are usually supplied by any prospective vendor. Packet delay through a typical packet switch node is 50 to 200 milliseconds due to packet processing. Typical packet processing per node is anywhere from 300 packets per second on the low end up to 10,000 packets per second on the high end. Look for switches that have constant packet per second processing at all levels of traffic throughput (although there is always a point where a switch is line rate limited). Also, the switch performance should be as near constant as possible irrespective of packet size. Some packet switches drastically degrade performance and packet processing as the packet size decreases below 64K or 16K. Larger packets are characteristic of batch processing, which uses packet-switched networks because dedicated or leased-line circuits cannot be cost-justified. This type of traffic absolutely requires good performance and constant throughput, regardless of packet size.

16.4.2 Protocol Implementation

Protocol conversion can degrade performance. The greater the amount of protocol conversion the switch or PAD has to perform, the lower the throughput and performance. This is primarily due to the CPU resources, memory, and resources required to perform the additional tasks. As the amount of PAD processing increases, the throughout of the service degrades because of the increased processing power required. These considerations will be discussed as they relate to the design of packet-switched networks in later chapters.

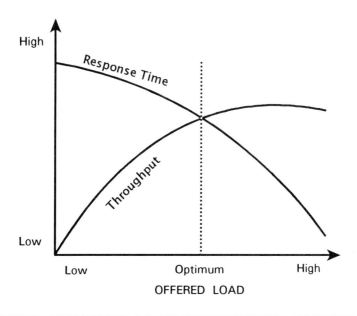

Figure 16.12 Typical Vendor Throughput versus Delay Curves

Performance can also degrade due to incomplete implementation of OSI layer protocols, insufficient use of transport layer protocols, and using remote access protocols rather than sharing resources on the local LAN. As the number of protocol layers increases, so does the amount of overhead, delay, and network complexity. It becomes increasingly more difficult to perform fault isolation and network (database) synchronization.

LAN and WAN protocol speeds must be compatible, and the network backbone speeds should be at least as fast as the offered LAN protocol speed, or else delay will occur. For example, the FDDI transport WAN must be as fast as the TCP/IP and other protocol conversions on the interface card. TCP/IP encapsulation overhead, packet reads, and CRC checking and verification will all add overhead — in fact all high-level protocols will add overhead which, in turn, decreases total data throughput. This overhead can be decreased by using larger packet transfers for large graphics files and bandwidth-intensive applications. This works best when performing bridging and routing of MAC-level protocols. For example, LAN packet sizes at least 512 bytes should be used, and faster filtering and forwarding will be needed to take advantage of FDDI 100M bps speeds.

16.4.3 Speeds and Connectivity

Speed mismatch on LANs/WANs can have drastic effects on performance. This can lead to congestion and loss of data, usually at the slowest link. The phrase, "the network is only as fast as its weakest link" applies here. In addition, switching between connection-oriented and connectionless services can cause performance problems if not managed properly.

16.5 DESIGNING FOR PEAKS

When calculating the arrival rate of calls (λ), a figure was used called "busy hour". Any network must be designed to handle the busiest traffic periods, or data will be blocked or lost. Take for example the IXC long distance carriers who need to design networks based on the nation's busy hour. How long would you keep your long-distance carrier if there was a good chance that your calls would be blocked from 9 to 10 AM each day (one of the nation's busiest hours for phone calls)? The IXCs specifically design their networks to handle periods of peak usage (such as Mother's Day — the busiest calling day of the year). Data networks must be designed with the same thought process.

16.5.1 Standard Busy Hour Calculations

When λ is calculated and applied to the formulas for traffic load, it is assumed that the designer will use a call arrival rate measured or estimated during the peak busy period. This period is called the busy hour, and an hourly measurement is fine for most voice networks.

16.5.2 Data Equivalent of Busy Hour

Data networks have their own equivalent of busy hour. Take, for example, the New York Stock Exchange, which each day after 4 PM Central Time transmits huge amounts of information about the day's transactions across the world. Or the data center which performs its switch/billing/file updates at 2 AM each day. These are the times when the highest throughput is required for these networks. Thus, a network is designed to accommodate the peaks of its particular "busy hour". Often a cost trade-off is used to balance the need of a larger network compared to buffering or delaying transmissions.

The other major component of data is its burstiness. The bandwidth allocated to a phone call is usually the same: 56K bps or 64k bps for a standard voice channel, and down to 8K for compressed voice. But in data, an average user may transmit anything from a 2K bit text file to a 2M bit graphics file. To complicate matters, multiple users may be simultaneously accessing a MAN which, in turn, uses the WAN, and peak traffic patterns could vary drastically with virtually no notice. Because of these characteristics, data traffic can be analyzed on a busy hour basis, but is more accurately represented with multiple snapshots on a "busy minute" or even "second" basis.

Now to take a look at some methods of calculating data busy hour, keeping in mind that if applications warrant, these calculations could be made in minutes or even seconds.

busy hour PPS = (busy hour packets sent + busy hour packets received) (3600 seconds / hour)

busy hour bps = (busy hour PPS * packet size) * 8 bits

link utilization = BH bps / Speed of link

for example,
BH PPS = (5 PPS + 5 PPS)(3600) = 36,000 PPS
BH bps = (5 PPS)(128 Bytes)(8 bits) = 5120 bps.
Link Utilization = 5120 bps / 9600 bps = 53.33% utilization

Chapter 19 will show how busy hour statistics fit into the access node design.

16.6 DELAY

When a network begins to slow down because of buffering, retransmissions, and/or any other time-affecting phenomenon, the network begins to experience delay. Delay will cause response time and throughput to degrade, application time-outs and retransmissions, and may even cause users to loose data. Delay can also affect performance.

16.6.1 Causes of Delay

While actual transmission path delay does contribute an element of delay to the network, the primary contributors to delay include:

+ propagation path length
+ line speed
+ number of access hops
+ hardware and software interfaces
+ load on every component across the path
+ elements traversed (each adds delay)
+ windows sizes
+ bit-setting selections (e.g., D-bit)
+ memory and buffers
+ pad functions
+ address database look-up
+ address verification
+ addition of new users
+ filtering, forwarding, processing packets, frames, cells

Most network design tools routinely model delay, but they also need to account for the components of delay listed above. There are also many ways to decrease delay by decreasing network load, capacity, or routing, as well as adjusting device-specific tuning parameters.

16.6.2 Circuit-, Message-, Packet-, and Cell-Switching Delay

In circuit switching, the delay incurred by a user is based upon the message size and the time it takes to transmit it over the available bandwidth. For example, if a graphics file of 2Mb is being transmitted over a 128K frame relay trunk, the total delay in transmission (assuming a zero network delay element) would be:

2M bits / 128K bps = 15.625 seconds

Message switching delay closely resembles that found in frame relay networks. This is where the total delay is calculated the same as in circuit switching, but multiplied by the number of nodes it must traverse or transit, minus one. For example, assume that a user is transmitting the same 2M bps file through the 128K bps port on the user device. The frames are being passed through a four-node frame relay network (not counting the origination and destination node), with the middle two nodes are passing the information at T1 speeds. The total delay would be [(2)(2M bits / 128K bps)] + [(2)(2M bits / 1.544M bps)] = 33.84 seconds. This calculation assumes that the user has the entire bandwidth of each link (and may vary for different implementations of frame relay).

Packet switching delay is based on many of the packet switching calculations examined above. The total message delay is calculated as:

$$[(p / c)(n + 1)] + (r / c)$$

where p = packet size of first packet
   n = number of nodes
   c = capacity available in the transmission medium
   r = remaining number of bits to be delivered in message

For example, if the 2M bits graphics file is transmitted over the four-node network connected by 56K bps trunks, using packet sizes of 1024 bits, total delay in transmitting the entire graphics file would be [(1024 / 56,000)(4 + 1)] + (1,998,976 / 56,000) = 35.79 seconds.

Cell-switching delay best resembles packet-switching delay. There is some queuing (as in packet switching) but with a *fixed* number of cells. Data exceeding the available throughput is discarded, with no retransmissions. The entire frame/cell needs to be completely read and processed before it can be transmitted out.

16.6.3 Data Services Delay

Sometimes delay components are not readily visible. For example, a PAD receiving asynchronous transmissions may packetize this traffic at 128 Bytes at a time, or it may wait until it receives 4K Bytes and then split the message into packets and transfer (store and forward). Thus, greater delay is incurred when an entire message must be read into a node before being transferred. This example is very important when designing frame relay networks. When a frame relay device receives a frame, it must read the entire frame into memory, calculate the CRC to verify the frame, and then transmit the entire frame on to the next node or to the user. The larger the frame size, the greater the delay incurred, and the more buffer space required.

Let us take a packet switch example and calculate total network delay. The total network delay is comprised of many elements, such as:

+ call setup delay (20 ms - Call RQ, Call CN, ENQ, Auth, Resp, Ack, Clear RQ, and Clear CF)
+ node delay
+ buffer filling time for characters (32ms per 4 asynch on 1200 bps line - so 8 ms for each access line)
+ packet formation time (50ms)
+ processor instruction execution (11-33 ms)
+ each hop across network (e.g., if utilization on a 56K was 10%, the delay would be 28 ms)
+ modem-to-modem (80 ms) initiate
+ modem-to-modem terminate (10 ms)

for a total network round trip delay on a three-hop network of $(2)[(20)+(8)+(28)] + (3)[(50)+(11)] + 80 + 10 = 385$ ms, best case. It is important to note that the modems in this example contributed to much of the delay, while workstations in a frame relay or LAN/MAN environment would contribute very little delay.

Delay will be handled differently in a frame relay network, where the result depends on the how the user applications higher-level protocols deal with delay. Most services, such as frame relay, will state a guaranteed figure for delay, such as the percentage of PDUs/frames which are delivered within a specified time (ms). For example, 90% of PDUs delivered within 1 ms, the rest delivered within 5 ms. This figure is driven by both what the user application requires and what the network devices can provide. Another example is buffering, which tells the user how many packets or frames/PDUs can be buffered and at what size (including CRC check), as in the last example on frame relay frame size and its effect on delay.

With larger frame sizes, the service which is providing the transport may segment or loose the frame. Also, delay may be incurred in storing portions of frames in buffers until they can be received and retransmitted in their entirety. In frame relay, the entire frame is received and verified (CRC) before being retransmitted. There is also the danger of large frames being segmented by other protocols, such as IP, SMDS, and even ATM. This causes additional overhead the larger the frame size. The effects of protocol conversions, encapsulations, and translations on delay must be accounted for by the network designer.

16.7 AVAILABILITY AND RELIABILITY

Two major measures of hardware and software must be known by the network designer: availability and reliability. These values are found through vendor-provided calculations such as Mean Time Between Failure (MTBF) and Mean Time To Repair (MTTR). MTBF is calculated based on stress tests, the results of which are projected into the future, as well as through theoretical model projections, possibly compilations based on the individual parts that make up the system. Other methods of measuring availability and reliability performance will be explored also.

16.7.1 Availability

Availability is the amount of time the system is working when compared to the entire lifetime of the system. Availability is calculated by:

$$A_i = \text{Availability} = \frac{\text{MTBF}_i}{\text{MTBF}_i + \text{MTTR}_i} = \frac{\text{time system is working}}{\substack{\text{time system exists} \\ \text{between failures}}}$$

For highly reliable systems, this number should be at least .999, or 99.9%. This is commonly referred to as "three nines". Networks that provide public service offerings often have availability figures of 99.99% or higher. Every additional nine added increases the order of magnitude by 10, thus an increase from 99.99% to 99.999% is a drastic increase in availability. What does it cost for an extra 9? It may be more than you are willing to pay, as the cost increase may also be tenfold. Availability is related to available capacity, whereas a user always wants to have enough capacity to transmit messages (regardless of the message size).

Another way to look at availability is through its complement: unavailability.

$$U = \text{Unavailability} = 1 - \frac{\text{MTBF}}{\text{MTBF} + \text{MTTR}} = \frac{\text{MTTR}}{\text{MTBF} + \text{MTTR}}$$

When unavailability is calculated, the time the system will be unavailable can be calculated or, in other words, its probability of failure. The system is unavailable MTTR hours out of every MTBF + MTTR hours.

Another way to look at the number of failures during a given period is by the formula:

$$\text{Average \# of failures in } t = t / (\text{MTBF} + \text{MTTR}) \cong t / \text{MTBF}$$

where "t" is the number of hours of operation. MTTR is included in the formula because each failure must be repaired. When MTBF is much greater than MTTR (MTBF >> MTTR), MTTR can be omitted from the equation. Thus, if the MTBF was 1,000 hours (a very poor MTBF) and the number of failures must be determined within a year (8760 hours), there would be the likelihood of 8.76 failures that year, or .024 failures per day (almost one failure each month). If there are 100 nodes in your network, with .024 failures per day, it means that there are 2.4 failures per day throughout the network. Now it can be seen why an MTBF of 1000 is quite poor! Consider a good MTBF of 40,000 hours, which would calculate to one failure every 4.57 years! Now this 100-node network has less than 2 failures per month.

For multiple network elements with different MTBFs, as in a hybrid network, availability (A_i) and unavailability (U_i) would be calculated as follows:

For a serial network as seen in Figure 16.13, the availability with two devices would be calculated as:

$$A_s = (A_1)(A_2) = [(\text{MTBF}_1) / (\text{MTBF}_1 + \text{MTTR}_1)][(\text{MTBF}_2) / (\text{MTBF}_2 + \text{MTTR}_2)]$$

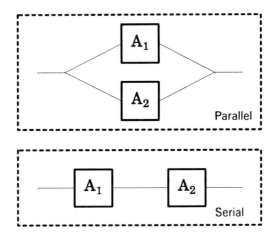

Figure 16.13 Availability

Thus, the greater the number of nodes, the greater the chance of network failure.

The unavailability would then be calculated as:

$$U_s = (U_1)(A_1) + (A_1)(U_2) + (U_1)(U_2)$$

For a parallel network as seen in Figure 16.13, the unavailability with two devices would be calculated as:

$$U_p = (U_1)(U_2) = (1 - A_1)(1 - A_2) = 1 - A_1 - A_2 + A_1A_2$$

The availability would then be calculated as:

$$A_p = (1 - U_p) = A_1 + A_2 - A_1A_2$$

Thus, the greater the number of nodes, the less chance of network failure. Meshing also has the potential to increase *network* or *system* reliability.

16.7.2 Reliability

Reliability is the distribution of time between failures. Reliability is often specified by the Mean Time Between Failure (MTBF) for Markovian failures. A high-reliability (MTBF) figure means that the system contains many reliable components which together constitute a reliable system. Reliability is specified as the probability that the system does not fail prior to "t" hours.

Reliability = e $^{-t / (MTBF)}$

Take a sample MTBF of 10,000 hours and a time interval (t) of one year (8760 hours). The reliability would be 41.64%. If we were to increase the MTBF to 20,000 hours, the reliability jumps to 64.53%. For multiple network elements with different MTBFs, as in a serial hybrid network (such as a bus environment), the total failures calculation would be:

R(total) = e $^{-t((1/MTBF1) + (1/MTBF2) + (1/MTBF3))}$

As an example, if we had MTBF figures of 20,000, 25,000, and 30,000 hours, respectively, measured over one year, the network would yield a 34% reliability! Thus, reliability can also be cumulative, but is always as weak as the weakest network element and always decreases as network devices are added. It can be seen that reliability is a trade off with redundancy when adding additional network resources.

Two additional performance measurements are:

Mean Time To Repair (MTTR). After finding a problem, how long does it take to isolate and repair. This figure is usually stated at less than 15 minutes.

Sometimes used: Mean Time Between Service Outages (MTBSO). How long (historically) it has been since the system has been down for a service outage. This figure is usually stated at greater than 2000 hours. Over 2000 hours translates to (250 days * 8 hours per day) 2.5 hours down time per year.

16.7.3 Additional Performance Measurements

In service or maintenance contracts, negotiated penalties apply if performance objectives provided by the vendor are not met. Also, each network performance parameter must be tested before placing users or applications on it. Not only must the level of error tolerability be determined, but also how much sharing of resources and capacity is tolerable. Some common methods of measuring performance on packet-, frame-, and cell-switched services include:

- less than X packets, frames, PDUs among Y are:
  - delivered without error
  - delivered to wrong customer
  - are lost
  - are duplicated
- total lost frames/packets/cells
- total network delay
- guaranteed aggregate throughput per channel based on characteristics of data (ex., 1.22M bps per T1 using 16-Byte frames)
- guaranteed overhead limits (based upon packet/frame/cell size)
- measure of DDS service in error free seconds (99.99% desirable)
- load balancing/load sharing limitations
- masking, filtering, and forwarding rates for access and backbone
- error detection/correction effects upon overhead, throughput
- level of redundancy built into network

16.7.4 Plan for Failures

Is there extra capacity to allow for failures in the network backbone nodes, access nodes, links, network hardware, and network software? Make sure that the system is designed to survive failures. Select hardware and software with high availability and reliability figures. Implement designs that minimize failure weak points by adding additional redundant subsystems. It may also be prudent to force a failure early in the life of the network (or before it goes "live") to determine if fault-tolerant system backups are working properly.

16.8 PLANNING FOR FUTURE CAPACITY

Capacity planning was once a 5- to 10-year cycle. It is now a 1- to 3-year cycle, and many times even shorter. There are two elements to the plan, a *short-term* objective and task-oriented plan usually revised each year, and the *long-term* three- to five-year plan, which should also take into account the data directions for the next 5 to 10 years. As the cost of computing hardware decreases, the entropy of capacity requirements increases. This makes capacity planning for data communications networks a challenging task. The results of short- and long-range capacity planning can provide the corporation with a competitive advantage, assuming that the plans fit in with the long-range corporate strategic plan. Each design must provide the flexibility to change technology as business needs change without major network restructuring. Both plans must take into account the business needs, customer needs, and the technologies.

One method is to migrate multiple network platforms and plan for a consolidated network platform that conforms to industry standards. Decide early on whether to maintain the existing architecture and build upon it, or plan to build a new integrated network architecture in the future. One of the best architectures is that which is built in a hierarchical nature and can be replaced in layers over time. Plan for port expansion, CPU capacity, hardware storage requirements, memory storage requirements, and many other critical growth areas discussed in this chapter. Many design tools have the capability to model future network changes and assist in both the short-term and long-range capacity planning process.

16.9 REVIEW

Going beyond the voice world and standard circuit-switched traffic modeling, the traffic modeling of data services and technologies such as packet switching, frame relay, DQDB (SMDS), and even ATM was introduced. The traffic matrix, or possibly even multiple traffic matrices, allow the calculations to determine requirements and sizing of access nodes. This has identified capacity needs, including how much capacity, what the utilization will be, how performance will be affected by these decisions, and how to improve performance. The network has been sized to accommodate peak traffic conditions, and to determine what level of delay will be tolerable. The availability, reliability, and other performance measurements required of vendor products has been identified with which to build the network. Finally, an organized plan was set forth which accommodates future growth in capacity. Next, the best technology must be chosen to suit network access requirements, and a choice of the vendor(s) to supply this technology established, and then the access network design can be completed.

17

Comparisons — Circuit vs. Packet vs. Frame vs. Cell

The next step in the design process is to choose the technology or hybrid of technologies to use in the access and backbone portion of the network. Previous chapters have covered a working knowledge of both circuit and packet switching technologies, the standards which define them, and the services which can be offered. This knowledge will be coupled with the user's requirements and the type of capacity required from the network. The major technologies have been discussed; circuit switching, X.25 packet switching, frame relay, IEEE 802.6 and SMDS, B-ISDN and ATM, and SONET. This chapter draws comparisons between these technologies and services, their protocols and interfaces, and details of each switching technique. The access and backbone portion of the network design can now be accomplished with the best mix of technologies to meet the user's requirements and to ensure the correct capacity requirements for both access and backbone.

TABLE 17.1 Technologies and Services Comparison

TECHNOLOGY	SERVICES
Circuit Switching	High-Speed Circuit Switching (HSCS), Digital Reconfiguration Service (DRS), other Switched Services (e.g., Switched 56/64K, FT1, T1), DRS, ISDN
Packet Switching	X.25 (network) and X.75 (internetwork and gateway services)
Frame Relay	Frame Relay, X.25
Cell Relay [DQDB (IEEE 802.6)]	SMDS (CL 802.6)
FDDI	FDDI, FDDI-II
ATM	B-ISDN, SMDS, ISDN

17.1 TECHNOLOGIES AND SERVICES

The previous chapters have given a good idea how the technologies and services defines in the last two sections relate. Table 17.1 shows the technologies and services that we will be comparing in this chapter. Note that many of the more advanced technologies support many of the earlier services (backward compatibility).

The major technologies, standards, and services discussed in this text derive from two major modes of switching: circuit and packet. Circuit switching evolved from manual operators, to basic patch panels and DXCs, to devices which perform software circuit switching for services such as Switched 56/64K. These services range up to the high-speed circuit switched (HSCS) services at T1 and T3 speeds. One example is the HSCS service offered by MCI Communications — VPDS 1.0.

The other major mode of switching is packet switching. Packet switching is broken down into traditional packet switching operating over X.25 protocols and "fast packet" technologies such as frame relay and cell relay. Figure 17.1 shows the evolution of circuit and packet switching services for data communications.

17.2 PROTOCOLS AND INTERFACES COMPARISON

Throughout the past few chapters, multiple interfaces and protocols were discussed for each of the switching technologies. Each protocol and interface has been related to the Open Systems Interconnect

Reference Model (OSIRM). Figure 17.2 combines the OSI protocol stack with the varied protocols and interfaces discussed so far.

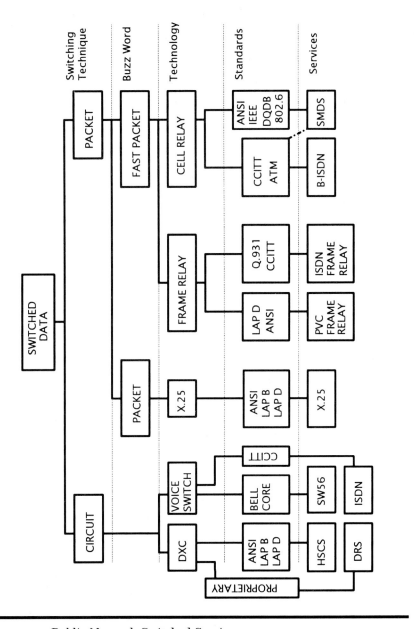

Figure 17.1 Public Network Switched Services

OSIRM	PROTOCOL OR MEDIA
Application	Operating Systems, Applications, File Management Protocols
Presentation	NCP, NETBIOS, NETBeui, FTP, NFS, FTAM
Session	SPX, NETBIOS, NETBeui, FTP, SMTP
Transport	TCP, XNS, SPX
Network	IP, IEEE 802.1, X.25, ISDN, ARP / IPX, XNS, Ethertalk
LLC	IEEE 802.2 LLC, HDLC, SDLC
MAC	IEEE 802.X (.3 Ethernet, .4 Token Bus, .5 Token Ring, .6 L2_PDU/L3_PDU), FDDI X3.139
Physical	Copper/Coax/Fiber Media systems and interfaces

Figure 17.2 Protocols and Interfaces of the OSI Reference Model

OSIRM	SERVICES AND SPEEDS
Application	Distributed File Services, Messaging, Multimedia, Applications
Presentation	Distributed File Services, Messaging, Multimedia, Applications
Session	Distributed File Services, Messaging, Multimedia, Applications
Transport	TCP
Network	X.25, IP LAN/WAN Interconnect, SMDS, B-ISDN
Data Link	FDDI, SMDS, Frame Relay, B-ISDN, High Speed Circuit Switching
Physical	DS0, FT1, T1, T3, High Speed Circuit Switching

Figure 17.3 Services and Speeds of the OSI Reference Model

17.3 SERVICES COMPARISON

Many services were discussed relating to and offered on each technology. These services are shown compared to the OSIRM in Figure 17.3. Each of these services meets the following criteria:

- ☞ carrier based
- ☞ global
- ☞ fast
- ☞ ubiquitous access
- ☞ based on gigabit bandwidth technologies

In Figure 17.4 these services are shown in relation to their ability to carry various types of traffic.

17.4 ARCHITECTURE COMPARISON

Throughout the past several chapters technologies have been covered that fall into one of multiple vendor and industry standard architectures. Figure 17.5 shows a summary of the architectures discussed in Chapter 3, for example.

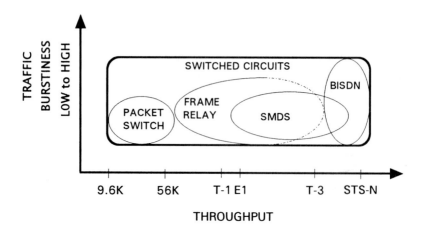

Figure 17.4 Switched Services Comparison

17.5 SWITCHING METHODS COMPARISON

This section is designed to compare circuit switching and packet switching (private-line circuit switching, X.25, frame relay, IEEE 802.6 (including SMDS implementations), and ATM (including B-ISDN). The primary categories of comparison are traffic types supported (Table 17.2), technology and service characteristics (Table 17.3), protocol functionality (Table 17.4), and hardware types (Table 17.5). Note that a "No" in any column does not necessarily mean that type of traffic cannot be carried, but it is not recommended. The headings for the table are as follows:

TrafficSup = Traffic Supported, Char = Characteristics, Prot = Protocols, Hdwr Types = Hardware Types, PL/CS = Private Line/Circuit Switch, X.25 = X.25 Packet Switching, FR = Frame Relay, SMDS/802.6 = SMDS/IEEE 802.6, BISDN = B-ISDN (ATM), Y = Yes - indicates support, N = No - indicates not supported, P = Poor, G = Good, E = Excellent, VL = Very Low, M = Moderate, H = High, VH = Very High.

17.6 ACCESS COMPARISON

The technologies and services reviewed provide various methods of access, but the predominant feature of network access is the capability to reduce the number of access circuits and thereby reduce network access costs — thus more effectively utilizing bandwidth. The number of access circuits required to support the same amount of bandwidth and traffic seems to decrease each year. This is primarily due to the evolution of technology. Figure 17.6 shows a comparison of the major technologies and services illustrating their ability to reduce access compared to private lines.

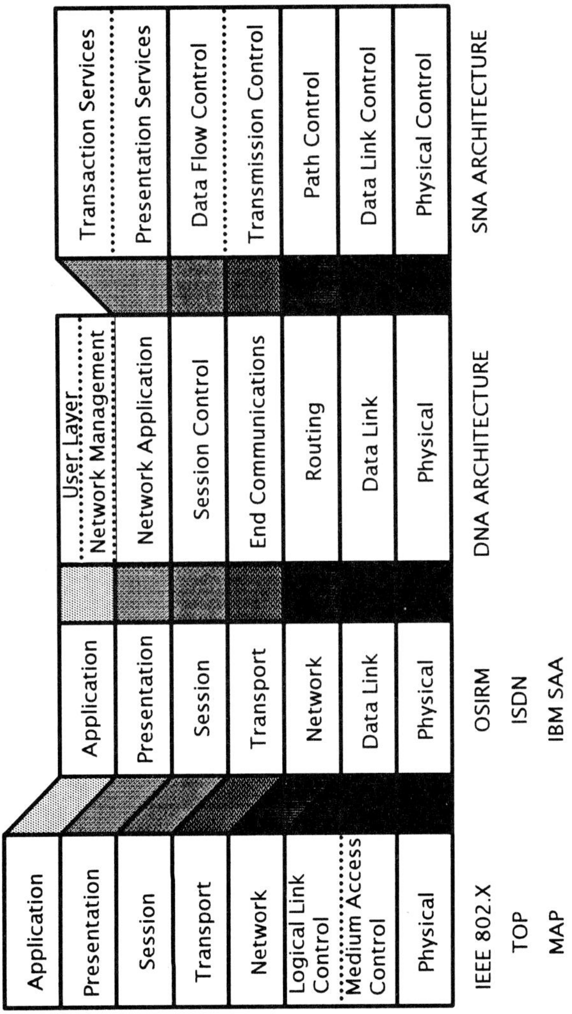

Figure 17.5 Architectures Summary

TABLE 17.2 Comparison of Traffic Types Supported

TrafficSup	PL/CS	X.25	FR	SMDS/ 802.6	BISDN
Voice	Y	N	N	N/Y	Y
File Transfer	Y	Y	Y	Y/Y	Y
CAD/CAM	Y	N	N/Y	Y/Y	Y
Transaction-Oriented	Y	G	G	E/E	E
Channel-to-Channel	Y	N	N	Y/Y	Y
Imaging	Y	N	Y	Y/Y	Y
Video	Y	N	Y	N/Y	Y
LAN-LAN	Y	Y	Y	Y/Y	Y
MAN-MAN	Y	N	Y	Y/Y	Y
Isochronous	Y	N	N	N/Y	Y
Bursty (w/small packets	P	G	E	G	G
Bursty (w/large packets	P	P	E	E	E
Primary Application	Inter-data center	PSDN	LAN-to-LAN	MAN, ESCON	WAN/ GAN HiPPI
Action Taken	passes	routes	forwards	routes	routes

TABLE 17.3 Comparison of Technology and Service Characteristics

CHAR	PL/CS	X.25	FR	SMDS	BISDN
Performance	VH	L	H	VH	VH
Speed	VH	L	H	VH	VH
Transport Delay	VL	H	M	L	VL
BW Efficiency	P	G	E	E	E
BW Usage	fixed	var	var	var	var
Throughput	E	P	E	E	E
Reliability	H	H	L	L	L
Nodal Processing Overhead	VL	H	L	L	L

TABLE 17.3 CONTINUED

CHAR	PL/CS	X.25	FR	SMDS	BISDN
International Connectivity	G	G	P	G	G
BW-on-demand	N	Y	Y	Y(true)	Y(true)
Circuit Types	dedicated	PVC, SVC	PVC, SVC*	CL	CL
Switching Unit	bit/Byte	packet	frame	cell	cell
Packet Type	-	LAP-B	LAP-D	IEEE 802.6	CCITT ATM
Packet Length	-	< 1024 bits	<8192 Bytes	53 byte	53 byte
Connect Time	2-4 seconds	none	none (PVC)	none (CL)	none
WAN Transport	DS0 - T3	56K	T1 (T3*)	T3,150M/ 300Mbps ATM	ATM
Physical Port Access	DS0 - T3	DS0, 128K, 256K	DS0, FT1, T1	DS1, DS3, SONET	DS3, SONET VTR, 155M bps
Transmit on copper	Y	Y	N	Y	N
Primary Problem	Expensive	Overhead/ Delay	Congestion Control	No Isochronous	Deployment
New Hardware Req'd	N	N	N	Y	Y
Suppliers	All	CPE/ CO	CPE/ CO	CO/ (CPE)	CO
Ubiquitous Access	N	Y	N	Y	Y
Available	Now	Now	Now	Now	1994
Orientation	CO	CO,CL	CO	CL	CL,CO
Evolution Capabilities	H	L	M	H	H
Point-to-multipoint	Y	Y	Y	Y	Y
CPE Cost Effective	Y	N	Y	Y	N

TABLE 17.4 Comparison of Protocol Functionality

FUNCTION	PL/CS	X.25	FR	SMDS	BISDN
Protocol Transparency	E	P	G	G	E
Call Setup	-	PVC	PVC	CL	B-ISDN Q.931
Primary OSI Layer	Layer 1	Layer 2	Layer 2	Layer 2	Layer 2
Service Layer	Layer 1	Layer 3	Layer 2	Layer 2	Layer 2
OSI Layers Used	Layer 1	Layers 1-3	Layers 1-2	Layers 1-2	Layers 1-2
Address Type	Q.931/ ASCII	X.121, E.164	DLCI	MAC (E.164)	VCI
Address Size	none	DTE: 56b+, VC:12b	2 octets+	60b	20-28b
Flow Control	N	Y	Y (some)	N	N
Congestion Control	P	E	P	P	P
Error Handling	user	node-node-user	user-user	user-user	user-user
Sequencing	N	Y	N	Some	Some
Logical connections per physical port	1	1000+	992	many	many
Error Tolerance	P	E	P	P	P

TABLE 17.5 Comparison of Hardware Types

HARDWARE TYPES	PL/CS	X.25	FR	SMDS	BISDN
Repeater	Y	N	N	N	N
Bridge	Y	N	Y	N	N
Router	Y	Y	Y	Y	Y
Gateway	Y	Y	Y	Y	Y
Packet Switch	N	Y	N	N	N
ATM Switch	N	N	N	Y	Y

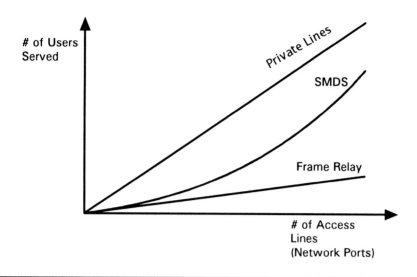

Figure 17.6 Reduction of Access Circuits Through the Use of Switched Services

17.7 TRANSPORT COMPARISON

Each of the technologies shown in the last table transports data differently. Figure 17.7 illustrates a comparison of private line, circuit, packet, frame, and cell transmissions. Notice in circuit switching that variable bandwidth is not available to the user, whereas in frame and cell switching, bursty data can be allocated the appropriate bandwidth-on-demand. Packet, frame, and cell technologies also allow for the capability to dynamically route around failures at the packet level, rather than at the circuit level, and circuit reconnect time is in milliseconds rather than seconds. Usually, in circuit switching, the entire call has to be reestablished in the event of a failure or disconnect.

Figure 17.8 shows a chart depicting how access and transport speeds have increased since 1988. The technologies discussed in recent chapters are providing the services to handle these increases. In 1990, the average transport speeds included DS0, T1, and some T3. In 1992, the speeds have increased up to STS-1. In 1993, average speeds may reach into the STS-1 range, and in 1994 STS-3 and STS-12 transmission speeds will be common.

TRANSPORT CHARACTERISTICS

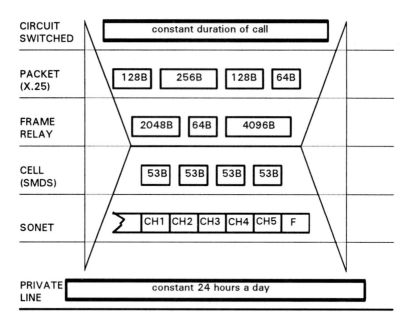

Figure 17.7 Private Line, Circuit, Packet, Frame and Cell Transmissions Comparison

These speeds can be compared to those of circuit switching, which are DS0 and T1 in 1990, T1 and T3 in 1991, and up to OC-3 in 1993. The greatest speeds will be achieved through packet, frame, and cell switching, which in 1990 accomplished T1 and T3 speeds through OSI and IP. In 1992 and 1993 IEEE 802.6/SMDS are reaching 150M bps and 300M bps, and will reach into the B-ISDN and SONET gigabit speeds after 1993 and 1994.

17.8 CHOICE OF LAN/MAN

Numerous comparisons of LAN and MAN protocols, interfaces, and services have been provided already. The choice of LAN/MAN technologies depends on many factors; some are listed below.

+ protocols to support
+ architecture to interface
+ traffic characteristics (steady or bursty)

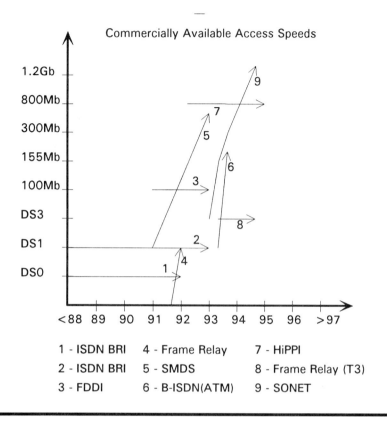

Figure 17.8 Access and Transport Speeds (1988 to Present)

- ✣ cost of hardware
- ✣ availability
- ✣ vendor
- ✣ speed
- ✣ line quality
- ✣ response time required
- ✣ network topology
- ✣ throughput requirements
- ✣ private or public connectivity

17.9   X.25 PACKET-SWITCHING vs. FRAME RELAY

Currently, there are three primary options for packet-switched services; X.25 packet switch networks, router-based LAN/MAN/WAN networks, and frame relay networks. Router-based networks can support many protocols, including X.25 and frame relay, but routers mainly provide encapsulation and translation through routing rather than true switching.

Now for a few comparisons of packet switching to frame relay. Most of these items are covered in the matrixes presented earlier in this chapter, but it is important to restate the major differences as they relate to network design. While some view frame relay as being only an extension of packet switching (merely a speed upgrade), retaining the dynamic capabilities of packet switching, packet switching can only truly be compared to frame relay at the LAPB/LAPD layer. X.25 packet switching covers all three layers of the OSIRM, while frame relay covers only part of the data link layer (LAP-D) in the Core Q.921 and upper Q.921 protocols. In packet switching, the intelligence (error checking, correction, recovery, retransmission, flow control) lies in the network nodes. In frame relay it resides at the user or network access hardware. While X.25 introduces more delay, it also provides for more accurate transmission and throughput. This accuracy is achieved through additional overhead, but is not required for transmission over fiber optic facilities. Also, while both technologies are ideal for bursty data, frame relay breaks the 56K, 128K, and 256K barrier with T1 and eventually T3 speeds, capable of handling delay-insensitive LAN-to-LAN, SNA, and image transfer types of traffic.

17.10 FDDI vs. IEEE 802.6/SMDS

The two primary MAN services currently available are FDDI (eventually FDDI-II) and SMDS. SMDS service was designed primarily for public service MANs, whereas FDDI was designed to provide for a campus backbone LAN environment. Shared FDDI MANs can be connected via OC-3 SONET pipes to form a WAN. SMDS is supported by the RBOCs and IXCs, whereas FDDI is supported primarily by CPE vendors. Also, FDDI does not provide address screening and security features found in SMDS, and also contains some distance limitations. FDDI-II may eventually correct the deficiency of isochronous services and allow FDDI to compete more effectively with SMDS.

17.11 FRAME RELAY vs. IEEE 802.6/SMDS

Comparing frame relay to IEEE 802.6/SMDS is like comparing an eagle to a penguin; they are animals of a different species, born and bred for different purposes. In this section a comparison of the more disputed topics of frame relay and IEEE 802.6/SMDS will be discussed.

Service. Frame relay is both a series of protocols forming a private network service and a public network interface, while SMDS (CL IEEE 802.6) is a public network service.

Access and Transport. SMDS access and transport speeds far exceed those of frame relay. Frame relay is an access technology, where IEEE 802.6 is both an access and a backbone technology. User information is packaged into variable-length frames (frame relay) or fixed cells (DQDB and ATM) with segmentation. This allows IEEE 802.6 to prioritize and switch isochronous data, which frame relay cannot do. Frame relay uses connection-oriented PVC statistical multiplexing, while IEEE 802.6 uses connectionless datagramlike statistical multiplexing. Frame relay provides a single access point, where IEEE 802.6 can offer shared access of DS3 access loops. IEEE 802.6 transport includes more overhead than frame relay.

Traffic. Frame relay can pass data traffic only, where IEEE 802.6 supports transport of voice, data, and video. The Bellcore SMDS does not support isochronous transport at this time. The DQDB cell relay technology can carry delay-sensitive traffic because it fixes the size of the cells and dedicates a portion of each cell to delay-sensitive isochronous traffic. Frame relay cannot support delay-sensitive traffic. IEEE 802.6 based services are cost distance-insensitive, provide ubiquitous connectivity and availability, and perform various protocol options and conversions, whereas frame relay cannot.

Congestion Control. Both frame relay and SMDS services have poor congestion control techniques. Frame relay uses the FECN and BECN and DE bits to notify the user to throttle back the data flow, whereas SMDS uses access classes to define maximum bursts of data. Both methods are inefficient.

Market Push/Pull. Frame relay service has been pushed by the vendors and users (user-driven) while SMDS service is being pushed by the standards committees, RBOCs, and IXCs (service- or industry-driven). Frame relay is now offered as a service by the IXCs and service providers, but primarily as an interface into a public data network. At DS1 speeds, they offer similar benefits. But they do not compete, but rather they are complementary.

Migration. Frame relay provides a good migration path to SMDS and B-ISDN, but the path is not a smooth one, and there is a much easier migration from SMDS to B-ISDN (ATM). Users moving from circuit or packet (X.25) to frame relay require few hardware and software changes, while the move to a cell-based service such as SMDS requires a much greater hardware and software CPE and network investment. Frame relay hardware is less expensive than SMDS and can be modified from existing hardware. SMDS access requires new hardware and software if confined to the CSU/DSU for L3_PDU to L2_PDU conversion.

Internetworking. Frame relay standards do not provide a standard for interswitch and internetworking, but SMDS allows it via the ISSI interface defined by Bellcore. This provision makes end-to-end interoperability easier with SMDS than with frame relay.

Hardware. Frame relay is found in routers and as multiplexer interfaces, whereas SMDS is primarily found in standalone 802.6 switches (Alcatel, Siemens, AT&T) or as part of a broadband ATM switch (NEC, Fujitsu, NTI). Nonintegrated 802.6 requires special CSU/DSUs for access, whereas frame relay uses existing CSU/DSU hardware. Integrated 802.6 implementations can use existing CSU/DSU equipment.

17.12 SMDS (IEEE 802.6) vs. B-ISDN (ATM)

The comparison of SMDS to B-ISDN is difficult, because SMDS is a service that can also be offered over ATM since it is technology independent. A technology-independent infrastructure is the basis of the primary mode of transport for B-ISDN services as defined in the standards. There seems to be more of a migration from IEEE 802.6 to ATM, with a parallel migration from SMDS to B-ISDN in the works. Thus, SMDS to B-ISDN has a migratory rather than a competitive relationship. While SMDS is primarily a MAN service, WANs can also be designed using SMDS. B-ISDN also fits this category. Both the 802.6 and ATM technologies package the user information into fixed cells, segmenting large chunks of user data into multiple cells. Isochronous data is prioritized and allotted a portion of each cell, which contains 48-octet data cells with 5-byte headers. These cells are then switched over the transmission network, which could be conventional asynchronous, ATM, or SONET transmission facilities.

Both IEEE 802.6/SMDS and B-ISDN are public network services, while SMDS provides an easy migration path to B-ISDN services (based upon similar cell structures). B-ISDN will support all media over the same channel, as will IEEE 802.6 implementations, and both are aimed at business and residential markets. In fact, the RBOCs and IXCs will support 802.6/SMDS, and are now evaluating ATM/B-ISDN (especially the IXCs). The need for multimedia and mixed services over a single access will allow IEEE 802.6 and B-ISDN to quickly surpass frame relay, with the additional cost of new hardware and software. IEEE 802.6/SMDS and B-ISDN can both provide a connectionless medium.

ATM is a connectionless technology, but supports both connection oriented and connectionless services. On the other hand, B-ISDN has the collection and distribution mechanisms. Either way, both SMDS and B-ISDN services over IEEE 802.6 and ATM fast-packet technologies provide switching, multiplexing, and transmission capabilities. In addition, they provide the high-capacity, packet-oriented dynamic call establishment, dynamic bandwidth allocation, and true switching that is now required by users.

Both SMDS-to-B-ISDN and X.25 to frame relay migrations should be smooth. Each have similar cell/frame/packet sizes, hardware, and many other synergism. Frame relay to SMDS, or to B-ISDN, will be much more difficult.

17.13 REVIEW

This chapter provides a summary of the previous technology and requirements chapters. The reader now has the tools to choose the appropriate technologies, interfaces, services, switching method, and hardware required for both access and backbone (transport). We also reviewed a few of the key comparisons of LANs to MANs, FDDI to SMDS, frame relay to SMDS, and SMDS (using IEEE 802.6) to B-ISDN (using ATM). The next step in the design process is to choose the vendor.

5

Choosing the Vendor

Choosing a vendor is like getting married. Proposals for a relationship are extended through solicitations (RFIs and RFPs) providing a list of requirements by which to build a long-lasting relationship. The vendor (the user's future partner) responds to the RFI/RFP with a list of what they can provide, when, and at what cost. The capabilities of each vendor are examined carefully, and ultimately one is chosen (or, as in some cases, multiple patrons are engaged) to build the network. The contract is written and the future (network) is committed to the vendor as the vendor commits their future services to the user. Both find out how they must live with each other. Vendor partnerships are often for the life of the network, although the capability of divorcing the network from the vendor may become necessary if service becomes poor or the vendor priorities change.

18

The Vendor Selection Process

Now that requirements have been defined, the technology chosen, and the high-level design process has begun, the vendors must be chosen. Vendor selection should be done with care. The network and network manager's job may directly relate to vendor performance. In this chapter, the Request for Proposal (RFP) process is summarized. Requests for Information (RFI) may have been issued as well, but it is now time to select vendors to propose specific solutions to specific network requirements. Conformance to standards, the business parameters of the company, and the industry as a whole is critical. The RFP is the one chance to "get it all in writing". If a requirement or conformance is not included in the RFP, the vendor does not have to provide it. The watchword is: be thorough in the selection and contract process.

Vendors must be chosen based on weighted technical and business needs: what trade-offs are critical, assured multiprotocol and multimedia support from the vendor, industry standardization in the equipment purchased, and how to ensure vendors will deliver on their promises. The vendor-user relationship must benefit the strategic and technical direction of both parties to be a lasting relationship. In this chapter, both end users and network designer are "users" — when considering vendor relationships. Vendor responsiveness to user needs is critical. One critical support structure which vendors must provide is network management. Vendor selection can affect the future business of the company.

The network is designed, the purchase orders are written, and the network sites are awaiting the equipment's arrival. Vendor XYZ promised that the hardware would arrive in six weeks. It has now been over three months and still no equipment. The recent trade magazine shows that vendor XYZ is having financial problems. A call is placed to the sales representative at vendor XYZ. They tell you, "no problem, the stuff is on its way, you are only hearing rumors". A week later the hardware arrives and the bill is paid. The next month, XYZ goes out of business. When finally the equipment you bought is up and running, it does not work properly. When it is interfaced to another vendor, it does not work because the interfaces are proprietary. The network is eventually scrapped, the network designer loses his job. This story, of course, is over-exaggerated to emphasize the point that a thorough vendor analysis must be completed prior to choosing a vendor. This chapter helps separate promises from reality. Don't be an XYZ fatality!

18.1 THE REQUEST FOR INFORMATION (RFI)

A Request for Information (RFI) is designed to get enough information from suppliers for the user to eliminate suppliers who cannot provide the needed requirements and thus arrive at a "short list" of suppliers from which a final candidate is chosen. The RFI simply offers prospective suppliers a version of the end result desired and allows them to provide their style of solution. This is the "feeling out" stage, and many vendors can be provided an opportunity to respond. There are usually parameters placed on the technology, and user guidelines are offered. The user will typically provide:

- traffic volumes and characteristics (as explained in this chapter)
- existing systems and technologies to replace or integrate
- network end objectives
- future plans and requirements
- evaluation criteria

The user will typically request from the supplier respondents:

- pricing
- vendor capabilities
- complete documentation

18.2 THE REQUEST FOR PROPOSAL (RFP) PROCESS

After the RFIs are evaluated, the number of vendors is narrowed down. It is now time to begin the Request for Proposal (RFP) process. Generally the rule is that RFIs go out to only 10 to 15 vendors, and RFPs to 3 to 5 vendors. RFIs are used

- to clarify buyer needs
- as a legal agreement of vendor provisions and capabilities
- to weed out competition
- to solve pricing discrepancies

RFPs are a means of requesting multiple vendors to formally commit detailed solutions, propose the appropriate technology, and quote services to satisfy designated requirements. Before issuing the RFP, the buyer must

- understand the technology in the bid
- understand all of the alternatives available
- properly define what is required from the vendor
- clearly define the technology, business, and financial issues with priorities
- clearly define support issues such as those defined in this chapter

Make sure that user requirements are clearly understood. What does the RFI from the vendor provide? All correspondence should be thoroughly documented. This may seem harsh for two businesses who are trying to establish a lasting relationship, but it is necessary to build a professional relationship and to avoid later finger-pointing.

During the RFP process, try to have vendors with sales, marketing, and technical contacts respond to the RFP. The intelligent user will work with vendors to complete RFPs. This stage is very important to avoid confusion and misrepresentation of a vendor who may actually turn out to be the best choice.

18.2.1 The RFP Structure

RFPs vary based upon the systems being bid, from simple requests for a specific piece of hardware or software to an entire communications system complete with vendor support and possibly even an outsourcing agreement. A general RFP format includes:

- Synopsis or objectives
- General terms and conditions (explanations/interpretations)
- Qualifications of bidder
- Current systems, technologies, architectures, protocols, services
- Hardware and software technical specifications
- Equipment specifics and environmental requirements
- Delivery schedules
- Vendor required engineering and installation
- Implementation plan
- Service support
- Equipment substitutions
- Warranty requirements
- Summary of financial bid
- Work and responsibilities
- Proposal assumptions
- Training
- Vendor credibility and references
- Legal considerations

18.2.2 Analyzing the RFP Response

Each respondent to the RFP should respond paragraph by paragraph. Responses will be varied, but this seems to be the best format. Beware of terms like "substantial compliance", "intend to comply", and "will be contained in future releases". Look for terms like "full compliance" with an explanation of how they comply and to what extent. Replies should contain all one-time and long-term costs identified, including costs for:

- price of product
- maintenance
- training
- learning curve
- on-going expenses

- tax financing
- contract termination fees
- warranties
- documentation
- purchase or lease decision

These factors should be included in the vendor response. In the next section each factor will be examined for a determination on which vendor to choose.

18.3 CHOOSING THE VENDOR(S)

There are many reasons for choosing one vendor over another. The vendor is not chosen only on compliance with industry standards or the ability to meet the users needs. Factors such as politics, business, financial, and operational needs, future position of the vendor, goals of both businesses, and a variety of others may outweigh the technical abilities of the vendor. This is particularly true of a larger corporation or user dealing with a large national or international vendor.

The relationship established with the choice of a vendor must be both practical for the business needs and affordable to both the user and vendor. It is sometimes termed as "sleeping with the vendor". You have to know who you are sleeping with, just as you would know who you were marrying. Look into the vendor's past for not only the product you are purchasing, but financial history, past customer base, and industry experience.

Some evaluation criteria for these hardware devices includes the vendor history, the product versatility, processor and packet-processing speeds and capabilities, LAN and WAN interface and protocol support, local or remote support ability, OSI layer and proprietary protocol support and operation, performance characteristics, features and functions (including network management and security), and price. Remember, price is often not the most important evaluation factor, because the manpower and resources committed to the network often exceed hardware costs.

18.3.1 The Requirements Matrix and Weighting Methods

Once a requirements matrix has been established, a weight is assigned to each requirement based on its importance. More important criteria are given a higher weight, less important a lower weight.

Some requirements such as price are often weighted heavier than others, such as security features.

Table 18.1 shows a sample weighting of ten typical requirements for an RFP specifying a three node router network. Each was provided a weight from 1 (1%) to 99 (99%), with a total weight of all requirements of 100 (100%). Then each requirement was provided a rating from one to ten, with ten being the highest score and one the lowest. In the table below the vendor scored a total of 650 points out of a possible 1000 points.

TABLE 18.1 Requirements Matrix Weighting

REQUIREMENT	WEIGHT	SCORE	TOTAL PTS
total costs	20	8	160
features and functions	10	10	100
vendor reputation	2.5	4	10
product reputation	2.5	8	20
architecture flexibility	5	6	30
system capacity and performance	10	7	70
reliability	15	4	60
availability	20	6	120
service aspects	10	4	40
installation	5	8	40
TOTAL	**100**	-	**650**

The weighting could be changed to provide an equal weighting for all requirements, or more requirements could be added. Either way, the vendor with the highest number of total points should be chosen. Another method of requirements analysis is the cost-benefit analysis, where the costs of each requirement are weighed against the benefits brought by the requirement. This is similar to the weighting method shown previously.

Some companies perform an exposure analysis, which matches the expenses if the contract is a total failure. This includes both the capital expenditures and the operational and personnel costs involved in the project from conception. Maximum exposure is usually measured from 1 to 2 years with the vendor stretching out delivery and then finally not delivering, or worse yet, going bankrupt after spending the user's prepaid money.

18.3.2 What Trade-offs are Critical?

The weighting method is an excellent way to determine which vendor to choose. But how can it be determined which trade-offs are critical and warrant higher weighting factors? In this section a few major topics are micro-analyzed. These lists are by no means comprehensive but do represent the most commonly asked questions.

COSTS INVOLVED

- price performance ratio (PPS and cost per PPS)
- modularity expansion costs
- price per port

FEATURES and FUNCTIONS (PROTOCOLS and INTERFACES)

- security
- filtering capability
- route policy enforcement
- bus speeds
- PPS processor speed
- packet forwarding rate for each protocol
- protocols that can be routed/bridged
- memory required for each protocol
- physical interfaces required/supported
- bridging/routing protocols supported
- local- and remote-load software for operation
- good mix of protocol support
- mix protocols on same circuit
- in-band network management
- multiple fully redundant, load sharing power supplies

VENDOR and PRODUCT REPUTATION

- industry leader or follower
- references and recommendations

ARCHITECTURE/PRODUCT FLEXIBILITY

- ability to migrate to future technologies
- proven architecture which will continue to evolve
- fault tolerance
- standards-based or proprietary

- future investment or concrete
- large range of internetworking products or single solution source
- future investment protection by standards-based
- "plug and play" compatibility with all modules
- future product line migration path (no disjointed software or hardware upgrades/revisions/fixes)
- expandability of chassis
- commonality of common logic across entire product line
- recent new card upgrades compatible with old
- cards work together in the same box/software release
- fewer types of cards and more reliance on software to change protocol (decreases sparing plan and maintenance costs)

SYSTEM CAPACITY and PERFORMANCE

- system architecture
- PPS performance for all interfaces
- packet loss rate (with and without acknowledgment)
- throughput
- PPS transmission over loaded ports, cards, and nodes

RELIABILITY and FAULT TOLERANT HARDWARE

- redundancy in all components and systems
- parallel backplanes/buses for reliability
- daughterboard expansion built in for future processor speed upgrades
- independent load sharing over data paths
- redundant power supplies
- redundant common logic boards
- redundant local media
- hot swap functionality (replace cards while on-line without service interruption)
- dynamic reconfiguration while on line
- no single point of failure
- isolation of failed components, cards, or buses
- for FDDI, support of optical bypass and redundant rings
- used EEPROMs increase boot time after failure

AVAILABILITY

- fault tolerant (no single point of failure)

- multiple processors per interface card (look at processor MIPS, bus speed, PPS processing, memory, etc.)
- symmetric multiprocessing (gives scalable performance which can be both sustained over long periods of time and routing tables and protocol conversion is done at port level and no bus transfer needed
- easy device configuration and customization
- maximum network availability

SERVICE ASPECTS

- service levels and customer support
- small vs. large corporation
- support staff size and skillset

INSTALLATION

- ease of software installation
- documentation
- capabilities

There could be a large equipment base already existing in the corporation that will need integration and interoperability. If this is the case, an integration requirement must be factored into the requirements analysis. This is one situation where multiprotocol or multimedia support will be required. In this instance, chose a vendor who has experience in both.

18.3.3 Adherence to Industry Standards

In today's age of data communications standardization and open systems, it becomes increasingly important to choose vendors who adhere to de facto standards and justify this commitment to standards by continuing to develop products around them. Beware of vendors who are banking on a single-standard product or especially a proprietary standard product, even though they may promise eventual standards compliance. These rules also apply to regulatory, protocols, and manufacturing compliance on both a national and international scale. Compliance to the standards listed above lends to interoperability at both the server and client level and at the user's and network interface. A complete guide to standards support is found in Chapter 3.

18.3.4 Vendor Promises and Support

Vendors will make many promises when courting a contract. Just make sure that any contract is fully explained in writing. To verify information, research the vendor. Ask around. Talk to other customers of the vendor, and even to their competitors (and weigh their responses accordingly). Documentation of all vendor correspondence should be compiled and stored for future reference.

One of the most hotly contested issues among vendors and buyers is support. When the vendor is first courting the user, look for vendors who both educate and consult your staff on the technologies and issues involved, rather than just trying to sell a product. Look for the following support from the vendor:

- maintenance services are on-site or central number
- hours/level of support
- how much will they assist with the installation, turn-up?
- percent of staff assigned to support
- built-in training support for entire staff
- size of vendor support staff
- training by the vendor (includes fixes to unknown problems, features and functions, and options)

18.3.5 Vendor Expertise

What expertise does the vendor hold in sales and marketing their product, in design, and in systems-level consulting? Have other account references been checked? Also check their manufacturer's position in the industry. A vendor may have to delay a product delivery because they have been short of power supplies from their overseas power supply vendor (who has gone out of business)!

18.3.6 Vendor Delivery Dates and Reality

Again, check the vendor's references for their delivery track record. Do they often miss deliveries, postpone deliveries? What have been some of their other customers' experiences with late deliveries? In the United States, it is common to have two-week to two-month average delivery cycles from the date the order is placed to the date the equipment ships from the vendor. In Europe, the average delivery is much longer, typically two to four months. This average is getting shorter, as many international vendors are now competing in national markets where project timelines and delivery dates are more competitive and much shorter.

18.3.7 Product Announcements and Product Futures

Be aware of all product announcements and future product direction of each vendor. Product announcements should be received free of charge through data sheets, advertisements, and electronic mail bulletins. If information is sensitive in nature and not yet public, a user-vendor mutual nondisclosure form can be used, which states that neither party will divulge information about the other. This contract is legally binding and strictly enforced. Nondisclosure agreements can be selective or blanket agreements.

Also ask for the future product direction of the company. Discern their dedication to R&D activities, and what percentage of their profit goes toward R&D and new product development. Make sure that the company's spending direction is the same as your vendor's development direction.

18.3.8 The Proprietary Vendor and Futures

While the same vendor can provide backward compatibility, they can also lock the user into a product unable to interface to the market. The vendor could get lost with technological change, and take the user with them. Avoid proprietary vendors. While they provide specialized solutions for the short term, they cause atrophy in the long term. Lastly, look at the overall future potential of the vendor. This includes market position, product futures, financial condition, and a myriad of other company health factors which will paint the true picture if the vendor is in for the short haul or the long haul.

18.4 THE VENDOR-DESIGNER RELATIONSHIP

A user-vendor relationship can be a rewarding experience. As mentioned before, this is similar to a marriage, where both parties benefit under good conditions. Both must clearly understand each other's business needs, and maintain good communications channels to head off any misunderstandings.

There are some rules to follow to achieve these goals. Don't let the vendor experiment with your network. This is especially true when you are trying to stabilize your network, and the vendor is experimenting with untried solutions such as new software revisions. Also, all upgrades should be planned to eliminate these surprises.

Maintain good interpersonal relationships. Make sure it is clearly understood what is expected of the vendor and of the user. Speak honestly and cooperatively with the vendor, and work together to achieve your goals. Understand the levels of peer-to-peer relationships and the roles and responsibilities of the account team. Frequent pre-sales and postsales support calls, as well as user visits to vendor facilities and labs are desirable. The vendor should make available an engineering staff contact, such as a knowledgeable technical consultant, and provide some level of continued on-site support personnel for large accounts; 7 day by 24 hour (7x24) maintenance support is left to maintenance contracts.

Make sure you are purchasing the products you want, not what the vendor wants you to purchase. When signing vendor development contracts, make sure you do not get locked in for long periods of time. Have the vendor produce good documentation before project/product hand-off to the user.

Look for vendors who are choosy about their business partners, and how the vendor relates the chosen customers to its long-term strategy. Does the vendor deal only with interexchange carriers, or does it also deal with local access providers? The primary customer base of a vendor can tell much about how the vendor is positioning for the future.

The relationship between vendor and user will usually last a long time; therefore, the vendor will have a profound impact on the user's business. The two are tied together. If the primary vendor for a communications network suddenly goes out of business, it could mean substantial budgetary impacts on the user's business to either replace the existing product base or migrate to a new one.

18.5 STRATEGIC USER-VENDOR AGREEMENTS

Strategic agreements between users and vendors take many forms depending on the level of commitment of each party. Flexibility is the most important aspect of the agreements listed below. Both user and vendor will learn through the process, and both will be required to adapt to each other's needs. Large users typically have enough buying clout to impact the R&D direction of the vendor to provide technology when needed. These agreements can offer many advantages, such as guaranteeing a revenue stream for both parties, providing the lowest prices, skipping the repetitive and resource-intensive and expensive RFP process, and building upon the strengths of both user and vendor. Often, the vendors achieve greater control, for without them there is no product. The following are some common terms of agreement defined:

Affiliate — no legal contract, but work together for business purposes

Alliances — nonexclusive contractual agreement, separate funding

Joint Venture — central funds and exclusive contract

Strategic Vendor-User Partnership — work together to solve complex problems, may involve any level of those relationships listed above

18.6 SERVICE LEVELS

It is important to define the service expectations of the vendor. This includes the vendor's service perspective, how service is defined, and how that service is provided. What are the categories of service provided, and what are their priorities? What priority is the user to the vendor — what size account — and how does the user compare to other users? What is the vendor's response time, parts and sparing plan, software and hardware support hours?

Vendors provide many service levels to users. There are four levels of service that can be provided, and the user must decide which best suits the business strategy and is most cost effective. The four levels of service are: nonparticipant, value added, full service, and joint development.

Nonparticipant vendors deliver their products to their customers and do not provide continued service after receipt of delivery. These are typically off-the-shelf products not requiring vendor installation or maintenance. They are also often low-cost items with sales based on price, availability, and speed of delivery. An example of nonparticipant vendors is Black Box, a company which specializes in delivery of well-known low-end and midrange data communications products.

Value-added vendors provide not only good products but include some level of service and support. These vendors are judged by factors other than price and availability, such as quality, feature function, and support staffs, and make it a part of their business to interface with each customer. These vendors offer some long-term commitment to their products and services, as well as to the customers who purchase them. While value-added vendors provide customer support, they draw the line when it becomes a financial burden to do so; customer service is an added value for their product, not part of their revenue-generating business.

Full-service vendors are differentiated by their dedication to and relationship with each customer. These vendors typically have large corporate accounts and support users who require a dedicated account team and are willing to pay for it. Full-service vendors interface much more with the end user, and may be involved with the actual implementation of the hardware purchased. Full-service vendors provide customer service and account teams and relate this service directly to revenue. These vendors often prefer long-term relationships with the user based on joint service offerings and future business potential.

The *joint development vendor* is the last level. This vendor not only works with the users, but also develops products in and around the users. The 1990s are the time for mergers and strategic partnerships among many businesses in the communications industry. Vendors which can partner up with large businesses can gain a strategic advantage in the marketplace. Small vendors can claim interoperability with major users and cut a niche into the market at the expense of time spent on development work with the user.

At most, a help line may be available during normal business hours. Sometimes a vendor will not be able to provide all the service or product that is required. A good design will incorporate alternate sources and strategies for network support which will not leave you out in the cold. It is the network manager's job to find and implement these alternatives.

18.7 NETWORK MANAGEMENT CAPABILITIES

Network management should be one of the most important decisions in the vendor analysis process. Networks are now moving toward a single network management platform for all equipment and elements present in the network. This incorporates the need for integrated network management into the RFP process. Some users may already have an existing hardware base which will be integrated with the new — mixing protocols, operating systems, billing, alarms, and operating platforms, making integrated network management all the more important. This creates a need for a network management platform that can provide global fault detection, analysis, and correction, measure network performance, provide security control and accounting measures, correlation, configuration and reconfiguration capability, and self-learning skills to a variety of bridges, routers, gateways, multiplexers, switches, concentrators, LAN servers, and other wide area network elements.

There are two network management platforms on the horizon which provide the integrated capability needed in today's computing networks; TCP/IP's Simple Network Management Protocol (SNMP) and OSI's Common Management Information Protocol/Common Management Information System (CMIP/CMIS). Both management schemes will enable integrated network management for network elements on a communications network. SNMP is available now for many devices, including ATM switches, and CMIP/CMIS, the OSI standard under development, will soon be available.

The vendor must also provide their plans for the transition from SNMP to CMIP/CMIS, and the migration period when both protocols will be active in the network. SNMP implementations must support Management Information Base (MIB) I and II plus extensions. In reality, both network management approaches will exist in the market for the next decade.

Much of the burden still lies on the user, who must both interpret the information provided by the network manager and develop interfaces for the translation of this data into a workable form. Some important network management requirements which should be incorporated include:

- nonproprietary implementations
- local hard drive/floppy
- multiple security levels
- data scope functions
- point and click graphics
- real-time geographical network status
- statistical collection
- accounting and billing collection (ASCII or binary)
- remote and local downloads
- remote and local nodal configuration
- remote and local board diagnostics
- remote and local inventory management capabilities
- real-time performance graphics

18.8 FUTURE BUSINESS OF YOUR COMPANY

It cannot be stressed enough that the future success or failure of a business can ride on the success or failure of the vendor(s) chosen for data and computer communications networks (not to mention the designer's job). For this reason, never sell the entire business to only one vendor. A multivendor environment is a healthy and safe one. It creates a competitive environment so that vendors work harder for the business and a secondary supplier is available in case the primary supplier falters. If all vendors meet the criteria presented in this chapter, especially standards compliance, there should be minimal difficulties in network integration and interoperability.

18.9 REVIEW

This chapter was designed to assist the reader in choosing a vendor capable of providing the features, functions, and services required, all based upon the criteria defined in previous chapters. After the RFI is analyzed and the choice of vendors is narrowed down to a "short list", the RFP is issued. The RFP is then analyzed based on a host of weighted criteria including adherence to industry standards, vendor promises and support, vendor expertise, and vendor futures. A look at real world user-vendor issues revealed the importance of items such as delivery dates, product announcements and features, and the vendor-designer and vendor-user relationships which are required for a successful vendor implementations and ongoing relationships. Finally, existing vendor network management capabilities were reviewed along with how the vendor becomes an integral part of the future network. The next phase of the design is the access and backbone network designs. The choice of vendors can and often does follow the access and backbone designs.

6

The Network Design

Now that the requirements have been analyzed, the capacity plan produced, and the vendors chosen, the time has come to perform the access and backbone network design. Thus far, each technology and protocol chapter has provided detailed design criteria which must be followed during the design process. These specifics will be used to accomplish the design. This chapter will step the reader through both the access and backbone portion of the network design. While it is impossible to focus on every aspect of network design for every protocol and technology discussed, the next two chapters provide insights into the key issues which may be encountered. Documentation should be completed at each step of the design process. After designing both the access and backbone portion of the network, the next step will be to implement the design. Although the access network design is often completed before the backbone and drives the backbone design, early versions and phases of network designs may be configured as an access network only, with a backbone built at a later date, and either step can be completed first as long as the rules presented in these two chapters are maintained.

19

Access Network Design

This first section deals with the access portion of the network design. It encompasses the point where the access network interfaces to the data transport network. Access may also include some portion of the actual data transport to the backbone network. The first step is to define access requirements as identified in Chapters 15 and 16.

Next, the capacity required in the total access network and in each access device should be defined. The style by which local access is mapped follows next: either ubiquitous access or a hierarchical scheme. An overall access network topology can then be planned to accommodate the applications, access devices, and access and network circuit design, and to allow interfacing either to access devices or the backbone network, which could take the form of a public network service. Finally, detailed methods of combining SNA and non-SNA traffic and local networking into a single wide-area network access to the backbone is explained.

19.1 ACCESS AND BACKBONE LEVEL DESIGN DEFINED

There are three possible access levels to any network design: the user or application, concentration or access, and backbone. These levels are shown in Figure 19.1 and are explained below. Generally, the user or application-layer design concentrates on all seven layers

of the OSI reference model (and corresponding levels of other architectures). The access and backbone designs concentrate on the first
three layers (physical, data link, network) with access also sometimes providing the fourth (transport) layer. Higher-layer support
(fifth, sixth, and seventh layers) are usually controlled by the application software.

User/Application Design. The first level defines the user or application
access to the network resources. For physical interface this level
typically includes the PC LAN (workstation and server) interface
card, FEP interface to the Token Ring LAN, or any other device
where data originates or terminates. Usually few "network" services
exist at this layer, but this depends on the level of protocol structure
operated by the user or application. This area provides the greatest
diversity of interfaces, protocols, architectures, technologies, and
standards of any network layer. The user layer of design is based
not only on protocols which access and transport data across the
network, but also on file transfer and session-oriented protocols.

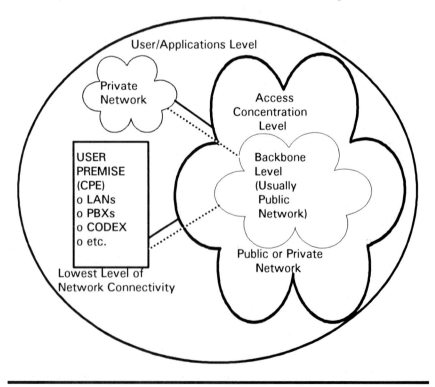

Figure 19.1 Network Access Levels

Access Network Design. The second level defines the user access point into the network. Typical access devices include routers, bridges, PBXs, switches, or any device that provides a focal point for standardization of interfaces, protocols, architectures, technologies, features, functions, and services required. Protocols are typically network, data link, and physical, and are similar, if not identical, between all network access devices. This is also the point where the user interfaces and interacts with switched-data services and the WAN. Access devices are sometimes called terminal or network devices, terminal adapters, or DCE. Terminal devices (DTE) terminate data, while terminal adapters (DCE) pass or modify data.

Backbone Network Design. The third level defines the backbone transport for the access network layer of the design. It is generally transparent to the access portion of the network, unless the user skips or omits the access network portion and interfaces directly with the backbone. The backbone provides a standard platform interface, protocol, architecture, and technology. It also provides user features, functions, and services. To the user, the backbone can be the carrier-provided switched service, becoming the "network cloud" to which the access devices send traffic. One example is SMDS service, where the user access device (a router) interfaces with the carrier-provided gateway into backbone service, or the user can directly access the IXC backbone cell switch. This method of interface will become more common as high-bandwidth data services proliferate.

19.2 NETWORK ACCESS REQUIREMENTS

What are the requirements for network access? The access design will be primarily based on the previous traffic analysis and capacity plan. The steps were provided in the last few chapters for requirements analysis and capacity planning and will not be reiterated here. However, it should be emphasized that there is a need to coalesce user requirements onto a common platform defining the interfaces, protocols, architecture, technology, features, functions, and services required by users.

19.2.1 Interfaces and Physical Connectivity

Perhaps the widest variance of any connectivity requirements will be at the physical media level. The access device design needs to accommodate both the user-side interfaces (determined in Chapter 15 and 16) and the network-side interfaces. If building a private network comprised only of access nodes, the network interfaces will be the point-to-point links between access nodes. These are often synchronous links utilizing the most efficient protocols to the devices chosen. When interfacing to a public network, the network side will often be determined by the public network access (gateway) and backbone interface.

The existing wiring configurations at each access site will dictate many connectivity options. The primary options include unshielded twisted pair (UTP), shielded twisted pair (STP), thin and thick coax, and glass (single mode or multi-mode) or plastic (multi-mode) fiber.

First, twisted pair. Twisted-pair wire (shielded and unshielded) can be used for both analog and data transmission. Many existing locations will probably have twisted-pair telephone wiring as the main building wiring. Recent advances have increased the data rates which can be transmitted over twisted-pair wires. Ethernet 10M bps, Token Ring 16M bps, and even FDDI 100M bps are tried and tested.

Coaxial cable was the second generation of wiring to proliferate. "Coax" cable is used extensively for Ethernet 10M bps. Many buildings have coax cable which has been used for telephone and cable TV, as well as security applications. Also, previous to fiber deployment, coax cable was used for interbuilding wiring. Coax can handle much higher bandwidths than twisted-pair wiring, up to 500 MHz.

Fiber optic cable provides the highest bandwidth available, ranging into the gigabits per second data transfer rates. Data is transmitted through light wave pulses converted from electrical signals. Advantages to fiber are the obvious high-bandwidth capability, security, the lack of copper-media-affecting disturbances, and the logistical advantages of smaller, lighter cable.

A variety of physical and electrical interfaces are found, ranging from the telephony/telecommunications world to the computer LAN arena. The physical telecommunications transmission media (facilities) are comprised of the electrical and physical access path. Common physical and corresponding electrical and mechanical interfaces include DB-25 (RS-232, RS-530 — touted as a replacement for RS-449), DB-37 (EIA-449), M-34 (V.35).

For computer and LAN communications, each LAN architecture provides multiple interfaces. These were discussed in detail in Chapter 5. Most personal computers contain interface cards that have RJ-45 phone jack outputs. A twisted-pair cable will run from the interface card to an RJ-45 wall jack. This wall jack is wired through the building to the telephone closet, where it is terminated on a 66 punch-down block (a row of wire prongs extended from the main distribution frame for "punching down" the wires, while stripping back the insulation to provide contact with the prong). This punch-down block interfaces the user to the destination device, which can be an IBM connector for a multistation access unit (if IBM Token Ring), an RJ connector on an Ethernet concentrator (hub), or some other type of terminal concentrator, controller, or hub. Most wiring closet designs will also include some form of patch panel before the concentrator for testing and ease of reconfiguration. Figure 19.2 shows an example of user connectivity to the network access device. The PC interfaces to the wall jack in the office via a LAN interface card with an RJ-45 interface plug. This wall jack is wired to the patch panel in the local wiring closet (the point at which everyone else on the LAN is also wired). The patch panel is wired to a 66 block, where patching capability is provided. The four wires are run down to a LAN concentrator or hub, where all four wire circuits terminate onto the LAN media. The hub is then passed to a port on the access device.

Other connection devices that may be required are an RS-232 line driver or a limited-distance modem (LDM), a balun for impedance matching, and external radio frequency modems or fiber optic inter-repeater links for distance connectivity. ARCnet uses coax shielded cable. These connections eventually require interfacing to the network access device, such as repeaters, CSU/DSUs, modems, bridges, or routers.

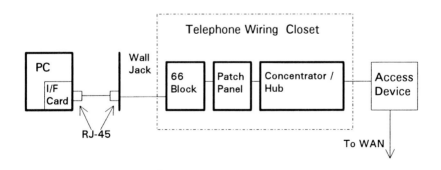

Figure 19.2 User Access Device Connectivity

19.2.2 Protocols

What protocol support is required? Protocol support can range from providing transparent transport to translating and converting multiple protocols from multiple architectures. First, each protocol must be defined that operates over each interface by identifying its syntax, semantics, timing, and proprietary implementations and idiosyncrasies. Next, determine what portion of each protocol is used. Identify and focus on protocols which are not physical media or hardware dependent.

Identify the characteristics of the file transfer protocols for impacts, specifically, their packet sizes in each direction (such as NFS, which transmits large packet sizes and receives small ones back). Determine each protocol's mean data unit size and ratio of transmit-to-receive data units. Then determine which protocols can be bridged, routed, or switched, and which cannot. Isolate the media access protocols, and determine if they will be passed transparent to the network, encapsulated (tunneled), translated, or converted by the network. Also, many protocols have interdependencies on other protocols which limit their handling and operation. Analyze which protocols will be affected by error detection, correction, notification, flow control, or buffering.

There will most likely be many dissimilar protocols, and most will not communicate directly with any other, or, if they do, they are doing so via IP, or in the SNA/SDLC world using gateways. These protocols should be mapped with their interdependencies and conflicts clearly understood. Also look at the protocol requirements from the perspective of inside the access node. What protocol requirements are expected from the access portion of the network? For example, packet switching protocol specifics would be the speed in bps or Bps, the number of packets, calls, resets, and interrupts allowed, the VC call memory allocations, packet size and window size per VC. All of these parameters are expected to be set by the access device (PAD) and the network switch and must be defined. Other technologies have similar protocol-specific parameters that must be identified.

This brings up the question of the amount of control that the access network needs to exert over the user or application, or more importantly the level of access network device to user interaction required. Again, these will be different depending on the protocol(s) being used. The network designer is required to clarify all of these interdependencies before progressing with the access design. It is obvious how important these details become for an accurate network design.

19.2.3 Architecture and Technology

Next determine what architecture the protocols interact with, and how these architectures themselves interact. Most of this design will be protocol-to-protocol specific. Again, determine what level of encapsulation, translation, bridging, routing, or switching is required. Also ask where these functions will be accomplished: at the physical input or output of the access device (transport media), within the access device, or in the network backbone.

Another question: Is the technology base a single technology or multiple technologies? Obviously, the more complex the mix of technologies the more complex the access design and the more limited the number of device options available. When doing the access design, refer back to the Chapter 17 technology comparison and make sure that the choice of technology has not changed. Unless the future plan or economics dictate, do not use a more complex device than what is required. For example, do not use a bridge when a repeater is required, and multiport repeaters are sometimes optioned rather than more complex devices. Do not use a router when a bridge is required, etc. Remember that the more meshed and dynamic the topology, the more intelligent the access devices must be.

19.2.4 Features, Functions, and Services Required

When the user interfaces to the WAN through an access device, certain features, functions, and services are required over and above the normal interface and protocol support. These features and functions are specific to the combination of interface and protocol selected, such as FECN, BECN, and DE bit usage with frame relay. The protocol may support these functions, but how does the user and access device use them to implement flow control? Another example is the access class of SMDS. Many of these questions can be answered when it is determined where the user traffic hits the service. The service interface point is often the place where features and functions operate. Does the access node provide the first advantages of the data service, or does it begin in transport to the backbone, or even at the backbone? Where do the value-added services begin; at the access or in the backbone? Once these questions are answered the initial loading phase of access design can begin.

19.3 ACCESS NETWORK CAPACITY REQUIREMENTS

In Chapter 16 the loading on each access device was calculated based on the known inputs. For many small or private network designs, the user applications and inputs are well known. If this is the case with the network design, the access node configurations are already known. If so, proceed to connect the access devices via one of the architectures defined in the next chapter.

However, when designing large private or public data networks, the user access specifics are usually not clear and well defined. This is especially true when designing a network from scratch, and the user inputs are estimations or, worse yet, speculations. In this case, some broad approximations of user traffic can be placed into a model to define the number of access and backbone ports required, their speed, and utilization. This assumes that the user protocol characteristics are either the same or similar, and that a single technology and set of protocols is used for internetworking these users. Also take into account service aspects, such as the CIR in frame relay, which provide logical constraints that may be exceeded by the user at any time.

19.3.1 Access Device Loading and Link Utilization

Now to calculate the access network design required when only raw access bandwidth numbers are provided. These calculations are protocol and technology independent, and would require modification for statistical multiplexing, queuing, or any other buffering or efficiency increases from the input to output of the access device. These calculations are most applicable to calculating multiplexer devices, as well as frame relay and ATM network access devices.

A certain user community will transmit a given number of MBytes per day of data over the data network (M) through a given number of access node ports (n). With these numbers a model for the access network with (n) user input ports and (T) backbone access trunks. A limit must be set on the size (in ports) of the access devices, as well as the access and trunk speeds available. Once this is done to model the access network, modifications can be made to these variables by adding factors which model queuing delays, statistical multiplexing, and internodal traffic.

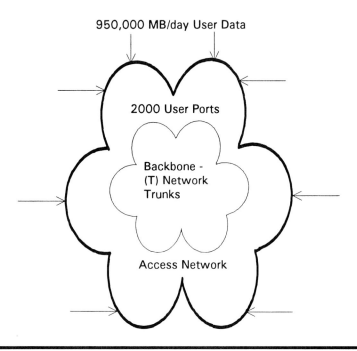

Figure 19.3 Access Port Sizing

Figure 19.3 shows an example where M = # MBytes/day/network = 950,000 and the number of access ports in the network = n = 2000. For simplicity designate that the access ports speeds = s = 1.544 M bps (T1). The average utilization for each port would be calculated as:

$$\frac{M}{n} * \frac{1 \text{ day}}{24 \text{ hr}} * \frac{1 \text{ hr}}{3600 \text{ sec}} * \frac{8 \text{ bits}}{1 \text{ Byte}} * \frac{1 \text{ port}}{s} = \text{avg \%U per port}$$

$$\frac{950,000}{1000} * \frac{1}{24} * \frac{1}{3600} * \frac{8}{1} * \frac{1}{1.544} = \frac{5.7 \%}{\text{Utilization}}$$

Now add a busy hour calculation where 20% of the total traffic occurs during the busy hour. The new calculation would include a factor of 5 hours per day to attain the 20% busy hour peak.

$$\frac{M}{n} * \frac{1 \text{ day}}{5 \text{ hr}} * \frac{1 \text{ hr}}{3600 \text{ sec}} * \frac{8 \text{ bits}}{1 \text{ Byte}} * \frac{1 \text{ port}}{P} = \text{avg \%U per port}$$

$$\frac{950,000}{1000} * \frac{1}{5} * \frac{1 \text{ hr}}{3600} * \frac{8}{1} * \frac{1}{1.544} = \frac{27.35 \%}{\text{Utilization}}$$

Notice that port utilization has increased from 5.7 to 27.35% during the busy-hour period. Take this one step further and look at the utilization during the busy minute and second. The busy minute or busy second could be calculated by reducing the number of seconds in the denominator of the hour to seconds conversion. When this is done, bursty traffic can have a drastic effect on the bandwidth utilization of a dedicated circuit.

19.3.2 Access Node Design

Now look at the average access node design. Since only a general idea what the traffic will be at each node is known, further assumptions can be made based on the fact that each access node will be configured in the same manner. Access design can start out this way, and when more detailed site-specific information becomes available, the access design can be further refined and modified.

To find the access node input-to-output utilization, a comparison is made concerning the number of ports, the speed of each port, and the port utilization to the output trunk utilization. The formula would read:

$$(T)(S)(\%UT) = (p_i)(s)(\%Up)$$

where the
\# of Trunks Required $= T$
\# of ports of type $i = p_i$
Speed of trunks $= S$
Speed of port $= s$
% utilization of trunks $= \%UT$
% utilization of port $= \%Up$

For example, if inputs and outputs to a router are designed with the following parameters: input utilization is 27% on eight frame relay T1s, you want an output utilization per trunk of no more than 50%, and the output trunks are also frame relay T1s.

$$(T)(1.544)(.5) = (8)(1.544)(.27) \text{ then } T = 4.32 \text{ or } 5 \text{ trunks required.}$$

This calculation can be taken one step further, where there are now multiple access speeds and port utilization, but the trunk speeds remain constant. The formula would now read:

$$(T)(S)(\%UT) = \sum (p_i)(s)(\%Up_i)$$

Here is another example where the inputs are five T1, ten FT1 (256K), and twelve DS0 (56K) circuits. Assume that the access utilization remains the same at 27% utilization, the total number of trunks needed would be:

$$(T)(1.544)(50\%) = (5)(1.544)(.27)+(10)(.256)(.27)+(12)(.056)(.27)$$

where T = 3.83, or 4 T1 trunks are required. As the input port utilization increases so would the number of trunks required to support the traffic.

Obviously, these numbers will vary based on statistical multiplexing efficiencies, time variances in utilization and busy time periods, protocols, and usage patterns, to name a few. LAN inputs and loading factors were explained in Chapter 16. It is important to note that not only would the bandwidth provided be required, but also the peak bandwidth during busy hour/minute/second will also be necessary.

19.3.3 Utilization, Loading Factors, and Anticipating Failures

The link utilization and link efficiency relies heavily on the protocols operating over the link and the technology used. These can vary from frame relay and HDLC which are very efficient, to BSC and asynchronous protocols which are relatively very inefficient. The utilization of a link is affected by the number of end-user devices using the same link, the propagation delay, packet sizes and overhead due to flow control, error control, window and buffer sizes, and many other protocol and technology proprietary factors.

The utilization of a link will range from the average utilization to the busy hour/minute/second and even millisecond peak period utilization. The network designer can plan for high utilization periods through three major methods of link loading. The first is to overbuild the capacity. This is the simple yet expensive method of throwing more capacity at the link (a link can constitute multiple circuits between the same two endpoints). The second is to simply add up the capacity required and double the number of trunks. Again, this is expensive and wastes bandwidth.

The most efficient method of link (trunk) loading to allow for the failure of one link without losing traffic is by loading the links at a percentage which allows for the failure of a single trunk by the following:

$$L_f = [\mu - (\mu/N)],$$

where L_f is the load factor of each trunk, N is the number of backbone access trunks of the same speed, and μ is the maximum percentage load on a trunk given a failure of another trunk in the same access node.

For example, a network where the trunk overhead accounts for 50% of the bandwidth (μ) is a drastic case! For two links of T1 speed, the loading factor would be 25% or 386K bps maximum traffic load (thus, one link could fail and the other link could carry the traffic at a 50% utilization). Now, take the same example for four links with the same overhead. This is shown in Figure 19.4. Each trunk is loaded with no more than 37.5% traffic, or 579K bps, because if one link fails the remaining links have to carry an additional 12.5% of the total traffic, resulting in a 50% utilization of the remaining three links. This formula could be modified based 12547*on the level of reliability of failure analysis and appropriate link loading. Typical 9600 bps loading for packet switching is at 60-65%, and for 56K bps is at 75-80%. The network access devices in packet networks are typically utilized at much lower levels than the network trunk side.

This example was drastic for an overhead of 50%, but some technologies actually provide much overhead. For example, an ATM channel over a DS3 can be considered. The maximum throughput of a DS3 (44.736 M bps) is 43.008M bps (at 4% overhead due to framing). The overhead of the ATM packet structure of a 53-Byte packet is 5 Bytes. Thus, the aggregate overhead is approximately 14.6%. This does not take into consideration the higher-layer protocol overhead within the cell. It is not uncommon to have over 40% overhead on the actual user data which originates as a file.

The utilization of LAN links is usually measured in packets. For example, on an Ethernet LAN it takes 3,000 PPS (of the smallest packets) to fully utilize a T1, and for 64 Byte packets, or larger, the throughput would be less. If the throughput begins to degrade, and the packet size remains the same, it may be necessary to reallocate LAN resources, or segment the LAN. Additional trunks may also be an option. Either way, define allowable margins and limit access to within those margins.

TRUNKS TO   ACCESS NODE   USER ACCESS
NETWORK

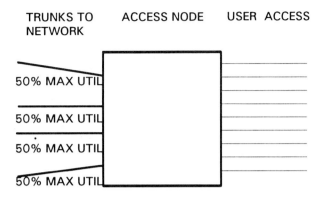

Figure 19.4 Network Trunk Sizing

19.3.4 Efficiencies of Statistical Multiplexing

As discussed before, statistical multiplexing can also increase the efficiency of available bandwidth use. Statistical multiplexing is found in packet switches, routers, cell switches, and of course statistical multiplexers. LAN and MAN operation can also be considered as a form of statistical multiplexing, because users who need the bandwidth get it when it is available. If one station on the LAN/MAN requires all of the bandwidth, that station gets most of it (again depending on the architecture). The ratios of statistical multiplexing are expressed in X:Y, where X is the user input and Y is the access device output. For example, if eight T1 user input circuits are transmitting an average of 12.5% utilization each, and a single output T1 trunk could be driven at 100% utilization, the ratio would be 8:1. On the other hand, if there were four inputs at an average utilization of 25% each, the ratio would be 4:1. In the same manner, the ratio could also change based on the load of each user input.

19.3.5 Future Capacity

Plan for providing both physical and logical future capacity in each network access device. This planning for growth is part of the short- and long-term capacity plan. Make sure resources are in the right place to begin with, as well as placed in areas of future growth. The smart network designer will anticipate areas of growth and build in extra capacity in those areas. This capacity could be additional hardware or ports, low initial trunk utilization, extra processing power, or just adding the capability to expand that area of the network.

19.4 STYLE

Style has been discussed previously in terms of network topological styles. This section discusses how the design of the geographically local environment may have a major impact on the WAN, both in loading and topology. A look will first be taken at the geography of user requirements as defined in the traffic matrix, and then at the distribution of those resources within each geographic region in either a hierarchical or matrix (ubiquitous access) style.

19.4.1 Based on Geography

The placement of local access nodes will be based primarily on geographical user and application distribution. Use the traffic matrix built in Chapter 16 to determine the placement of access nodes. Organize access by CPE, LATA, nation, and continent.

19.4.2 Ubiquitous Access (Matrix)

There are two major styles of providing connectivity to local area resources (subnets): *hierarchical* and *ubiquitous* access. The ubiquitous access approach allows all users to transmit and receive data from all areas of the network. This is also called the "matrix approach". One example of ubiquitous access is shown in Figure 19.5, where a single intelligent hub provides access for the entire building's LAN connectivity. Any user on any LAN can pass data through the hub, as well as access any server in the network. This style of access is good for organizations that are spread out and non-structured (nonhierarchical) in nature. The ubiquitous access style provides a one-dimensional, flat network and is often used with smaller networks. This type of network becomes difficult to manage as it grows in size, especially when multiple interfaces, technologies, architecture, protocols, and vendors are involved. Problems with performance degradation (see throughput versus user curves in previous chapters) are not easily diagnosed and the network becomes difficult to manage.

UBIQUITOUS ACCESS w/ HUBS

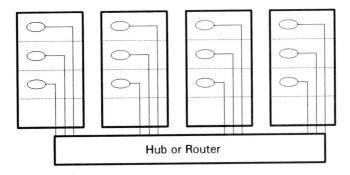

Figure 19.5 Ubiquitous Access Example

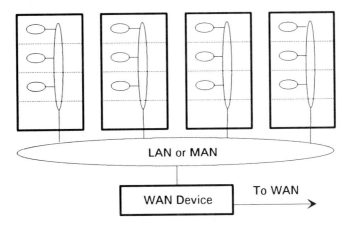

Figure 19.6 Hierarchical Access

19.4.3 Hierarchical

Hierarchical designs provide a user access hierarchy where traffic destined for its own local, metropolitan, and wide area remains in that geographical area, rather than accessing the common switching or routing point as in the ubiquitous access approach. Figure 19.6 illustrates a building with a four-tier hierarchy; each user is connected to a floor LAN, each floor LAN is connected to a building LAN, the four-building LANs are connected to one campus MAN, and the campus MAN is connected to the WAN. In this example, connectivity between segments, or subnetworks, is accomplished via bridges or routers. The hierarchical design is often used in larger networks warranting the segmentation of user traffic. In this design, any user should not have access to every server in the network, instead going through his local server for most applications. Thus, servers are generally more specialized.

A deviation of the hierarchical design would be to use intelligent hubs. Traffic would be aggregated into building routers, and then passed to centralized hubs as in Figure 19.7. This design is taken one step further in Figure 19.8, where three building hubs access a single WAN router. The routers would perform routing within a building, and the hubs would act as concentration points for the facility. The routers would then balance the load and perform the routing function, and the hubs would perform pure hubbing. Since routers often have a limited number of ports, the role of the router and hub could also be reversed.

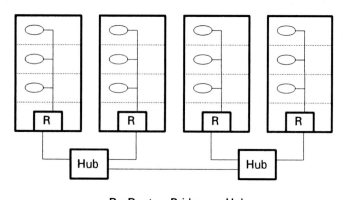

R - Router, Bridge, or Hub

Figure 19.7 Hierarchical Access with Hubs

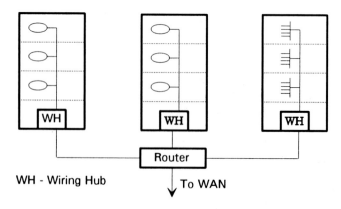

WH - Wiring Hub

To WAN

Figure 19.8 Hierarchical Access with Router

The segmentation of hierarchical networks starts at the work group subnet. This unit provides interconnection of resources (workstations) between LANs — (e.g., RIP protocol routers using TCP/IP and multiple operating systems). The next higher level is the departmental subnet where routers or gateways connect a local area. This provides the LAN-to-LAN internetworking using OSPF to efficiently provide wide area network routing. The next higher level is MAN and WAN networking. Departmental subnets should be arranged to provide flows of connectivity which form the enterprise backbone. The enterprise backbone could be the network as portrayed in the last figure, or many such networks connected over the WAN. Typically, the enterprise network is the latter, with many local designs connected via access nodes (the routers or the hubs in these examples) to the carrier providing the service (e.g., SMDS (DQDB) or B-ISDN (ATM) backbones).

The hierarchical access style will allow the capability of connecting lower-speed LANs to high-speed MANs, without the degradation of performance. The rule that the transport is only as fast as its "slowest link" still applies, and with a hierarchical style the bottlenecks can be eliminated by assuring that each level of the hierarchy is at least as fast as the previous one. Hierarchical style networks help to protect users from broadcast storms and allow administrators to regulate traffic flows between segments. Some other tips to follow when deploying this style are that servers on local LAN segments should never route packets, but instead routers should be used. And regardless of which style is chosen, centralized support of all elements of the WAN is required.

19.4.4 Hierarchical vs. Ubiquitous Access

The hierarchical approach to local access design provides many advantages over the ubiquitous access style, including:

- ⮡ cost-efficient use of network media
- ⮡ keeps performance from degrading
- ⮡ protects from broadcast storms
- ⮡ allows for hierarchical address schemes (see Chapter 20)
- ⮡ uses access control filters in routers to police segments
- ⮡ ease of security administration
- ⮡ isolation and diagnosis of problems made easier

Obviously, the ubiquitous access method causes much more internetwork traffic, because much of the traffic could be limited to the area of transmission and reception of data. In large WAN designs, this could be disastrous for WAN throughput and performance. The major drawback of the hierarchical style is the additional cost of network equipment and servers to localize the traffic. Also, ubiquitous access is sometimes required when many remote LANs need to communicate.

Regardless of which style is used, the servers of each LAN should be designed in line with hierarchical or ubiquitous access layers. In the hierarchical style, more servers are needed to ensure local access of server-based information and segmenting of that access. Putting servers and clients on the same LAN segment also keeps backbone WAN traffic reasonably low. In ubiquitous access all servers can talk to all other servers so all resources can be shared across the entire local access network. The network or subnetwork addressing should also correspond to the style chosen. In hierarchical networks using routers, parts of the network can be isolated or fire-walled to provide fault isolation and keep data contained to specific areas to prevent data storms. Both routers and bridges can perform filtering of data to eliminate excess data across the WAN. Cost and network control (turf) issues also become factors in choice of style, and hybrids of both are not uncommon, but rather the norm.

19.5 COMPLETING THE ACCESS NETWORK DESIGN

The steps for completing the access network design include verifying the user application intelligence, confirming design and choice of the access device(s), choosing the number and type of access devices, completing the circuit design to and from the access node, and verifying the total access topology. Most of these steps may have already been fixed from design practices of previous chapters, but are presented here in a wrap-up fashion to assure that every step is summarized in completing the access design portion of the network.

19.5.1 User/Application Intelligence Verification

When the access network design was performed, it was assumed that there was some level of intelligence in the user devices and applications. The greater this level of intelligence (and the more expensive), the less is required from the network and the more the network can concentrate on passing data. If not done in the initial requirements phase, some time should be spent reanalyzing the user application intelligence to confirm the intelligence expectations of both the user and the network. Also confirm that the required intelligence is in the access device selected, and that this device will be able to communicate that intelligence to the backbone network. One example of misconceptions that may arise is operating system support from LAN to LAN. Most routers can convert between LAN protocols, such as Ethernet to Token Ring, but cannot allow LANs with different operating systems to interoperate. This process is complex because of the different framing and addressing formats. Some users may not look at this level of detail, and just expect the higher-level protocols to interoperate because the lower-level LAN/MAN/WAN protocols do.

19.5.2 Access Device Level

The next level of verification is of the access device itself. The access device should be the feeder portion of the network design, where user applications, devices, and protocols of a defined geographic area are fed into a single device. This single-access device will then communicate with other access devices or directly with the backbone. The typical access node aggregates and feeds voice, data, and video traffic from many different interfaces, protocols, architectures, and technologies to the backbone via a single protocol and technology. Access devices, therefore, should reduce the number of access ports, protocols, conversions, and data formats fed to the backbone. With technologies such as frame relay, access to the WAN is in terms of multiple logical channels rather than single physical channels per port.

One often overlooked benefit is that access devices such as multiport bridges and routers can also serve as a LAN or MAN. This is when the internal bus of the device exceeds that of a single LAN or MAN but provides the same functionality at faster bus speeds. Take the Wellfleet BCN node for example. This device has four 250M bps backplanes. Ten LANs can access the router through access ports and have in effect access to four MANs at 250M bps or a single MAN at 1G bps! (assuming the router can route packets at these speeds). Figure 19.9 illustrates the evolution of a router from (a) the role of strictly providing WAN access to (b) the role of actually becoming the LAN (or MAN) itself. The same arrangement can be built with a multiport bridge or intelligent hub, but without the capability to route. Eventually, the router and hub market will tend to merge, and the choices will not be as clear.

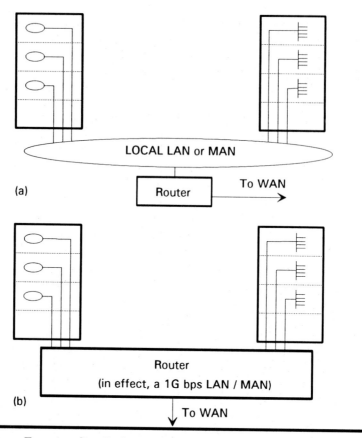

Figure 19.9 Functionality Evolution of Routers

19.5.3 Number and Type of Access Nodes

The number of access nodes is determined by both the traffic matrix and the optimized grouping of traffic and the placement of access nodes. Fewer nodes tend to cause longer access circuits, but decrease total network delay. More nodes increase the chance of network failure (see availability calculations in Chapter 16). The network design should always allow for at least a single node failure scenario, as well as alternate trunks for a minimum of a single access trunk failure. Loading factors which accommodate for trunk failures have been discussed already, and the same formulas can be used to calculate the number of nodes required to survive a single node failure.

By now, the reader should be able to choose the type of access node required. Here are a few extra tips on the selection of access hardware. Using a single router with multiprotocol support may preclude buying two bridges that each support a single protocol. This would reduce the number of network access points and equipment costs while providing maximum functionality. Also, with large amounts of heavy client-server, multiprotocol traffic, a router is always a better choice over a bridge. Since client-server traffic (TCP/IP, NetBIOS, and other protocols) usually constitutes the majority of user traffic (rather than SNA), this is a major consideration in LAN/WAN access design.

Frame relay access devices can be any normal access device discussed so far: multiplexers, packet switches, bridges, or routers. Access devices for the DQDB architecture (SMDS access) are more limited, and include routers, Macintosh PCs, and even some multiplexers, but these devices require hardware/software upgrades to support this new service. SMDS operation also presently requires special DSUs specifically designed to perform L3_PDU to L2_PDU segmentation. But SMDS interfaces are beginning to proliferate to all types of LAN/MAN/WAN access device, even the CRAY supercomputer. ATM and SONET access devices are still specialized and details are found in their respective chapters. ATM is now appearing on the user/customer premises, but it may be some time before CPE access devices provide SONET interfaces.

Host-to-host connectivity has been through the channel-to-channel bus-and-tag adapter at 3M *Bytes* per second. Channel extension can be accomplished through either high-speed dedicated T1 and T3 circuits, or through high-speed LAN/MAN/WAN access. The latter is beginning to predominate with FDDI or IEEE 802.6. More vendors, including IBM, are now designing frame relay and SMDS interfaces for their processors and host computers, and these interfaces may soon replace the older distance-limited methods described above.

19.5.4 Access Circuit Design

The best method of designing the access and backbone circuit sizing is with a network design tool. Barring access to such a tool, circuits can be chosen and prioritized on a variety of factors including:

- ★ shortest path
- ★ maximum cost
- ★ maximum congestion
- ★ minimum delay
- ★ maximum loss
- ★ minimum/maximum throughput

★ utilization
★ spanning tree minimum hop count

Once the circuit topology is decided, the placement of nodes should be reviewed. Based on the total ingress/egress traffic into/out of a node, the decision to add/delete nodes or change nodal configurations can be made. This is an iterative process, and takes multiple iterations to achieve an optimum design. This is especially true with multiple protocol layers, each affected by changes in the routing. The best advice is to understand the limits of the protocols being used (e.g., hop counts, overhead, etc.) and make sure they are not violated by changes in topology.

The procedure goes something like this: assign the capacity in circuits between access nodes, taking care to account for every constraint. Manually fail parts of the network and determine where additional capacity may be required. Add the capacity and fail another part of the network. Continue until the design meets the correct criteria for availability and reliability. Now optimize the network wherever possible. Analyze backhauling (the amount of distance extension required to reach the nearest access node) and determine if it is cost-justifiable to eliminate any nodes or links (keep original user requirements in mind). After optimization is complete, return to failure analysis and repeat the process. A good design tool will perform these steps and many more (design tools will be discussed in Chapter 24).

All access to backbone circuits should have diverse homing based on the availability and redundancy required. If at least two diverse circuits are used, they should have the capability to load share and load balance. Test all calculations on circuit failure and survivability. Make sure that the network contains the desired level of redundancy. Circuits between access and backbone nodes look like access circuits to the backbone, and the shorter the access circuit distance the better. Long-access circuits incur more delay, are more expensive, decrease reliability, and decrease total capacity of the network. Run all access circuits through a backhaul analysis to determine if an additional access node should be used. Typical 1992 circuit costs for backhaul were $10 per T1 mile and $50 per T3 mile per month.

19.5.5 Access Network Topologies

Access network topologies can vary:

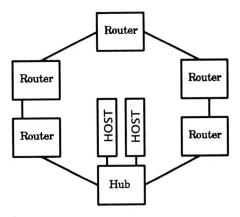

Figure 19.10 Cascaded Router Configuration

☆ direct connect to backbone nodes *without* direct trunking between access nodes

☆ direct connect to backbone nodes *with* direct trunking between access nodes

☆ ring (loop) connected to backbone

☆ trees with "root" connected to backbone

☆ multipoint with bus connected to backbone

☆ star with hub connected to backbone

☆ hybrid meshed configuration

Many of these topology styles have been defined in Chapter 4, and will be covered in more detail in the next chapter.

Access nodes can also be cascaded, but the reliability decreases when this is done. Figure 19.10 shows five routers cascaded and then joined into a common hub which accesses some centralized computing elements. Delay and survivability would be the primary problem factors with this configuration, but the wiring and connectivity constraints may prohibit interrouter connectivity. The delay would also increase if packets had to transit many routers to reach the hub and beyond the hub. The best technology for this access arrangement would be a MAN ring architecture such as IEEE 802.6 SMDS or FDDI.

Also, when building bridged networks (or non-routed networks) beware of creating loops. As point-to-point connections grow across the WAN, the likelihood of loops increases. Loops can cause duplication of packets. Protocols such as Spanning Tree Protocol (STP) works for local bridges, but remote bridges can cause loops and multiple non-unique paths. STP will place some links (if loops found) on standby mode and the subsequent routing may cause increases in delay and hops. As stated before, be aware of the effects of all protocols used in the access network architecture, as well as the architecture impact on the protocols.

19.6   IBM SNA AND WAN ACCESS

It is a well-known fact that SNA networks and traditional non-SNA WANs do not mix well. How should SNA traffic be moved onto the common multiprotocol backbone? There are many methods of tying SNA networks into the non-SNA WAN environment, and several will be discussed. Why combine SNA traffic with non-SNA traffic? One of the advantages of placing SNA traffic over a WAN is that broadcast packets can be eliminated, similar to a more dynamic method of filtering. Broadcast packets cause excessive overhead with source routing; the larger the network, the more overhead used.

Many users are upgrading SNA devices with Token Ring interfaces. The device then interfaces to the WAN through the Token Ring LAN. Some of the older IBM devices, such as the 3X74s, S/36 and S/38s, 3600s, 4700s, and 5520s, cannot be upgraded with Token Ring capability. This isolates them from the LAN internetworking arena. SDLC is another option, but acts as a point-to-point protocol, where Token Ring (LLC) can be either matrix or hierarchical. When multiple SDLC lines are placed on a Token Ring network, much bandwidth is wasted. Also, the SDLC address space is not large enough to support LAN networking, so other methods have been developed to pass SDLC traffic over the WAN. Therefore SDLC must be encapsulated and passed through the WAN. There are other methods of passing SNA traffic through a common non-SNA access and eventually through the backbone network.

Synchronous pass-through. Point-to-point port mapping that provides transparent pass-through using IP encapsulation is called tunneling. This method adds additional overhead and creates larger packets which affect user response times, may affect time-outs, and requires priority schemes similar to those of video over data or overbooking of bandwidth (4 to 1) to prevent variations in response times. There is also a problem with constant and excessive polling (up to 50% overhead); information passes over the network from primaries to secondary. Figure 19.11 shows an example of synchronous pass-through.

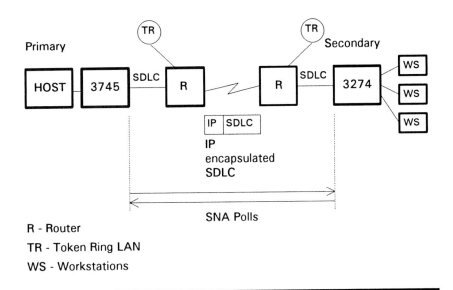

R - Router
TR - Token Ring LAN
WS - Workstations

Figure 19.11 Synchronous Pass-Through

Remote SDLC/3270 polling with retransmission. This eliminates polling overhead with a technique called spoofing. In this case, the access device passes only blocks containing SNA data over the dedicated SNA line. Polling is done locally with both primary or secondary modules performing the polling functions. Spoofing provides the same benefits as synchronous pass-through, yet eliminates some overhead and time-out problems. Figure 19.12 shows an example of spoofing.

Remote SNA switching with host pass-through. This replaces the primary and secondary polling nodes with primary and secondary SNA nodes in the router, where the secondary device-connected routers emulate SNA Type 5 nodes and the primary device-connected routers emulate SNA Type 2/2.1 nodes. This provides dynamic path routing rather than the SNA-specified routing, and eliminates the need to establish SNA cross-domain host sessions. Remote SNA switching forms a virtual SNA environment, but does not provide response time advantages over remote polling. Figure 19.13 illustrates remote SNA switching. Figure 19.14 shows this method in a strictly IBM environment.

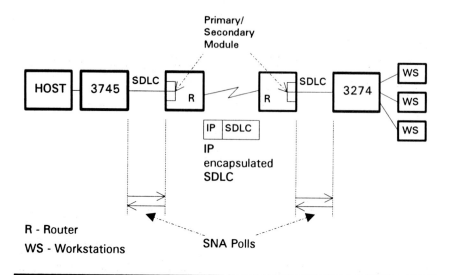

Figure 19.12 Spoofing Example

SNA Routing. There are two major methods of SNA routing. SNA 37XX or APPN Type 4 routing establishes an optimum path between routers for host communications through router emulation of SNA Type 4 routing. This method retains the benefits of remote SNA switching, but does not eliminate cross-domain session setup overhead. Type 4 SNA routing is shown in Figure 19.15. The second method is SNA cross-domain Type 5/4 host/FEP routing. This method is similar to the Type 4 routing, but more manageable.

Both methods are new and still extremely vulnerable to changes in any of the SNA domains using SNA routing.

Other methods of passing SNA traffic over the public network exist. In Chapter 5 the SNAC/TR method of connecting older 3274 communications controllers was covered, which do not have Token Ring interfaces, to the Token Ring LAN. This can be seen in Figure 19.16. Many router vendors are implementing the capability to convert SNA/SDLC to the Token Ring LLC format, and many routers also support both the synchronous pass-through of SDLC and HDLC from origin to destination through the IEEE 802.1 spanning tree protocols. The same method is used by many routers to route IBM LU6.2, NetBIOS, IPX, and TCP/IP protocols. IBM uses a proprietary source-routing protocol, or implements true SNA Type 5/Type 4 host/FEP node emulation previously discussed. If IP encapsulation is used to transport SNA traffic, up to 50 Bytes of overhead per message will be added. The smaller the size of the message, the worse the throughput. Another router option is to simply add to the SDLC address. This, however, will only increase the message overhead by up to 10 Bytes or so.

The IBM Network Control Program (NCP) Packet Switching Interface (NPSI) is also available. NPSI encapsulates SNA traffic into X.25 packets. X.25 then rides inside the routed TCP/IP environment. This allows access to a packet switched network or a WAN capable of routing X.25 protocol.

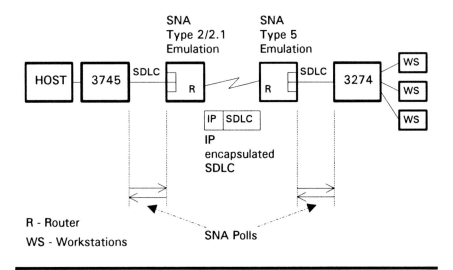

Figure 19.13 Remote SNA Switching

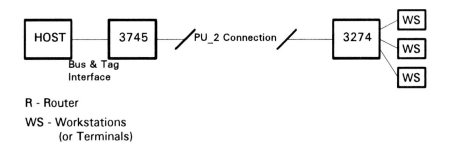

Figure 19.14 Present IBM Solution

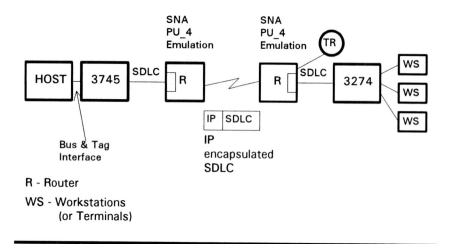

Figure 19.15 Router Solution with PU_4 Emulation (Type 4 SNA Routing)

Implementing a full SNA to OSI gateway is also an option, albeit an expensive one. Implementations of this type will vary, but are almost always proprietary. Discussions of TCP and LU6.2 gateway conversions are left for further reader study.

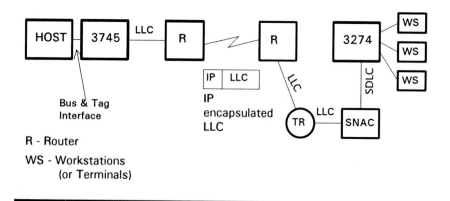

Figure 19.16 Router Solution with SNAC Conversion

19.6 CHOICE OF CARRIER TECHNOLOGY AND SERVICE

Carrier technology and service decisions should be based on the decisions made in the access portion of the network design. The primary focus should be on the future of the network. In the next chapter the backbone network design will be covered and a look taken at the carriers that provide the required backbone technology or service. All of the methods discussed are offered by the carriers. It is up to the reader to analyze each option based on the information provided in this text. Then choose the best fit for network integration.

19.7 REVIEW

In this chapter, requirements explained in Chapter 15 and 16 were built into the access portion of the network design. First, the requirements in the categories of interfaces, protocols, features, functions, services, and technologies were modeled. The importance of building the access portion of the network to accommodate these requirements was emphasized while assuring that the requirements were indeed supported by the design. It was found that there are many methods of modeling network capacity on a site-specific level and a global level. The importance of building future capacity into the network was discussed. It was seen that the hierarchical style of topology is best for large, complex-access networks and that ubiquitous access is best for small, heterogeneous designs. The access design was verified by taking into account the user device and application intelligence, the device type and quantity to adequately support the users, and the circuit design to support optimal performance with room for future growth. Finally, methods of integrating IBM SNA traffic into the non-SNA WAN environment were outlined. Now to move to the backbone portion of the network design.

Chapter

20

Backbone Network Design

Now that the access network design is completed, the decision must be made whether to interconnect the access design, build a backbone network, or interface to an existing backbone such as a public data network. Another review of the interfaces, protocols, architectures, features, functions, and services is required in the same manner as done for the access network design. The capacity required to support the access network was determined primarily from the outputs of that design stage. Next, a topology and style consistent with the services and technology must be chosen. The resultant topology will in turn determine the number and placement of backbone nodes, the circuit design, and the overall integrated network design. After the topology is established, there is a necessity to go into great detail on network addressing. Multiple techniques are explored. The examples used take this exercise to the leading edge of IP address design. Both network management and network timing are discussed, and this chapter closes with a discussion of methods of fine tuning the entire network design. Many of the same principles discussed in access design also apply to backbone design and will not be explained in detail but the results given and applied.

20.1 BACKBONE REQUIREMENTS

Why is a backbone needed? If an existing meshed point-to-point network is composed of network concentrators or access devices, what economies and benefits are derived that would drive the design of a backbone? Backbones provide many efficiencies not achievable from a meshed-access network, including:

- ☞ intelligent routing
- ☞ dynamic bandwidth resource allocation
- ☞ self-healing architecture
- ☞ flexible topologies and styles of design
- ☞ distributed or centralized network management
- ☞ flexibility
- ☞ traffic consolidation
- ☞ eliminate multiple paths with different types of traffic
- ☞ sharing of equipment and facilities by multiple locations
- ☞ economies of scale
- ☞ reroute and redundancy
- ☞ high-bandwidth switched-services platform

Public switched-data networks can also act as backbone networks to an access network in a hybrid fashion. Thus, the backbone design aspects of this chapter will refer to public switched-data network designs. When viewed from the backbone network, the network perspective must be global. As LAN interconnectivity grows in size, possibly company-wide, it is called an enterprise network. Either the enterprise network keeps a matrix structure (as in SNA networks), or subnetworks begin to form a hierarchical structure (as in many non-SNA or OSI multiprotocol networks). As the number of point-to-point trunks grows between access nodes, it becomes time to design the backbone network layer.

Figure 20.1 shows a WAN with six access/concentrator nodes and private lines between them. This is compared to the same network, this time with a three-node backbone providing a switched service to the access nodes. The access nodes could be packet, frame relay, DQDB IEEE 802.6, or ATM switches, bridges, routers, gateways, or any other network access or switching device discussed so far. This achieves many of the efficiencies listed above. Now to look at the characteristics of backbone networks and the requirements which drive their design.

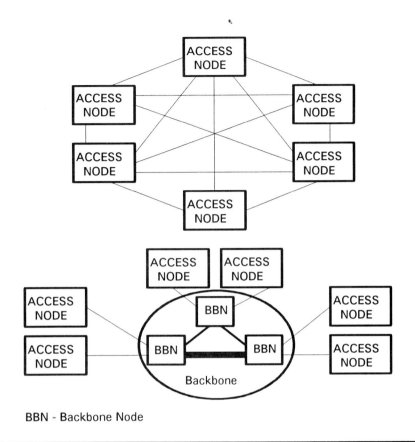

BBN - Backbone Node

Figure 20.1 Private Line WAN Backbone Compared to Switched Services WAN Backbone

20.1.1 Interfaces

The primary interfaces for the backbone design will either be access circuits from the access design portion (see Chapter 19 results) or direct-user access into the public switched-services backbone. The primary speeds for backbone interfaces are 56K bps, T1, T3, 100M bps, OC-N. Many of the specifics on these interfaces have been discussed in previous chapters.

20.1.2 Protocols

Many of the user protocols will be transparent to the backbone, but the design of the backbone may still have an affect on them. Performance degradation, delay, and other protocol factors may ripple back from the backbone to the user protocols. The backbone design must be flexible enough to accommodate multiple protocols and operating systems, whether switching them transparently or actually becoming involved in the intelligent operations of the protocol.

The backbones of the '90s are moving toward more intelligence in the control of the network and less intelligence in the actual handling of the traffic. Backbones must have the capability to interconnect diverse sites, switch or route all the protocols of the access network, and accommodate both in-band and out-of-band protocol transmissions.

TCP/IP remains the most common backbone network and internetwork protocol, and can be deployed in more ways than one. IP addressing schemes are used at both a network level and at the router information exchange level. Many new technologies offered by carriers and service providers will at some point be encapsulated within an IP packet. In fact, many commercial offerings are designed around IP networking. The IP protocol has also enabled private network protocols such as frame relay to be offered as a public network service.

It is very important to take into account the protocols passed over the backbone, even though they may be passed transparently. From the previous study of network protocols, it is known that routable protocols such as TCP/IP and DECnet can be either bridged or routed and are the preferred WAN protocols, while protocols such as LAT, IBM SRT, and NetBIOS (NetBEUI) can only be bridged. In fact, most methods of passing bridged protocols involve IP encapsulation or translation bridging. Determine whether the protocols are operating half or full duplex. Half-duplex protocols generate large amounts of overhead because the sender has to wait for acknowledgments from the receiving station before more information can be sent. Thus, over half of the available bandwidth is wasted with turnarounds. Also, many routing protocols such as RIP add extra overhead to the WAN through their constant exchange of routing table information. Newer routing protocols such as OSPF take up drastically less overhead (about 1% of RIP) than older routing protocols. This demonstrates the importance of minimizing and localizing the routing tables in each router. Some access designs will contain proprietary protocols within a local area. These instances show that the backbone designs should always attempt to use standard protocols for wide area networking.

20.1.3 Architecture and Technology

Usually the backbone is the same or one generation further advanced than the access network technology, and is almost always based upon a single architecture. Consider, for example, MCI's Virtual Private Data Service (VPDS). MCI offers frame relay access devices with an 802.6 SMDS backbone architecture and service. This will probably be upgraded to an ATM/SONET backbone as the SMDS hardware moves closer to the customer premises. This shows that the provider is one step ahead of the user in technology. Backbone designs should also be faster than the access devices, at least on a maximum burst basis. Historically, the WAN has been slower than the LAN because multiple LANs were connected via point-to-point bridges. 56K and 1.544M bps circuits could not keep up with multiple 10M bps and 16M bps LANs. Performance would degrade, file transfer would take longer over the WAN, and bottlenecks would occur. New MAN and WAN technologies such as SMDS and B-ISDN are breaking that bottleneck and providing the high-capacity backbone bandwidth users require.

Long-term planning is also easier with the backbone than with access, because capacity additions and technology changes can be easily migrated into the network in layers. In the access configuration, protocol, service, and technology changes will require replacement of the core switching equipment. This is one of the primary reasons many businesses are building access networks and using public switched-data services as their backbone. Why pay for the technology and be concerned with keeping it current when the network provider can incur most of the costs? Figure 20.2 illustrates this concept of a layered migratory backbone, where "Userland" is where the users reside, and each successive ring represents the backbone layer and then future successive access layers.

Another prime decision when designing backbones is whether to build a connection-oriented or a connectionless service, and their impact on each other if layered. Connectionless services such as cell-based IP, SMDS, and B-ISDN are actually more efficient than connection-oriented services such as frame-based frame relay. There is less node-to-node overhead involved, and therefore throughput is better. Connectionless services also avoid the delays inherent in setting up a connection. When a connection-oriented service such as frame relay is placed over a connectionless service such as SMDS, higher-level protocols are required to assure end-to-end reliability. Since TCP/IP, OSI, or other transport and network protocol schemes are used with frame relay, this is not considered to be a problem.

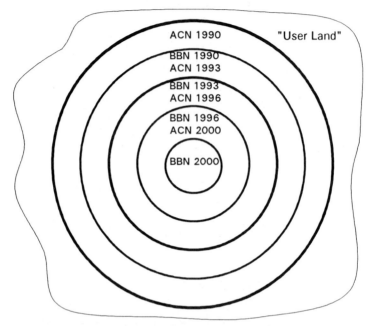

ACN - Access Node In Year 19XX/20XX

BBN - Backbone Node In Year 19XX/20XX

Figure 20.2 Access and Backbone Network Layer Migration Concept

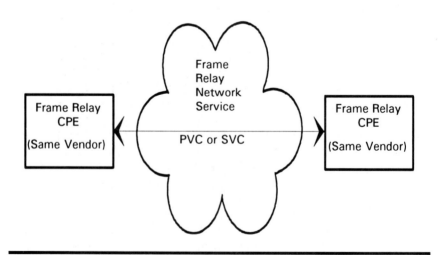

Figure 20.3 Frame Relay Network with Proprietary CPE

The primary concern in this case would be optimizing the packet size to ensure the amount of overhead generated is at a minimum.

20.1.4 Features, Functions, and Services

The features and functions of the backbone will depend upon the technology chosen. The real benefits derived from using a network backbone are the public network value-added, switched-services available. These services include circuit switching, packet switching, TCP/IP, frame relay, SMDS, ATM, and B-ISDN. Many of these services reside on the backbone and are fed by user premises access nodes such as PADs and FRADs. With these newer ultra-high-speed services offered by the IXCs and LECs, it is less expensive for the user to simply access them rather than build their own corporate backbone. These service backbones must also support many of the features and functions that operate on the access devices. Also, any new feature or function offered must be fully tested so that it does not degrade or affect the access portion of the network.

Remember the many idiosyncrasies of each technology and service presented in previous chapters, and plan on how to support them. Take frame relay for example. In fact, with frame relay, many CPE devices must be from the same vendor and have the same protocol settings to communicate, but the frames can be sent across the backbone of a different vendor. This is shown in Figure 20.3. This is due to the proprietary framing technique and passing of proprietary information between the common frame relay CPE devices. Common PID usage will eventually alleviate this problem. Congestion control is also a problem with frame relay, because each vendor implements it differently. Also, FECN, BECN, and DE may be ineffective if there is no intelligence in the CPE device to make use of it, or if there is no intelligence in the end device to translate to higher-level transmission protocols which are able to then regulate the flow. Even if they do get around to regulating the flow, the congestion may already be gone. Hitches like this must be planned for and contingency plans developed. Bottom line: understand the deficiencies of each service offered on the backbone and make sure the customer and user also understand them.

20.2 BACKBONE NETWORK CAPACITY REQUIRED

Backbone capacity is measured in both processing and bandwidth. Nodes are selected based on the amount of local, remote, and pass-through traffic they need to process. Their capacity is also measured by the amount of ports for access devices, trunks to other backbone nodes, and the bandwidth on the trunks between those ports. Backbone nodes are designed within utilization and loading parameters similar to access node design. Once the access loading is determined, the total backbone capacity can be found. It is then up to the designer to apply good judgment for each backbone node design — this ensures that the loading can handle both normal operating conditions and a single node or link failure analysis.

20.2.1 Backbone Node Selection

First, determine what percentage of the traffic will remain in the access design and what needs to be passed over to the backbone. This is found by interpreting the traffic matrix one step further: perform a backbone traffic matrix in the same manner as the access matrix, but this time instead of user application traffic use the access node traffic passed to the backbone. Then determine what percentage of traffic that is passed to each backbone node goes in and back out the same backbone node. Finally, determine how much traffic enters the backbone node and leaves toward another backbone node.

Figure 20.4 shows an example of traffic patterns for a twelve node network. Notice that each access node is trunked to a single backbone node. Forty percent of the user traffic remains local (to and back out of the same access node), 30% of the traffic remains within the same state or province (to and back out of the same backbone node), and 30% of the traffic must transit the backbone. Take this one step further. Assume that each access node is dual-trunked: with one trunk to two separate backbone nodes. This is shown in Figure 20.5. The same amount of traffic accesses each access and backbone node, however, only half the previous amount actually transits the backbone links. This takes the utilization of the backbone links down to 15% and allows either less links to be used (reducing the design by two backbone-to-backbone trunks) or keeps the links for even greater diversity and redundancy. These considerations will be used together with topology styles defined later in this chapter to choose the number and location of the backbone nodes.

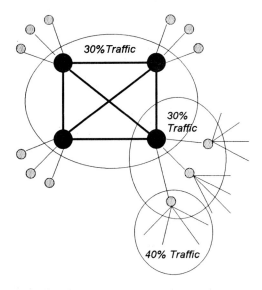

Figure 20.4 Typical Traffic Patterns — Single Trunk Access Nodes

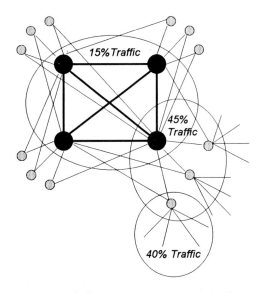

Figure 20.5 Typical Traffic Patterns — Dual Trunked Access Nodes

20.2.2 Utilization, Loading Factors, and Anticipating Failures

The same calculation from Chapter 19 for access networks can be used in backbone networks to calculate node and link utilization and loading. These calculations are made for trunks into the backbone nodes (labeled as "user ports" in Chapter 19) and the trunks between the backbone nodes (labeled as "trunks" in Chapter 19). Yet these calculations will be more precise because the traffic patterns between the access nodes are exactly known, and thus the backbone nodal design will be more accurate. There is also the theory that if the initial calculations were off, the errors will be multiplied in the backbone design! Also, at this stage loading factors should be in the same units of measure, for example, packets, frames, or cells per second.

It is easier to change the utilization and loading of the backbone if required. It is much harder to change utilization and loading on access nodes (as it directly effects the user there). Performance problems can also be isolated and fixed more easily on a backbone network than on the access side. It is also easier to throw more capacity (circuits) at the backbone. To make these changes requires the point at which additional capacity is required in the nodes and trunks. Response times at the beginning of service must be good, and not degrade during service. Sometimes the service (backbone) provider will introduce artificial delay into the initial design so that additional users will not degrade the service.

Some styles of backbone designs provide for high reliability during failure conditions by over-building the network and providing up to twice the capacity required at each network node. This way, half the links out of a node could fail without limiting the capacity of the network. The calculations for node failure are the same as those for link failure defined in Chapter 19.

20.2.3 Total Backbone Capacity

Given the backbone access traffic patterns, backbone capacity can be calculated in two ways. It can be calculated as the total capacity of the backbone network with a chosen number of nodes as a given, or it can be calculated by the total number of backbone nodes required, given the capacity required. The next set of calculations performs the first method, and the second is left to the reader (a simple manipulation of formulas).

Take the case where users are single-trunked to a network backbone, once a topology configuration is decided upon, and if the number of nodes in the backbone (N) and capacity of each node (c) is known, the total capacity of the backbone (T) based on the type of traffic it will carry can be determined. There are four major types of traffic patterns.

All traffic that enters a node leaves the same node. There are times when a majority of traffic that enters a node will leave the same node. The trunking from this node is minimal, and many good network designs operate by this concept, where the backbone nodal trunks are primarily used for backup and redundancy. The formula for calculating the total backbone capacity in this arrangement is: $T = (N)(c)$. Figure 20.6 shows a four-node backbone network with twelve access nodes. Each backbone node processes 50 units (packets/frames/cells) per second (ups). For this network,

$T = (4)(50) = 200$ ups.

Traffic originating on a backbone node is transmitted symmetrically to every other backbone node. This is the case for broadcast or public networks. The backbone nodes' trunks are primarily used for switching and the links between them are heavily used. The formula for calculating the total backbone capacity in this arrangement is: $T = (N+1)(c)/2$. Figure 20.7 shows a four-node backbone network with twelve access nodes. Each backbone node processes 50 units (packets/frames/cells) per second (ups). For this network,

$T = (4+1)(50)/2 = 125$ ups.

All traffic patterns are asymmetrical and are divided into user classes such as terminal-to-host and LAN-to-LAN communications. The backbone nodes' trunks are again primarily used for switching and the link usage varies. The formula for calculating the total backbone capacity in this arrangement is $T = (N^2)(c)/(2N-1)$. Again, refer to Figure 20.7. Each backbone node processes 50 units (packets/frames/cells) per second (ups). For this network, $T = (4^2)(50)/((2)(4)-1) = 114$ ups.

Users never talk to nodes on the same backbone (this is a multiple backbone scenario with backbone nodes are connected via WAN links). The applications for this are varied, but again this relates to a public network service. The formula for calculating total backbone capacity in this arrangement is: $T = (N)(c)/2$. Again, refer to Figure 20.7.

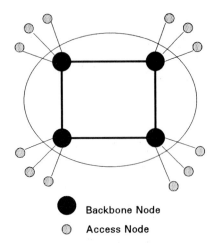

Figure 20.6 Backbone with Single Trunked Access Nodes

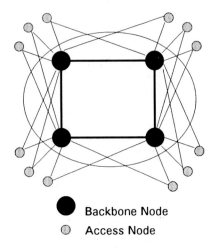

Figure 20.7 Backbone with Dual Trunked Access Nodes

Each backbone node processes 50 units (packets/frames/cells) per second (ups). For this network, $T = (4)(50)/2 = 100$ ups.

As the traffic patterns become more distributed, the network capacity decreases. This is because there are more options for traffic to use the limited bandwidth resources. This is particularly a problem with the next-generation networks covered in this text. Network designers are finding that the only way to add extra capacity to this type of network is to overengineer to begin with.

20.2.4 Route Determination

WAN elements such as broadband and baseband LANs, MANs, PABXs, data-circuit switches and packet (packet, frame, and cell) switches, bridges, routers, and other associated devices all route or switch data based on planned routes over access and backbone links. These routes are either static or dynamic, and are either pre-programmed (as PVCs) or decided dynamically (as SVCs). Some are routed "best way" as long as they get to the required destination. This choice of routing is accomplished node-to-node, hop-by-hop, and is based on a variety of variables (hop count, cost, bandwidth, priority, quality of line). Once these variables have been defined, then document and keep consistent route definitions. For example, the priority of traffic flow may be (from low to high) copper, coax, short-range microwave radio, VSAT, wireless communications, or fiber optics, and, after these priorities are met, traffic will be routed via the minimum hop count.

Do not let the users control the routing on the network. A prime example is the throughput class in X.25 packet switching. Users could misuse this parameter and cause loss of network control. Another example is the SMDS throughput class, which will be defined by the backbone.

Some technologies allow the traffic to route around network problems or congestion, the user is then notified of the condition, requiring use of a different route. For example, frame relay uses FECN and BECN bits to notify users of congestion. The user can then choose to slow down the transmission, take a different route, or ignore the notification. Other technologies such as ATM and X.25 can lose a trunk between backbone switches and the backbone network will automatically route individual packets or cells around the failure transparent to the user (except for the loss of the packet or cell, which would be noticed by an ATM user but not an X.25 user).

20.2.5 Future Capacity

Since user requirements on the average double annually (number of users, number of segments, and bandwidth required), extra capacity should be built into the backbone. Loading factors should be low early in the life of the network, and the network should have the capability to quickly add capacity when and where required. Again, this means both increased processing power and extra trunking bandwidth capacity at each backbone site. When turning up a new service, make sure that this extra capacity is available. For example, in frame relay, measure all the CIRs of each user input to the switch, calculate the total burst that may reach the backbone, and determine if the backbone can handle that burst or drop traffic. Then plan on providing just a little bit more than what is required. If the design practice of building a backbone with a higher level of technology and larger bandwidth pipes than the access is followed, the design will prove effective.

What will the future data communications network look like? Figure 20.8 shows a possible outcome. Notice that all voice and data ride over this network, and the traditional voice network hierarchy of the LECs and IXCs disappears and is replaced by an all digital data access network built upon a SONET infrastructure. Also notice the layering of the network which allows for expanding (adding) and contracting (reducing) layers of technology. Access can direct connect to any layer of the network, even to the SONET core.

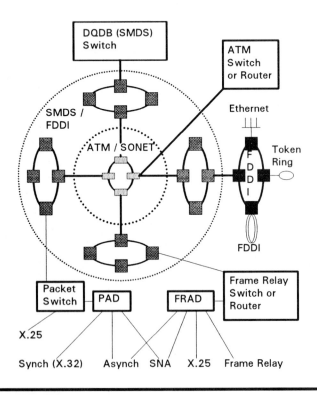

Figure 20.8 Future Data and Computer Communications Network

20.3   STYLES OF TOPOLOGIES

Backbone topologies come in two styles — those planned and those that just grow.   Recent trends of private and public data networks seem to fit these stereotypes.   Private networks, especially LANs that grow into MANs and then into WANs, tend to take on strange and asymmetrical shapes.   On the other hand, public networks respond to user needs and are often designed based on the procedures outlined in this book.   It is easier to plan a backbone design and then build the network than to try and modify an existing mesh of WAN connectivity.   Either way, the backbone network topology should be a function of the user applications, the access network topology, traffic volume, and range of connectivity (local to global).   Try not to tie yourself into a single technology or protocol suite.   Now we examine some of the more popular backbone topologies and the styles of design that go with them.

20.3.1   Star

The star design is similar to the star topology, where there is a central node serving as the hub node and all other nodes are connected via point-to-point circuits to the central node.   All communications pass through the central node.   (N-1) links are required for N node(s).   This star style is often used in an environment such as a LAN hub or ATM switch/hub.   The central node is often a multiport device that can handle large amounts of concentration, bridging, switching, or routing.   While this configuration provides a maximum of two hops, it is unreliable and susceptible to an entire network failure when the hub node fails.   However, network management has certain advantages.   Figure 20.9 illustrates a star-style backbone configuration.

A special version of the star topology is the distributed star.   This is typically used in LAN environments which use hubs as concentrators and tie the hubs together.   Figure 20.10 shows an example of the distributed star topology linking three hubs and nine Ethernet LANs.   This topology can also be called the star wired ring.

20.3.2   Loop

The loop design is similar to the loop or ring topology.   Each network node is connected to two other network nodes.   (N-1) links are required for N nodes.   This style is often used for distributed networks where nodes primarily talk to local nodes, or point-to-point communications are required over short distances (or cannot operate over extended distances such as MAN links). There is no maximum to the number of hops across this network, but it is reliable to the point of two link failures, which would then isolate the network into two pieces. The DQDB loop configuration is one example of this backbone topology. Figure 20.11 illustrates the loop style.

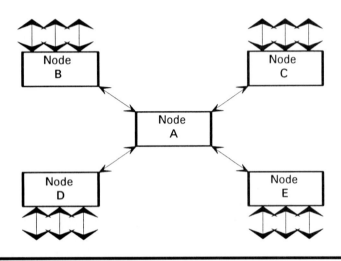

Figure 20.9 Star Backbone Topology

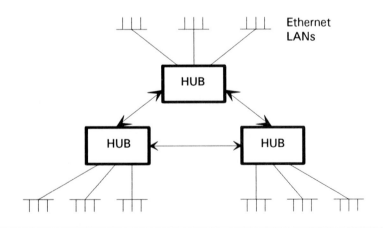

Figure 20.10 Distributed Star Backbone Topology

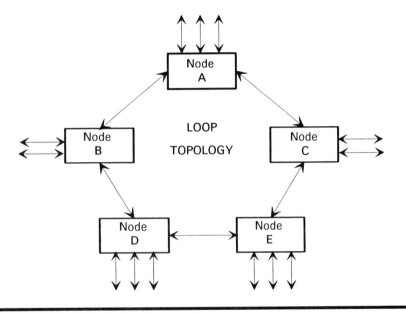

Figure 20.11 Loop Backbone Topology

20.3.3 Meshed and Fully Meshed

While the meshed network was discussed in Chapter 4, the degree to which a mesh is built depends upon the hardware and software expense of the ports, incremental cost of the links, network resource availability, and the amount of reliability and redundancy required. The number of links required for a fully meshed design is N(N-1)/2. The number of links required drastically increases with the number of network nodes. Obviously, a fully meshed network is highly desirable, but often cost prohibitive and rarely required. Again, design tools have the capability of modeling various scenarios of meshing. Figure 12.12 shows both a partially meshed and fully meshed backbone.

20.3.4 Daisy-Chained Access Nodes

Consider, for example, the packet network depicted in Figure 20.13(A). All concentrator PADs are dual homed to two high-capacity packet switches. While this provides for high availability, it wastes bandwidth if the applications are regional or their processing

624 Network Design

is distributed. Figure 20.13(B) shows an alternative, where each
PAD also acts as a switch through a daisy chain.

In this example the number of links required has been decreased
from 54 to 32. This has lowered the equipment costs of the larger
centralized switches, while retaining the connectivity requirements.
The distance between nodes will also be shortened. This network
will easily evolve into a frame relay network by replacing or upgrading the PADs to FRADs, and the same figure could be used to represent a frame relay or cell-switched distributed network.

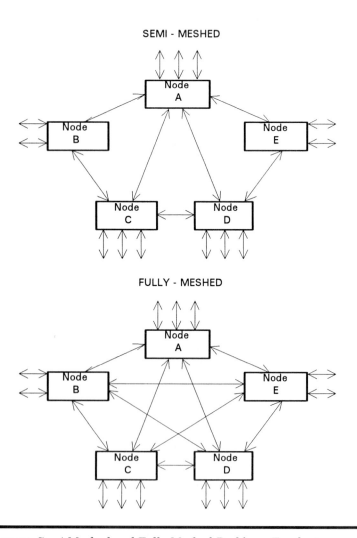

Figure 20.12 Semi Meshed and Fully Meshed Backbone Topologies

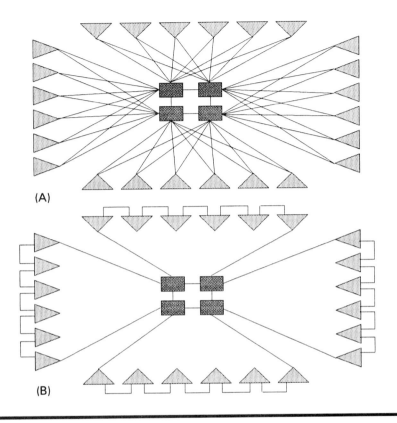

Figure 20.13 Daisy Chain Access to Backbone Network

20.3.4 Backbones Within Backbones

In the last section it was assumed that the location of network backbone nodes were dictated by the access nodes. In many cases, however, the backbone nodes and their topology are separate from the access node topology. In fact, many network providers often build multiple backbones within backbones in a hierarchical nature. This approach has both advantages and disadvantages, as we will soon see.

Figure 20.14 shows an example of a network where ten access nodes receive user CPE traffic. Each LATA is served by two access nodes. These network access nodes are configured in a loop-access topology, where each access node is connected to two other access nodes. These access nodes are then, in turn, connected to two different backbone nodes. Up to this point, all circuits are T1 speed. The backbone nodes are also configured in a loop backbone topology (with an additional link which limits network hops) and provide a majority of the inter-LATA high-bandwidth transport. The backbone provides high-capacity DS3 circuits between backbone nodes. This network topology allows each user a maximum of three hops to get to any destination, or two if the destination is in the same or adjacent LATA.

Figure 20.15 shows the same network, this time with an additional high-speed (level 2) backbone within the existing (level 1) backbone. Assume that the distance between access (level 0) nodes is increased, and the LATAs are now countries. The original backbone now has a network node at each country and spans a continent. The addition of a higher-level backbone will be built with three nodes in a fully-meshed configuration. Each level 1 backbone node will then connect to two level 2 backbone nodes. The new level 2 backbone will then provide high-speed (e.g., ATM) switching at gigabit levels.

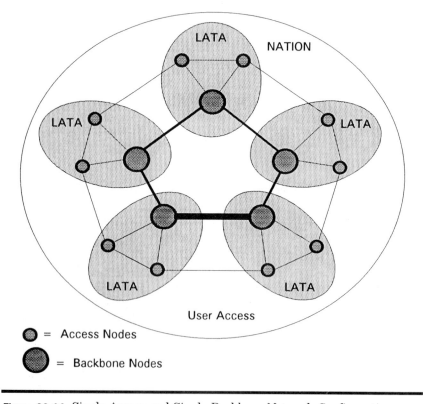

Figure 20.14 Single Access and Single Backbone Network Configuration

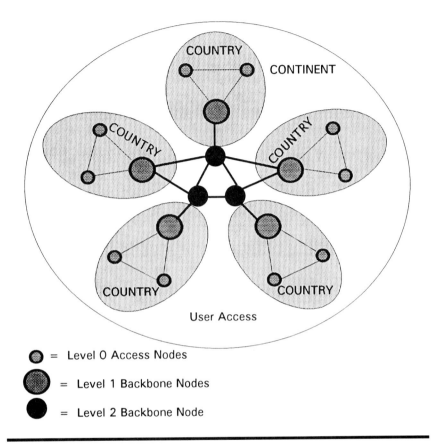

Figure 20.15 Single Access and Multiple Backbone Network Configuration

From this analysis, it is plain that networks are built using hierarchical structures offering not only redundancy and availability, but also a reduced hop count and an increased speed of traffic that transits long distances across the network. One example of this approach is the MCI Communications frame relay/HyperStream frame/cell switched network, which employs a frame relay access node arrangement with DQDB (SMDS) cell switches as the backbone network. MCI also plans to deploy an ATM and/or SONET backbone which would operate comparably to the last example.

20.4 BACKBONE TOPOLOGY STRATEGIES

As discussed in the last chapter, there are two styles of connecting local LANs; hierarchical and ubiquitous access. Now to look at the entire WAN, or enterprise backbone. The enterprise backbone provides not only interconnection of the departmental subnets, but also LAN/MAN/WAN interconnectivity through routers or switched networks (e.g., SMDS, B-ISDN).

The bottleneck in WANs is the point-to-point T1 facilities. This is the first place to look when the LAN-to-LAN traffic throughout begins to drop off, as the network device throughput is now usually much higher than the transport medium (routers have up to gigabit bus structures). When compression and concentration techniques fail, WAN services such as SMDS and B-ISDN take over. These services even exceed the expensive T3 facilities available, and utilize the bandwidth much more effectively. A few desirable topologies of each technology should be examined.

20.4.1 Desirable Topologies per Technology

With constant bit-rate traffic, the WAN connection of choice is often a multiplexer or a DXC offering circuit switching, with the private line being the most expensive solution (yet often used the most). For traffic of a variable bit rate nature, WAN connections include packet switches, bridges, and routers offering either point-to-point solutions (HDLC, frame relay, or SMDS), or accessing switched-data services. Combining the right topology with the right technology, and vice versa, will avoid design flaws. Make sure the device designed grows with the network, rather than limiting it through proprietary implementations or older multiplexer technologies.

Some devices can exceed the physical limitations of T1 and T3 circuits, but only in a local (limited distance) area. For example, routers can handle the new HiSSI interfaces, which allows allocation of bandwidth in 20M bps and 30M bps chunks, rather than the T1 constraints imposed by multiplexers. Also, some devices may be more desirable than others because of their value-added services, while others may save expenses. One example is using bridges as distance extenders rather than running coax cable.

For packet switching, the topologies available are quite flexible. For frame relay, all links are point-to-point unless they ride another switched-service platform. Also consider using integrated circuit and packet switch equipment, which provides good integration of packet and circuit switching at T1 speeds.

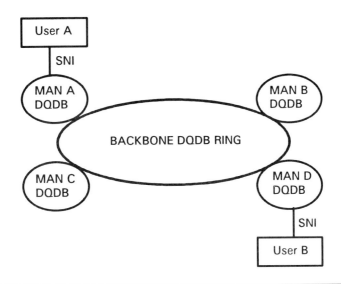

Figure 20.16 DQDB Access Rings Utilizing Single DQDB Backbone

With SMDS 802.6 networks, it may be desirable to connect each MAN cluster via point-to-point buses, rather than with an open bus. With the open bus configuration, all stations on the ring are limited to the bandwidth of the entire ring. If the ring is lightly loaded, or the bandwidth of the backbone far exceeds that of the access network, there should not be problems. If this is n true, and the backbone is heavily utilized, there may be the chance of high contention. With point-to-point bus, there is no contention problem, and the total bandwidth is shared between only two users. Also, in the open bus (ring) configuration, two bus failures could cut the network in half, although two simultaneous bus failures are highly unlikely. The point-to-point solution is more robust but also more expensive.

Another DQDB design methodology is to build DQDB access rings (MANs) and then connect these via a backbone ring. Figure 20.16 shows this concept, where user A connects to the network via an SNI interface to MAN ring A. The E.164 address is switched over the local MAN, across the WAN, and into the remote MAN ring B. The cells are then switched to the remote user B SNI.   s shown, there are many styles within each technology. Any of thes can be used as long as the access and backbone traffic requirements are fulfilled.

The next generation of backbones will be the ATM DXCs and switches. These devices together with SONET switches will be used to cross connect and switch electrical speeds to optical speeds in the gigabit range. In fact, there are now ring architectures in metropolitan areas of MAN rings built on SONET fiber transmission facilities.

20.4.2 Requirements Drive the Topology

One method for designing the topology of a network is to add the links which are absolutely required first, and proceed to add more links until all possible links are added where capacity is required for data flow. The designer then eliminates links and combines traffic over other existing links based on many factors such as shortest path, link cost, and quality of facilities. This method can also be performed by a design tool.

This method is shown in Figure 20.17. Figure A shows six links that have been added in order of the largest flow required to the smallest flow required (links 1 through 6, respectively). Figure B shows the selection of links 1, 2, 3, 4, and 5 to remain, and 6 to be deleted because it was too expensive not to route the traffic through node B (over links 1 and 3) rather than directly from node A to node D and vice versa. Figure C shows the final iteration where link 2 was eliminated due to only analog transmission facilities being available (traffic from B to C was rerouted from B to D to C and vice versa) and link 5 was eliminated due to its excessive distance (and traffic rerouted from A to B to D to C and vice versa). While this is a simplistic view of a network topology design, it does provide some insight into the method of selection for topologies of backbone links.

Never lose sight of the original user requirements during the backbone design. Understanding the effects that your backbone design will have on the user can help optimize the backbone design. This is a two-way process. Note that high-bandwidth access and backbone transport may not be required if the data can stand a bit of delay during times of congestion. But, if the opposite is true, and a backbone is designed with too few links, the extra network transport delay incurred across the backbone may have an adverse effect on requirements. When signing up for a carrier-offered service, the user needs to fully understand the carrier's access and backbone design and its potential affects upon the applications. For example, if a carrier is quoting 500 millisecond total delay across their frame relay network, how does this effect your file transfer sessions? Will this delay, when added to your access and egress delay to their service, affect your applications? These questions and many more from previous chapters must be asked of the carriers.

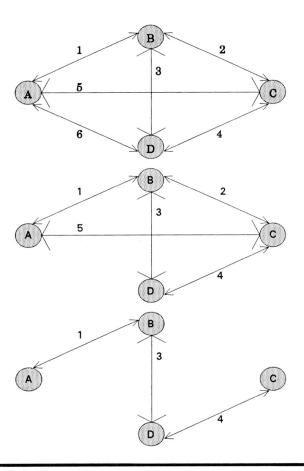

Figure 20.17 Shortest Path Design Methodology

20.4.3 Tariff Concerns

Where is the cost break-even point for a full-time private line circuit versus a switched circuit of lower speed? This varies based on tariffs and pricing schedules, but if the total daily usage is greater than three hours per day for the entire bandwidth, the calculations should be performed. Remember that switched services and frame relay are designed for bursty data periods and statistically multiplexed traffic, not dedicated-bandwidth-using-applications. If you are passing large volumes of constant bit-rate traffic, start looking at circuit-switch access. If passing large volumes of variable bit-rate traffic, start looking at switched services. Over the next decade, the United

States will be moving from voice and data private networks toward integrated voice and data networks. The network of the near future is a hybrid network where the user sends integrated voice and data to the carrier over a service such as SMDS/802.6 or B-ISDN, and the carrier splits out the voice and data for transport and services. The intra-LATA and inter-LATA tariffs are handled by the carrier, although many data services are not based on tariffs.

20.4.4 Hybrid Topologies

Many users today mix both leased facilities and switched services. Switched services often are used as the backup to the leased facilities. But this is changing. Services such as frame relay are providing users with cost efficiencies for bursty data, which is quickly overtaking the voice traffic at many corporations. Thus hybrid topologies must be dealt with. Many of the IXC service offerings span many technologies. The rule today is that when designing a private network, make sure it can be integrated (hybridized) with a public network-switched services. Public and private network hybrids allow control of critical resources to be retained while taking advantage of public services functionality and pricing.

When designing a hybrid backbone network, attempt to keep all speeds constant. Keep uniform bandwidth as much as possible across the wide area. The throughout constrictions will come from speed mismatches, such as an FDDI MAN which tries to communicate over the WAN through a T1 frame relay interface. Also, beware of multiple encapsulation schemes of hybrid networks, and understand that similar network devices often must be used at both network ingress and egress points.

Once a hybrid private/public network is built, it may require additional bandwidth out of the public side. Some carriers offer a bandwidth subscription service. Users can set up bandwidth accounts with the carrier service, where users just dial-up more bandwidth as required. This spare bandwidth is often offered by the carriers to multiple customers, at a lower priority and at lower cost to the customer, although it may not always be there when needed.

The Bay Bridge project at University of California-Berkeley created a bridge-router device which can bridge between SMDS and FDDI. While this provides a much needed WAN solution, it is still an encapsulation scheme where one or the other protocol is encapsulated and passed through. This can incur the same danger found with all multiple encapsulation designs — the problem of excess overhead in packetizing, segmentation, and the reassembly process.

20.4.5 Topology of the Future — Distributed Design

The primary attributes of distributed processing networks is the push to place the intelligence to process and switch data into the access devices, rather than the backbone devices. As seen previously, most backbone designs are either point-to-point trunks or access to a public network service.

Services such as SMDS and B-ISDN will offer greater efficiencies to networks with distributed designs. Before a company enlists the aid of ATM for isochronous transport, however, it must take a long look at SONET. SONET is the perfect means of isochronous transport, and may justify the wait. SONET may be the service for the next decade, but for now backbone support of isochronous services still finds its best fit with ATM (B-ISDN) — B-ISDN/ATM over DS3.

20.5 ADDRESSING

Addresses are used to direct traffic to a given destination. Addresses differ from names in that they form a communications system data structure defining the specific physical or logical location of an entity, device, or single access point, whereas names provide a human recognition symbol of a network entity or device. Addresses can be assigned to entities or devices, and a route defines the other devices with which a given address can interface (thus specifying the from and to). Addressing schemes should be nonproprietary when possible and should be used consistently across the network. In this section we will discuss various types of common addressing schemes and how they influence the network design.

20.5.1 Levels of Addressing

Addresses can be assigned in either a flat or a hierarchical scheme corresponding to the access and backbone style of design. For example, TCP/IP addresses are arbitrary in nature within the scheme of the Internet address classes, while E.164 are hierarchical with geographic significance (individual IP address assignments can also be hierarchical with geographic significance). It is obviously much easier to control and filter hierarchical addresses, but there is also the disadvantage of requiring change of addresses every time a user changes his location.

TABLE 20.1 Addressing by Technology

Technology	Type	Size
Circuit Switching	none	none
X.25 Packet Switching	LCI	12 bit
X.121	LCI	14 BCD digits
Frame Relay (CCITT)	DLCI	4 octet
Frame Relay (ANSI)	DLCI	2 octet
802.6	E.164/D15	16 bit
802.6	E.164/D15	48 bit
802.6/SMDS	E.164/D15	Individual 60 bit
802.6/SMDS	E.164/D15	Group bit
802.6/SMDS	E.164/D15	NPA/NXX
802.X LAN	MAC	16 to 48 bit

20.5.2 Types of Addresses

Throughout the technology chapters, many types of addressing have been discussed. A summary is found in Table 20.1.

Each network architecture has its own type of addressing. One example of a predominant architecture is DECnet. DECnet Phase IV addressing has both an area level 2 address and a node-specific level 1 address, and can support up to 63 areas with each area supporting up to 1023 systems. System names can be up to 6 alphanumeric characters each. These architecture-specific addressing schemes must be mapped to any network addressing scheme (such as network IP address design, which will be discussed next) to eliminate any addressing discrepancies. The other architectures have similar address designs.

20.6 WAN IP ADDRESS DESIGN

Internet Protocol (IP) was designed by the Internet Engineering Task Force (refer to Chapter 3), and has formed the basis for inter-networking computers (called hosts). The Internet is a huge world-wide network, using IP as the layer 3 global network addressing scheme. Each user is assigned a unique IP address of 32 bits, and corresponds to the following format:

XXXX.XXXX.XXXX.XXXX,

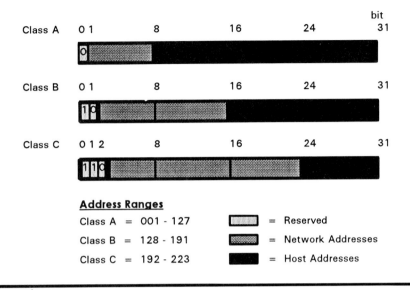

Figure 20.18 Internet Assigned Address Classes

where the pattern can range from:

00000000.00000000.00000000.00000000
to
11111111.11111111.11111111.11111111

A more concise way of writing this is 0.0.0.0 to 255.255.255.255 where there are 2^{32} or 4,295,000,000 possible addresses on the entire network. These addresses are allocated by the Internet committee, and conform to address classes A, B, and C. These network classes are provisioned according to network size. Each class is presented in Figure 20.18 along with the appropriate address assignments. A Class A address could be 20.0.0.0 with the possibility of 16 million host addresses (2^{24}). A Class B address could be 150.250.0.0 with a possibility of 64,000 (2^{16}) hosts. A Class C address could be 150.250.004.0 with a possibility of 255 (2^8) hosts. IP routing becomes more efficient when a large number of user machines are connected to a smaller number of network machines, which is typical of distributed computing environment designs.

IP addresses must remain unique. This is why their distribution and control is strictly administered by the Internet committee. If users on two separate networks with the same IP addresses are internetworked there could be network problems unless the network devices filter the address and translate it to another protocol-addressing scheme.

IP works with TCP for end-to-end reliable transmission of data across the network. TCP will control the amount of unacknowledged data in transit by reducing either the window size or the segment size. The reverse is also true, where window or segment size values can be increased to pass more data if error conditions are minimal. Chapter 6 contains a detailed discussion of TCP/IP operation.

20.6.1 IP Address Topologies

There are many methods of implementing an IP address scheme. The first determination is whether the network will follow a hierarchical or matrix topology. With a matrix topology, each user (workstation, host, server) will be assigned a single IP address. Communications will be peer-to-peer. As discussed previously, this style of addressing scheme would be fine for a small network with few users, but becomes unmanageable with large networks.

For larger networks, a more manageable design is the hierarchical topology. IP addresses are assigned in blocks based upon geographic region and where (by level) the user resides in the hierarchy. Boundaries are drawn by what are called subnet masks. These subnet masks provide a method of segmenting and reducing the number of network addresses that network devices (access and backbone) need to know.

For example, the matrix IP addressing would resemble the old style of sending mail, where the sender would simply place a person's name, address, city, and state on the letter. Each post office or stop would pass the mail along to the city, where the person would then be located by name and address. This was fine for a small amount of cities with few people where everyone knew each other, but would not work today. The hierarchical mail scheme adds a zip code. Today, each city has a minimum of one zip code, and larger cities have multiple zip codes. In fact, some companies have their own zip code (XXXXX-YYYY). Now each city can sort mail based on zip code (XXXXX), and filter out any mail not destined for that city. Then each post office in any area of a city can sort based on the last four digits of the code (YYYY) and is able to filter all addresses for that given zip code extension. The (XXXXX) portion of the zip code acts as the subnet mask, and the (YYYY) acts as the specific user zip code extension. This is an example of what subnet masking does for IP addressing. Let's look at some specific examples.

20.6.2 Addresses Based On Topology

Now that the style of topology for the network has been determined, it must be determined how many addresses will be allocated at each site for each level of the hierarchy. For example:

❶ each workstation receives 1 host address
❷ each LAN segment receives 64 host addresses
❸ each floor receives 256 host addresses
❹ each building receives 1024 host addresses

These numbers indicate a sample number of addresses per unit of measure. This allocation will be accomplished, along with segmentation and hierarchical design practices, through the use of subnet masks. Open Shortest Path First (OSPF) routing, an Interior Gateway Protocol (IGP), can be used in an autonomous system where subnet masks are required. Routers can then forward and filter traffic based on subnet addresses. They can forward traffic by mapping the IP host address to the router's subnet address. Now to discuss subnet masks.

20.6.3 Subnet Masks

A subnet mask is a method of segmenting the addresses to various subnets across the network. These masks should be applied as follows: one mask for the network, one for each geographical location (e.g., building), one for each floor, and one for each device (e.g., router). A subnet could be an entire network, or an individual LAN segment with X workstations. Some examples include a subnet mask of:

- 255.255.255.255, or 11111111.11111111.11111111.11111111, would include all nodes (broadcast)
- 255.255.255.252, or 11111111.11111111.11111111.11111100, would allow 64 subnets, with four assignable addresses per subnet
- 255.255.255.248, or 11111111.11111111.11111111.11111000, would allow 32 subnets, with eight assignable addresses per subnet
- 255.255.255.240, or 11111111.11111111.11111111.11110000, would allow 16 subnets, with sixteen assignable addresses per subnet

and so on until a final subnet mask of 255.255.255.128, or 11111111.11111111.11111111.10000000, would allow 2 subnets, with 128 assignable addresses per subnet.

Now take a look at the next level of address scheme, on a more local level. These subnet masks are used with a small number of locations with many users at each location. A subnet mask of:

- 255.255.252.000, or 11111111.11111111.11111100.00000000, would allow 64 subnets, with 1,024 assignable addresses per subnet

- 255.255.248.000, or 11111111.11111111.11111000.00000000, would allow 32 subnets, with 2,000 assignable addresses per subnet

- 255.255.240.000, or 11111111.11111111.11110000.00000000, would allow 16 subnets, with 4,000 assignable addresses per subnet

and so on until a subnet mask of 255.255.128.0, or 11111111.11111111.10000000.00000000, would allow 2 subnets, with 32,000 assignable addresses per subnet.

Appendix D shows all available subnet addresses and their corresponding number of assignable addresses per subnet for a Class C address. Next is an example of IP WAN design, so that the interdependencies of subnet masks, IP addressing, routing, and the summarize capability of routers will be understood.

20.6.4 Sample IP WAN Design

Now to build a WAN using IP routing and addressing schemes. First, look at the entire network IP addressing scheme. Consider a backbone subnet mask of 255.255.255.252. This mask has been chosen because it can have up to 64 locations, with each location having up to four subnets. An example is shown in Figure 20.19. This figure will be followed for the entire example.

When addressing the internodal links, a subnet mask is required to identify the addresses on each link between nodes. Subnet masks are placed on both access circuits and trunks. For example, a subnet mask of 255.255.255.252 would allow for 3 ports to be addressed on each internodal link. In this example, the link between ports on Node A and Node B is address 150.250.004.004 and 150.250.004.005, respectively. The summary address of that link is 150.250.4.4 (we will discuss summary later). This IP subnet address, and all others discussed, are programmed into the router during installation. While point-to-point connectivity meets this criteria (only requiring two end points, while this example uses 3 total per link), a multidrop or MAN connection may not.

Therefore, a subnet mask of 255.255.255.248 may be used to identify up to 6 ports per link. The designer may ask: Why not simply use a 255.255.255.128 mask and allow up to 127 ports per link? This should not be done, because the designer is limited to a certain number of addresses, and each address used by a trunk could have been used for a workstation or host. This relationship becomes clearer as we progress. In this example, the entire four-node router subnetwork can be summarized with the address 150.250.4.0, which takes into account links 150.250.4.4 through 150.250.4.12.

Now look at the ports on the user side of the network. Each location or site now has four subnets worth of addresses to use. Taking the example shown, where four subnets are routers A, B, C, and D, assume these subnets are a floor of a building. Based on the class C address obtained, site A uses addresses 150.250.008.000 through 150.250.011.255. Site B uses addresses 150.250.012.000 through 150.250.015.255, and so on. Each address corresponds to a single workstation or host port.

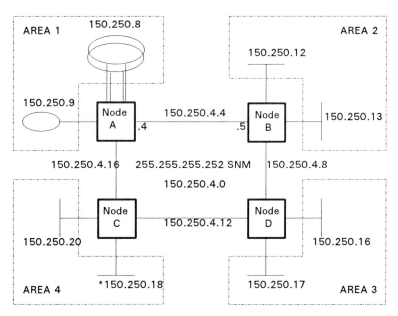

Figure 20.19 IP Network Addressing Example

Subnet addresses can be "summarized" by routers to reduce the number of addresses (and address passing) across the WAN. Through summarization, all subnets in area 2 are summarized by router B as address 150.250.12.0, even though there are many other addresses present in that area. In the routing tables of the routers, only the summary entry is used. If the entire network were to be summarized, a mask of 255.255.0.0 for the network address 150.250.0.0 would be used. Note that network summaries cannot be used when IP addresses are assigned at random, such as in a matrix design.

Applying subnet masks to this address scheme will segment user communities within each area. The summary feature can also be applied to reduce the number of addresses broadcast between network devices (routers). These network summaries can hide the complexities of the local areas.

Now to explain a subnet mask for each area. If a subnet mask of 255.255.248.0 is used the 32 subnets at that location would have eight subnets each.

Next, look at each router's area. If the subnet mask of 255.255.255.192 is used and the addresses 150.250.008.000, 150.250.008.128, 150.250.008.128, and 150.250.008.192, this would yield four subnetworks (LANs), each capable of 62 nodes (workstations with individual IP addresses). In our example, two LANs have been assigned to each subnet. At router B, the first Ethernet LAN address is 150.250.12. Sixty-four workstation addresses should be more than enough for each LAN. Any LAN exceeding 64 workstations should be segmented into small segments.

Now consider using the address "summary" feature of the routers. Using address summary, router B would broadcast only the address 150.250.12 and the addresses 150.250.13, .14, and .15 would be summarized (and assumed to be off that router). This is done because the subnet mask dictates that only four subnets can exist for each device. If this were the case, only four LANs could be segmented off router A.

Now to analyze this example. This example uses subnet masking on both the trunk and access side of the router. Also the summary feature of IP addressing has been used in conjunction with subnet addresses. The Class C address 150.250.XXX.XXX is used in this example, with addresses that range from 150.250.000.000 to 150.250.255.255.

For Area 1, the summary entry of 150.250.8 allows only this address to be advertised to the network, but the address also includes 150.250.9-10-11-12. The subnet mask is 255.255.252.0, allowing for 64 addresses on that subnet (or LAN).

For Area 2, the summary entry of 150.250.12 allows only this address to be advertised to the network, but the address also includes 150.250.12-13-14-15. The address 150.250.18 in Area 4 falls within the subnet defined in area 3 (150.250.16-17-18-19). Any IP packets destined for an address within the mask 150.250.18 would end up in Area 4. This is called a "routing hole". Routing holes are hard to troubleshoot after a network is operational, and using a strict hierarchical addressing scheme such as the one defined here will decrease the chance of routing holes.

Remember that the subnet is transparent to the higher layer protocols. If TCP is running on two hosts connected by one or more local and/or wide area subnets using IP, the TCP/IP does not really care about the subnet employed. The subnet is, in effect, transparent. The subnet could be T1, X.25, frame relay, SMDS, or FDDI and it is still transparent to the networking device (e.g., the router). Delay may become a problem as higher-level protocol windows and timers interact, but that problem falls in the area of network management.

20.6.5 OSPF vs. RIP and Default Gateways

Every IP addressing scheme should be designed in conjunction with the routing protocol. Note that it is more efficient to use RIP routing protocol in the local area subnets and OSPF in the WAN. The fact that summaries have been used in OSPF reduces the amount of address updates which transit the network. RIP is more efficient in the local area subnet where users know who the other users are and the environment is less dynamic, and OSPF is not implemented on most hosts. If a single router is the only interface to the LANs within an area, and if there is only one interface per LAN, RIP routing is not required. A default workstation gateway entry will allow the workstations to find problems with the network through a default gateway address (in our example, the router port to which the LAN connects).

If the packet has a remote destination, the router will send it to a specified IP address on the router (default gateway IP address). In this example, the subnet mask on the router would be the same as on the workstations. Thus, if the network ID, the gateway, and the subnet mask is known, the local areas are not required to run RIP. The default gateway function can be used since each LAN has only one interface. Remember that with RIP, each LAN device only advertises its status, but does not listen to advertisements. One other point, the domain name server, if used, is the same as the LAN server. Try not to use domain name servers, however, as the tables become too large.

IP routes effectively between subnets, but not within a subnet. Thus, an optimized network has different routing protocols within the subnet than those used outside the subnet. Keep RIP off IP backbones, because it generates too much overhead (and the larger the network, the more the overhead) due to transmission of entire routing tables as updates every 30 seconds. Also, RIP does not have password authorization. OSPF and EGP use filtering or subnet masking to filter out what addresses are broadcast to the entire network. Effective filtering and masking can create efficiencies in the network design.

Global addressing varies based on the service, technology, and protocol being used. In a mixed environment, IP addresses are individually mapped to each technology's specific addressing scheme (such as DLCIs for frame relay).

20.7 TOTAL NETWORK MANAGEMENT

Network management is much easier to administer at the network level than at the access concentrator or even user-device level. A single WAN management platform could manage the addressing, bandwidth management (e.g., congestion control, throughput classes, synchronization), administration, and performance management activities. Network management for the entire network is covered in detail in the next chapter.

20.8 TOTAL NETWORK TIMING

Timing is always an important aspect of each design, especially with technologies such as SONET, whose success relies on error-free synchronization. Timing is also important between the CPE and network access devices. External timing should be used when possible, because internal timing sources, while accurate to the originating device, may not be accurate to all network elements. Also, the higher the transmission speeds, the more critical timing becomes. Timing problems can cause problems such as line interference, data unit slips (frame slips in frame relay) which lead to loss of data, interruption of service, or a general decrease in the reliability of transmission.

Clock sources are rated in Stratum levels from one to five, where Stratum one is the most accurate and five the least accurate. Some Stratum one examples include the Basic Synchronous Reference Frequency (BSRF) and the DOD Loran-C (Global positioning system). Clock sources such as "Stratum one" assure no more than one frame slip per day.

20.9 TUNING THE NETWORK

Network tuning should be performed after the network design is finished, as well as at specific intervals in the lifetime of the network. Four specific areas, when tuned, will increase the efficiency and throughout of both the access and backbone design. These include optimizing packet, frame, and cell size, limiting the segmentation of user data by lower level protocols, decreasing the overall port-to-port transfer delay, and using window size changes to flag potential network degradation. Frame relay CIRs must also be checked for how they may be used and possibly abused. Not mentioned are the many hardware, software, and protocol-specific tuning parameters covered in previous chapters.

20.9.1 Packet/Frame Size

In packet-switched networks (including frames and variable length cells) there is a trade-off between large and small packet sizes. When small packets are used (and each packet has a small amount of data), the amount of overhead increases. This causes a disproportionate amount of overhead generated versus data passed, and the data throughput of the line degrades. Remember the definition of throughput — how much *data* a user can pass across a given circuit or device. Small packets have the advantage of better response time and less chance of errors, and therefore a reduced chance of retransmission.

Consider the case of larger packet sizes. A higher throughput can be achieved by placing larger blocks of data within each packet/frame/cell and thus creating a larger packet/frame/cell. While this improves throughput, there is a point where the retransmissions due to lost packets, errors, or buffer and transmission delays will cause an actual degradation in throughput.

Figure 20.20 shows a U-shaped curve depicting the phenomenon of small versus large packet sizes based on achievable throughput.

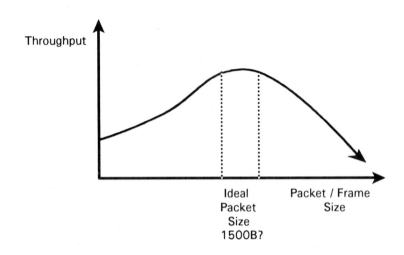

Figure 20.20 Ideal Packet Size Curve

This example uses an IP packet network, where the throughput decreases with very small packets (too much overhead) and with very large packets (IP begins to segment packetized user files less than 1500 Bytes in length). This does not show the effect of increased use of buffers and the delay imposed on the transmission for large packet sizes. In the example the optimal packet size is around 1500 Bytes.

The packet size is a balancing act, where additional factors influencing the size include:

* how long it takes to read-in a packet
* buffer space taken by held packets
* Packets per second dropped
* time to forward packet to next device or destination
* mix of protocols to bridge/route

among many others. In addition, each technology and service has its own operational factors, such as frame relay, where each node must read in the entire frame before transmitting. In this case mid-size and smaller frames process faster through the network. The best method for achieving optimal packet size is to tune the network based on the performance observed.

When vendors rate their packet processing capability, they often do so at the smallest packets which can be passed and sometimes containing no data. FDDI is one example where the PPS processing should be around 15,000 PPS regardless of packet size. Some vendors support this speed, but only at small packet sizes (64 or 128 Byte). Some vendors perform well at smaller packet size (<64 Byte), but worse at higher packet sizes (> 256 Byte).

20.9.2 Segmentation

Try to reduce the amount of segmentation at both the user file transfer level and at the access and backbone technology transport level. Let us review our IP example. When IP addressing is used on higher level protocols with large file or message sizes, segmentation occurs. Thus, if an NFS file of 6000 Bytes is encapsulated in IP, it is cut up into four data units of 1500 Bytes each and encapsulated with the IP header. This fragmentation causes even more overhead in the transmission. X.25, with an average packet size of 128 Bytes could add an exorbitant amount of overhead to this file transfer. Ethernet is similar to IP, with a maximum packet size of 1500 Bytes. Thus, when many layers of encapsulation are encountered across the network, packets may have to be broken down at each new protocol level.

20.9.3 Port-to-Port Data Transfer Delay

If data is routed at a portion of the maximum data rate, such as 12,600 PPS of a maximum 14,840 PPS (of an Ethernet LAN), the throughput is down to 85%, or 15% overhead is introduced by the device. This delay becomes more appreciable the larger the file size. Much of this delay and overhead depends on internal architecture and software protocol handling of the device. Actually, large file transfers should not use the minimal size packets for which the 14,840 PPS figure applies, but much larger packets for higher data throughput. Many devices implement multiple ports per interface card, thus reducing this effect. Also, the higher the bus speed, the faster the data rate between interface boards and the central (or distributed) processor(s).

20.9.4 Window Sizes

Window sizes can be tuned at each level of the X.25 packet switch network, or in the TCP portion of the protocol stack (usually in the user or access device). These windows provide transmission flow control. The window size determines how many packets can be outstanding on the network before an acknowledgment is received from the last unacknowledged packet sent. This procedure is fully explained in Chapter 8. In packet switching, high error lines should use a window size of 2 and low error full-duplex lines should use a window size of 7. TCP, on the other hand, will adjust its window size based on the current network throughput. If there are more errors in the network, it will *decrease* window size, if less errors, *increase* window size. Increased window size provides increased throughput but requires more memory and buffers in the network hardware and software, and can cause more problems that it reduces. Network performance problems can be determined by trending the window size changes and determining where additional throughput may be required in the network.

20.9.5 CIR and Bursts

Frame relay was designed to allow network access for multiple variable-bit rate users. These are the same users who once paid for low-speed private lines to connect local computing and the communications environment. When these users have a dedicated 56K bps private line, their bursts of data would take three seconds to transmit. The rest of the time their line would remain at low utilization. This is shown in Figure 20.21(a). Now, they can attach their LAN to a router and share a frame relay access to a public switched network service with four other LANs. Each of their bursts now only take 200 milliseconds to transmit the entire file, because during these bursts there is a good chance that they can get all the bandwidth available on the T1 up to their CIR (shown in Figure 20.21(b)). In fact, they may even be able to burst *above* the CIR during times of low network utilization (such as peak #3). In this case, their file transfer takes even less time (150 milliseconds).

The concept just described works well as long as the user does not try to use the frame relay interface for cheap bandwidth. If the user overloads the service and places many users on the same frame relay interface, the user will have to contend for the same resources at the same time, against many more users. Let us say that the owner of the frame relay interface and router decides to place sixteen LANs on the interface. If the increased traffic load of each LAN is proportional, our user will still be able to burst, but chances are he will only receive 128K worth of shared bandwidth during his file transmission. Now he is back to the original problem of the old 3 second delay during bursts. These calculations all assume a constant total network load. A different scenario needs to be analyzed for usage-based service.

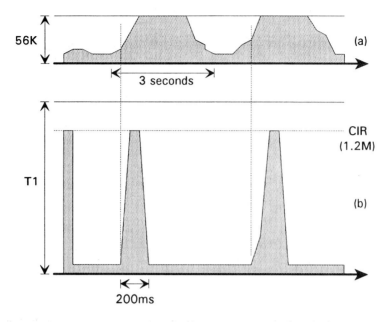

Figure 20.21 Effect of Frame Relay Service CIR on User Traffic

This example shows that while users can intelligently use the CIR and maximize throughput, their savings are passed off to the users in the form of delay. CIRs are allocated to allow the maximum burst on the line and still guarantee transmission. They were not designed to be the average transmission speed and definitely not the baseline speed of the circuit. Two measures of bursts are the committed burst size for forward and backward traffic and the excess burst size for forward and backward. Each user should be mapped according to these criteria to get an accurate picture of what the circuit can expect during peak periods. Ideally, users should move toward usage-based service with fully meshed PVC configurations. This will allow the greatest user flexibility, along with the greatest cost savings (if the service is priced cost effectively compared with CIR-based services). At the time of printing, only one service provider offered usage-based frame relay service — MCI Communications.

20.10 REVIEW

In this chapter the backbone network design is completed. This network design supports enough capacity to meet existing user access requirements, as well as future network growth. The backbone style and topology was chosen based on many factors, such as user requirements, the access and backbone technologies, and future networking concerns. Multiple techniques of network addressing were covered, including an example of a global IP addressing scheme utilizing subnet masks and the new router "summary" feature. The need and importance of global network management and network timing was stressed, and the chapter was ended with tips on tuning the entire network through packet, frame, and window sizes, avoiding segmentation, and minimizing port-to-port delay.

21

Operations and Maintenance

After the access and backbone network design is completed, the next important step is documentation. This requires both an engineering plan which steps a user or manager through the entire network, and an operations and maintenance manual which provides the support structure for both implementing and maintaining the network. The responsibilities of each department must be identified, and documenting each will help to make implementing the design a success. A detailed analysis of network management is presented, covering the much disputed Simple Network Management Protocol (SNMP), as well as CMIP/CMIS and OSI network management plans. Billing and security are briefly discussed, as is training. Finally, some soul searching must be done to decide to exercise some constraint. It is now time to stop designing and start implementing the network.

21.1 DOCUMENTATION

Documentation is one of the most important steps in the network design process. Documentation helps people understand the design, and allows them to remember the assumptions made during the design. Documentation should never be an afterthought. Both designer and user will suffer from poor documentation. It must be developed during the project.

The four primary documents required to support a network design are the business case document, the user specifications and acceptance document, the engineering plan, and the operations and maintenance (O&M) document. It is assumed that the first two documents were completed before the design began. The engineering plan will be developed throughout the design and distributed when the design is completed. The O&M document will also be distributed after network design completion, but requires specific inputs from the engineering plan.

21.1.1 The Engineering Plan Document

The engineering plan explains the history behind your present connectivity, the reasons for the network design, what the network will support and why it needs support, and the entire design from application to backbone connectivity. Of course, it should include everything discussed in this book referencing design. Capacity planning could also be included in the plan, or it could become a separate document.

Specific information in the engineering plan includes :

- ✦ node configuration
- ✦ circuit diagrams
- ✦ requirements matrix
- ✦ physical requirements
- ✦ locations
- ✦ bridge/router/gateway/ID addressing
- ✦ segment/port name/ID addressing
- ✦ serial numbers
- ✦ model numbers
- ✦ LAN/MAN/WAN addressing (e.g., Ethernet, E.164, TCP/IP)
- ✦ current software version level
- ✦ current hardware version level

21.1.2 THE O&M DOCUMENT

The Operations & Maintenance (O&M) document presents all information required to install and maintain the network. Many sections of the O&M are taken directly from the engineering plan, and many sections represent standard operating procedures for installing and maintaining network equipment. Some of the topics represented in the O&M document include:

* System/Network Description
* Physical/Topological Description
* Logical/Functional Description
* Design of Network (reference design document)
* Budgetary Impacts
* System/Network Administration
* OE/OP Procedures
* Network Management Procedures
* Installation and Testing
* Support Structure and Escalation Procedures
* Hardware / Software Maintenance and Sparing
* Test Equipment Requirements
* Training
* Vendor Specific Documentation
* Glossary
* Acronyms
* References

At the site-specific level, the details provided include:

* installation power and grounds
* implementation and cut-over
* naming and addressing conventions
* hooking up all devices
* testing all devices
* floor plan construction
* layout of LAN/MAN
* topology
* device locations
* wiring
* shared resources locations
* network management/administration locations
* maintenance schedules

21.2 RESPONSIBILITY MATRIX

Many groups within the company participate in the network design and implementation. Active participation from all groups insures a smooth implementation and continued operation of the network.

These group responsibilities are only guidelines, as many of these groups fall under a single engineering and support organization.

Planning — provides future architecture direction, plans on how future services will integrate and on what technology platforms they will ride, and design concepts to engineering.

Network Engineering — provides all engineering from the user requirements through the access design (and backbone design), initial design and continued engineering support after network roll-out, and customer interface for engineering applications.

Service Engineering — provides the same service as engineering, but with the software, control systems, and network management of the network, with more of a software and services orientation.

Order Entry/Order Provisioning (OE/OP) — orders the required access and backbone circuits, tracks circuit orders to completion, and provisions the required circuits.

Network Administration — provides all node, circuit, address, and other administrative responsibilities for the network, configuration management, database management, security administration, and performance reporting.

Billing — handles either billing of the customer or paying the bills for the network transport, access, hardware, software, and any other cost aspect of the network.

Operations — installs the network devices (hardware and software), controls the operations and maintenance of the network once it is installed, monitors the network management, support structure for trouble reporting and trouble management, provides performance monitoring, customer support — both centralized and field operations, handles network outages, and performs preventive maintenance,

Testing — tests and troubleshoots all applications on the network access devices, as well as over the backbone, provides component to system to end-to-end network testing, and certifies vendor hardware and software.

Training — ensures all personnel are trained on all network systems, and trains users on the network applications and services.

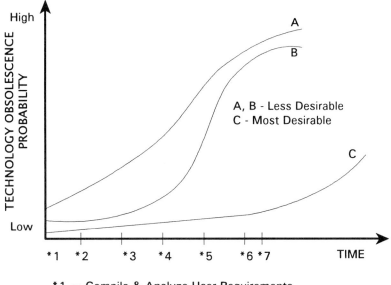

* 1 = Compile & Analyze User Requirements
* 2 = Perform Traffic Analysis / Sizing
* 3 = Choose the Technology
* 4 = RFI, RFP, Vendor Selection
* 5 = Design the Access & Backbone Network
* 6 = Order, Delivery, & Implementation of Network Equipment
* 7 = Network Operational

Figure 21.1 Project Life Cycles

Program Management — manages the project resources from design conception through implementation, maintains overall responsibility for the project, manages project cycles, and the budget. Most importantly, program management gets the buy-in and cooperation from all support groups listed above early on in the project, and begins the cycle early. Figure 21.1 shows average project cycles, from technology which rapidly becomes obsolete (curve A and B) to newer technologies (curve C).

21.3 DEFINITION OF NETWORK MANAGEMENT

Ambrose Bierce, an American author in the 19th century, published a book called *The Devil's Dictionary*, in which he gave satirical definitions to various words. One wag recently defined Network Management (NM) in Biercian terms as:

"...the art of making a network appear to be running smoothly when in actuality it is teetering on the brink of disaster."

21.3.1 Users Demand Better Network Management

There is some reality and truth behind the above definition. Customers are demanding more control of networks with assurances of enhanced reliability and disaster recovery. Vendors of equipment and services have traditionally furnished some type of control and management. As technology has improved and become more complex, users are now demanding better ways to control and manage their systems. The network must appear to be more than just running smoothly with the "brink of disaster" lurking somewhere beneath.

The distribution of computer systems and the "flattening" of the corporate organizational structure has caused more networking to take place. Users are concerned that network management become better, faster, more accurate, less expensive, more integrated, more human-factored, easier to use, and available when they want it and at a price they are willing to pay.

21.3.2 Evolution of "Network Element" Management

In the 1970s and 1980s the cost of embedded processing was high, so network operations were usually controlled from one main center. The centralized network was vulnerable to a single point of failure or catastrophe. This center monitored and sent commands and messages to the network elements that were "out there". This approach was somewhat similar to the centralized mainframe-oriented computing environment. Modern-day PCs and workstations show the opposite trend. More computer power is passing to the end user of which the "client-server" concept is an example. NM is following this trend too. As processing costs have fallen and as distributed processing has become more economical and practical, distributed element management (DEM) is becoming feasible. Each piece of equipment passes its own processing power to manage operation and status.

With frame relay, cell relay, and ATM and SONET services, the emphasis is on a blend of intelligent CPE and advanced transmission elements that monitor themselves using distributed element management (DEM) systems. Standards committees such as ANSI T1 and X3, IEEE 802.X, and various CCITT/ISO committees are now creating standards for interelement communications management. While the world waits for standards, the real world of vendors and users are usually implementing de facto approaches. This solves today's problems to stay in business long enough to maybe use official standards later. The Simple Network Management Protocol (SNMP) is one good example of an interim de facto standard. SNMP has met the short term need of users by its simplicity and availability. SNMP was developed by a handful of academics in the research and development community and has captivated many commercial corporate users searching for a way to harness their mushrooming, multivendor TCP/IP and multiprotocol data networks.

In the area of NM, the CCITT NM standards body has recently gone beyond the Common Management Interface Protocol (CMIP) approach by proposing a subset of CMIP called "CMIP Over Logical Link Control", or CMOL. Although CMIP is standardized to run over all seven layers of the OSI reference model, CMOL does not follow the seven-layer OSI standard but runs only on the first two layers (physical and data-link layers). In effect, the success of SNMP has caused the original CMIP standards proposal to be scaled back to the sub-set of CMOL. While CMIP uses elegant object-oriented programming approaches, SNMP, in contrast, uses a remote-debugging paradigm. Remote-debugging defines its objects, such as a routing table in a gateway device, in terms of simple management variables, called scalars, and also defines the simple tables constructed from scalars.

Regardless of whether CMIP (or CMOL) or SNMP is used, NM has to be implemented in two places: the management or "manager" station, and the managed device. Since object-oriented programming is "more elegant" and high-powered, it incurs more implementation and run-time costs for both the "management" and the "managed" station.

21.3.3 ISO Network Management

The International Standards Organization (ISO) has proposed and published the Open Systems Interconnection (OSI) structure for network management which uses two main protocols: Common Management In-Service protocol (CMIS) and Common Management Information Protocols (CMIP). These two protocols are for managing multivendor, multiprotocol networks. CMIS is the vehicle for collecting information from and sending information to network nodes. It has a set of service primitives for reporting and retrieving information, controlling the setting of parameters and initiating actions.

CMIS allows network managers to solicit network information, such as the value of a parameter, from a transport protocol layer. It also provides the ability to transfer certain control commands, such as setting a retransmission timer or initiating a diagnostic test.

CMIP, on the other hand, is a set of rules governing how information is exchanged between separate network management applications. In the multivendor environment, since multiple systems use different protocols, CMIP is envisioned as the "common protocol" to interface multiple vendors' systems. With CMIP, information is exchanged between network management systems, while CMIS is the grammar and vocabulary of CMIP. CMIS identifies the services or functions to be communicated under CMIP between network management applications.

Interoperability is a key issue for users with multivendor networks. Over 90% of major commercial users have equipment from at least two different vendors and 50% from five different vendors. These are "data" networks, composed of modems, T1 multiplexers, packet-switching nodes, LANs, and other computer resources. What happens with the addition of "voice" into the area of network management? There is no one easy answer to this question. The answer lies in knowing that the "hybrid" networking environment consists of voice/data, public/private network elements, CPE/POP-based switching systems, and varying degrees of user/vendor control.

21.3.4 Vendor Network Management vs. OSI (CMIP (CMOL)/CMIS)

AT&T has a nine-function Unified Network Management Architecture (UNMA). However, AT&T's products actually cover only five major functions of the nine: configuration management, fault management, performance management, accounting management, and network planning. By comparison, MCI Communications has a six-pronged approach consisting of configuration management, operations management, trouble management, performance and planning management, billing and cost management, and order entry and tracking management.

The Open Systems Interconnection/Network Management Forum has proposed a "model" that represents a third major view with five possible network management system functions; fault management, configuration management, performance management, security management, and accounting management.

Users need applications-specific network management, including disaster recovery, load balancing, and corrective maintenance. Fault management represents a classic area of confusion. Evolving customer requirements have innovatively added several branches to this category, such as: restoration, trouble ticket, and inventory management.

The following ten areas expand the common five-fold OSI definition of network management functions to what various vendors are actually implementing:

Configuration Management — provides the ability to monitor and reroute on a dynamic basis to meet changing needs. It allows on-demand allocation of bandwidth, preprogrammed configuration diagrams based on traffic patterns, electronic mail, and management-center personnel activity.

Restoration Management — relates to, but is different from, configuration management. The restoration management function tracks disaster recovery plans and other plans that have been preprogrammed to be executed when a line outage or some other disaster occurs.

Data Base Management — provides the user with a means to update and maintain the basic logical information contained in the user's data bases that define the network.

Operations Management — provides information concerning the present operation of the system, including network alarms. It also ties into restoration management in disaster situations.

Trouble Management — provides trouble ticket management functions in the case of line problems, and historical analysis of past trouble ticket problems and their resolution.

Administrative Management — allows for general administrative chores, file indexing and transfer, system configuration, security access, and administrative messaging via E-mail.

Billing Management — provides information concerning present bills due, billing formats, alternative billing methods, hierarchy definitions and bill delivery specifications.

Inventory/Provisioning Management — provides information concerning services inventory, order entry, order status, and delivery dates.

Traffic Management — provides information on present traffic patterns, via real-time transmissions, and past traffic patterns, allowing trunking analysis, cost-performance data analysis, and so forth.

Performance Management — provides information and reporting on present performance, via real-time transmissions, and historical trends in network performance.

21.3.5 OSI CMIS/CMIP Upstaged by Simple Network Management Protocol (SNMP)

The CMIS and CMIP components of the ISO network management scheme have been under development for over 10 years and have as of yet not been implemented on very many systems. Users have elected instead to go with the Simple Network Management Protocol (SNMP). SNMP was developed by a group of users in the Internet community (academia, government, military, etc.) for managing their mushrooming Transmission Control Protocol/Internet Protocol (TCP/IP) multivendor data networks. Once considered only as an interim "ad hoc" protocol to be eventually replaced by the OSI CMIP protocol, SNMP has taken on a life of its own since the ISO CMIS/CMIP protocols have been slow to market.

Over 7,000 "generic" SNMP systems have been installed at US. user sites. Projections are: 14,000 networks by 1993, and 30,000 by 1995. The going price for hardware and software is around $20,000 and this will drop to about $10,000 in 1993. Most users are choosing two main SNMP management platforms: Hewlett-Packard's Open-View and Sun Microsystems' SunNet Manager. The international market for SNMP is also growing.

Many users are finding that the standards process is too time consuming, lacks agreement as to what should be developed, tries to be too comprehensive, is too costly, and is too late. The real world demands user solutions in less time, at less expense, with less over-head, and that work.

21.3.6 New Technology Outpaces Official Standards

The ISO-OSI NM protocol concepts were developed in an era of multikilobit-per-second networking around 1979. However, the era of multimegabit-per-second networking is already here via local network and T1 technology. Soon, with the Fiber Distributed Data

Interface (FDDI) and T3, the stakes will be even higher. By the time ISO-OSI is mature, the brink of multigigabit-per-second networking will be reached. So ISO-OSI originally planned for kilobit speeds must now contend and deal with gigabit speeds.

The ISO-OSI NM forum tacitly admitted the weakness of CMIP by their actions. As noted, they have scaled back from CMIP to the simpler subset of CMIP called CMOL or "CMIP over Logical Link Control": it runs over fewer than the seven-OSI-layer stack; manages only devices on the same LAN network; and requires the use of proxy agents for dissimilar LAN connection — with the resultant increased overhead, reduced performance, decreased interoperability, and lessened transparency.

Most major hardware and software vendors will announce support for standards, especially as their customers see a possible solution to business problems, but will delay delivering them to the market-place. For example, IBM will preserve its Systems Network Architecture (SNA) as long as possible, being IBM's proprietary data networking architecture, since it "locks" customers into the IBM platform. Some of IBM's standards implementations still tie into the embedded SNA architecture with proprietary calls for actions to be taken.

21.3.7 OSI NM Standards Will Follow SNMP's Example

Presently the SNMP protocol for TCP/IP data networks has 30 or so vendors building to its specs, while the OSI CMIP camp has 2. Even AT&T and others who have implemented so-called OSI protocols have done a proprietary version of the original OSI protocol (what happened to standards?), so that proprietary "hooks" are built into the equipment to favor the entrenched vendor. SNMP is nonstandard without OSI or any other blessing, but: it works, it is cheap, and it is available. Marshall Rose, chair of the Internet's Network Management Protocol working group, gives his opinion:

> SNMP uses a remote-debugging paradigm. Remote-debugging defines the objects, such as a routing table in a gateway device, in terms of simple management variables (scalars), and simple tables, which are constructed from these scalars. Although this technique lacks the flair of an object-oriented approach (CMIP) it has a no-nonsense attitude that has proven quite effective in promoting the definition of all kinds of managed objects. ...the products that are out there are doing yeoman's work in the field and providing a robust, competitive market. Talk to your colleagues, see what they are using, and find out what their gripes are. ...start managing real networks with real products and draw your own conclusions.

Recently a new twist has been added to SNMP — SMP. Simple Management Protocol (SMP) is a proposed specification that plans to offer capabilities not currently available with SNMP. It promises to overcome some of the limitations of SNMP and move closer to some of the appealing aspects of the OSI Common Management Information Protocol (CMIP).

The main new functionality proposed is that it would offer event-driven management of applications as well as network services, unlike SNMP which now polls devices for information. The new SMP was occasioned by a request issued by the Internet Engineering Task Force (IETF) for an improved version of SNMP. The SMP specification is being positioned to provide management support for more than only networks, by allowing users to manage applications on host and personal computers, while providing improved network element management.

SMP is proposed to offer a means for bulk transfer of management information and will support communications among management stations, which is now missing from SNMP. It will also offer event-driven management rather than just polling. It also includes an upcoming security standard which the IETF is planning to announce in its finalized version soon. It will also be backward compatible with SNMP and SMP will contain a proxy agent to translate between SNMP and SMP commands and messages.

The TCP/IP orientation of SNMP will be removed in SMP so that SMP can work over TCP/IP, Apple Computer's AppleTalk and Novell's Internetwork Packet Exchange (IPX) and OSI networks. SMP is presently a proprietary protocol, and it remains to be seen if it becomes an Internet standard. It appears that the developers of SMP are cleverly not immediately submitting SMP for Internet standard approval, but seem to be attempting to introduce it to the marketplace via existing SNMP vendors and with market acceptance can then approach the lengthy and specifications changing approval process. It is expected that SMP will further cool interest in CMIP. If SMP proves to be successful it could derail the still existing thrust to provide an object-oriented systems and network management platform via CMIP.

In summary, SNMP will continue to evolve technically. SNMP started in 1988 and CMIP started in 1979 — SNMP since 1988 has 30 vendors fitting it to their systems with an installed base of about 7,000 users who like the simplicity, low cost, and "now" availability of SNMP, while CMIP has 2 sponsoring vendors and few installations and has recently been scaled down to CMOL, a 'subset' of CMIP operating only on the first two layers of the OSIRM.

21.3.8 IBM Network Management

In 1986, IBM took its four or five standalone data network management products and combined them into a new offering under the banner of NetView. IBM positioned NetView as running on the IBM mainframe using MVS or VM operating systems. It was positioned as an enterprisewide focal point for managing large networks and data centers. All management data has to be translated into IBM SNA protocols before being sent to the focal point at the mainframe for processing. Over 50 vendors have pledged to interface to NetView. More than 6,000 copies of NetView have been installed around the world. It is a marketing success. This shows the marketing trend for the integration of different vendors' NM systems.

21.3.9 Network Management Elements and New Technologies

To understand the complexity of network management, one must have a bird's-eye view of the private and public network elements that must be managed. Integrated network management must be provided for these hybrid mix of elements:

- ✪ CPE/premises elements (voice and data)
- ✪ Central office switching/DXC elements
- ✪ Tandem level switching/DXC elements
- ✪ International switching/DXC/gateway elements

Networks are growing larger and more complex. Around 1970, the average data network consisted of 20 to 25 terminals connected across a network to the Front End Processor (FEP) of a mainframe. Today, data networks are much larger on average — sometimes consisting of thousands of terminals applying to multiple FEPs and hosts for access to multiple applications. New technologies that must be managed include Local Area Networks (LANs) using various transmission media: twisted pair (unshielded and shielded), coaxial cable, and fiber, all using multiple LAN protocols (Ethernet, Token-Ring, etc.) and internetworking protocols (IP, IPX, etc.). A whole new family of new communications devices has been interposed between LANs and WANs — bridges, routers, brouters, and gateways. These value-added services must also be managed.

Increasing volumes of alarms and status messages are being received from complex networks composed of: larger systems, more complex systems, higher-paced applications (such as just-in-time delivery systems), real-time inventory management, CPU-to-CPU communications, Computer Assisted Design (CAD)/Computer Assisted Manufacturing (CAM), large file transfers, increased graphic dumps between workstations, and exploding LAN-to-LAN connectivity among similar and dissimilar LANs and, finally, LAN-to-WAN connectivity: across the street, the city, the state, the country, and the world.

Local Area Network (LAN) to Wide Area Network (WAN) connectivity is the leading force in forcing change and has spawned the introduction of new technologies into data networking: frame relay, cell relay, ATM, and Synchronous Optical Networks (SONET) will come into their own chiefly as LAN/WAN connectivity agents. To help the network manager cope with this veritable explosion of network alarms and status indicators, important new technologies are being employed to assist to manage modern networks.

CRTs with color graphics are being used instead of printers and CRTs that report status by "text" trouble alarms. Text, unfortunately, is highly mnemonic and often requires documentation to explain the meaning of what went wrong. Network management is now starting to use enhanced tools. NM is beginning to use graphical displays with well-designed icons and symbols to display configuration and status for key network elements under distributed element management (DEM) systems. New visual displays and programming paradigms are being used to compress and summarize (exception management) the volumes of data constantly flowing across the network.

Sifting information by new programming methods and using color visual displays of the overall network with icons and network layouts, is allowing NM operators to better address strategic networking needs, trend analyses, and planning issues. Rapid mouse-clicking on icons to indicate and/or change their status is also proving helpful and meets human-factors engineering.

21.3.10   Around-the-Clock Operations Increasing

Off-hours or "graveyard shift" traffic is becoming more important with the globalization of networks. Just as the local supermarket is now open 7 days a week and 24 hours a day in many cases, so too networks are becoming more available too. Extended hours and around-the-clock operation of world stock exchanges and economic markets from Tokyo to New York to London shows this. Data critical for global decision making is being passed over networks at all hours of the day/night. Not only are multinational (or now so-called transnational) businesses run monthly from afar but are now making daily decisions from afar. Older human-based systems of NM must be increasing automated to meet these challenges.

Recent trends in new NM technology applications include:

+ High-performance engineering workstations
+ Improved graphical user interfaces (GUIs)
+ Windows, menus, icons, color graphics, etc.
+ Rapid mouse-clicking on icons to display or change status
+ Enhanced human factors engineering (ergonomics)
+ "Exception" report to prioritized alerts, alarms, status
+ New emerging technologies:
+ Display management
+ Neural networks
+ Object-oriented coding
+ Object-oriented databases
+ Self-healing nets/disaster recovery via fiber rings

The sheer volume of network status, alerts, and alarms flowing over a network demands that network operators have all the help they can get to keep control. The information provided to the operator must be summarized, condensed, and made quickly digestible. It should be graphically oriented so the network operator can make operations decisions by observing symbols, color changes, and various condition states on color-CRT screens rather than being forced to decipher, with constant reference to paper documentation, the cryptic messages of network status, alarms, and alerts rapidly scrolling past in text format on a CRT.

Information, too, should be displayed in hierarchical fashion, with the ability to "telescope" from the macro view down to the micro view — for example, looking at the overall network of 100 nodes represented by icons, and then further isolate a problem by rapid mouse-clicking down to the individual icon level of, let us say, a multiplexer in Denver, with the icon showing the multiplexer's status lights in color on the workstation's windows. Action can then be taken based on status. Icon clicking with the mouse can take the network operator from the macro view to the micro view in 3 to 6 seconds — from the overall network down to the level of a specific location, unit, and status.

21.3.11 Multivendor, Multitransport Media Networking Growing

Multiple vendors usually implement incompatible products and services (not always on purpose). Proprietary designs evolve even within the implementation of standards, such as AT&T's Network Management Protocol (NMP) which is OSI-oriented, but has AT&T's proprietary additions. Often there is dissimilar hardware, programming, unique command structures, and displays. As multiple NM platforms have developed, multiple display terminals have also developed.

To address this multivendor environment, the leading vendors have developed "umbrella" platforms or architectures to deal with the multivendor problem. This approach is called "managing the managers". An integrated network management system (INMS) is developed which monitors existing disparate NM systems through some common interface. However, the problem is not just multivendors but also multi-products *within* a vendor. For example, IBM presently uses more than 14 incompatible operating systems in its business products. AT&T has tried to bring over 15 NM systems in under the UNMA umbrella. Several other AT&T systems are so different that they have not even been addressed.

Two basic NM platforms have developed. The data world, dominated by IBM, uses IBM's NetView. The voice world, dominated by AT&T, uses AT&T's Universal Network Management Architecture (UNMA). The fact that a recent IBM-AT&T press release announced a link between these two NM architectures emphasizes the increasing "hybrid" nature of voice and data networking. Data traffic, for instance, over public switched networks will approach nearly 50% of traffic volume by the mid-1990s.

AT&T and IBM have agreed to interconnect their NM systems (IBM's NetView and AT&T's Accumaster Integrator — UNMA) because their largest customers have demanded it. The IBM/AT&T NM agreement includes:

- Four interfaces to be developed for information sharing
- First interfaces use existing interfaces; eventually OSI
- No off-the-shelf product available until 1994
- Initial implementations will employ LU6.2 protocols (IBM)

The issue is more marketing positioning than standards developing. When IBM and AT&T connect, they plan to use the IBM LU6.2

interface, thus using a "de facto" IBM standard for NM. All major vendors are moving to standards but each in his own manner and method. As one wag has observed: "If you like standards, hang around, there will be more of them to chose from in the future".

Most existing NM systems are "reactive" management techniques instead of "proactive" ones. Existing NM systems tend to flood the network operator with alarms and status information. And the operator tries to react as quickly as possible. As networks grow more complex, however, the operator is flooded with still more information. Rapid decisions must be made on *more* information, in *less* time, and with *escalating penalties* for mistakes.

Human factors must step in to help. NM systems must be built to adapt to the human mind's ability to process information. People are not good at parallel processing or handling multiple streams of information simultaneously. People are mono, serial processors — one thing at a time, although they can be interrupt-driven. NM systems must take information coming in from the network in parallel fashion and adapt it to the human mind's ability to serially process events and actions.

NM systems are beginning to interface to the parallel incoming information sources streaming in from the network. By neural networks and rule-based expert systems, NM systems can reconfigure incoming parallel information streams for "mono" presentation to the human network operator — thus automating the analysis of information in a human warm and fuzzy manner.

Some basic requirements for adapting new technologies to NM include:

- Improve display management
- Implement rule-based and neural Artificial Intelligence (AI) systems
- Employ object-oriented coding and databases
- Handle growing networks with more network elements
- Summarize network data for decision making
- Move rapidly from macro- to microviews of net elements
- Automate the control process
- Improve data presentation and alarm annunciation
- Accommodate emerging technologies and standards
- Replace reactive NM with proactive NM

21.3.12 Improvements in Display Management

Improved graphic-display management software allows the network operator to now scan detailed network diagrams and determine the general health and status of a several-hundred node network in seconds. This is a major improvement over the chore of reading text-related messages for each network element. Improvements have come largely because of developments in X-Windows, a low-level window manager that runs on UNIX, and with higher-level graphic toolkits such as Open Look, Motif, and Graphic Modeling Systems (GMS). X.11 is the latest version of the popular MIT-developed X-Windows display manager. It runs on virtually all UNIX workstations and on many other platforms, from mainframes to PCs. It provides a programming interface to a variety of library functions, called Xlib, that can be used to create and manage windows and their basic functions, such as scroll bars and window resize buttons.

Open Look and Motif are two toolkits for creating X Windows-based Graphical User Interfaces (GUI). They implement calls to Xlib and present mechanisms that allow the end user operate the application. Open Look was developed by Sun Microsystems with the support of AT&T. Motif was originally developed by Digital Equipment Corporation (DEC) and Hewlett-Packard and, since both companies are sponsors of the Open Software Foundation (OSF), Motif receives strong support from the OSF.

Open Look and Motif both provide programmers with tools to reduce the coding effort needed to produce the basic GUI framework. But neither provides direct support of the high-level graphics needed for network maps or performance charts. Nor do they let the user interact with graphical elements, such as modem front panels or network nodes, to monitor and control network elements.

User interfaces will become more mouse-, menu-, and windows-oriented as more detailed graphics become available, thus allowing the operator to interact with the actual equipment represented by the icons to program or take in and out of service. Using toolkits and window managers, users can draw complex network diagrams to allow an operator to view the status of hundreds of network nodes and elements simultaneously. Yet with the click of a mouse-button the operator can, within a few seconds, focus on a particular icon and observe its alarm lights for the state of the actual element represented by the icon.

The network operator's attention can be enhanced by the animation of icons to represent "living" systems: turning red, green, blue or yellow, as well as blinking, showing happy or sad faces, or using speech processing to literally cry out for "Help!" in a strident manner like some video games. Workspace on the CRT screen can be maximized by the ability to resize and open/close windows on the network management workstation. Closed windows, however, continue in "real time" to monitor the status of the network elements for later recall and display if required. When a window is reopened later, the current, cumulative network status is displayed in near real time. This timeliness and availability of information makes network decision making easier and faster. This allows quick resolution of problems with less downtime. This equates to reduced costs and increased revenues.

21.3.13 Artificial Intelligence (AI)/Neural Networks

Since voluminous amounts of network management information must be processed, and network information arrives at the control point in parallel streams as status/alarms originating from perhaps thousands of network elements — such as a modem, multiplexer, or CSU/DSU — the parallel streams must be analyzed and patterns and statuses recognized. The "real" alarms, or the "trigger" alarms must be isolated by exception methods from "sympathy" alarms set off by the original "trigger" alarm. Otherwise, thousands of alarms would have to be dealt with.

To move from reactive to proactive NM, rule-based expert systems and neural networks will allow for the detection of deteriorating conditions on the network before a full-fledged outage or breakdown. This allows for "proactive" effecting of changes before the faults actually occur. Artificial intelligence monitors hundreds of points for traffic flow rates, user transaction response times, and equipment health and status. These data streams are correlated to recognize trends that can predict problems by extrapolation.

The knowledge of experienced network operators can be captured by AI methods by storing various corrective procedures. It also stores "memory" for things **not** to be done. These procedures can be stored and referenced later as holiday or after-hours outages occur so that, in a sense, network operations experience is available almost real-time. Rule bases are produced over time (the experiences of expert human network operators) and these rule bases predict ranges of decisions of things "to do" or "not to do" in certain situations in the network.

To release this overload, a neural network is used. A neural network is a new form of computer processor that gathers and correlates high volumes of measurement data for input to the rule-based expert systems. Unlike traditional computers which must be programmed to produce certain outputs based on specific input data, neural networks are "trained" to recognize patterns by running sample data. Neural networks can also process many inputs simultaneously, that is, perform "parallel processing" while conventional computers are limited to serial-like processing, performing one operation at a time. Neural nets can correlate multiple measurement data streams against preprogrammed measurement trend-data ranges that could cause network faults. This is proactive network management! Neural network output could also be used by the rule-based expert systems to select corrective courses of action. Thus, "self-healing" networks are possible not only by physical design but by software design as well.

21.3.14 Voice Processing and Network Management

The "Speak & Spell" computer from Texas Instruments introduced the era of voice processing. Automobiles now use voice processing to supplement visual signals and indicators. The "gas empty" indicator, whether an analog needle that travels to the "empty" position or a digital red light going from "off" to "on", has now been augmented by a human voice speaking to the driver. Network operators, even though they may have existing "visual" indicators, can now also be reminded at critical points by voice processing messages so their networks will not run out of gas on some lonely road.

21.3.15 Network Management Industry Summary

Users must have the potential to pull multiple element management systems (EMS) together into a single-vendor platform like AT&T's Accumaster Services Workstation (ASW) or MCI's Integrated Network Management Services (INMS) Workstation.

MCI's INMS workstation, based on two IBM compatible PCs running in a client-server relationship, provides a window on trouble tickets for all services. It provides enhanced management services for the MCI's public virtual private network offering (VNET) and cut-through/terminal emulation access to other private network-selected Element Management Systems (EMS).

MCI's VNET Configuration Manager (CM) allows users to set, change, activate and deactivate calling privileges and codes. MCI has opted to consolidate much of its outbound functionality into the

Configuration Manager and a second service geared to performance management — the Integrated Network Management Service (INMS), which provides historic reports on calling statistics, call attempts, traffic profiles and active configurations.

On the public network side, MCI provides an element management system that AT&T does not. It is called MCIView. This service forwards information about VNET access lines directly to NetView and the information can be viewed on NetView screens. MCI's link to IBM is much more direct than AT&T's.

On the private network side, MCI has taken on AT&T's dominance in dedicated services management with two releases of its Dynamic Reconfiguration Service (DRS). DRS allows for either dynamic or fixed network reconfiguration and competes with AT&T's Customer Controlled Reconfiguration (CCR) and Accunet Bandwidth Manager (ABM) products. MCI also offers trouble ticket service through the INMS workstation and NetView links via MCIView.

MCI's Global Communication Service (GCS) acknowledges that with the globalization of networks NM is taking place. GCS is focused on meeting customer needs for private networks and offers services to facilitate the design deployment and maintenance of these networks. MCI outsourcing services for GCS includes both GCS Network Monitoring Service and GCS Network Management Service, and is made available on an individual case basis. MCI will take on the responsibility of monitoring or managing a customer's network by configuring a private line between the customer's Element Management System (EMS) and an MCI umbrella network management system, or by positioning the customer's EMS at the Global Network Management Center (GNMC). Network elements to be monitored include customer premises equipment, collocated equipment, and GCS circuits.

The GNMC monitors and performs network administration on customer circuits as well as the customer premises equipment, located at the customer site or collocated at a GCS node. If the customer so chooses, the GNMC will become the network administrator and perform services ranging from monitoring the network, to fault isolation and vendor dispatch, to network management for adds, deletes, changes and configuration management.

Announcements by MCI regarding Integrated Network Management Services (INMS) at the September 1991 TCA trade show in San Diego revealed the development path for MCI INMS Network Management products. MCI has provided real-time monitoring of the MCI network for MCI data and inbound services. Expanded operations management capabilities are available for MCI terrestrial data (TDS), digital data (DDS), digital private line (DSO) and MCI 800 &

900 services. Customers can monitor their portion of the MCI network for conditions affecting data traffic as well as operational alarms and traffic status.

New features include digital cross-connect and extended superframe (ESF) data monitoring, which provides alarms and real-time trending reports. Increased performance and planning management for TDS and inbound services. This capability allows customers to generate reports based on traffic and alarm data on these services. ESF-based performance reports will enable customers to identify network degradation before it affects service. This information also permits customers to anticipate and plan service upgrades or reconfigurations by trending long-term data. It includes new access to MCI's Digital Reconfiguration Service (DRS) for configuration management of MCI data services. Current support is for VNET and 800 CM services only. New workstation dynamic topology maps are added for greater visibility into real-time network status. INMS will also soon be integrated into FocusNet.

MCI Communications has announced the trend for "dynamic bandwidth on demand" for data using the evolving technologies of frame-relay, cell relay, ATM/B-ISDN, and SONET, MCI's overall strategy of complementing the public VNET and the private network services shows a good integrated approach using a single INMS platform for network management.

21.4 BILLING

The billing scheme will be based on many factors:

- technology used
- specific implementations of protocols
- time the circuit is used
- number of data units passed (packets, frames, PDUs) during a certain time period
- total number of Bytes

Billing schemes fall into three categories: fixed cost, usage based, or a hybrid of both. The switched services discussed in this text (with the exception of circuit switched) are primarily billed based on usage. This is because the user wants to pay for only the data sent, not the entire bandwidth. If he wanted to pay for the entire bandwidth, he would have bought a private line! The service provider who offers switched services will establish a pricing structure that makes services like frame relay and SMDS cost effective for bursty variable-bit rate traffic, and not cost effective for users who abuse the service and try to realize cheap bandwidth or to pass constant bit rate traffic. Each tariff, upon analysis, clearly shows the break-even point between buying a private line as opposed to a switched service.

21.5 SECURITY

Network security should be clearly defined in the network engineering plan, and should be strictly administered. Breaches of security can be classified as query or information gathering, modification, permanent change, or theft. The easiest way to loose control of the network is through insufficient security. There are two primary modes of network security: access and transmission.

Transmission security can be implemented through password-protected modems, encoded transmission devices, and at a variety of other network access points. This is to prevent line monitoring, passive wiretapping, and stolen data from an active tap. Technologies such as ATM and SONET contain their own form of transmission security because even if the circuits are tapped and "read" the data flow is meaningless. The way in which ATM and SONET frames are formed distorts the original shape of the data so much that it could not be interpreted.

Security can also be implemented at many protocol layers within the network, within hardware, software, and computer services. Access control filtering is done primarily at the data link, network, and application layers. These layers contain the protocol feeds which make security control possible. Perhaps the best security control is at the user or application level. Here, protection from illegal use of the network is gained through:

- user names
- passwords
- password life of days, not months or years
- software access levels
- user lockouts
- user ID card/key
- no user echo of password
- imbedded non-alphanumeric character
- tight administration and control
- one time user passwords
- fingerprint or eye retina scanner

▣   tissue sample or DNA match
▣   first born inserted into scanner

Combinations can also be used. Some of these techniques go overboard on security, but use whatever is required based upon the sensitivity and mission-critical nature of the traffic. Also, educating users on how to use the security system will help. Security can be implemented through the access control mentioned above, hazard protection, and as personnel practices clearly defined in policy and procedures. Remember, network security is only as strong as the weakest link.

Once the network is up and running, have the engineer most familiar with the design attempt illegal entry. If he can get in, someone else can also. Beware of security holes during system updates and software upgrades.

21.6 TRAINING

The training process should begin as soon as the network design is complete, and can actually begin when the vendor is chosen. Also, allow the vendor to provide materials for and actually do training as practical. The key point to stress is to begin training as soon as possible. Training should cover at a minimum:

📖 user training as early as possible
📖 concepts and architecture
📖 basic operations
📖 hardware and software (including database and addressing)
📖 installation
📖 network management
📖 advanced techniques
📖 documentation

Training is most critical for those who will actually install and configure the hardware and software. This should be the focus of training before network implementation.

21.7 WHEN TO STOP DESIGNING AND START IMPLEMENTING

The design process is an iterative one, but the designer must know when to stop designing and start implementing.   There is new technology emerging on the market every day. Don't give in to the temptation to stop the design process based on a new product some vendor says will be out next quarter.   The design will never be perfect, and modifications can always be made after the network is up and running. If the network is designed based on the guidelines provided in this book, it will be easier to integrate new technology into the network when required.

21.8 REVIEW

In this chapter the importance of documenting both the network design (engineering plan) and the method by which it will be implemented (O&M document) was covered. It takes teamwork among many departments to design and implement a network. The importance and the capabilities of network management were noted: the responsibility to provide thorough network management to the user, and to make best use of the protocols defining network management. The issues of security, training, and billing are also important. Finally, the design is reviewed one final time before beginning the implementation phase.

22

Ancillary Issues —
Design Tools
and the Job Market

This chapter provides the reader with insight into the selection of network design tools. Design tools provide an accurate method of modeling and designing networks. In this chapter the criteria by which to choose a design tool will be explained. The chapter also provides an overview of the current network design and management job market for the mid and late 1990s.

22.1 DESIGN TOOLS

Over the past decade, many companies have written software design tools. The capabilities of these tools include analyzing existing networks, providing meshed and hierarchical network designs, modeling design change and failure scenarios, and performing clocking analysis. While these tools exist, few are actually used by most network designers and managers. This is in part due to the lack of tool flexibility and inherent constraints in limited options, and also in part by the sheer difficulty of modeling distributed hybrid networks. Some available tools are also well suited for modeling specific types of networks, such as intelligent multiplexer networks. But, conversely, when an intelligent multiplexer network is meshed with a multiprotocol router-based backbone which operates over an FDDI MAN, the tools hit their limits.

Design tools analyze two types of data networks: circuit switched and packet/frame/cell-switched networks. When choosing a design tool, certain criteria should be present in the tool. At a minimum, it should support the basic user requirements defined in the last chapter. Before a tool can be successful, the data fed into it must be accurate. Hence the importance of the detailed requirements analysis of Chapters 15 and 16 so that good input will produce good outputs.

Design tools use either mathematical modeling or packet level simulation. Mathematical modeling uses statistics to predict user traffic flows and model their characteristics. Packet-level simulation is the most common method, however, where the tool inputs are a snapshot of the network at a *single point in time*. Since the traffic on packet networks is often bursty in nature, multiple snapshots are required. This type of modeling works well for small networks where the user traffic characteristics are well known, but processing times and the sheer computing power required to perform large, dynamic network design often makes this process very time-consuming, resource insensitive, and difficult. Some design tools, such as the MAKE Systems tool, use neither of the above approaches. Instead, they model each user input as a single demand, based on their file transfer protocol characteristics. The specifics of the MAKE Systems modeling tool will be looked at later in the chapter devoted to case studies. The author has found that the best network design tools focus on heuristic routing algorithms based on user-definable parameters and queuing theory, rather than any one form of mathematical simulation of theory.

22.1.1 User Inputs

Network design tool user input parameters vary widely, based on the type of network to be modeled. Some of the more important parameters include:

- ✱ packet size (average or peak)
- ✱ number of packets/frames/cells per second (mean and in each direction)
- ✱ minimum/maximum bandwidth per user access
- ✱ time between packets (burstiness)

* file size per transmission
* links/trunks required (length, speed, priority, etc.)
* cost and priority factors
* overhead statistics per protocol
* network device (node) specifics
* protocol specifics

This information is typically input into the tool through an application profile input file. This file also includes application specifics such as name, description, traffic type, coordinates origin(s) and destination(s), etc.

22.1.2 Tool Support

While the features and functions of design tools vary widely, below are listed some of the more important items to be supported:

◎ multiple protocols (HDLC, X.25, Frame Relay)
◎ multiple architectures (OSI, SNA, SAA, ARPANET)
◎ layered view of each protocol and architecture
◎ handle multimedia traffic (voice, data, video — time sensitive/insensitive)
◎ end-to-end view upon selection
◎ support for both hierarchical (LAN/MAN) and meshed (multiplexer) network designs
◎ user-friendly interface — Graphical User Interface (GUI) and Applications Program Interface (API), CAD/CAM capabilities, multilevel graphics, international
◎ location inputs in Vertical and Horizontal (V&H) of serving CO, Latitude and Longitude, NPA/NXX, LATA, and CRT screen coordinates
◎ conversions for the location finders above
◎ display utilization
◎ powerful high-level user language accessible by user (user able to make coding changes to tool)
◎ on-line editing of links, nodes, and other properties
◎ flexible data structures (circuits/packets/frames/cells)
◎ flexible input parameters (distance, cost, quality, bandwidth)
◎ flexible parameters for defining link and node placement
◎ capability to upgrade/add new technologies, protocols, and parameters
◎ modular design (run each step of design separately)

◎ general-purpose design criteria
◎ capability to input national and international tariff and network specifics
◎ dial-up capability
◎ varied node and link types
◎ survivability analysis
◎ unlimited number of nodes
◎ configuration and clocking design
◎ runs on industry-standard platform
◎ current software revision
◎ multiple homing of links to multiple or same node
◎ variable equipment costs
◎ fast, efficient algorithms
◎ data export including laser printing and plotting
◎ multiple save file formats: text and graphics
◎ sensitivity analysis
◎ tariff manager
◎ multivendor, multicarrier device libraries
◎ short run and processing times
◎ add in for growth
◎ thorough debugger, editor, journalizing
◎ performance analysis simulator
◎ query capability for all of the above
◎ traffic generator
◎ least-cost topology design
◎ assess network delay and throughput
◎ applications profiler
◎ variable decision matrix
◎ price-performance model

22.1.3 Reporting Capability

Design tools need to report their findings to the user so the user can: understand what the tool has modeled, what can now be model, and how the modeling can be translated to management. Some of the more important reporting characteristics include:

➲ end-to-end response time
➲ topology options
➲ throughout, delay, cost (per link and node)
➲ effect of link or node failure
➲ individual link and node views
➲ bandwidth usage per user/port/protocol

- ⊃ what changed? analyses (for topology, node/link demand, link load, node throughput, routing differences, and packet throughput)
- ⊃ topology changes
- ⊃ link and node upgrades
- ⊃ protocol type changes (routing, switching)
- ⊃ maximum hop count
- ⊃ delay and throughput analysis
- ⊃ traffic type changes
- ⊃ traffic pattern changes
- ⊃ link and node characteristics changes (PPS processing)
- ⊃ busy hour/minute/second changes
- ⊃ black box support for unrecognized devices
- ⊃ constraints in link and node demand
- ⊃ hybrid topologies
- ⊃ redundancy

22.1.4 User Functionality

User functionality defines the level of control the user can exert over both the design and the design tool. The functions the tool provides to the user should include:

- ☑ complete design
- ☑ incremental design
- ☑ multinetwork design
- ☑ subnetwork design
- ☑ built-in redundancy
- ☑ specific rules of routing
- ☑ layered (protocol or architecture) design
- ☑ user-specified specific parameter design
- ☑ manual additions
- ☑ time interval design
- ☑ specialized routing
- ☑ forced homing
- ☑ connectivity constraints
- ☑ import traffic statistics recorded by data analyzers
- ☑ macro functions
- ☑ zoom-in capability
- ☑ printing capability

The tool should also have the capability to import current configuration files in ASCII format, ranging from database downloads of existing equipment to entire network maps of available routes and the tariff information for each. All parameters should be displayed and configurable in both domestic and international measurements, distances, tariffs, maps, etc.

22.1.5 WAN Design Tools Available

Some examples of WAN design tools can be found in Table 22.1.

TABLE 22.1 WAN Design Tools

Vendor	Design Tool Product
Network Management Inc.	Modular Interactive Network Design(MIND)
Network Design & Analysis Corp.	AUTONET Designer
Connections Telecommunications	NETConnect
MAKE Systems Inc.	NETOOL (Workbench)
Network Equipment Technologies, Inc.	Network Design & Analysis System
IBM	INTREPID

Remember that the design tool process does not stop with the initial design. It is a cyclical process to be completed every few months by the network engineering organization. Design tools can drastically improve the efficiency of an existing network, provide significant cost savings through reducing access and thus reducing costs, and recover underutilized hardware from the network. Network design tools can optimize efficiency from the local LAN to the entire WAN. These tools can help perform the fine tuning discussed in Chapters 19 and 20. Try to avoid the use of *proprietary* tools.

While tools may be valuable for modeling current network topologies and technologies, they may quickly become obsolete when new technology becomes available in the network. And, not only should the tool itself be considered, but on-going support and updates are also important.

22.2 THE JOB MARKET

Where is the job market going today and in the future? With each furtherance of automation and the emergence of the postindustrial information society, the people who do repetitious non-value-added jobs will be attrited slowly and replaced by computerization. This statement should be taken with a grain of salt, however, since it may end up like the so-called "paperless office" promised ten years ago, so that with more automation there seems to be more paper.

However, the communications people who find a niche and add value to their position will be better set for survival than those who "go with the flow". To stay on top of the wave and not be lost in the swirling backwash demands that the communications professional first identify what newer technologies are replacing the older technologies — that is, what is being "cost-justified" into the business (even though the old may be working well) but the economic indicators are irresistible to upper management for cost cutting considerations.

The smart telecommunications professional will make a concerted effort to prepare himself as a subject-matter expert on that emerging new technology. The electrical engineering or computer science degree now should be coupled with some type of business and management degree, such as an MBA, to provide a combination to cover both skill areas of data processing technology and business skills. However, someone coming from a liberal arts background with management/people skills (although it is much harder than moving in the opposite direction) can add to the business acumen aspect the new telecommunication graduate programs now available and earn an Master of Science degree from an engineering school without the rigors of a heavy mathematical engineering curriculum and achieve the same end.

Trends show that data managers generally command higher salaries than voice managers. This difference is even higher if the manager knows both voice and data. Also, studies show that data managers tend to be higher educated than voice managers.

22.3 REVIEW

This chapter has provided us with the capability to choose a network design tool which meets the requirements of user input parameters, user functionality, feature and function support, and reporting capability. The leading design tools on the market were summarized and the main features and functions required were outlined. Part 7 will include two case studies reviewing MAKE Systems' "NETOOL" and IBM's "INTREPID", two of the leading design tools in the industry for multiplexer and IP routed environment design and modeling. Finally, a review was made of the 90s job market for network design engineers and managers and of how communications professionals must plan for change by pursuing a mix of technical and business skills targeted to emerging technologies that can cost-displace older technologies.

7

International Networks and Case Studies

This section is designed to provide the reader with an overview of the international data communications arena. The changing role of the PTTs and international network and service providers will especially be delved into. This includes an analysis of what it takes to compete in these arenas. With an understanding of these aspects of international networking, a network design can be performed by adjusting priorities to suit the situation. The book then finishes with the presentation of two case studies presenting analysis of two predominant network design tools — MAKE Systems' Netool and IBM's Intrepid — which model multiplexer and packet/frame/cell switched networks.

23

International Network Design

This chapter highlights both the business and technology aspects of international network design. The focus will be on the changing role of many national communications systems and regulations. The range will cover a sole PTT ownership to the competitive environment resembling the United States divestiture. This will give an understanding of the advantages and drawbacks of providing international network services using both public and virtual private network services. By studying the interest of foreign service providers in the new unification of European and Asian communications infrastructures, it can be determined who are the key worldwide players and how they shape the structure of worldwide data communications. Next, a quick look at frame relay, SMDS, B-ISDN, and SONET services and their potential in the world markets will be made, and the chapter ends with a discussion of European opportunities emerging from the formation of the European Community, as well as other worldwide attempts at privatization and divestiture.

23.1 TERMS DEFINED

This chapter discusses many network configurations and services provided by international carriers and dominant public providers. Now to define these terms and their relation to providers.

PDN. Public data networks provide a common access network to allow many users to share a common network facility, as well as the services it offers (also could refer to a private data network, where a set of common users share private facilities and services). PDNs support voice, data, or both, and provide access for users who cannot cost-justify building a private data network of their own. The international arena is moving from a single PDN provider (historically the PTT) to a competitive environment with multiple providers.

PSN.   Packet switch networks are shared-data networks where customers are charged based on how much data (measured in packets) they send across the network. The new frame relay services also fall under this category.

IVAN. International value added networks are privately owned service providers who offer public services not available through the local service (or local PDN). Typical services include basic file or voice transfer, and enhanced services include electronic data interchange (EDI), electronic mail (E-Mail), and protocol interpretation and conversion. Users interface to IVANs through direct connect or dial-up access. IVANs provide a good alternative to private lines, and their popularity is evidenced by a 25% growth rate. If the country's communications are PTT owned, IVANs will interface to monopoly providers through the local PDN gateway.

IVPDN. International virtual private data network, an international public data network (IPDN) service where facilities and services are offered to the customer in a manner where it appears he is operating his own private data network.

IRC. International record carriers, prior to 1980 were the only international telex and packet switch providers that worked with the PTTs through gateways to each country PDN.

Figure 23.1 shows the evolution of these various service providers.

23.2 THE CHANGING ROLE OF PTTs

The role of the PTT has drastically changed in the past decade. Worldwide divestiture is causing open competition in both industrialized and developing countries. Voice and data networks are being modernized at an exponential growth rate, and public and private network service providers are scrambling to take advantage of new international markets created by these tides of change. The players in the worldwide markets are changing and the number of players increasing. The new role of these PTTs and service providers in the world communications market will prove interesting.

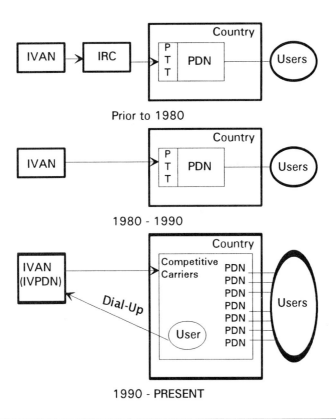

Figure 23.1 Service Providers

23.2.1 Worldwide Divestiture

Until recently, the Post Telegraph & Telephone companies (PTTs) of many countries worldwide held control of the voice and data communications service providers within an entire country. Government owned and influenced, the PTTs operated in a monopolistic environment, primarily for profit.

As in the United States, carrier monopolies are now collapsing and free competition on communications is occurring in the same manner as deregulation occurred in the United States — driven by the regulatory climate and the end user. The same exorbitant costs for communications which drove users in the United States to rebel against the monopolies has taken place all over the world. Now the companies who built business empires out of divestiture in the

United States are profiting from European and Pacific Rim divestiture.

As previously mentioned, this move away from PTT monopolies and toward divestiture has been driven by the customers. These customers demanded:

- ☎ regulatory change and competition
- ☎ separation of regulatory from operating company
- ☎ value added services
- ☎ cost-based tariffs
- ☎ guaranteed service levels
- ☎ further liberalization and competition
- ☎ strategic alliances with major global carriers such as MCI, Sprint, & AT&T
- ☎ privately-owned companies who compete, are customer driven, and respond to the market

The PTTs are now changing their role. Historically, their primary interest was to generate government revenues. Customer service suffered, and the cost of that service was high. Now with privatization, PTTs will reverse their historical priorities and focus on customer satisfaction, offering better quality services at lower prices. To date, international providers have had to deal with each PTT, or local competition, one-on-one to form partnerships. This remains true as the PTTs still represent the driving factor in the overseas markets, and basic services remain regulated. As privatization spreads through PTT-owned countries, PTTs are selling government assets to the private industry sector in an effort to obtain the substantial investments required to fully restructure. The PTTs have also tried to gather funds by selling portions or the entire PTT holdings, by offering public stock, and through employee buyout options. These trends will continue with privatization.

23.2.2 Dominant Public Providers

Table 23.1 shows the dominant public providers (or PTTs) of the major international voice and data service providers in Europe. Also listed is their international value-added network service. The same table is provided for Asia and Australia, also called Pacific Rim (Table 23.2), South America (Table 23.3), and other countries (Table 23.4).

TABLE 23.1 Dominant World Public Data Providers

Country	Dominant Carrier(s)	Service
Austria	Radio-Austria, A.G. (RADAUS)	SI
	Gendion OE P.T.V.	Datex-P
Belgium	Regie des Telegraphes et Telephones (RTT)	DCS
Denmark	Kopenhagen Telecom AS (KTAS)	DATAPAK
Finland	Tele	DATAPAK
France	France Telecom	TRANSPAC
Germany	Deutsch Bundespost Telekom	DATEX-P
Greece	Hellenic Telecommunications Organization	HELPAC
Netherlands	Administrations des Postes et Nether-lands	DATANET
Ireland	Telecom Eireann	EIRPAC
Italy	Societa Italian per I'Esercizio delle Telecommunicazioni	ITAPAC
Italy	Italcable	-
Luxembourg		LUXPAC
Norway	Teledirecktoratet	DATAPAK
Portugal	Companhia Portuguesa de Radio Marconi	TELEPAC-P
Spain	Telefonica de Espana	IBERPAC
Switzerland	General Swiss PTT	TELEPAC
Sweden	Televerket Sverige	DATAPAC
U.K.	British Telecom PLC (BT)	IPSS
U.K.	Mercury Communications Ltd.	IPDS

The major United States based IVAN service providers include BT Tymenet, Inc., GE Information Services, IBM Information Network, Infonet Services Corp., Sprint International, and MCI International.

23.2.3 The PTTs and Global Players After 1992

The predominant carriers mentioned in the previous tables will be the key players in the globalization of data and computer communications. Take a look specifically at Europe. This area has undergone the most radical changes and will continue to revolutionize rather than evolutionize most of their data communications.

TABLE 23.2 Pacific Rim Data Service Providers

Country	Dominant Carrier(s)	Service
Australia	Australian and Overseas Telecommunications Corp. (AOTC)	Data Access
Australia	AAP Communications Pty., Ltd. (AAP)	-
Japan	Nippon Telegraph & Telephone Corp.	DDX
	Kokusai Denshin Denwa Company Ltd. (KDD)	VENUS-P
	International Telecom Japan, Inc. (ITL)	-
	International Digital Communications Inc. (IDC)	-
	Daini Denden, Inc.	-
	Teleway Japan Corp.	-
	Japan Telecom Co.	-
Hong Kong	Hong Kong Telecom International Ltd. (HKTI)	DATAPAC
	Cable & Wireless Hong Kong Ltd.	IDAS
India	Videsh Sanchar Nigam, Ltd.	GPSS
Indonesia	PT Indonesian Satellite Corp.	Paksatnet SKDP
Korea	Korean Telecommunications Authority (KTA)	-
	Data Communications Corporation of Korea (Dacom)	-
Singapore	Singapore Telecom (ST)	TELEPAC

TABLE 23.3 South American Data Service Providers

Country	Dominant Carrier(s)	Service
Argentina	Empresa Nacional de Telecommunicaciones (ENTEL)	-
Brazil	Empressa Brasileira de Telecomunicacoes, S.A.	INTERDATA
Chile	Empresa Nacional de Telecomunicaciones, S.A. (ENTEL)	VTRNET
Columbia	Ministry of Communications	COLDAPAQ
Peru	Ministerio de Transportes y Comunicacion	TELEACCESO
Venezuela	CANTV	-

TABLE 23.4 Other World Data Service Providers

Country	Dominant Carrier(s)	Service
Canada	Telecom Canada	DATAPAK
Canada	CNCP Comm.	Infoswitch
Mexico	Direccion General de Telecomunicaciones	TELEPAC
	Telemex	-
People's Republic of China	Ministry of Posts and Telecommunications and Beijing Telecommunications Administration	CNPAC
Saudi Arabia	Cable & Wireless Riyadh, Ltd.	IDAS
South Africa	Department of Posts & Telecommunications	SAPONET

There are three primary communities of interest in Europe:

The European Community (EC) consists of Belgium, Denmark, England, France, Germany, Greece, Ireland, Italy, Luxembourg, Netherlands, Portugal, and Spain. This group of nations represents the most diversified group of communications evolving in the world. While this represents the greatest challenge, it also represents the greatest potential. An EC-wide common network would provide the keystone to EC success in the 1990s.

The European Free Trade Association (EFTA) consists of Austria, Finland, Iceland, Liechtenstein, Norway, Sweden, and Switzerland. This group of nations has some of the most advanced communications infrastructures in Europe, yet they are also highly diversified on standards and interconnectivity. While the possibility of an EFTA-wide common network would greatly enhance the power of the EFTA, a combined EC-EFTA-wide network could signal the beginning of the world's largest economic power — all with the keystone of integrated computer and data communications.

The Eastern Bloc (now defunct) consists of Albania, Czechoslovakia, Bulgaria, Estonia, Hungary, Latvia, Lithuania, Poland, Romania, and Yugoslavia. While these countries lack the infrastructure that exists in Western Europe, they have the capability to build a modernized network from the ground up. To do this they will require funding assistance from Western Europe or private business moving in to fuel their stagnant economies. But they can quantum leap from old technologies to cellular and digital networks.

23.3 TRANSMISSION NETWORKS

Many factors are hindering the modernization of European networks. The primary causes are the quality of transmission networks and the expense of quality facilities. Location also influences the cost and availability of quality transmission. With today's technology, fiber optic transmission facilities are required to move data around the globe, and many satellite facilities no longer provide sufficient quality. Government stagnation and bureaucracy are also factors.

23.3.1 Expensive, Poor Facilities

The problem plaguing many Eastern Bloc countries is the lack of quality transmission facilities. There is still no fiber optics or digital communications facilities (all old analog). In fact, the wait time for residential phone connectivity is still a 10 to 15 year waiting list. Fiber is now the predominant digital service media over satellite. Fiber began in 1988 with TAT-8 and has now overtaken satellite. Many under-developed areas make it impossible for fiber to reach. It is these areas that are still served by satellite systems.

The other problem is getting the new data communications services and facilities past the local regulation. Private lines are still very expensive. Some satellite opportunities are now opening up, but there are few packet switching services and almost no ISDN.

23.3.2 Importance of Location

The pervasiveness of data communications in all countries is based on location of facilities, and areas of high-growth markets. These are the locations where global network connectivity supports global business. Many European and Asian cities have been labeled as "hubs" or focal points for inter- and intra-country traffic. These locations serve as the primary origination and termination points for international fiber and satellite. Some examples include: London, Paris, Frankfurt, Zurich, Munich, Dusseldorf, Brussels, and Amsterdam. They are also prime locations for international network management centers. The quantity and quality of service degrades drastically as the distance from these hubs increases.

23.3.3 Costs

International bandwidth is considerably more expensive than in the United States. Private line costs within Europe and Asia average about $7,000 per month for a 64K bps. Lines from the United States to Europe are slightly more expensive.

IVANs can help reduce this private line cost while providing better connectivity. Compare the private line solution to accessing a carrier-provided service. The VAN would charge per access port, not per circuit. While the United States to Europe connection would cost about the same, the intra-Europe ports would cost less than half the price of dedicated circuits, and provide the same bandwidth, but now have ubiquitous access to each other. This is why the IVANs are quite popular in the international market. Ultimately, it is users driving the communications industry, choosing their providers, and insuring their success.

23.4 THE IVAN/IVPDN MARKET

The IVAN market exists for both voice and data, although the two are merging — if ISDN standards follow through as planned. The IVAN data market can be segmented into four major technologies and services. These include:

➊ router and bridge WANs (including IP networks)
➋ X.25 packet switching networks
➌ frame relay networks (equipment varies)
➍ SMDS and FDDI MANs

These technologies and services will be offered by either the local predominant carrier, local competition, or foreign service providers. The first two have been discussed, foreign service providers are next.

23.4.1 Foreign Service Providers

One alternative to using the local dominant carrier services is foreign service providers. These companies provide IVAN service when the local PDN cannot adequately service the user's needs. Companies like MCI Communications and Sprint International offer complete end-to-end communications between the United States, European, and Asian locations. The IVANs provide international virtual private data network services (IVPDNs) to the foreign country. This connects local data networks to an international data network, as well as offering the services customers require. Often, these services will predominate over the local or domestic virtual network services. The voice and data advanced services include features such as time of day routing, ISDN, and switched services which are very cost effective for users.

These providers also offer IVANs and IVPDNs. These IVANs and IVPDNs let the international network and service provider (e.g., MCI or Sprint) handle all the foreign PTT or dominant carrier dealings, provide the appropriate levels of technical support and network management, upgrades when required, and the people required to operate it. This is often the most cost-effective solution as opposed to building an international private network, especially when switched services are offered. Just make sure that connectivity and service uses worldwide industry standards.

One of the largest players in the European data services market is Infonet which bills itself as "The World Network". Infonet offers packet switched and frame relay service to many countries in Europe. This is offered through an existing base of NET IDNX multiplexers. Many of the services offered use IP routing. Infonet is owned by MCI, Deutsch Bundespost Telekom, and France Telecom, as well as 11 other PTTs. Another example of international gateway services is the MCI International Digital Gateway. This service helps multinational companies combine national and international traffic over a single T1 or T3 circuit. Siemens of Germany has also shown an interest in the nationalized phone companies of the Eastern Bloc countries, as have other large European providers.

23.4.2 International Outsourcing and Joint Partnerships

Many of the leading service providers and predominant carriers are forming joint partnerships. The focus of these new units are to provide a complete end-to-end international outsourcing package to multinational corporations. For example, there is a partnership between MCI Communications, KDD, and BT PLC called Commax designed to provide end-to-end international communication services from voice to data to video, and including both dedicated and switched services. The caveat to understand, however, is that there must be a financial incentive to *both* parties of a partnership for the relationship to successfully continue.

23.5 Technology Effects On International Traffic

Frame relay over routers and multiplexers, SMDS over DQDB switches, B-ISDN over ATM switches, and SONET infrastructure switches will soon shape international communications.   Each country is pursuing its own avenues of network infrastructure development. Switched services are very popular in Europe. Many European countries are embracing SMDS (DQDB), and a few are investing in frame relay.   The Deutsch Bundespost Telekom is installing a SONET transmission network across Germany, and the Japanese NTT plans to roll out B-ISDN services by 1995, while Fujitsu and NTT are working on ATM network applications and services. Development is moving along, but with very little common direction, somewhat resembling United States IXC, RBOC, and LEC lack of development and subsequent support for these new technologies.

23.5.1 Common ISDN Platform

The 25 European carriers and PTTs have come to an agreement on a common ISDN implementation. This standardization is currently being implemented. The agreement provides a common set of ISDN services, standards, and applications, defining a common, non-proprietary BRI (NET3) and PRI (NET5) standard through CEPT, and also on a common CCITT Signaling System 7 (SS7) for interconnection of ISDN services. There is also a future direction toward B-ISDN standardization based on ETSI and CCITT standards. The target date for EC B-ISDN standardization is 1995.

ISDN has had almost no customer interest of penetration in the Pacific Rim countries, particularly not in Japan. Japan's six local carriers and two international carriers have shown minimal interest in ISDN to date.

23.5.2 International Frame Relay

Frame relay is ideal for transoceanic data communications, utilizing international private lines more efficiently and thus reducing the cost of both access and transmission. This was the primary cause of the move from private networks to accessing international public network data services. But these international transmission facilities must have low bit-error rates to carry frame relay traffic. Widespread international frame relay deployment remains constrained by the lack of fiber optic facilities required to carry frame relay traffic, as well as the lack of PTT interest. Frame relay may be given a boost when the CCITT completes its frame relay standards by 1994.

Another hindrance is that there exists no common method of implementing frame relay. Each CPE vendor provides their own version of the ANSI or CCITT standard, and supports various local management implementations. At present, there are many carriers and service providers who are offering international frame relay service. These services come with the caveat that implementations of international frame relay require the access devices to perform the same implementation of frame relay protocol sets.

23.6 INTERNATIONAL TRAFFIC PATTERNS, ARCHITECTURE, AND TOPOLOGIES

International traffic patterns will vary and are difficult to predict. This is primarily due to the matrix design which has grown from the conglomerate of private line connectivity and packet-switched services. The topologies tend to be flatter, more matrix than hierarchical, yet still conform to those presented in Chapter 20. A WAN usually employs gateways between VPNs in each country.

When designing an international network, all of the steps discussed in this book apply and should be followed in the same sequence and order. Designing an international network is similar to a national design, with the priority shifted toward reducing the cost of facilities and transport and away from a "balanced", reliable design. Such conveniences are often sacrificed in international designs. All architectures discussed may be present, although the OSIRM and SNA predominate throughout the world. OSI and TCP/IP are the predominant internetworking protocols, with the new move toward OSI. Soon, OSI may be the common standard across the entire world, and any international design would conform to standard OSI protocols. The topologies follow those presented in Chapter 20.

Look for business overlaps when planning traffic patterns. One example is the early morning overlap of business hours between the U.S. and Europe. This yields high voice traffic volumes. Similar patterns can be found with bank transactions, stock market information transfers, and other industrial and financial traffic. Data is not as susceptible to these patterns, as it can be transferred when required (computers don't have business hours).

23.7 EUROPEAN OPPORTUNITIES AFTER 1992

There are two efforts driving European "network unification": the realization of the EC in 1992 and the EC Commission Green Paper. The Single European Act of 1987 calls for the establishment of a common European market where goods, people, services, and capital may move freely among member countries. The second important catalyst is the EC Commission Green Paper. This document spells out the objectives of the EC to jointly develop a strong telecommunications infrastructure based on recent technological developments and to offer services commensurate with that network. This document will give the EC consumer what they want and demand.

These two efforts will pull together the varying network infrastructures of each country and provide a common technology platform. It will promote advanced services and electronic data interchange (EDI) across national boundaries. With the European Community working together with the European Free Trade Association, service lead times will be decreased, OSI standardization will become a reality, and a new economic power may emerge on the basis of a modernized European communications infrastructure. Through major EC projects such as PROJECT, RACE, STAR, and APOLLO the EC and EFTA will become one huge distributed data network, and the rest of Europe will be forced to play catch-up.

23.7.1 Developing Countries — Worth the Risk?

International arenas such as the Commonwealth of Independent States (former 11 Soviet republics), South America, New Zealand, Australia, and Africa have antiquated systems which could easily be upgraded to new technologies. These markets are wide open. To enter these markets, providers must know the size of the market, type of service required, current organizational structure and partnerships, pricing structure in place and expected, and especially the competition (both national and international — did you think you were alone?). Only then can the obstacles be surpassed of competing in international communications markets, such as the culture, business ethics, politics, rules and regulations, laws, geographical restrictions, industrial stage of development, new distribution patterns, new price structures, and value-added tax or VAT.

23.7.2 The Top Players

Can we speculate on who will be the top players in the international arena? The key players seem to be system integrators, IVAN providers, and many more. They are the companies who look beyond one area of communications, forming partnerships with other key worldwide players — in the United States, AT&T, MCI Communications, Sprint, Infonet, and IBM; in Europe, Alcatel, Siemens, Philips, British Telecom, and the Deutsch Bundespost; and in the Pacific Rim, Fujitsu, NEC, NTT, and KDD. There are many other key players, but these are the ones who control the dominant share of the worldwide data communications market.

23.8 REVIEW

This chapter has allowed us to step into the shoes of both the PTTs and the service providers competing for market share in the move toward privatization. The changes in PTT posture have been observed from that of a government owned monopoly to that of a profit center devoted to customer service satisfaction. With this change has come opportunities for international network and service providers. Together with entities such as the EC and EFTA, these providers are helping to shape a new world for communications based on OSI standards and devoted to customer service satisfaction. Many opportunities have been identified for these providers, as well as new ground rules that they must follow to compete in the international arena. International network designs resembled national designs with a shift in priorities toward cost and quality of facilities. These factors also influence the base technologies that can be used, and each was explored.

24

Case Studies

This last chapter presents an analysis of two industry-leading design tools — MAKE Systems' Netool and IBM's Intrepid. These tools perform circuit, packet, frame, and, soon, cell switched network modeling and design. Both case studies have been provided by the respective vendor. Anyone wishing to obtain additional information about these tools should contact the marketing/sales representative listed at the end of the case study.

24.1 NETWORK DESIGN CASE STUDY #1 — MAKE SYSTEMS' NETOOL OVERVIEW

Netool is a network asset management system which applies the benefits of decision-support technologies to the complex tasks of network management so users can make better decisions about their networks. Netool helps with daily operations as well as long-term strategic planning, providing the ability to make decisions based on accurate, up-to-date information.

Unlike network management systems, which perform real-time *on-line* data collection and monitoring, Netool does its work *off-line*. By capturing current information from an on-line element management system, then using vendor-certified "device libraries" to accurately model operating characteristics of specific devices on the network,

Netool is able to simulate changes then report their effects on network characteristics (e.g. routing, bandwidth allocations, priority and clocking schemes, node throughput and performance) — *all without impacting real-time network operations.*

Through accurate simulation of complex communications equipment, users can see exactly what will happen if they bring a new application on line, install new nodes or trunks, make a call priority change or reroute a particular circuit. Users can "dry run" a cutover, fail a node or nodes or try out any change imaginable - all off line. They can test the change, determine effectiveness of a decision, then generate management reports, safely and accurately. This simulation capability is extremely valuable in implementing network change, in managing reliability and in controlling costs.

24.1.1 Vendor-Specific Device Libraries

Through technology licensing agreements with key hardware manufacturers, Make Systems has developed vendor-validated device libraries which accurately model the behavior of specific devices on a network. These device libraries are fundamental to the quality and accuracy of data provided by Netool. Each device library captures and models characteristics such as capacities, performance, protocols supported and other proprietary features so the user is assured of reliable simulation results.

Applications such as design and simulation work in conjunction with device libraries to ensure designs are optimized to specific network equipment. In addition, users can explore vendor features in a simulated network prior to implementation; they can determine the value of implementing product features, and ascertain the real world results, prior to actual implementation.

24.1.2 TDM Support

Initially, Make Systems developed device libraries for multiplexer (TDM) devices, building relationships with Ascom Timeplex, N.E.T., Newbridge Networks Corp. and Tellabs. The company also is developing a device library for the ADAPTIVE SONET Transmission Manager™ (STM). Netool helps users of these TDM devices to better understand their current network and to plan and design future networks.

24.1.3 Internetworking Support

Most recently, Make Systems introduced products for the LAN internetworking arena, through agreements with 3Com Corp., Cabletron and Wellfleet. Future device libraries will model the characteristics of X.25, frame relay and SMDS products, among others. Users of internetworking products can use Netool to help with planning, analysis of network utilization and simulation of various effects on networks.

24.1.4 Applications and Benefits

Netool provides a number of valuable benefits related to network management:

Optimized network design. Netool allows for a network design which ensures capacity, handles peak loads, provides for future expansion, and ensures that operations can handle the growing workload. To avoid overbuilding the network, Netool lets users explore various network designs, as well as vendor equipment and technologies, for maximum performance *before* implementing them.

Simpler network management. Netool can assess every aspect of "what if" scenarios with a series of mouse clicks, as opposed to the hours or even days required using manual methods. Netool's simulation capability can create a growth template that simulates and tests steps to implement the change, without disruption to the actual network.

Greater flexibility of analysis. Netool provides flexibility in analyzing the network through "what if" questions to accommodate unexpected changes, and through the performance of non-disruptive testing for the addition of new applications and other network expansions. Users can manually explore failure scenarios or use automated survivability analysis features to explore overall network survivability, and they can plan for maintenance outages while minimizing service disruption.

Improved cost control. Automating network analysis increases productivity. The analysis capability results in more effective use of existing facilities, a better design in support of additional users and demand, improved performance-to-price ratio and reduced network costs through design improvement, network optimization and better choice of tariffs and/or vendors. Other cost savings come through reducing or eliminating the cost of network outages. These include user and application costs as well as lost opportunity costs.

Improved productivity. Netool's management tools increase staff efficiency by automating difficult tasks which require specialized expertise. For example, load planning and service priority planning improve service quality and availability. In addition, Netool's support tools help alleviate problems caused by the shortage of skilled personnel. Finally, the use of Netool helps to enhance personnel development by providing more challenging tasks and by allowing staff (both experienced and new personnel) to learn more demanding network operations, analysis, planning, and design skills — *all in a nondisruptive environment.*

24.2 NETOOL ARCHITECTURE

Netool is designed specifically for network management. Its architecture consists of three main subsystems:

✣ graphical user interface
✣ data management
✣ simulation and analysis

24.2.1 Graphical User Interface (GUI)

Intrinsic to the Netool concept is the ability to empower the human decision maker; for this reason, its user interface is both approachable and intelligent. Netool features a graphical user interface that allows users consistently easy access to enterprisewide network information. Further, the modular architecture on which Netool is built provides an expandable set of tools for displaying, manipulating, and visualizing complex network data.

Network query and navigation. The Netool interface provides a unique query and navigation tool that helps users organize and view both physical and logical network data. Users can create any number of specific and customized network views, and they can continuously modify or refine these queries until they have accessed only the information they need. They can specify both the type of view (geographical, logical or textual) and the kind of information in the view (topology, link loading, clocking, etc.).

Object-oriented interface. Key to the network navigation concept is an object-oriented approach which allows users to "drag and drop" icons representing devices, demands, routes, etc. from one portion of the network to another, or from one view of the network to another. For example, if a user wants routing information for a specific demand — an application requiring networking bandwidth — the demand object can be "dragged and dropped" into a window that reports on the end-to-end connectivity of any demand dropped into it.

Automatic layout. While geographic views provide useful information, users also often want to view their networks from a logical perspective. Netool's automatic schematic layout capability creates logical displays which optimize representation of the network data. Clutter is reduced, space is used more efficiently, and overlapping of objects is avoided. Auto layout lets users concentrate on functional relationships between network objects, many of which are obscured in a geographical view.

Consistent "look and feel" across platforms. The Netool user interface is built in the OSF/Motif™ environment, which is based on the X Window System from MIT. This technology provides the foundation for distributed (client/server) applications and, because it is the UNIX standard, allows delivery of a consistent "look and feel" across many different platforms. Figure 24.1 shows the Netool graphical display.

24.2.2 Data Management

The data management component of Netool's architecture supports both proprietary and standard data interfaces that bridge to multiple sources of network data. Netool provides data bridges as interfaces to vendor network management systems to extract updated network information in the form of network "snapshots". These data bridges rest on a data engine which internally manages and manipulates all data, and provides query and report generation.

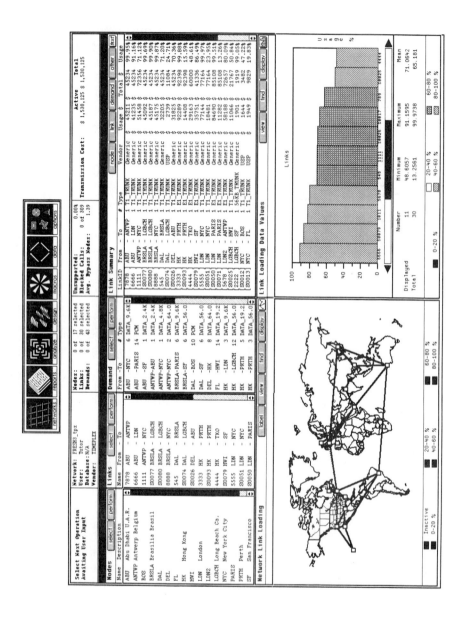

Figure 24.1 Netool Graphical Interface

In addition, a new traffic analysis facility helps users define application profiles by importing and exporting network data between Netool's common data model and external ASCII files. The common data model is a published format containing all Netool network data, and provides three primary functions:

★ export of Netool common data to external files for reporting

★ import of network configuration from vendor NMSs

★ import of tariff database

The tariff manager facility allows users to override the Netool tariff database and supply their own tariffs. This tariff database may replace or augment Make Systems' tariff database service. Capabilities supported include:

✦ User-defined tariffs on a point-to-point basis, or by local exchange (LEC) and interexchange (IXC) portions.

✦ Volume discounts applied to the Netool tariff subscription service and user-defined tariff database.

✦ File import facility for user tariffs through the data import/export utility.

✦ User addition of countries, cities, POPs and international gateways. The user can add countries and cities to the latitude and longitude database for countries outside the U.S. and Canada. If the user adds a country, then tariffs are calculated either as "generic" or based on the Tariff Manager.

24.2.3 Simulation and Analysis

The Netool simulation and analysis subsystem provides a variety of information to assess the quality of network design and performance. Simulation results show:

✥ what applications are not achieving maximum throughput, and what level of throughput the applications are achieving,

✥ what applications are not meeting minimum performance requirements, and

✤ node and link utilization levels, providing an indication of highly loaded parts of the network as well as potential trouble spots.

Netool determines and highlights on an individual demand (i.e. application) basis which nodes or links are causing application throughput problems. Netool also provides a list of unmet demands at each node and link where applications are not achieving minimum performance, identifying major congestion points.

Using the survivability analysis feature, Netool automatically fails each node and link in the network, reroutes traffic and determines new network loading, calculates performance costs and identifies unmet demands. Users can quickly determine network vulnerability, where weak points are located and what applications will be most impacted.

24.3 NETOOL LAN/WAN INTERNETWORK SUPPORT

Netool's demand-level simulation is unique. Two commonly used network analysis methods are mathematical modeling and packet-level simulation. Mathematical modeling utilizes formulas and techniques derived from queuing theory. It attempts to model network behavior abstractly, using basic assumptions about the capacities (bandwidth) of links and nodes. Packet-level simulation involves discrete event simulation of large numbers of individual packets. While practical in certain situations, the two methods have significant limitations. The mathematical model does not provide the accuracy needed to make correct decisions in complex real world situations, and packet level discrete event simulation techniques are overwhelmed by detail and complexity.

Netool's demand-level approach to simulating network traffic is an extension of the techniques first developed for the TDM environment, with added features to provide accurate, detailed information on which to base key design decisions.

24.3.1 Application Profiles

Netool provides a complete methodology for characterizing network applications and traffic, through network design and implementation. Application profiles of network traffic are characterized by the following:

Network and MAC protocol. Identifies protocol overheads and routing constraints.

Packet sizes and ratio of packets each direction. Determines capacity and bandwidth requirements.

Burstyness or probability of activity. Differentiates between applications such as NFS which may be mounted but only infrequently transfer data, and applications such as FTP which continuously transmit data while active.

Maximum and minimum throughput required. User defined range within which the application's performance is considered acceptable.

Application profiles can be defined manually, by using a network monitoring device or by using the default application profiles within Netool. For new applications, characteristics should be determined during development. Table 24.1 shows sample application profiles.

TABLE 24.1 Sample application profiles provided with Netool

Characteristic	NFS	FTP	Telnet	Transaction Processing	Measured Flow
Maximum bandwidth	4 Mb/s	4 Mb/s	1.5 kb/s	1 kb/s	200 kb/s
Minimum bandwidth	300 kb/s	30 kb/s	1.2 kb/s	500 b/s	200 kb/s
Probability of activity	.01	1.0	1.0	0.1	1.0
Mean packet size, each direction	1500 to, 100 from	1080 to,64 from	64 to, 70 from	200 to, 70 from	140 to, 110 from
Ratio of packets to/from	1.25	1.25	1	1	1

24.3.2 Topology Design

Netool provides an object-oriented design capability that allows considerable user definition and control over the design process. The topology design toolset is intended as an aid to users who are interested in investigating alternative topologies and design strategies in their networks. Topology design supports multivendor, multiclass networks for the integration of feeder multiplexers and switching multiplexers, as well as internetworking products. Design tools balance theoretical design goals with real-world constraints, such as maximum hop limits and redundancy requirements. With real-world design as a goal, Netool makes judicious use of the simulation engine in determining feasibility.

A set of design strategies allows users to perform network design without knowledge of the design primitives or the need to create their own design strategies. However, if they desire, users can create their own strategies, or they can edit Netool-supplied strategies, which include a variety of least cost, ground-up and incremental design capabilities. For more information on Netool contact Beverly L. Dygert, Corporate Communications Manager, at MAKE Systems, Inc., 201 San Antonio Circle, CA 94040. (415)941-9800.

24.4 NETWORK DESIGN CASE STUDY #2 — IBM'S INTREPID

INTREPID is an integrated tool for the design and analysis of communication networks. The tool is interactive and runs on a PS/2™ (PS/2, OS/2, and PM are trademarks of the IBM Corporation) under OS/2TM and Presentation Manager PM™. It is capable of selecting backbone nodes, finding both local access and backbone link topologies, assigning capacities to links, creating a class of service table for routing, and analyzing delay and throughput. It also includes interactive graphics which allow the user to make presentation quality displays and edit the network graphically. Finally, it includes generic input/output modules allowing the tool to freely interact with other tools. It is comprised of independently compiled and spawned processes which communicate via shared memory. Thus, it is easily expanded to include new network architectures.

24.5 STRUCTURED DESIGN OF INTREPID

INTREPID (Integrated Network Tool for Routing, Evaluation of Performance, and Interactive Design) is an integrated tool capable of handling different network architectures and allows the designer to choose among them. The integrated structure of modules relies upon features of the operating system and systems libraries whenever possible, reverting to IBM generated generic utilities for much of the rest of the code. By virtue of this modular and adaptive design, a substantial tool was created in a relatively short time which is able to add modules to handle new network architectures comparatively easily, allowing the tool to keep pace with changes to its environment. At the same time, features unique to OS/2 have been avoided so that the system will be able to migrate to other comparable operating systems, most notably, UNIX. Structurally, the tool is divided into several parts.

24.5.1 Command Shell

The command shell module recognizes user commands to invoke functional modules. In fact, this capability is not based in the tool itself but rather in a utilities library which comes with the computing platform the tool runs on. Specifically, the initial implementation of INTREPID has been done in the C Language on a PS/2 running under OS/2 and Presentation Manager (PM). PM includes the ability to create menus and user dialogs. While the menus and dialogs are specific to the tool, these are essentially tables or tabular code (CASE statements) and the bulk of the shell is generic. The next implementation of INTREPID will run under UNIX™ (UNIX is the trademark of the ATT Bell Laboratories) and XWindows which have similar capabilities. IBM believes that the bulk of the implementation will port in a straight forward way to X-windows, especially if the capabilities of Motif are used.

24.5.2 Task Management

The task management tool functions in a multitasking environment, with computer intensive tasks running in the background while user interaction is taking place in the foreground. In this environment, the user is rarely idle waiting for computation to finish. INTREPID relies heavily on the operating system for these capabilities.

24.5.3 Shared Memory Management

All modules communicate with one another via shared memory. The shared memory module allows other modules to allocate, reference, and deallocate blocks of shared memory.

24.5.4 Common Input and Output Modules

These modules, including the input processor and reader/writer modules, control the transfer of information between disk and shared memory. The input processor is an interactive module which facilitates preparation of an input file by allowing the user to specify parameters and component input sources. This simple format facilitates exchange of information among modules and also between this tool and others.

24.5.5 Network Design Programs

These modules do the design of the network, selecting node types, homing nodes at one level to another, and selecting a link topology. Most of the network design algorithms are based on fundamental network design principles. The design actually proceeds in two major phases. The first, where the bulk of the combinatorial optimization takes place, uses generic algorithms. In the second phase, algorithms more specific to the particular network architecture are used. These latter algorithms need only consider a relatively small number of alternatives. The generic algorithms bear most of the computational load.

24.5.6 Network Analysis Procedures

These include modules to analyze throughput, delay, loss, routing, and reliability. These modules are specific to individual types of networks.

24.5.7 Network Display

This module graphically displays candidate networks. It is capable of displaying on a CRT, printer, or plotter. One of the capabilities of PM (and other modern graphics libraries) is device-independent graphics. Thus, essentially the same program is used to display on all devices. The display includes an optional background map which the user can toggle on and off. The display module includes routines for transforming between map coordinates, telephone company vertical and horizontal (V and H) coordinates and screen coordinates. The user can control the display, moving nodes and node labels, zooming, and adjusting which nodes and links are visible. Links are color-coded to indicate multiplicity or utilization, again, adjustable under user control. Thus, the view can be refined for use in reports and management presentations.

24.5.8 Network Editor

This module allows the user to manually modify a network, adding and deleting links and altering their properties (e.g., speed and multiplicity). It even allows the user to input an entire design manually. One can also test variations on the basic design heuristics and verify the quality of the designs produced by the algorithms in the tool. Finally, one can do incremental designs.

24.6 DESIGN FLOW PROCESS

The tool is meant to be used interactively, with the user moving freely from one module to another, iterating among the modules as the design is refined. There is, however, a natural flow for the design process. The user begins by preparing an input file containing information about the locations, link capacities, link costs, and traffic requirements associated with the network. The telephone company V and H coordinates and all the link costs are retrieved automatically by the tool from a tariff database. The user also has the option of obtaining this information independently.

An important feature of this tool is that it accepts inputs in a fairly general form which facilitates using files prepared for other purposes as input. After an input file has been prepared, the user makes a number of preliminary runs of the design algorithms to get a general idea of the structure of the network. Specifically, one may wish to get an idea of a reasonable number of backbone nodes, the speed of most of the links, a reasonable level of delay or loss, and a reasonable reliability. The tool produces alternative designs so that one can reasonably set the parameters for the constraints and guide the optimization process. Often, (e.g., if a proposal or feasibility study is being done) such preliminary designs are all that one needs. Other times, these designs form the basis for other more detailed design work.

During this initial phase, the designer is usually not concerned with details. In this phase one gets a picture, or several different pictures, of the overall topology. Given this, one can make important decisions about the overall design, specifically about the architecture of the network. For example, it might become clear via such an analysis that a private network does not make sense at all.

This first phase of the design process can be carried out primarily using the design algorithms in the tool. After the designer has obtained one or more viable candidate topologies, the second phase is entered. In this phase, design parameters are set to concentrate on the areas which proved fruitful in the first phase. Thus, the designer may force certain nodes to be backbone nodes, limit the choice of line speeds, select an appropriate algorithm from among those available, and set parameters to produce denser, more reliable designs. The number of candidates the designer works with is dramatically reduced. Analysis procedures are run to test the feasibility of the candidates and local optimization is done, both manually and automatically, to refine the most promising candidates. The final phase produces presentation quality output for each of the designs — hard copy of network diagrams and the analysis modules to produce detailed reports.

The natural flow of decision making within the design and analysis processes includes many steps. The first step is to determine which nodes are backbone nodes and which are not. The user may specify some nodes as mandatory backbone nodes, others as mandatory end nodes and the rest as free for the algorithm to assign a type to. Part of the backbone node selection process is a determination of which backbone node to home each end node to. Both the node selection and homing processes can be modified using the network editor. The homing process decomposes the overall design problem into a backbone network design problem and one or more local access network design problems.

Local access topologies and the backbone topology can be determined in any order, once the homing has been done. The first step, finding a backbone topology, is to produce a compressed traffic matrix between backbone nodes. With this, it is possible to determine a backbone link topology satisfying the requirements using the design algorithms. After this is done, the designer may wish to modify the topology using the network editor.

Most routing procedures are essentially shortest path algorithms. The next step in the design process is to assign a length to each link. We refer to this as class of service (COS) determination. An algorithm may be run to optimize the COS. Such algorithms are specific to a particular network architecture. Given a COS for the links, it is now possible to do a routing using the specific algorithm suitable for the actual network architecture at hand and then to do detailed delay, loss and throughput analysis. Given link flows, output by the routing process, it is possible to do a capacity assignment; i.e., to reassign capacities to the links.

At any time during this process, once a link topology is available, it is possible to do reliability analysis. The basic idea is to do as much optimization as possible using fast, simple algorithms and models and to allow the user to interact closely with the process, providing manual input throughout. This methodology relies heavily on the ability to get reliably good designs using only approximate models. As can be seen from the discussion above of the design process, it is essential for the designer to be able to move freely from one module to another.

24.7 COMMON INPUT/OUTPUT

A simple, table-based structure for data files allows the tool to communicate freely with other software, obtaining inputs and producing outputs. The Common I/O modules provide an effective mechanism for moving data between disk and memory.

24.8 DESIGN ALGORITHMS

Several of the most important algorithms are listed below, including highlights of the basic principles upon which they were built.

Node selection. Two types of algorithm are used — a threshold algorithm and a median finding algorithm.

Link topology selection. The MENTOR algorithm is used to select a backbone link topology. This algorithm, while fast and simple, has worked well in practice in the design of many types of networks because it relies on network design principles which are widely applicable. It seeks to create relatively direct paths for traffic, make reasonable link utilizations and aggregates traffic to take advantage of economies of scale. While these goals are in general conflicting, the algorithm balances among them and realizes most of the benefit of all three principles.

COS determination. COS determination is particular to the network architecture. IBM describes the algorithm implemented for IBM's Advanced Peer-to-Peer Network (APPN) architecture.

Capacity assignment. Given a link topology, link flows, the cost of links of various capacities and a function relating link capacity to delay, the tool will assign a capacity to each link, minimizing cost while keeping the average network delay within a given constraint.

24.9 EXPERIENCE WITH TOOL AND CONCLUSIONS

The tool is very fast and can produce a design for a 250-node backbone in less than 1 minute, and for backbones of less than 20 nodes (which are currently most common) in less than 1 second. In reality, it produces multiple designs which the user can then explore interactively. The multitasking capabilities of OS/2 are utilized to allow the tool to continue producing alternative designs while the user examines a candidate and to analyze one design while the user examines another. Thus the designer and the computer work efficiently together, neither spending much time waiting for the other. The tool relies heavily on heuristics (and simple ones at that!) that it produces reliably good results when used by designers with modest training.

For more information on INTREPID contact Robert S. Cahn, Pao-Chi Chang, Parviz Kermani, or Aaron Kershenbaum, at the Thomas J. Watson Research Center, 30 Saw Mill River Road, Hawthorne, New York, 10532. (914)784-7367.

Acronyms and Abbreviations

AAL	ATM Adaptation Layer
ABM	Asynchronous Balance Mode (HDLC)
AC	Access Control (IEEE)
ACF	Access Control Field (DQDB)
ACK	Acknowledgment
ADM	Add/Drop Multiplexer
AIS	Alarm Indication Signal (SONET)
ANS	American National Standard
ANSI	American National Standards Institute
APS	Automatic Protection Switching
ARM	Asynchronous Response Mode (HDLC)
ARP	Address Resolution Protocol
Asynch	Asynchronous
ATM	Asynchronous Transfer Mode
AU	Access Unit (DQDB)
AUI	Attachment Unit Interface (Ethernet 802.3)
BCC	Block Check Characters
BCD	Binary Coded Decimal
BECN	Backward Explicit Congestion Notification (FR)
Bellcore	Bell Communications Research
BER	Bit Error Ratio or Rate
BGP	Border Gateway Protocol
B-ISDN	Broadband aspects of Integrated Services Digital Network
BO	Bit Oriented (SONET)
BOC	Bell Operating Company
BOM	Beginning of Message (DQDB)
bps	Bits per second

Bps	Bytes per second
BRI	Basic Rate Interface (ISDN)
BSC	Bisynchronous Communications
CAD	Computer Aided Design
CBEMA	Computer & Business Equipment Manufacturers Assoc.
CCI	Carrier to Carrier Interface
CCITT	Consultative Committee Int'l Telegraph & Telephone
CD	CountDown counter (DQDB)
CE	Connection Endpoint
CEPT	Conference on European Post & Telegraph
CIR	Committed Information Rate (FR)
CL	Connectionless (SONET)
CLLM	Consolidated Link Layer Management (FR)
CMIP	Common Management Interface Protocol (ISO)
CMIS	Common Management Information Service (ISO)
CMISE	CMIS Element (ISO)
CMT	Connection Management (FDDI)
CO	Central Office
COAM	Customer Owned and Maintained
COCF	Connection-Oriented Convergence Function (DQDB)
COM	Continuation of Message (DQDB)
CPE	Customer Premises Equipment
C/R	Command/Response Indicator or bit
CRC	Cyclic Redundancy Check or Test
CS	Convergence Sublayer (DQDB)
CSMA/CD	Carrier Sense Multiple Access w/ Collision Detection
DA	Destination Address field
DAL	Dedicated Access Line
DAS	Dual-Attach System connection (FDDI)
DCE	Data Communications Equipment
DCS	Digital Cross-connect System
DDD	Direct Distance Dialing
DE	Discard Eligibility (FR)
DEC	Digital Equipment Corporation
DH	DMPDU Header (DQDB)
DLCI	Data Link Connection Identifier (FR)
DMPDU	Derived MAC PDU (DQDB)
DPG	Dedicated Packet Group (FDDI)
DOD	Department of Defense
DQDB	Distributed Queue Dual Bus (IEEE)
DS 0	Digital Signal Level 0
DS 1	Digital Signal Level 1
DS 3	Digital Signal Level 3
DSAP	Destination Service Access Point (LLC)

DSG	Default Slot Generator (DQDB)
DT	DMPDU trailer (DQDB)
DTE	Data Terminal Equipment
DTMF	Dual Tone MultiFrequency
DXC	Digital cross-connect (X) System
EA	Extended Address
ECN	Explicit Congestion Notification (FR)
ECSA	Exchange Carriers Standards Association
ED	End Delimiter (IEEE 802)
EGP	Exterior Gateway Protocol
EGRP	Exterior Gateway Routing Protocol
EMA	Enterprise Management Architecture (DEC)
EOM	End of Message
EOT	End of Transmission
ESF	Extended Super-Frame
ES-IS	End System to Intermediate System protocol (OSI)
ETB	End of Transmission Block
ETX	End of Text
F	Flag
FC	Frame Control field (FDDI)
FCS	Frame Check Sequence (FR)
FDDI	Fiber Distributed Data Interface (ANSI)
FDDI-FO	FDDI Follow-On
FDM	Frequency Division Multiplexing
FEC	Forward Error Correction
FECN	Forward Explicit Congestion Notification (FR)
FM	Frequency Modulation
FOIRL	Fiber Optic Inter-Repeater Link (Ethernet 802.3)
fps	Frames per second
FR	Frame Relay
FRAD	Frame Relay (frame) Assembler/Disassembler
FRAD	Frame Relay Access Device
FS	Frame Status field (FDDI)
FT1	Fractional T1
Gb	Gigabits (billions of bits)
G bps	Gigabits per second (10^9)
GFID	General Format Identifier
GFI	General Format Identifier (X.25)
GGP	Gateway-Gateway Protocol (DOD)
GOS	Grade of Service
GOSIP	Government Open System Interconnection Profile
HCS	Header Check Sequence (DQDB)
HDTV	High Definition Television
HDLC	High-Level Data Link Control (ISO)

HOB	Head of Bus (DQDB) A or B
Hz	Hertz or cycles per second
ICF	Isochronous Convergence Function (DQDB)
ICMP	Internet Control Message Protocol
IEEE	Institute of Electrical and Electronics Engineers
IGP	Interior Gateway Protocol
IGRP	Interior Gateway Routing Protocol (cisco™)
ILMI	Interim Local Management Interface
IMPDU	Initial MAC Protocol Data Unit (DQDB)
IMSSI	Inter-MAN Switching System Interface (DQDB)
I-MAC	Isochronous Media Access Control (FDDI)
intraLATA	intraLocal Access and Transport Area
ISN	Initial Sequence Number (DOD)
IP	Internet Protocol (DOD)
IPCP	Internet Protocol Control Protocol (DOD)
IPX	Internetwork Packet Exchange protocol (Novell)
IS	Intermediate System (OSI)
ISDN	Integrated Services Digital Network
ISDU	Isochronous Service Data Unit (DQDB)
IS-IS	Intermediate System-to-Intermediate System (OSI)
ISO	International Standards Organization
ISSI	Inter-Switching System Interface (SMDS)
ISU	Isochronous Service User (SMDS)
kb	Kilobit (thousands of bits)
k bps	Kilobits per second (10^3)
km	Kilometers (10^3)
LAN	Local Area Network
LAP-B	Link Access Procedure - Balanced (X.25)
LAP-D	Link Access Procedure - D (ISDN/Frame Relay)
LAT	Local Area Transport protocol (DEC)
LATA	Local Access Transport Area
LB	Letter Ballot
LCGN	Logical Channel Group Number
LCP	Link Control Protocol
LEC	Local Exchange Carrier
LLC	Logical Link Control (IEEE 802.X)
LME	Layer Management Entity (DQDB)
LMI	Local Management Interface (FR)
LTE	Line Terminating Equipment (SONET)
LU	Logical Unit (SNA)
m	meters
MAC	Media Access Control (IEEE 802.X)
MAN	Metropolitan Area Network (DQDB, FDDI)
Mb	Megabits (millions of bits)

M bps	Megabits per second (10^6)
MCF	MAC Convergence Function (DQDB)
MCP	MAC Convergence Protocol (DQDB)
MIB	Management Information Base (SNMP)
MIC	Media Interface Connector (FDDI)
MID	Message Identifier (DQDB)
MIPS	Millions of Instructions Per Second
MMF	Multimode Fiber
MOP	Maintenance and Operation Protocol (DEC)
ms	millisecond (one-thousandth of a second, 10^{-3})
MSAP	MAC Service Access Point (SMDS)
MSDU	MAC Service Data Unit (SMDS)
MSS	MAN Switching System (SMDS)
MTU	Maximum Transmission Unit
MUX	Multiplexer
NCP	Network Control Protocol or Point (SNA)
NE	Network Element
NetBIOS	Network Basic Input/Output System protocol
NFS	Network File Server
nm	Nanometer (10^{-9})
NMP	Network Management Process (SMDS)
NNI	Network Node Interface (SONET)
NNI	Network-to-Network Interface (FR)
NOS	Network Operating System
NRM	Normalized Response Mode (ISO)
NRZ	Nonreturn to zero
NRZI	Nonreturn to zero invert ones
ns	Nanosecond (10^{-9})
NTx	Network Termination x (where X=1, 2, ...)
OC-1	Optical Carrier Level 1 (SONET)
OC-n	Optical Carrier Level n (SONET)
OH	Overhead
OOF	Out Of Frame
ONA	Open Network Architecture
OS	Operating Systems
OSI	Open Systems Interconnection
OSI CLNS	Connectionless Network System (OSI)
OSIRM	OSI Reference Model
OSPF	Open Shortest Path First
OTC	Operating Telephone Company
PA	Prearbitrated segment or slot (DQDB)
PABX	Private Automatic Branch Exchange
PAD	Packet Assembler/Disassembler (X.25)
PAF	Prearbitrated Function (DQDB)

PBX	Private Branch Exchange
PDC	Packet Data Channel (FDDI)
PDN	Public Data Network
PDS	Packet Driver Specification for public domain
PDU	Protocol Data Unit (IEEE)
Ph-SAP	Physical Layer SAP (DQDB)
PHY	Physical Layer standard (FDDI)
PIR	Protocol Independent Routing
PL	PAD Length (DQDB)
PLCP	Physical Layer Convergence Protocol (DQDB)
PM	Performance Monitoring
PMD	Physical Layer Medium Dependent standard (FDDI)
POH	Path Overhead (SONET)
POI	Path Overhead Identifier (DQDB)
POP	Point of Presence
PPP	Point-to-Point Protocol
PPS	Packets Per Second
PRI	Primary Rate Interface (ISDN)
PSPDN	Packet-Switched Public Data Network
PTE	Path-Terminating Equipment (PTE)
PTT	Postal, Telegraph & Telephone Ministry/Administration
PU	Physical Unit (SNA)
PVC	Permanent Virtual Circuit
QA	Queued Arbitrated (DQDB) segment, slot, access function
QAF	Queued Arbitrated Function (DQDB)
QOS	Quality Of Service
QPSX	Queued Packet and Synchronous Exchange
RBOC	Regional Bell Operating Company
RCP	Remote Console Protocol (DEC)
REJ	Reject frame
RIP	Routing Information Protocol
RISC	Reduced Instruction Set Computer
RJE	Remote Job Entry
RMT	Ring Management (FDDI)
RNR	Receive Not Ready
RQ	Request Counter (DQDB)
RR	Receive Ready frame
RTMP	Routing and Management Protocol (Apple)
s	second
SA	Source Address field
SAP	Service Access Point (ISO)
SAPI	Service Access Point Identifier (ISO)
SAS	Single-Attach System connection (FDDI)
SD	Start Delimiter

SDLC	Synchronous Data Link Control (IBM)
SDH	Synchronous Digital Hierarchy (CCITT)
SDU	Service Data Unit (DQDB)
SES	Severely Errored Seconds
SF	SuperFrame
SIP	SMDS Interface Protocol (SMDS)
SMDS	Switched MultiMegabit Data Service
SMF	Single-Mode Fiber
SMT	System Management protocol (FDDI)
SNA	System Network Architecture (IBM)
SNI	Subscriber Network Interface (SMDS)
SNMP	Simple Network Management Protocol (DOD)
SOH	Section Overhead
SONET	Synchronous Optical Network (ANSI)
SPE	Synchronous Payload Envelope (SONET)
SPF	Shortest Path First protocol
SPM	FDDI-to-SONET Physical Layer Mapping standard (FDDI)
SREJ	Select Reject frame
SRT	Source Route Transparent protocol
SS	Switching System (SMDS)
SSAP	Source Service Access Point (LLC)
STE	Section Terminating Equipment (SONET)
STM	Synchronous Transfer Mode or Station Management (SDH)
STM-n	Synchronous Transport Module level n (SDH)
STP	Shielded Twisted Pair
STP	Spanning Tree Protocol (IEEE 802.1d)
STS-1	Synchronous Transport Signal Level 1 (SONET)
STS-n	Synchronous Transport Signal Level n (SONET)
STS-Nc	Concatenated Synchronous Transport Signal Level N
SVC	Switched Virtual Circuit
SYN	Synchronous Idle
TA	Terminal Adapter (CCITT)
TCP	Transmission Control Protocol (DOD)
TCP/IP	Transmission Control Protocol/Internet Protocol (DOD)
TDM	Time Division Multiplexing
TDMA	Time Division Multiple Access
TP	Transport Protocol (CCITT)
TP4	Transport Protocol Class 4 (ISO)
TR	Technical Report
UDP	User Datagram Protocol (DOD)
UNI	User Network Interface
UNMA	Unified Network Management Architecture (AT&T)

UTP	Unshielded Twisted Pair
VC	Virtual Channel or Virtual Call
VCI	Virtual Channel Identifier (DQDB)
VLSI	Very Large Scale Integration
VT	Virtual Tributary (SONET)
VTx	VT of size "x" (currently x = 1.5, 2, 3, 6)
VTx-Nc	Concatenated Virtual Tributary (SONET)
WAN	Wide Area Network
XNS	Network Services protocol (XEROX)
ZIP	Routing and Management protocol (Apple)
μs	Microsecond (10^{-6})

*NOTE : Additional comments in ( ) are a clarification or refer to the standard from which the term is derived. Many acronyms are used by multiple standards and only the most prevalent are mentioned.

B

Standards Sources

Alpha Graphics
 10215 N. 35th Avenue, Suite A&B, Phoenix, AZ 85051
 Ph:602-863-0999 (IEEE P802 draft standards)

American National Standards Institute — ANSI — Sales Department
 1430 Broadway, New York, NY 10018
 Ph:212-642-4900/FAX:212-302-1286 (ANSI and ISO standards)

Association Francaise de Normalisation, Tour Europe — Cedex 7
 92080 Paris La Defense, FR
 Ph: 33-1-4-778-13-26; Telex:611-974-AFNOR-F; FAX:33-1-774-84-90

Bell Communications Research — Bellcore Customer Service
 60 New England Ave, Room 1B252, Piscataway, NJ 08854-4196
 Ph:908-699-5800 / 1-800-521-CORE (1-800-521-2673)
 (Bellcore TAs and TRs)

British Standards Institution
 2 Park St., London, WIA 2BS England
 Ph:44-1-629-9000; Telex:266933 BSI G; FAX:+44-1-629-0506

Canadian Standards Association
 178 Rexdale Boulevard, Rexdale, ON M9W 1R9 Canada
 Ph:416-747-4363; Telex:06-989344; FAX:1-416-747-4149

Comite Europeen de Normalisation
 Rue Brederode 2 Bte 5, 1000 Brussels, Belgium
 Ph:32-2-513-79-30; Telex:26257 B

Computer and Business Equipment Manufacturers Association (CBEMA)
 311 First Street N.W., Suite 500, Washington, DC 20001-2178
 Ph:202-626-5740; FAX: 202-638-4299, 202-628-2829 (ANSI X3
 secretariat)

Dansk Standardiseringsrad
 Aurehojvej 12, Postboks 77, DK-2900 Hellerup, Denmark
 Ph:45-1-62-32-00; Telex:15-615 DANSTA DK

DDN Network Information Center - SRI International
  333 Ravenswood Avenue, Menlo Park, CA 94025
  Ph:415-859-3695 / 1-800-235-3155 / e-mail: NIC@NIC.DDN.MIL
  (Requests for Comments [RFC] documents)

Deutsches Institut für Normung
  Burggrafenstrasse 4-10, Postfach 1107, D-1000 Berlin 30, Germany
  Ph:49-30-26-01-1; Telex:184-273-DIN D; FAX:49-30-260-12-31

Electronics Industries Association (EIA)
  Standards Sales, 2001 Eye Street NW, Washington, DC 20036
  Ph:202-457-4966; Telex:710-822-0148 EIA WSH; FAX:202-457-4985

European Computer Manufacturers Association (ECMA)
  Rue du Rhone 114, CH-1204 Geneva, Switzerland
  Ph:41-22-735-36-34; Telex:413237 ECMA CH; FAX:41-22-786-52-31

European Conference of Postal and Telecommunications Administrations — CEPT
  CEPT Liaison Office, Seilerstrasse 22, CH-3008 Bern, Switzerland
  Ph:41-31-62-20-78; Telex:911089 CEPT CH; FAX:41-31-62-20-78

Exchange Carriers Standards Association (ECSA)
  5430 Grosvenor Lane, Bethesda, MD 20814-2122
  Ph:301-564-4505   (ANSI T1 secretariat)

Global Engineering
  2805 McGaw Ave., Irvine, CA 92714
  Ph:1-800-854-7179   (ANSI, IEEE, US. Federal and Military
  standards and drafts)

Institute of Electrical and Electronics Engineers (IEEE) — Standards Office/Service
Center
  445 Hoes Lane, Piscataway, NJ 08855-1331
  Ph:908-564-3834; FAX:908-562-1571 (IEEE standards)

International Organization for Standardization
  1 Rue de Varembe, Case Postale 56, CH-1211 Geneva 20,
  Switzerland
  Ph:41-22-734-1240; Telex:23-88-1 ISO CH; FAX:41-22-733-3430

International Telecommunications Union — General Secretariat — Sales Service
  Place de Nation, CH 1211, Geneva 20, SWITZERLAND
  Ph:41-22-730-5860; Telex:421000 UIT CH; FAX:41-22-730-5853
  (CCITT and other ITU recommendations)

Japanese Industrial Standards Committee
  Standards Department, Agency of Industrial Science & Technology
  Ministry of International Trade and Industry
  1-3-1, Kasumigaseki, Chiyoda-ku, Tokyo 100 Japan
  Ph:81-3-501-9295/6; FAX:81-3-680-1418

National Institute of Standards and Technology
  Technology Building 225, Gaithersburg, MD 20899
  Ph:301-975-2000; FAX:301-948-1784

National Standards Authority of Ireland
Ballymun Road, Dublin 9, Ireland
Ph:353-1-370101; Telex:32501 IIRS EI; FAX:353-1-379620

Nederlands Normalisatie-Instituut
Kalfjeslaan 2, P.O. Box 5059, 2600 GB Delft, Netherlands
Ph:31-15-61-10-61

Omnicom, Inc.
115 Park St. SE, Vienna, VA 22180-4607
Ph:703-281-1135; Telex:279678 OMNI UR; FAX:703-281-1505

Omnicom International, Ltd.
1st Floor, Forum Chambers, The Forum, Sevenage, Herts, United
Kingdom SG1 1EL
Ph:44-438-742424; Telex:826903 OMNICM G; FAX:44-438-740154

Saudi Arabia Standards Organization
P.O. Box 3437, Riyadh 11471, Saudi Arabia
Ph:9-661-4793332; Telex:201610 SASO

SRI International
333 Ravenswood Avenue, Room EJ291, Menlo Park, CA 94025
Ph:800-235-3155 (Internet Protocol RFCs)

Standardiseringskommissionen i Sverige
Tegnergatan 11, Box 3 295, S-103 66 Stockholm, Sweden
Ph:468-230400; Telex:17453 SIS S

Standards Association of Australia — Standards House
80-86 Arthur Street, North Sydney N.S.W. 2060 Australia
Ph:61-2-963-41-11; Telex:2-65-14 ASTAN AA

Suomen Standardisoimisliitto
P.O. Box 205, SF-00121 Helsinki 12, Finland
Ph:358-0-645-601; Telex:122303 STAND SF

US Department of Commerce — National Technical Information Service
5285 Port Royal Road, Springfield, VA 22161
Ph:703-487-4650 (CCITT recommendations, US Government and
Military standards)

United Nations Bookstore
United Nations General Assembly Building, Room GA 32B, New
York, NY 10017
Ph:212-963-7680 (CCITT recommendations)

Appendix

Standards Reference

CCITT Standards

CCITT Recommendation E.164, Numbering Plan for the ISDN Era, Vol.II, Fas.II.2, Blue Book, ITU, 1988.

CCITT Recommendation G.703, Physical/Electrical Characteristics of Hierarchical Digital Interfaces, Vol.III, Fas.III.4, Blue Book, ITU, Geneva, 1988.

CCITT Recommendation G.707, Synchronous Digital Hierarchy Bit Rates, Vol.III, Fas.III.4, Blue Book, ITU, Geneva, 1988.

CCITT Recommendation G.708, Network Node Interface for the Synchronous Digital Hierarchy, Vol.III, Fas. III.4, Blue Book, ITU, Geneva, 1988.

CCITT Recommendation G.709, Synchronous Multiplexing Structure, Vol.III, Fas. III.4, Blue Book, ITU, Geneva, 1988.

CCITT Recommendation I.121, Broadband Aspects of ISDN, Blue Book, ITU, Geneva, 1988.

CCITT Recommendation I.122, Framework for Providing Additional Packet Mode Bearer Services, Blue Book, ITU, Geneva, 1988.

CCITT Draft Recommendation I.3xx, Congestion Management for the Frame Relay Bearing Service.

CCITT Recommendation Q.920, ISDN User - Network Interface Data Link Layer — General Aspects, Blue Book, ITU, 1988.

CCITT Recommendation Q.921, ISDN User - Network Interface Data Link Layer Specification, Blue Book, ITU, 1988.

727

CCITT Draft Recommendation Q.922, ISDN Data Link Layer Specifications for Frame Mode Bearer Services, April, 1991.

CCITT Draft Recommendation Q.922, Annex A, Core Aspects of Q.922 for Use with Frame Relay Bearer Service, April, 1991.

CCITT Recommendation Q.931, ISDN User - Network Interface Layer 3 Specifications for Basic Call Control, Blue Book, ITU, 1988.

CCITT Recommendation X.1 - X.32, Data Communications Networks: Services & Facilities Interfaces, Blue Book, November, 1988.

CCITT B-ISDN Standards — Summarized

B-ISDN Standards (All are Draft Recommendations for the 1992 books)
I.113 - Vocabulary of Terms for broadband aspects of ISDN
I.121 - Broadband aspects of ISDN
I.150 - B-ISDN ATM functional characteristics
I.211 - B-ISDN service aspects
I.311 - B-ISDN General Aspects
I.321 - B-ISDN Protocol Reference Model and its applications
I.327 - B-ISDN functional architecture
I.361 - B-ISDN ATM layer specification
I.362 - B-ISDN ATM Adaptation Layer (AAL) functional description
I.363 - B-ISDN ATM Adaptation Layer (AAL) specification
I.413 - B-ISDN user-network interface
I.432 - B-ISDN user-network interface — Physical layer specification
I.610 - OAM principles of B-ISDN access

ANSI Standards

ANSI T1.102-1987, Digital Hierarchy — Electrical Interfaces.

ANSI T1.105-1990, Digital Hierarchy — Optical Interface Rates and Formats Specifications (SONET). - Draft revision May 1990.

ANSI T1.106-1988, Digital Hierarchy — Optical Interface Specifications (Single Mode).

ANSI T1.107-1988, Digital Hierarchy — Format Specifications.

ANSI T1.403-1989, Carrier-to Customer Installations — DS1 Metallic Interface, February 22, 1989.

ANSI T1.602, Telecommunications — ISDN — Data Link Layer Signaling Specifications for Application at the User-Network Interface, 1990.

ANSI T1.605-1989, ISDN — Basic Access Interface for S and T Reference Points (Layer 1 Specifications) Appendix K, October 17, 1988.

ANSI T1.606-1990, ISDN — Architectural Framework and Service Description for Frame Relaying Bearer Service, 1990.

ANSI T1.607 (ANSI T1S1/90-011) — Digital Subscriber Signaling System No. 1 - Layer 3 Signaling Specifications for Switched Bearer Service, 1990.

ANSI T1.617 (old T1.6fr, ANSI T1S1/90-213R1) - DSS1 - Signaling Specification for Frame Relay Service, December 1990.

ANSI T1.617-1991 (old T1S1/91-352) - ISDN - DSS1 - Signaling Specification for Frame Relay Bearer Service, 1991.

ANSI T1.617 Annex D (ANSI PM91-1) — Additional Procedures for PVCs Using Unnumbered Information Frames, May, 1991.

ANSI T1.618 (old T1.6ca, ANSI T1S1/90-214R1), DSS1 — Core Aspects of Frame Relay Protocol for Use with Frame Relay Bearer Service, November, 1990.

ANSI T1.618-1991 - ISDN - Core Aspects of Frame Protocol for Use with Frame Relay Bearer Service, May, 1991.

ANSI T1S1/90-051R2, Carrier to Customer Metallic Interface — Digital Data at 64K bps and Sub-rates, May, 1991.

ANSI T1S1/90-175R4 (T1.606 Addendum) - Frame Relay Bearer Service — Architectural Framework and Service Description, May 1991.

FDDI ANS Standards

ANS X3.139-1987, Fiber Distributed Data Interface (FDDI) — Token Ring Media Access Control (MAC). (ISO 9314-2, 1989)

Draft ANS X3.139, Fiber Distributed Data Interface (FDDI) — Media Access Control (MAC-2), Draft maintenance revision 4.0, October 29, 1990.

ANS X3.148-1988, Fiber Distributed Data Interface (FDDI) — Token Ring Physical Layer Protocol (PHY). (ISO 9314-1, 1989)

Draft ANS X3.148, Fiber Distributed Data Interface (FDDI) — Physical Layer Protocol (PHY-2), Draft maintenance revision 4.0, October 25, 1990.

ANS X3.166-1990, Fiber Distributed Data Interface (FDDI) — Physical Layer Medium Dependent (PMD). (Draft ISO 9314-3, 1990).

Draft ANS X3.184-199X, Fiber Distributed Data Interface (FDDI) — Single-Mode Fiber Physical Layer Medium Dependent (SMF-PMD), Rev. 4.2, May 18, 1990. (Draft ISO 9314-4)

Draft ANS X3.186-199X, Fiber Distributed Data Interface (FDDI) — Hybrid Ring Control (HRC), Rev. 6.2, May 14, 1991. (Draft ISO 9314-5)

ANS X3T9.5/84-49, Fiber Distributed Data Interface (FDDI) — Station Management (SMT), Rev. 6.2, May 18, 1990.

D

Class "C" IP Network Addressing Reference

This reference contains a guideline for choosing IP subnetwork addressing. It is based on the assumption that the user owns a Class C address.

255.255.255.252

```
xxx.xxx.xxx.000 network  1
xxx.xxx.xxx.004 network  |
xxx.xxx.xxx.008 network  |
xxx.xxx.xxx.012 network  |
xxx.xxx.xxx.016 network  5
xxx.xxx.xxx.020 network  |
xxx.xxx.xxx.024 network  |
xxx.xxx.xxx.028 network  |
xxx.xxx.xxx.032 network  |
xxx.xxx.xxx.036 network 10
xxx.xxx.xxx.040 network  1
xxx.xxx.xxx.044 network  |
xxx.xxx.xxx.048 network  |
xxx.xxx.xxx.052 network  |
xxx.xxx.xxx.056 network 15
xxx.xxx.xxx.060 network  |
xxx.xxx.xxx.064 network  |
xxx.xxx.xxx.068 network  |
xxx.xxx.xxx.072 network  |
xxx.xxx.xxx.076 network 20
xxx.xxx.xxx.080 network  |
xxx.xxx.xxx.084 network  |
xxx.xxx.xxx.088 network  |
xxx.xxx.xxx.092 network  |
xxx.xxx.xxx.096 network 25
xxx.xxx.xxx.100 network  |
xxx.xxx.xxx.104 network  |
xxx.xxx.xxx.108 network  |
```

```
                              xxx.xxx.xxx.112 network  |
                              xxx.xxx.xxx.116 network 30
                              xxx.xxx.xxx.120 network  |
                              xxx.xxx.xxx.124 network  |
                              xxx.xxx.xxx.128 network  |
                              xxx.xxx.xxx.132 network  |
                              xxx.xxx.xxx.136 network 35
                              xxx.xxx.xxx.140 network  |
                              xxx.xxx.xxx.144 network  |
                              xxx.xxx.xxx.148 network  |
                              xxx.xxx.xxx.152 network  |
                              xxx.xxx.xxx.156 network 40
                              xxx.xxx.xxx.160 network  |
                              xxx.xxx.xxx.164 network  |
                              xxx.xxx.xxx.168 network  |
                              xxx.xxx.xxx.172 network  |
                              xxx.xxx.xxx.176 network 45
                              xxx.xxx.xxx.180 network  |
                              xxx.xxx.xxx.184 network  |
                              xxx.xxx.xxx.188 network  |
                              xxx.xxx.xxx.192 network  |
                              xxx.xxx.xxx.196 network 50
                              xxx.xxx.xxx.200 network  |
                              xxx.xxx.xxx.204 network  |
                              xxx.xxx.xxx.208 network  |
                              xxx.xxx.xxx.212 network  |
                              xxx.xxx.xxx.216 network 55
                              xxx.xxx.xxx.220 network  |
                              xxx.xxx.xxx.224 network  |
                              xxx.xxx.xxx.228 network  |
                              xxx.xxx.xxx.232 network  |
                              xxx.xxx.xxx.236 network 60
                              xxx.xxx.xxx.240 network  |
                              xxx.xxx.xxx.244 network  |
                              xxx.xxx.xxx.248 network  |
                              xxx.xxx.xxx.252 network 64
---------------------------------------
255.255.255.248               xxx.xxx.xxx.000 network  1
                              xxx.xxx.xxx.008 network  |
                              xxx.xxx.xxx.016 network  |
                              xxx.xxx.xxx.024 network  |
                              xxx.xxx.xxx.032 network  5
                              xxx.xxx.xxx.040 network  |
                              xxx.xxx.xxx.048 network  |
                              xxx.xxx.xxx.056 network  |
                              xxx.xxx.xxx.064 network  |
                              xxx.xxx.xxx.072 network 10
                              xxx.xxx.xxx.080 network  |
                              xxx.xxx.xxx.088 network  |
                              xxx.xxx.xxx.096 network  |
```

```
                              xxx.xxx.xxx.104 network  |
                              xxx.xxx.xxx.112 network 15
                              xxx.xxx.xxx.120 network  |
                              xxx.xxx.xxx.128 network  |
                              xxx.xxx.xxx.136 network  |
                              xxx.xxx.xxx.144 network  |
                              xxx.xxx.xxx.152 network 20
                              xxx.xxx.xxx.160 network  |
                              xxx.xxx.xxx.168 network  |
                              xxx.xxx.xxx.176 network  |
                              xxx.xxx.xxx.184 network  |
                              xxx.xxx.xxx.192 network 25
                              xxx.xxx.xxx.200 network  |
                              xxx.xxx.xxx.208 network  |
                              xxx.xxx.xxx.216 network  |
                              xxx.xxx.xxx.224 network  |
                              xxx.xxx.xxx.232 network 30
                              xxx.xxx.xxx.240 network  |
                              xxx.xxx.xxx.248 network  |
-----------------------------------
255.255.255.240               xxx.xxx.xxx.000 network  1
                              xxx.xxx.xxx.016 network  |
                              xxx.xxx.xxx.032 network  |
                              xxx.xxx.xxx.048 network  |
                              xxx.xxx.xxx.064 network  5
                              xxx.xxx.xxx.080 network  |
                              xxx.xxx.xxx.096 network  |
                              xxx.xxx.xxx.112 network  |
                              xxx.xxx.xxx.128 network  |
                              xxx.xxx.xxx.144 network 10
                              xxx.xxx.xxx.160 network  |
                              xxx.xxx.xxx.176 network  |
                              xxx.xxx.xxx.192 network  |
                              xxx.xxx.xxx.208 network  |
                              xxx.xxx.xxx.224 network 15
                              xxx.xxx.xxx.240 network  |
-------------------------
255.255.255.224               xxx.xxx.xxx.000 network  1
                              xxx.xxx.xxx.032 network  |
                              xxx.xxx.xxx.064 network  |
                              xxx.xxx.xxx.096 network  |
                              xxx.xxx.xxx.128 network  5
                              xxx.xxx.xxx.160 network  |
                              xxx.xxx.xxx.192 network  |
                              xxx.xxx.xxx.224 network  |
-------------------------
255.255.255.192               xxx.xxx.xxx.000 network  1
                              xxx.xxx.xxx.064 network  |
                              xxx.xxx.xxx.128 network  |
                              xxx.xxx.xxx.192 network  |
```

```
-----------------------------
```

| 255.255.255.128 | xxx.xxx.xxx.000 network | 1 |
| | xxx.xxx.xxx.128 network | 2 |

```
=====================================
```

255.255.255.000	xxx.xxx.000.xxx network	1
	xxx.xxx.004.xxx network	|
	xxx.xxx.008.xxx network	|
	xxx.xxx.012.xxx network	|
	xxx.xxx.016.xxx network	5
	xxx.xxx.020.xxx network	|
	xxx.xxx.024.xxx network	|
	xxx.xxx.028.xxx network	|
	xxx.xxx.032.xxx network	|
	xxx.xxx.036.xxx network	10
	xxx.xxx.040.xxx network	|
	xxx.xxx.044.xxx network	|
	xxx.xxx.048.xxx network	|
	xxx.xxx.052.xxx network	|
	xxx.xxx.056.xxx network	15
	xxx.xxx.060.xxx network	|
	xxx.xxx.064.xxx network	|
	xxx.xxx.068.xxx network	|
	xxx.xxx.072.xxx network	|
	xxx.xxx.076.xxx network	20
	xxx.xxx.080.xxx network	|
	xxx.xxx.084.xxx network	|
	xxx.xxx.088.xxx network	|
	xxx.xxx.092.xxx network	|
	xxx.xxx.096.xxx network	25
	xxx.xxx.100.xxx network	|
	xxx.xxx.104.xxx network	|
	xxx.xxx.108.xxx network	|
	xxx.xxx.112.xxx network	|
	xxx.xxx.116.xxx network	30
	xxx.xxx.120.xxx network	|
	xxx.xxx.124.xxx network	|
	xxx.xxx.128.xxx network	|
	xxx.xxx.132.xxx network	|
	xxx.xxx.136.xxx network	35
	xxx.xxx.140.xxx network	|
	xxx.xxx.144.xxx network	|
	xxx.xxx.148.xxx network	|
	xxx.xxx.152.xxx network	|
	xxx.xxx.156.xxx network	40
	xxx.xxx.160.xxx network	|
	xxx.xxx.164.xxx network	|
	xxx.xxx.168.xxx network	|
	xxx.xxx.172.xxx network	|
	xxx.xxx.176.xxx network	45
	xxx.xxx.180.xxx network	|

```
                          xxx.xxx.184.xxx network  |
                          xxx.xxx.188.xxx network  |
                          xxx.xxx.192.xxx network  |
                          xxx.xxx.196.xxx network 50
                          xxx.xxx.200.xxx network  |
                          xxx.xxx.204.xxx network  |
                          xxx.xxx.208.xxx network  |
                          xxx.xxx.212.xxx network  |
                          xxx.xxx.216.xxx network 55
                          xxx.xxx.220.xxx network  |
                          xxx.xxx.224.xxx network  |
                          xxx.xxx.228.xxx network  |
                          xxx.xxx.232.xxx network  |
                          xxx.xxx.236.xxx network 60
                          xxx.xxx.240.xxx network  |
                          xxx.xxx.244.xxx network  |
                          xxx.xxx.248.xxx network  |
                          xxx.xxx.252.xxx network 64
-------------------------------------
255.255.248.000           xxx.xxx.000.xxx network  1
                          xxx.xxx.008.xxx network  |
                          xxx.xxx.016.xxx network  |
                          xxx.xxx.024.xxx network  |
                          xxx.xxx.032.xxx network  5
                          xxx.xxx.040.xxx network  |
                          xxx.xxx.048.xxx network  |
                          xxx.xxx.056.xxx network  |
                          xxx.xxx.064.xxx network  |
                          xxx.xxx.072.xxx network 10
                          xxx.xxx.080.xxx network  |
                          xxx.xxx.088.xxx network  |
                          xxx.xxx.096.xxx network  |
                          xxx.xxx.104.xxx network  |
                          xxx.xxx.112.xxx network 15
                          xxx.xxx.120.xxx network  |
                          xxx.xxx.128.xxx network  |
                          xxx.xxx.136.xxx network  |
                          xxx.xxx.144.xxx network  |
                          xxx.xxx.152.xxx network 20
                          xxx.xxx.160.xxx network  |
                          xxx.xxx.168.xxx network  |
                          xxx.xxx.176.xxx network  |
                          xxx.xxx.184.xxx network  |
                          xxx.xxx.192.xxx network 25
                          xxx.xxx.200.xxx network  |
                          xxx.xxx.208.xxx network  |
                          xxx.xxx.216.xxx network  |
                          xxx.xxx.224.xxx network  |
                          xxx.xxx.232.xxx network 30
                          xxx.xxx.240.xxx network  |
```

```
                                    xxx.xxx.248.xxx network  32
----------------------------------
255.255.255.240                     xxx.xxx.000.xxx network  1
                                    xxx.xxx.016.xxx network  |
                                    xxx.xxx.032.xxx network  |
                                    xxx.xxx.048.xxx network  |
                                    xxx.xxx.064.xxx network  5
                                    xxx.xxx.080.xxx network  |
                                    xxx.xxx.096.xxx network  |
                                    xxx.xxx.112.xxx network  |
                                    xxx.xxx.128.xxx network  |
                                    xxx.xxx.144.xxx network 10
                                    xxx.xxx.160.xxx network  |
                                    xxx.xxx.176.xxx network  |
                                    xxx.xxx.192.xxx network  |
                                    xxx.xxx.208.xxx network  |
                                    xxx.xxx.224.xxx network 15
                                    xxx.xxx.240.xxx network 16
------------------------
255.255.255.224                     xxx.xxx.000.xxx network  1
                                    xxx.xxx.032.xxx network  |
                                    xxx.xxx.064.xxx network  |
                                    xxx.xxx.096.xxx network  |
                                    xxx.xxx.128.xxx network  5
                                    xxx.xxx.160.xxx network  |
                                    xxx.xxx.192.xxx network  |
                                    xxx.xxx.224.xxx network  |
--------------------------
255.255.255.192                     xxx.xxx.000.xxx network  1
                                    xxx.xxx.064.xxx network  |
                                    xxx.xxx.128.xxx network  |
                                    xxx.xxx.192.xxx network  4
----------------------------
255.255.255.128                     xxx.xxx.000.xxx network  1
                                    xxx.xxx.128.xxx network  2
```

Glossary

access unit - In DQDB, the functional unit within a node that performs the DQDB layer functions and controls access to both buses.

address - A station identifier on the network or a logical identifier of a service access point within network stations. DQDB supports IEEE 802 addresses as well as 60-bit ISDN addresses per CCITT Recommendation E.164.

address translation - A method of converting an address of a user protocol into the address standard format of the network protocol, and vice versa.

American National Standard (ANS) - An ANSI-sanctioned US. national standard; X3-series standards deal with information processing and T1 series standards address telecommunications.

American National Standards Institute (ANSI) - A private, non-governmental, non-profit national organization which serves as the primary coordinator of standards within the United States. CBEMA, ECSA, and the IEEE are accredited by ANSI, and ANSI is an active participant in the ISO.

analog - Voice or data signals which are continuously variable and posses an infinite number of values (compared to digital, which has discrete variables).

Application Layer (OSI) - Layer 7 of the OSIRM. Provides the management of communications between user applications. Examples include e-mail and file transfer.

Asynchronous Transfer Mode (ATM) - A high-speed connection-oriented data transmission method that provides bandwidth on demand through packet-switching techniques using fixed-size cells. ATM supports both time-insensitive and time-sensitive traffic, and is defined in CCITT standards as the transport mechanism for B-ISDN services.

asynchronous transmission - The transmission of data through start and stop sequences without the use of a common clock.

B-channel - An ISDN bearer service channel which can carry either voice or data at a speed of 64K bps.

Backward Explicit Congestion Notification (BECN) - Convention in frame relay for a network device to notify the user (source) device that network congestion has occurred.

bandwidth - The amount of transport resource available to pass information (passband), measured in Hz for analog and bps for digital carriers.

bandwidth balancing - A DQDB scheme where a node that is queued for access will occasionally *not* seize an empty QA slot. This helps to ensure effective sharing of QA slots.

Basic Mode - An FDDI mode of ring operation that supports packet switching services only where MAC PDUs are transmitted directly by the PHY protocol.

basic rate interface (BRI) - An ISDN access interface type comprised of two B-channels each at 64K bps and one D-channel at 16K bps (2B+D).

Bell Operating Company (BOC) - One of the 22 local telephone companies formed after the divestiture of AT&T (e.g., Illinois Bell, Ohio Bell).

bisynch (BSC) or Binary Synchronous Communications Protocol - An IBM proprietary bit-oriented protocol.

bridge - A LAN/WAN device operating at Layer One (physical) and two (data link) of the OSIRM.

broadband - While broadband once represented bandwidths in excess of the voice channel (3 kHz), or in basic data communications using analog, modulated signals, it now refers to channels supporting rates in excess of T1 or E1.

Broadband ISDN (B-ISDN) - ISDN service requiring a broadband channel operating at speeds greater than PRI. This service is defined in CCITT standards as operating over ATM (CCITT Recommendation I.121).

broadcast - a transmission to all addresses on the network or subnetwork.

broadcast address - A predefined network address that indicates all possible receivers on a network.

brouter - A device which serves some elements of both bridging and routing.

busy slot - A DQDB slot which is "in use" and not available for access by the QA access functions.

cell - A fixed-length transmission unit used in DQDB and ATM transmissions; similar to a packet. In DQDB the cell is often referred to as a fifty-three octet "slot".

Central Office (CO) - Telephone company switching office providing local user access to the local switched telephone network and its services; often the first interface to interexchange carriers.

Central Office vendors - A reference to vendors who provide switching equipment conforming to central office standards, such as DQDB switch vendors Siemens and Alcatel.

client server architecture - The distribution of network control across many computing elements within the network. Thus, some elements act as servers, controlling the transfer, and some as clients which transmit and receive the information. Servers can do all three functions, and are often the workhorse computing elements (multi-MIP machines), while the clients are typically workstations and terminals.

circuit switching - Switching based on dedicated paths between terminating devices, and adding no intelligence to the transmission. Bandwidth is dedicated and delay is minimal.

collocated - Devices near one another at the same site.

Committed Information Rate (CIR) - Holds various definitions depending on the service provider, but in general represents, in a frame relay service, the maximum bandwidth a user can "burst" and still be guaranteed delivery of all data. Transmissions exceeding the CIR are subject to discard depending upon use of the DE bit.

Concatenated Virtual Tributary (VTx-Nc) - A combination of VTs where the VT envelope capacities from N VTxs have been combined to carry a VTx-Nc that must be transported as a single entity (as opposed to transport as separate signals).

concentrator - A device providing a single network access for multiple user devices. In FDDI, a device which has additional ports beyond what is required for its own attachment to the ring.

congestion - The condition where network resources (bandwidth) are exceeded and additional information cannot be passed (in frame relay, data is dropped during congestion).

Consolidated Link Layer Management (CLLM) - In frame relay, an ANSI-defined method of sending Link Layer management messages over the last DLCI (1023). These messages are used to identify the exact cause of congestion and modify transmissions based on each DLCI.

convergence function - A DQDB protocol layer which interfaces service-specific interfaces to higher-layer protocol functions.

countdown counter (CD) - A queued arbitrated access method for determining how many empty slots must pass before a node has access to the DQDB bus.

customer premises equipment (CPE) - Equipment which resides and operated at a customer site.

cycle - The Protocol Data Unit used in FDDI-II.

cyclic redundancy check (CRC) - A mathematical algorithm which detects bit errors after data transmission.

D4 - AT&T-defined framing and synchronization format for T1 transmission facilities.

D-channel - The ISDN out-of-band (16K bps or 64K bps, depending on BRI or PRI, respectively) signaling channel which carries the ISDN user signals or can be used to carry packet mode data.

Data Communications (or Circuit Termination) Equipment (DCE) - Data communications equipment defined by the standards as a modem or network communications interface device.

Data Link Connection Identifier (DLCI) - A frame relay address designator for each virtual circuit termination point (or port).

data link layer (OSI) - Layer 2 of the OSIRM. Provides for the error-free communications between adjacent network devices over a physical interface. Examples include the LLC and MAC layers which manage LAN and MAN operation.

datagram - A packet mode of transmitting data where there is no guaranteed sequential delivery (connectionless service).

Data Terminal (or Termination) Equipment (DTE) - Data processing equipment defined by the standards as interfacing to the communications link.

default slot generator function - In DQDB, the function defining the identity for each bus in the dual bus network. In the looped bus topology, this function also provides the head of bus function for both buses.

delay-insensitive - *see time insensitive.*

delay-sensitive - *see time sensitive.*

derived MAC protocol data unit (MAC-PDU or DMPDU) - In DQDB, single 44-octet portion of the original IMPDU, comprised of 4 overhead octets and a 44-octet segmentation unit.

digital - Voice or data signals which have discrete values, such as binary bit streams of 0s and 1s (compared to analog digital which has a continuously variable signal).

digital cross-connect systems (DXC) - Breaks down a T1 into individual DS0s for testing and reconfiguration.

digital signal 0 (DS0) - One 56K bps framed channel out of the 24 contained in a T1 channel.

digital signal 1 (DS1) - One T1 channel (1.544M bps).

digital signal 3 (DS3) - One T3 channel which combines twenty-eight DS1s onto a single facility at 44.736M bps.

Discard Eligibility (DE) bit - Used in frame relay, this bit signals (when set to 1) that the particular frame is eligible for discard during congestion conditions.

distributed processing - Sharing of applications, data, and the tasks operating among several small or mid-range processing devices, as opposed to a single mainframe in centralized processing.

distributed queue - The operation of the DQDB Queued Arbitration MAC scheme, where all nodes keep track of the number of stations queued for access in their request counter; when a station queues itself for access, it keeps track of its position in the queue using its countdown counter and it counts the number of stations behind it in the queue in the request counter.

Distributed Queue Dual Bus (DQDB) - The IEEE 802.6 MAN architecture standard for providing both circuit-switched (isochronous) and packet-switched services.

DQDB layer - The lower portion of the DQDB link layer which provides the connectionless MAC data service, connection-oriented data service, and an isochronous service with the help of physical layer services.

dual-attachment station (DAS) - A workstation that attaches to both primary and secondary FDDI MAN rings which enables the capability for network self-healing.

dual bus - Bus A and Bus B, dual DQDB bus structure. The dual bus supports both the open dual bus and the looped dual bus.

E1 - The European T1 CEPT standard operating at 2.048M bps.

E1 carrier - Part of the European and Asian (excluding Japan) digital TDM hierarchy: a single multiplexed 2.048M bps channel.

E.164 - A CCITT Recommendation for defining addresses in a public data international network, varying in size up to 15 digits (carried as 60-bit addresses in DQDB).

empty slot - In DQDB, a Queued Arbitrated slot not currently in use which may be seized by a node queried for QA access.

enterprise network - A network which spans an entire organization.

entity - In the OSIRM, a service of management element between peers and within a sublayer or layer.

Ethernet - A LAN which uses CSAM/CD media access method and operates at 10M bps usually over coax medium.

Explicit Congestion Notification (ECN) - In frame relay, the use of either FECN and BECN or CLLM messages to notify the source and destination of network congestion (as opposed to implicit congestion notification).

fast packet - The generic term used for advanced packet technologies such as frame relay, DQDB, and ATM.

FastPacket™ - StrataCom Corporation's trademark for their proprietary switching technique which uses 192-bit packets and packetized voice.

Fiber Distributed Data Interface (FDDI) - Fiber optic LAN operating at 100M bps.

FDDI-II - FDDI standard with the additional capability to carry isochronous traffic (voice/video).

FDDI Follow-On (FDDI-FO) - Future ANSI standards for extending the speed of FDDI up to 600M bps.

fiber optics - Plastic or glass fibers which transmit high data rates through optical signals.

filtering - The selection of frames not to remain at the local LAN but to be forwarded to another network by a network device (i.e., router).

flag - Character which signals a beginning or end of a frame.

Forward Explicit Congestion Notification (FECN) - Convention in frame relay for a network device to notify the user (destination) device that network congestion is occurring.

fractional T1 (FT1) - The transmission of a fraction of a T1 channel, usually based in 64K bps increments but not less than 64K bps total.

frame - An OSI data link layer defined unit of transmission whose length is defined by flags at the beginning and end.

Frame Check Sequence (FCS) - A field in an X.25, SDLC, or HDLC frame which contains the result of a CRC error-checking algorithm.

frame relay - An ANSI and CCITT defined LAN/WAN networking standard for switching frames in a packet mode similar to X.25, but at higher speeds and with less nodal processing (assuming fiber transmission).

frame relay assembler/disassembler (FRAD) - A device which acts as a concentrator and protocol translator from nonframe relay protocols (e.g., SDLC, SNA) to a standard frame relay transmission.

frequency division multiplexing (FDM) - The method of aggregating multiple simultaneous transmissions (circuits) over a single high-speed channel by using individual frequency passbands for each circuit (for example, RF Broadband LANs).

full-duplex - The simultaneous bidirectional transmission of information over a common medium.

gateway - A network device which interconnects dissimilar types of network elements through all seven layers of the OSIRM.

global addressing - A frame relay addressing convention where a single, unique DLCI value is given to each user device on the network.

half-duplex - The bidirectional transmission of information over a common medium, but where information may only travel in one direction at any one time.

head of bus (HOB_A and HOB_B) - In DQDB, the node responsible for generating empty slots and management information octets.

host - An end-communicating station in a network; also an IP address.

implicit congestion notification - A congestion indication which is performed by upper layer protocols (e.g., TCP) rather than network or data link layer protocol conventions.

individual address - The address of a specific network station or node. In IP, the format is XXXX.XXXX.XXXX.XXXX.

Initial MAC Protocol Data Unit (IMPDU) - In DQDB, the PDU formed by the DQDB layer providing a connectionless MAC service to the LLC.

Integrated Services Digital Network (ISDN) - CCITT I-series Recommendation defined digital network standard for integrated voice and data network access, services, and user-network messages.

integrated switching - The method of performing multiple switching techniques with one device or within a single hardware architecture, including consolidated configuration and network management.

Interexchange Carrier (IXC) - The provider of long distance (inter-LATA) service in the United States. Also provider of worldwide switched data services.

interface - In OSI, the boundary between two adjacent protocol layers (i.e., network to transport).

internetwork - A master network made up of multiple smaller networks, or the concept of bridging, routing, switching, or gateway between homogeneous network devices, protocols, and standards.

interoperability - The ability of multiple, dissimilar vendor devices and protocols to operate and communicate using a standard set of rules and protocols.

InterSwitching System Interface (ISSI) - The interface between two MAN switching systems for purposes of MAN extension beyond the metropolitan area or for wide area inter-networking.

intraLATA - LEC-defined geographic areas (Local Access Transport Area). LEC must pass cells to IXC to go inter-LATA.

isochronous - The circuit-switched transmission service offered in DQDB and FDDI-II. This allows a consistent timed access of network bandwidth for time-sensitive transmission of voice and video traffic.

latency - The minimum amount of time it takes for a token to circulate around the LAN Token Ring or FDDI ring in the absence of a data transmission.

layer management - Network management functions which provide information about the operations of a given OSI protocol layer.

Layer Management Entity (LME) - In DQDB, the entity within the protocol layer responsible for performing local management of the layer.

Layer Management Interface (LMI) - In DQDB, the interface between the LME and network management systems.

Line-Terminating Equipment (LTE) - A device which either originates or terminates an OC-n signal and which may originate, access, modify, and terminate the transport overhead.

Link Access Protocol on the D-channel (LAPD) - CCITT Recommendations Q.920 (I.440) and Q.921 (I.441) defined standards for the data-link layer operation of ISDN "D" and frame relay channel.

local area network (LAN) - A MAC level data and computer communications network confined to short geographic distances.

local bridge - A high-throughput, collocated LAN-to-LAN interconnectivity device.

local exchange carrier (LEC) - In the United States, a local phone service provider (cannot provide long distance service).

Local Management Interface (LMI) - A set of user device-to-network communications standards for frame relay as defined by the "Gang of Four".

Logical Link Control (LLC) - The upper half of the OSIRM data link layer, Layer 2, as defined by the IEEE 802.2 standard. This layer provides a common LAN platform for all IEEE 802.X protocols.

logical ring - The circular closed set of point-to-point links between network stations on a token ring and FDDI network.

looped dual bus - A DQDB bus configuration where the head of bus functions for both A and B bus are contained within the same node.

MAN Switching System (MSS) - A single metropolitan area network comprised of many MAN switches, usually linked by a common DQDB bus.

media - The plural form of medium, or multiple mediums (twisted-wire pair, coax cable, fiber, etc.).

medium - The single common access platform, such as a copper wire, fiber, or air in the case of microwave.

Medium Access Control (MAC) - IEEE 802 defined media specific access control protocol.

message identifier (MID) - In DQDB, a value used to identify all DMPDUs that together comprise the same IMPDU.

metropolitan area network (MAN) - A MAC level data and computer communications network which operates over metropolitan or campus areas, and recently has been expanded to nationwide and even worldwide connectivity of high-speed data networks. A MAN can carry voice, video, and data, and has been defined as both the DQDB and FDDI standards sets.

multicast - A frame relay option defined in the LMI specifications allowing frame relay devices the capability to broadcast to multiple destinations on the network (not necessarily to *all* destinations on the network).

multimode fiber (MMF) - 50- to 100-μm core diameter optical fiber with many propagation paths for light, and is typically used for lower speed or shorter distances (as compared to single mode optical fiber).

multiplexing - The technique of combining multiple single channels onto a single aggregate channel for sharing facilities and bandwidth.

network - A system of autonomous devices, links, and subsystems which provide a platform for communications.

Network Layer (OSI) - Layer 3 of the OSIRM. Provides the end-to-end routing and switching of data units (packets), as well as managing congestion control.

network management - The process of managing the operation and status of network resources (e.g., devices, protocols).

node - A device which interfaces with the transmission medium through the physical layer (and often the data link layer) of the OSIRM. This device is sometimes called an access unit in DQDB.

octet - An 8-bit long transmission unit of measure.

open dual bus - In DQDB, a non-fault-tolerant subnetwork configuration where the head of bus functions for Bus_A and Bus_B are in different nodes (can also be the configuration after a failure and subsequent network self-healing).

Open Systems Interconnection Reference Model (OSIRM) - A seven-layer model defining the international protocol standards for data communications in a multiple architecture and vendor environment. Both the OSI and CCITT define standards based on the OSIRM.

Optical Carrier level n (OC-n) - The optical carrier level signal in SONET which results from an STS-n signal conversion. In SONET, the basic transmission speed unit is 58.34M bps.

Packet Assembler/Disassembler (PAD) - A concentration and network access device which provides protocol conversion into X.25 packet format.

Packet Switch Public Data Network (PSPDN) - A public data network utilizing packet switching technology (X.25, SMDS, ATM).

packet switching - A method of switching which segments the data into fixed or variable units of maximum size called packets. These packets then pass the user information (addressing, sequencing, error control, and user controlled options) in a store-and-forward manner across the network.

packet type - Identifies the type of packet and its use, such as for user data, call establishment and termination, and routing information.

path overhead (POH) - Overhead transported with the SONET payload and used for payload transport functions.

payload pointer - Indicates the starting point of a SONET synchronous payload envelope.

Permanent Virtual Circuit (PVC) - A logical dedicated circuit between two user ports in a point-to-point configuration.

Physical Layer Convergence Protocol (PLCP) - In DQDB, the part of the physical layer that adapts the actual capabilities of the underlying physical network to be able to provide the services required by the DQDB layer.

Physical Layer Medium Dependent (PMD) - In FDDI, the medium-specific layer corresponding to the lower sublayer of the OSIRM Physical Layer.

Physical Layer (OSI) - Layer 1 of the OSIRM. Provides the electrical and mechanical interface and signaling of bits over the communications medium.

Physical Layer Protocol (PHY) - In FDDI, the medium-independent layer corresponding to the upper sublayer of the OSIRM physical layer.

plastic optical fiber (POF) - A low-cost, low-distance plastic alternative to glass fiber.

Presentation Layer (OSI) - Layer 6 of the OSIRM. Identifies the syntax of the user data being transmitted and provides user service functions such as encryption, file transfer protocols, and terminal emulation.

primary rate interface (PRI) - An ISDN T1 access interface type comprised of twenty-three B-channels each at 64K bps and one D-channel at 64K bps (23B+D). The European version will operate at 2.048M bps (30B+D).

primary ring - The main ring for PDU transmission in FDDI, and the only attachment for SAS FDDI stations.

private network - A network providing interorganizational connectivity only.

private {automatic} branch exchange (PBX/PABX) - An {automatic} customer-site telephone switch, with some capability to integrate data.

protocol - The rules and guidelines by which information is exchanged and understood between two devices.

protocol data unit (PDU) - The unit of information transferred between communicating peer layer processes.

public data network (PDN) - A network designed to provide data transmission value-added services to the public.

queued arbitrated access - In DQDB, packet data users contend for access to the bus by queuing their requests; since all of the nodes know the length of the queue and their position in the queue, the access scheme is referred to as distributed queue.

Regional Bell Operating {or Holding} Company (RBOC or RBHC) - One of seven US. regional holding companies formed by divestiture of AT&T (e.g., Ameritech, Southwestern Bell). The RBOCs also manage the 22 BOCs.

remote bridge - A high-throughput bridge which provides remote LAN-WAN connectivity.

ring - A closed-loop, common bus network topology.

router - A LAN/WAN device operating at layer one (physical), two (data link), and three (network) of the OSIRM. Distinguished from bridges by its capability to switch and route data based upon network protocols such as IP.

secondary ring - In FDDI, the ring which carries data in the opposite direction as the primary ring; primarily used for backup to the primary ring.

section - A transmission facility between a SONET Network Element and regenerator.

segment - In DQDB, the payload (user data) portion of the slot.

segmentation unit - The 44-octet unit of data transfer in DQDB.

self-healing - The ability for a LAN/MAN to reroute traffic around a failed link or network element to provide uninterrupted service.

service - The relationship between protocol entities in the OSIRM, where the service provider (lower-layer protocol) and the service user (higher-layer protocol) communicate through a *data service*.

service access point (SAP) - The access point at a network node or station where the service users access the services offered by the service providers.

service data unit (SDU) - Unit of information transferred across the OSI interface between service provider and service user.

Session Layer (OSI) - Layer 5 of the OSIRM. Provides the establishment and control of user dialogues between adjacent network devices.

simplex - One way transmission of information on a medium.

single-attachment stations (SAS) - In FDDI, stations which are attached only to a single ring (primary ring).

single-mode fiber (SMF) - 8- to 10-µm small core diameter optical fiber with a single propagation path for light; typically used for higher speeds or longer distances (as compared to multimode optical fiber).

slot - The basic unit of transmission on a DQDB bus.

SMDS Interface Protocol (SIP) - The three layers of protocol (similar to the first three layers of the OSIRM) which define the SMDS SNI user information frame structuring, addressing, error control and overall transport.

SNA - IBM's communications networking architecture.

source routing - A routing scheme where the routing of packets is determined by the source address and destination route in the packet header.

station - An addressable logical or physical network entity, capable of transmitting, receiving, or repeating information.

Station Management (SMT) - FDDI station management entity.

subnetwork - The smaller units of LANs (called LAN segments) which can be more easily managed than the entire LAN/MAN/WAN.

Subscriber-Network Interface (SNI) - A DQDB user access point into the network or MAN switch.

Switched MultiMegabit Data Service (SMDS) - A MAN service offered at present over the IEEE DQDB bus.

switched virtual circuit (SVC) - Virtual circuits similar to PVCs, but established on a call-by-call basis.

Synchronous Digital Hierarchy (SDH) - The CCITT original version of a synchronous digital hierarchy; based on optical fiber; called SONET in ANSI parlance.

Synchronous Optical Network (SONET) - A United States high-speed fiber optic transport standard for a fiber optic digital hierarchy (speeds range from 51.48M bps to 2.5G bps)

synchronous transmission - The transmission of frames which are managed through a common clock between transmitter and receiver.

Synchronous Transfer Mode (STM) - The T1 carrier method of assigning time slots as channels within a T1 or E1 circuit.

Synchronous Transport Module level N (STM-N) -The SDH line rate of "N" STM-1 signals.

Synchronous Transport Signal level N (STS-N) - SONET transmission signal created with byte interleaving of "N" STS-1 (51.84M bps) signals.

Synchronous Transport Signal level Nc (STS-Nc) - Concatenated SONET synchronous payload envelope.

T1 or T-1 - A circuit operating at 1.544M bps.

T1 carrier - The TDM digital T1 hierarchy used in North America and Japan, with 24 voice channels constituting a single 1.544M bps T1 trunk.

T3 or T-3 - A circuit operating at 45M bps.

telecommunication - The transmission of voice, video, data, and images through the use of both computers and a communications medium.

time division multiplexing (TDM) - The method of aggregating multiple simultaneous transmissions (circuits) over a single high speed channel by using individual time slots (periods) for each circuit.

time-insensitive - Traffic types whose data is not effected by small delays during transmission. This is also referred to as delay-insensitive.

time-sensitive - Traffic types whose data is affected by small delays during transmission and cannot tolerate this delay (e.g., voice, video, real-time data).

token - A marker which can be held by a station on a token ring or bus indicating that station's right to transmit.

Token Ring - A LAN which uses a token-passing access method for bus access and traffic transport between network elements, where bus speeds operate at either 4M bps or 16M bps.

Transmission Control Protocol/Internet Protocol (TCP/IP) - The combination of a network and transport protocol developed by ARPANET for internetworking IP-based networks.

Transport Layer (OSI) - Layer 4 of the OSIRM. Provides for error-free end-to-end communications between two "host" users across a network.

transport overhead - In SONET, the line and section overhead elements combined.

twisted pair - The basic transmission media consisting of 22 to 26 American Wire Gauge (AWG) insulated copper wire. TP can be either shielded (STP) or unshielded (UTP).

unshielded twisted pair (UTP) - A twisted-pair wire without the jacket shielding, used for short distances but subject to electrical noise and interference.

user channel - Portion of the SONET channel allocated to the user for maintenance functions.

User-to-Network Interface (UNI) - The point where the user accesses the network.

user-to-user protocols - Protocols which operate between the users and are typically transparent to the network, such as file transfer protocols (e.g., FTP).

virtual channel identifier (VCI) - In DQDB, a field within the segment header which determines whether a node is to read or write, or copy the segment payload.

virtual circuit - A virtual connection established through the network from origination to destination, where packets, frames, or cells are routed over the same path for the duration of the call. These connections seem like dedicated paths to the users, but are actually network resources shared by all users. Bandwidth on a virtual circuit is not allocated until it is used.

virtual tributary (VT) - An element that transports and switches sub-STS-1 payloads or VTx (VT1.5, VT2, VT3, or VT6).

wide area network (WAN) - A network which operates over a large region and commonly uses carrier facilities and services.

window - The concept of establishing an optimum number of frames or packets which can be outstanding (unacknowledged) before more are transmitted. Window protocols include X.25, LAP, TCP/IP, and SDLC.

X.25 - CCITT recommendation of the interface between packet switch DTE and DCE equipment.

Parts of this glossary were taken from Gary Kessler's book *Metropolitan Area Networks* with the author's permission.

Bibliography

Bertsekas, Dmitri, and Gallager, Robert, *Data Networks*, 2nd Ed., Prentice-Hall, Inc., Englewood Cliffs, NJ, 1991.

Black, Uyless, *Computer Networks - Protocols, Standards, and Interfaces*, Prentice-Hall, Inc., Englewood Cliffs, NJ, 1987.

Black, Uyless, *Data Communications & Distributed Networks*, Reston Publishing Co., Reston, VA, 1987.

Blyth, W. John, and Blyth, Mary M., *Telecommunications, Concepts, Development, and Management*, Glencoe Publishing Co., Mission Hills, Ca., 1985.

Chorafas, Dimitris N., *TRP, Handbook on Data Communications & Computer Networks, 2nd Ed.*, 1991.

Computer Communications Review, articles 1990-1991.

Coulouris, George F., and Dollimore, Jean, *Distributed Systems*, Int'l Computer Science Series, Addison-Wesley, Reading, MA, 1991.

Data Communications Magazine, articles 1990 - 1991.

Datapro Information Services Group, *Data Pro Reports on Communications*, McGraw-Hill.

Dayton, Robert L., *Guide to Integrating Digital Services*, McGraw-Hill, New York, 1989.

Dayton, Robert L., *Telecommunications, The Transmission of Information*, McGraw-Hill, New York, 1991.

Dromard, Horlait, and Pujolle, Seret, *Integrated Digital Communications Networks, Vol. 1*, John Wiley & Sons, New York, NY, 1988.

Freeman, Roger L., *Telecommunications Systems Engineering, 2nd Ed.*, John Wiley & Sons, New York, NY, 1989.

Held, Gilbert, *Digital Networking*, John Wiley & Sons, New York, NY, 1990.

Helmers, Scott A., *Data Communications*, Prentice-Hall, Englewood Cliffs, NJ., 1989.

IEEE Communications Magazine, articles 1990 - 1991.

IEEE Std 802.6-1990, *Local and Metropolitan Area Networks - 802.6*, IEEE, Inc., New York, 1991.

Kessler, Gary C., *ISDN*, McGraw-Hill, New York, 1990.

Kessler, Gary C., and Train, David A., *Metropolitan Area Networks, Concepts, Standards, & Services*, McGraw-Hill, Inc., New York, 1992.

Kim, B. G., *LANs, MANs, & ISDN*, Artech House, Norwood, MA., 1989.

Knightson, Keith G., Knowles, Terry, and Larmouth, John, *Standards for Open Systems Interconnection*, McGraw-Hill, New York, 1988.

LAN Magazine, articles 1990-1991.

LAN Times articles, 1990-1991.

McClimans, Fred J., *Communications Wiring and Interconnection*, McGraw-Hill, New York, 1992.

Minoli, Daniel, *Telecommunications Technology Handbook*, Artech House, Norwood, MA, 1991.

Netrix books

Network World, articles 1990-1991.

Powers, John T., and Stair II, Henry H., *Megabit Data Communications, A Guide for Professionals*, Prentice Hall, Englewood Cliffs, NJ., 1990.

Ranade, Jay, and Sackett, George C., *Introduction to SNA Networking*, McGraw-Hill, New York, 1989.

Rose, M. T., *The Simple Book: An Introduction to Management of TCP/IP-based Internets*, Prentice-Hall, Englewood Cliffs, NJ, 1991.

Rosner, Roy D., *Packet Switching*, Van Nostrand Reinhold Co., New York, NY, 1982.

Metropolitan Area Networks, Siemens Stromberg- Carlson, Boca Raton, FL, 1991.

Schwartz, Mischa, *Computer-Communications Network Design and Analysis*, Prentice-Hall, Englewood Cliffs, NJ, 1977.

Schwartz, Mischa, *Telecommunications Networks*, Addison-Wesley Publishing Co., Reading, MA., 1987.

Sherman, Kenneth, *Data Communications, 2nd Ed.*, Reston Publishing Co., Reston, VA, 1985.

Spragins, John D., *Telecommunications: Protocols and Design*, Addison-Wesley Publishing Company, Inc., New York, 1991.

Stallings, William, *Data and Computer Communications, 3rd Ed.*, Macmillan, New York, NY., 1991.

Stallings, William, *Handbook of Computer Communications Standards, Volume 1 and Volume 2*, Howard W. Sams & Company, Indianapolis, IN, 1987.

Tannenbaum, A. S., *Computer Networks, 2nd Ed.*, Prentice-Hall, Englewood Cliffs, NJ, 1988.

US Sprint, *US Sprint Frame Relay Service Interface Specification*. Document No. 5136.03, July 12, 1991.

Williams, Gerald E., *Digital Technology*, Science Research Assoc., Inc., Chicago, 1982.